COSPAR COLLOQUIA SERIES

VOLUME 12

SPACE WEATHER STUDY USING MULTIPOINT TECHNIQUES

SPACE WEATHER STUDY USING MULTIPOINT TECHNIQUES

Proceedings of the COSPAR Colloquium
held in Pacific Green Bay, Wanli, Taipei, Taiwan
27-29 September 2000

Edited by

Ling-Hsiao Lyu

Institute of Space Science, National Central University,
Chung-li, Taiwan, R.O.C.

2002

PERGAMON
An imprint of Elsevier Science

Amsterdam – Boston – London – New York – Paris – San Diego – San Francisco – Singapore –
Sydney – Tokyo

ELSEVIER SCIENCE Ltd
The Boulevard, Langford Lane
Kidlington, Oxford OX5 1GB, UK

First edition 2002

Library of Congress Cataloging in Publication Data
A catalog record from the Library of Congress has been applied for.

British Library Cataloguing in Publication Data
A catalogue record from the British Library has been applied for.

ISBN: 0 08 044057 6

♾ The paper used in this publication meets the requirements of ANSI/NISO Z39.48-1992 (Permanence of Paper).
Printed in The Netherlands.

PREFACE

Magnetic storms may cause damage to satellites, radiation hazard to astronauts, disruptions of radio communications, and interruption of ground electric power lines. Space weather prediction becomes an important issue to be addressed in the twenty-first century. International Solar Terrestrial Program (ISTP) employs five satellites to probe the solar wind and magnetosphere, providing valuable information for space weather prediction. The Asia-Pacific region is becoming one of the economic centers in the world. The continuous drive for scientific and technological progress in parallel is evidenced by the establishment of many space research organizations in many countries of this area. In Taiwan, the National Space Program Office (NSPO) established her third satellite program -- COSMIC (Constellation Observing Systems for Meteorology, Ionosphere and Climate), which is a science experiment to demonstrate the utility of atmospheric radio limb soundings from a constellation of six low-earth orbiting satellites in operational weather prediction, space weather monitoring, and climate monitoring and research.

In order to provide a forum to discuss the many new results in this rapid-moving field and to forge international collaborations, a three-day COSPAR Colloquium on “Space Weather Study Using Multi-Point Techniques” was held in Pacific Green Bay, a resort area in Wanli, Taipei, Taiwan from September 27 to September 29, 2000.

This colloquium has provided a forum for experts from the international community to present new results on the timely topic "space weather". In the meeting, new techniques, models, theories, and strategies for space weather research were addressed. In addition, the scientific programs and results of ROCSAT-1, 2, 3 were also discussed in this colloquium. This colloquium were attended by 119 scientists from 11 countries. There were 79 invited papers and 33 contributed papers presented in this meeting.

This proceedings consists of 40 reviewed papers, most of them are invited papers, that have been presented in this COSPAR Colloquium. Two papers are in extent length, one of them is the paper from Professor Akasofu, who give a keynote speech in this meeting. The other one is the paper by Tsurutani et al., in which observational results from Cassini are presented for the first time. In addition to papers presented in this proceeding, the colorful Power Point-2000 presentation: "Satellite Anomalies, Recent Events, and Possible Causes," by Dr. Joe Allen given in the summary session, is available on-line at the SCOSTEP website homepage. The URL address is:

http://www.ngdc.noaa.gov/stp/SCOSTEP/scostep.html

This COSPAR Colloquium was cosponsored by Committee on Space Research (COSPAR), Academia Sinica, Ministry of Education (MOE), National Science Council (NSC), National Space Program Office (NSPO), National Central University (NCU), National Cheng Kung University (NCKU), and Scientific Committee on Solar-Terrestrial Physics (SCOSTEP)

This COSPAR Colloquium was convened by J. K. Chao and L. C. Lee. They were advised by the following Scientific Organizing Committee: C.T. Russell (USA), J. Allen (USA), B. Fraser (Australia), R. A. Heelis (USA), H. Kamide (Japan), D. H. Lee (Korea), R. P. Lepping (USA), H. Nishida (Japan), A.

A. Heelis (USA), H. Kamide (Japan), D. H. Lee (Korea), R. P. Lepping (USA), H. Nishida (Japan), A. Richmond (USA), J. Roettger (Germany), I. Sandahl (Sweden), P. Song (USA), B. Tsurutani (USA), S. Wang (China), F. S. Wei (China), S. T. Wu (USA), and K. Yumoto (Japan).

The Local Organizing Committee (LOC) consisted of: W.H. Ip (NCU), chair, F.B. Hsiao (NCKU), L.H. Lyu (NCU), C. Y. Huang (NCU), Y.H. Chu (NCU), L.-N. Hau (NCU), C.M. Huang (NCU), H.J. Huang (PCCU), K.H. Lin (NSYSU), C.C. Liu (Academia Sinica), J.Y. Liu (NCU), C.J. Pan (NCU), S.Y. Su (NCU), W.H. Tsai (NCU), C.L. Tseng (NCKU), and C.T. Wang (NCU).

The editor and LOC members would like to express their special thanks to Miss B. Y. Kuo and Miss P. C. Yang, whose efforts made this COSPAR colloquium a memorable success. Finally, we would like to acknowledge and thank the following scientists, among others, who served as referees and provided insightful, critical reviews for the papers submitted to this volumn.

J. Bürchner
J. K. Chao
K. S. Chen
M. Q. Chen
Y. H. Chu
H. Fukunishi
L. N. Hau
R.-R. Hsu
Cheinway Hwang
C. M. Huang
B. Inhester
B. Inhester
W. Ip
J. R. Kan
C. A. Lin
K. H. Lin
Yuei-An Liou
J. Y. Liu
L. H. Lyu
C. J. Pan
J. Roettger
J. H. Shue
Y. J. Su
T. Tanaka
W. H. Tsai
D. Y. Wang
K. Wilhelm
B. Wilken
J. Wock
Ming Yang
K. Yumoto

Ling-Hsiao Lyu
Editor

Wing H. Ip
Chair of Taiwan National COSPAR Committee

Institute of Space Science
National Central University

OPENING ADDRESS OF THE COSPAR PRESIDENT TO THE COSPAR COLLOQUIUM ON SPACE WEATHER STUDY USING MULTI-POINT TECHNIQUES

This message comes in replacement of my personal presence. I apologize sincerely for the last-minute change forced upon me by requirements from the new job that I accepted recently. I regret very much not to be with you at the opening of this COSPAR Colloquium.

COSPAR values highly the space weather issue. This is documented by the fact that at the Nagoya Assembly in 1998 a Space Weather Panel was established. I also remember with great pleasure the interdisciplinary lecture by Terry Onsager at that Assembly. This year, in Warsaw, Vice-President Lou Lanzerotti continued the tradition with a well-attended and well-received public lecture on storms in space weather. I do not have to underline the importance of space weather, even in a country that because of its geomagnetic location does not experience damages in power transmission systems. I just like to express the observation that phenomena that a few decades ago seemed to belong entirely to the world of science have now acquired a human dimension, economic values in space, hazards for astronauts, risks for telecommunications, even for railway traffic, radiation exposure during air travel, power failures and pipeline corrosion. I had a recent personal experience with the loss of Equator-S during Walpurgis night 1998 when a combination of space weather and sensitive electronics led to the loss of this spacecraft.

But there is another aspect to space weather research. All of its basic elements include fundamental space plasma physics. They give new stimulus to this research and increase the demands for more quantitative understanding, for better agreement of measurement, theory and simulation, for more complete and global measurements. Multipoint measurements are basic to geophysics since Carl-Friedrich Gauss. The IGY was the first culmination of this strategy to be followed by the IMS, the International Magnetospheric Study and, still ongoing, by the ISTP. In this context, we are especially reminded of the latest contribution to the ISTP, the four Cluster spacecraft which are at present in the commissioning phase. What was conceived by us scientists as a mission dedicated to the investigation and understanding of mesocale structures and turbulence to transport mechanisms in collisionless cosmical plasmas is now sold to the public simply as a space weather mission. This demonstrates the appeal of this issue to the public, but does not imply that the original objectives have come out of focus of the scientific community. All of those involved look forward with excitement and great hope to a new era of space plasma physics. Of course, multipoint observations are badly needed on much more global scales than the Cluster quartet can offer. Many proposals of this nature have been submitted and, undoubtedly, there is much future in constellation approaches. But also on Earth cooperation and coordination of widely separated geophysical observatories have not lost their importance.

Finally, I regret that I miss the opportunity to visit Taiwan once more. I have wonderful remembrances from earlier visits, of the great hospitability, and of an impressive exposure to the natural beauty, the old culture, and modern technology. I would have enjoyed to see the progress of Taiwan's space program, and receive first-hand information of other novel plans, new instrumentation and recent insights relevant to the space weather and space plasma topics. To all those who have the pleasure to attend this colloquium I convey the greetings and best wishes from the COSPAR Bureau and Executive.

G. Haerendel
COSPAR President
September 27, 2000

Sponsored by

Committee on Space Research (COSPAR)
Academia Sinica, Ministry of Education (MOE)
National Science Council (NSC)
National Space Program Office (NSPO)
National Central University (NCU)
National Cheng Kung University (NCKU
Scientific Committee on Solar-Terrestrial Physics (SCOSTEP)

Convener

J. K. Chao
L. C. Lee

Scientific Organizing Committee

C.T. Russell (USA)
J. Allen (USA)
B. Fraser (Australia)
R. A. Heelis (USA)
H. Kamide (Japan)
D. H. Lee (Korea)
R. P. Lepping (USA)
H. Nishida (Japan)
A. Richmond (USA)
J. Roettger (Germany)
I. Sandahl (Sweden)
P. Song (USA)
B. Tsurutani (USA)
S. Wang (China)
F. S. Wei (China)
S. T. Wu (USA)
K. Yumoto (Japan)

Local Organizing Committee

W.H. Ip (NCU), chair
F.B. Hsiao (NCKU), L.H. Lyu (NCU)
C. Y. Huang (NCU), Y.H. Chu (NCU)
L.-N. Hau (NCU), C.M. Huang (NCU)
H.J. Huang (PCCU), K.H. Lin (NSYSU)
C.C. Liu (Academia Sinica), J.Y. Liu (NCU)
C.J. Pan (NCU), S.Y. Su (NCU)
W.H. Tsai (NCU), C.L. Tseng (NCKU)
C.T. Wang (NCU)

COSPAR Colloquium 2000 Taipei

CONTENTS

Keynote Speech

PREDICTING GEOMAGNETIC STORMS AS A SPACE WEATHER PROJECT

S.-I. Akasofu

International Arctic Research Center, University of Alaska Fairbanks, 930 Koyukuk, Fairbanks, Alaska 99775-7340, USA

ABSTRACT

The study of solar-terrestrial relationship has developed into four major disciplines: solar physics, interplanetary physics, magnetospheric physics, and ionospheric physics (aeronomy). Researchers have made considerable progress within each field of study during the twentieth century. It is pointed out, however, that many major problems in advancing solar weather prediction projects have been left behind. To be successful in this particular effort, space weather researchers need to establish a new discipline that synthesizes and integrates the four major disciplines and their subdisciplines. It is pointed out that we can learn all the missing components in space weather prediction only when we attempt to synthesize the four disciplines. Several missing components are pointed out.

INTRODUCTION

Geomagnetic Storms

A typical geomagnetic storm begins with a sudden increase of the horizontal component (H) approximately simultaneously over the whole Earth (Figure 1). The phenomenon is called the storm sudden commencement (SSC). It is caused by the impact of an interplanetary shock wave on the dayside magnetopause. Their impact generates Alfvén waves on the magnetopause, which propagate through the magnetosphere. The SSC is often, but not always, followed by the initial phase, a quasi-steady period of a few hours and then by a large decrease of the H component. It is generally considered that the initial phase is the period when the Earth is embedded in the region between the shock wave and the coronal mass ejected at the time of coronal mass ejection (CME).

It is also inferred that the main phase begins at the time of arrival of coronal mass. The coronal mass is considered to be a magnetic cloud which is ejected at the time of CME or a magnetic loop which expands as a result of CME. However, we know little at this time about how the coronal mass ejections are related to magnetic clouds or expanding magnetic loops. Such a fundamental problem has not been studied in any detail in the space weather project.

Geomagnetic Storms and Epsilon Function

Although the nature of interplanetary disturbances caused by various solar activities is not well understood, we have learned how the magnetosphere responds to changes of solar wind and of the interplanetary magnetic field (IMF) at the front of the dayside magnetopause. Three important parameters in this regard are the solar wind speed V, the IMF magnitude B, and the polar angle of the IMF θ. The power ε generated by the interaction between the solar wind/IMF and the magnetosphere is given by:

$$\varepsilon\,(\text{M watts}) = 20\,[V(\text{km/s})]\,[B(\text{nT})]^2\,\sin^4(\theta/2)$$

It can be seen in Figures 2a and 2b that if $\varepsilon = f(V,B,\theta,t)$ is given, it is possible to determine and predict characteristics of geomagnetic storms, which are given by the two geomagnetic indices $Dst(t)$ and $AE(t)$.

In terms of ε, geomagnetic storms can be grouped as follows:

Weak storms $\varepsilon \sim 0.25 \times 10^6$MW (e.g. $V = 500$ km/sec, $B = 5$ nT)
Moderate storms $\varepsilon \sim 1.4 \times 10^6$MW (e.g. $V = 700$ km/sec, $B = 10$ nT)
Very intense storms $\varepsilon > 8.0 \times 10^6$MW (e.g. $V = 1000$ km/sec, $B = 20$ nT)

Thus, the purpose of geomagnetic storm prediction is to predict the three parameters $V(t)$, $B(t)$, $\theta(t)$, and $\varepsilon(t)$ as a function of time at the front of the magnetosphere soon after a particular solar event. For this purpose, the following six steps should be taken:

1. Modeling of the background solar wind flow
2. Parameterizing solar events on the source surface
3. Modeling the propagation of shock waves (including the simulation of past events)
4. Estimating the velocity, density, and IMF of the solar wind at the Earth
5. Characterizing geomagnetic storms in terms of epsilon function
6. Predicting the size of the auroral oval

In this paper, a method developed by Hakamada and Akasofu (1982) and Akasofu and Fry(1986) is summarized. This method is called the HAF model and its general theme is shown in Figure 3. A brief summary of the HAF model is also given by Akasofu (2001).

MODELING THE BACKGROUND SOLAR WIND FLOW

The solar wind exhibits considerable variation even without any specific solar events, such as solar flares, CMEs, and sudden filament disappearances. This is particularly the case when high-speed streams flowing from long-lasting coronal holes, and the resulting interplanetary sector boundary structures corotating with the Sun, are present. Since any effects of specific solar events propagate into the existing solar wind structures and interact with them, it is important first of all to devise a simple way to model the background flow.

Conditions on the source surface, the imaginary spherical surface of 2.5 solar radii, are important in modeling interplanetary conditions. One of the most important aspects of the source surface is the magnetic equator (or the so-called neutral line). The axis of the dipolar field on the source surface rotates from 0° to 180° (or from 180° to 0°) during the Sun's 11-year cycle variations (Figure 4). We found that we can infer most of the main features of solar wind variations during the whole sunspot cycle at the Earth or at any point to about a distance of 2 AU by assuming that the solar wind speed is minimum at the sinusoidal magnetic equator and increases toward higher latitudes.

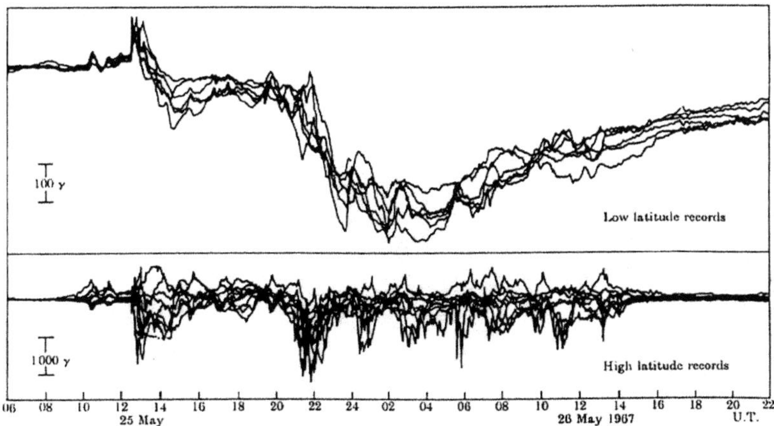

Fig. 1. Geomagnetic storm of May 25-26, 1967. Upper diagram: Superposition of six magnetic records from stations, widely separated in longitude in low latitude stations. Lower diagram: Superposition of nine magnetic records from auroral zone stations. Note the difference of the scale in the upper and lower diagrams.

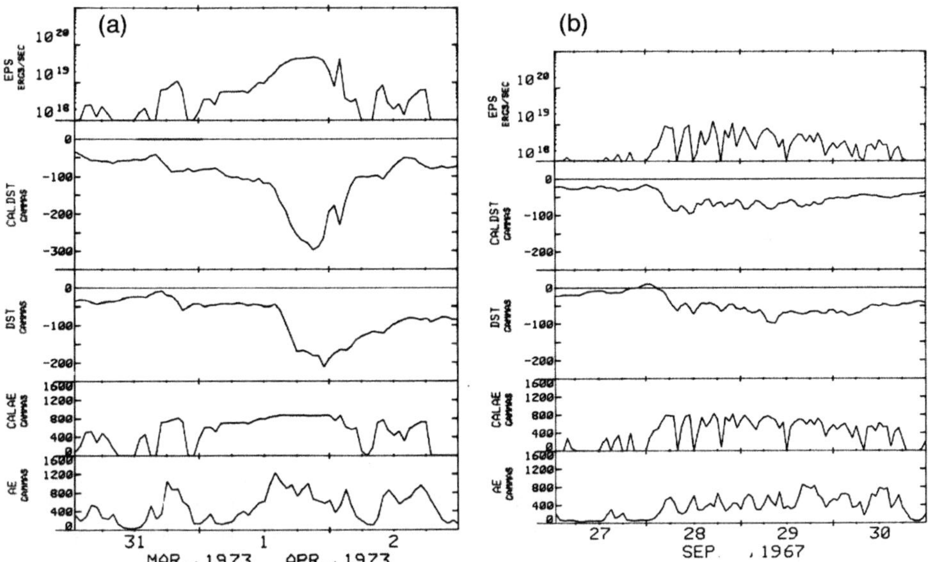

Fig. 2. From the top, the epsilon function (EPS), the computed *Dst* index (CALDST), observed *Dst* index, the computed *AE* index (CALAE) and observed *AE* index for (a) the March 31–April 2, 1973 storm, and (b) the September 27-30, 1967 storm. The main features of the storm represented by the observed *Dst* are well reproduced (CALDST).

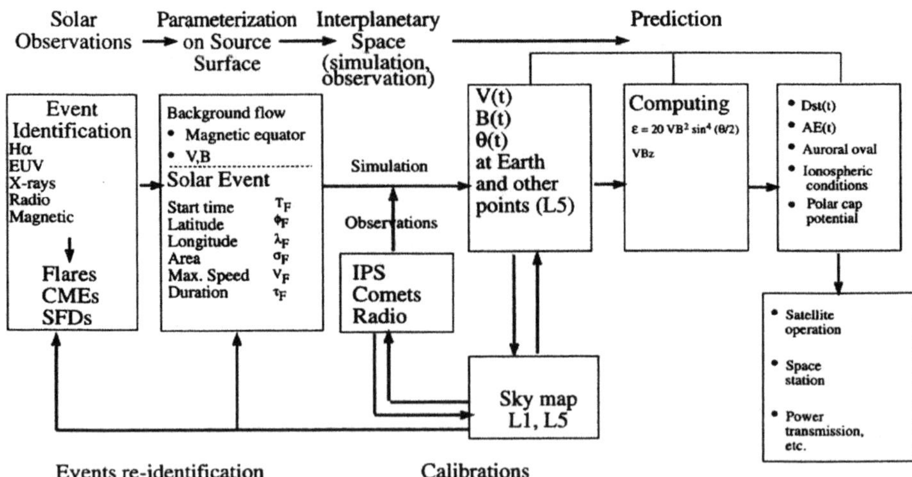

Fig. 3. The physically-based geomagnetic storm prediction scheme in a block-diagram form.

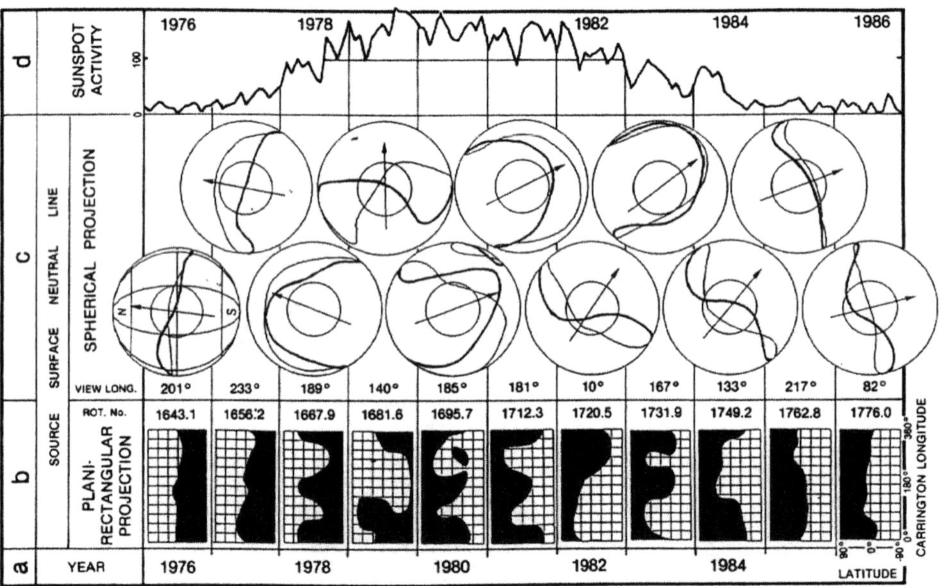

Fig. 4. Sunspot cycle variation of the orientation of the dipole field and the magnetic equator (neutral line) on the source surface.

As the Sun and its source surface rotate every 25 days, a fixed point in space (not on the source surface) at a distance of 2.5 solar radii scans horizontally the velocity field from solar longitude 360° to 0° in one solar rotation along a heliographic latitude line (e.g., 0° at June and December solstices). Figure 5 shows an example of model distribution of solar wind speed on the source surface and the resulting wind speed at a particular point (fixed in space) on the source surface. Solar wind particles leave radially from this particular point one by one with different velocities as the Sun rotates; the point depicts a sinusoidal variation of solar wind particles during one rotation.

As a result, each solar rotation sends out two waves. Subsequent changes of the radial speed of individual particles can be modeled by adopting a kinematic solution in a method developed by Hakamada and Akasofu (1982). A faster flow of particles interacts with a slower flow of particles to form a shock wave and a reverse shock. Thus, by integrating the velocity as a function of time graphically, one can determine the distance traveled by individual particles.

A magnetic field line originating from the source surface can be traced by following particles leaving a particular point on the source surface (not a fixed point in space). The resulting interplanetary magnetic field structure is the familiar Parker spiral, together with the corotating interaction region produced by the formation of the shock wave structure (Figure 6). The computed velocity (V), density (n), and IMF magnitude B agree reasonably well with the observed ones (compare Figures 7 and 8). It is important to realize that such a simple scheme can reproduce reasonably well the observed 27-day variations of the solar wind observed at the Earth. The arrival of the fast wind is associated with a sharp change of the azimuth angle (PHI) of the IMF (from toward to away or away to toward), indicating the crossing of the heliospheric current sheet. It is for this reason that the corotating structure is called the sector boundary.

PARAMETERIZING SOLAR EVENTS

One of the most difficult problems we face is how to parameterize individual solar events. All the models developed so far are nothing but *tools* for the prediction purpose. The tools cannot produce a desired result unless individual solar events can be properly parameterized. It is very surprising that little effort has so far been made in this respect, in spite of the fact that space weather forecasting has become such an important issue these days. It appears that solar physicists are interested in what happens on the Sun, while magnetospheric physicists begin their research on the basis of their satellite observations at the front of the magnetosphere. It is for this reason that the parameterization effort has been left behind.

In the HAF scheme, a solar event is represented by a high-speed source area on the source surface, which is superposed on the background structure described in the previous section. The source area is represented by a circular area (or an elliptical area); the speed is highest at the center and has a Gaussian distribution. The speed at the center varies in time in a characteristic way, which is parameterized by:

$$V = V_F e^{(1-t/\tau_F)}(t - T_F)/\tau_F$$

Thus, a solar event is parameterized by

1. V_F: the maximum speed at the center of the event;
2. τ_F: the time variations;
3. T_F: the start time of the event;
4. λ_F longitude and ϕ_F latitude; and
5. σ_F: the area size (standard deviation of the Gaussian distribution).

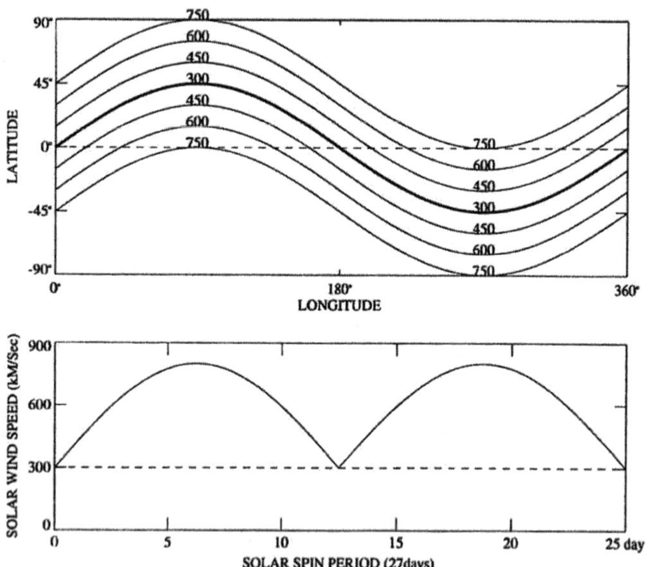

Fig. 5. The distribution of solar wind speed on the source surface and the resulting solar wind speed at a point on the equator, which is fixed in space at a point at a radial distance of 2.5 solar radii.

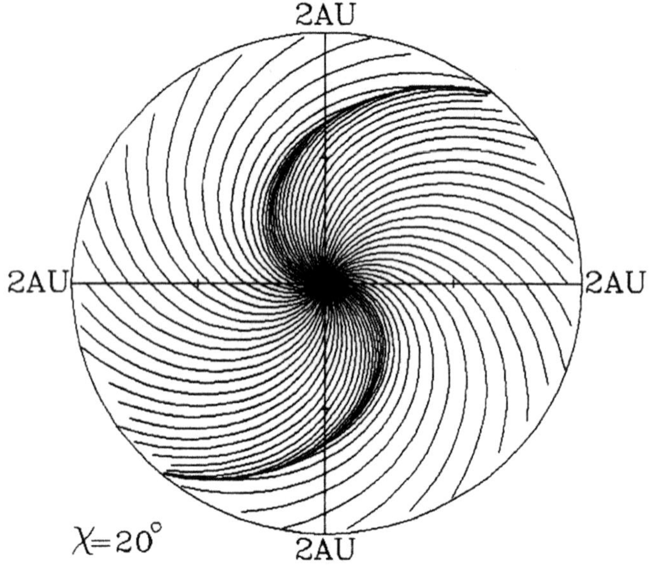

Fig. 6. The spiral structure of the interplanetary magnetic field lines on the equatorial plane. The magnetic equator on the source surface is assumed to be sinusoidal, the amplitude being 20° in latitude (see Figure 5).

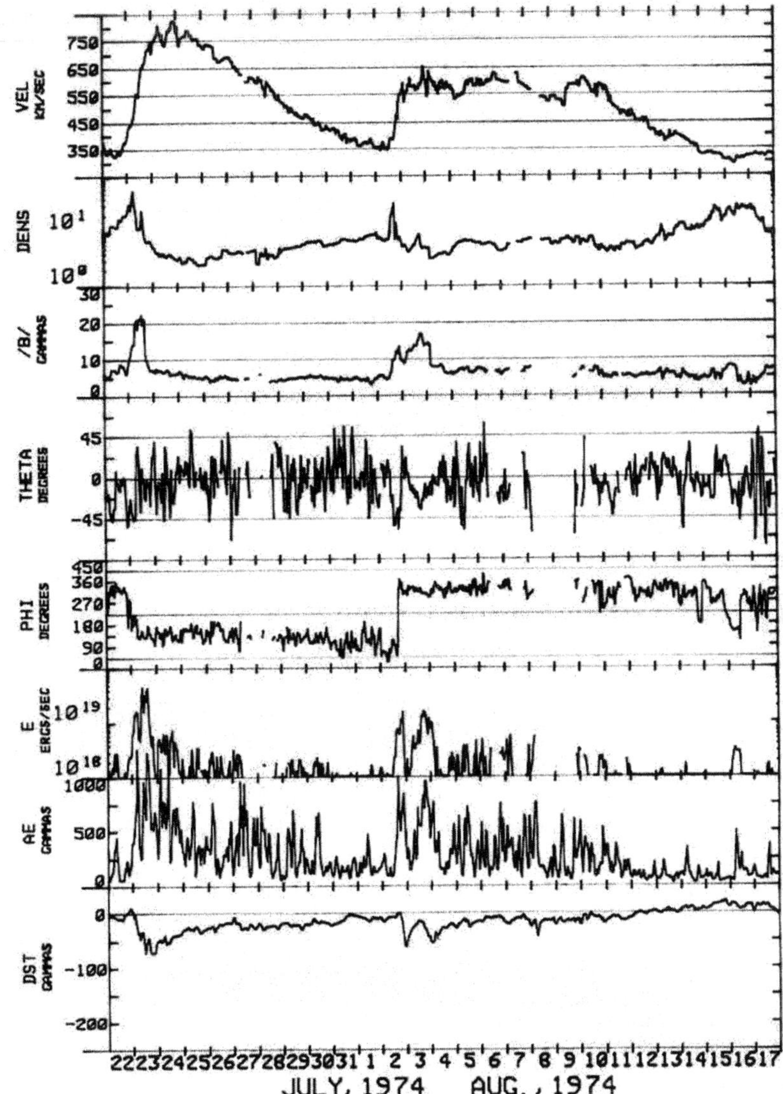

Fig. 7. Observed solar wind variations in one solar rotation (27 days), velocity (VEL), density (DENS), IMF magnitude (B), latitudinal angle (THETA), azimuth angle (PHI), the epsilon function (E), and the two geomagnetic indices AE and Dst.

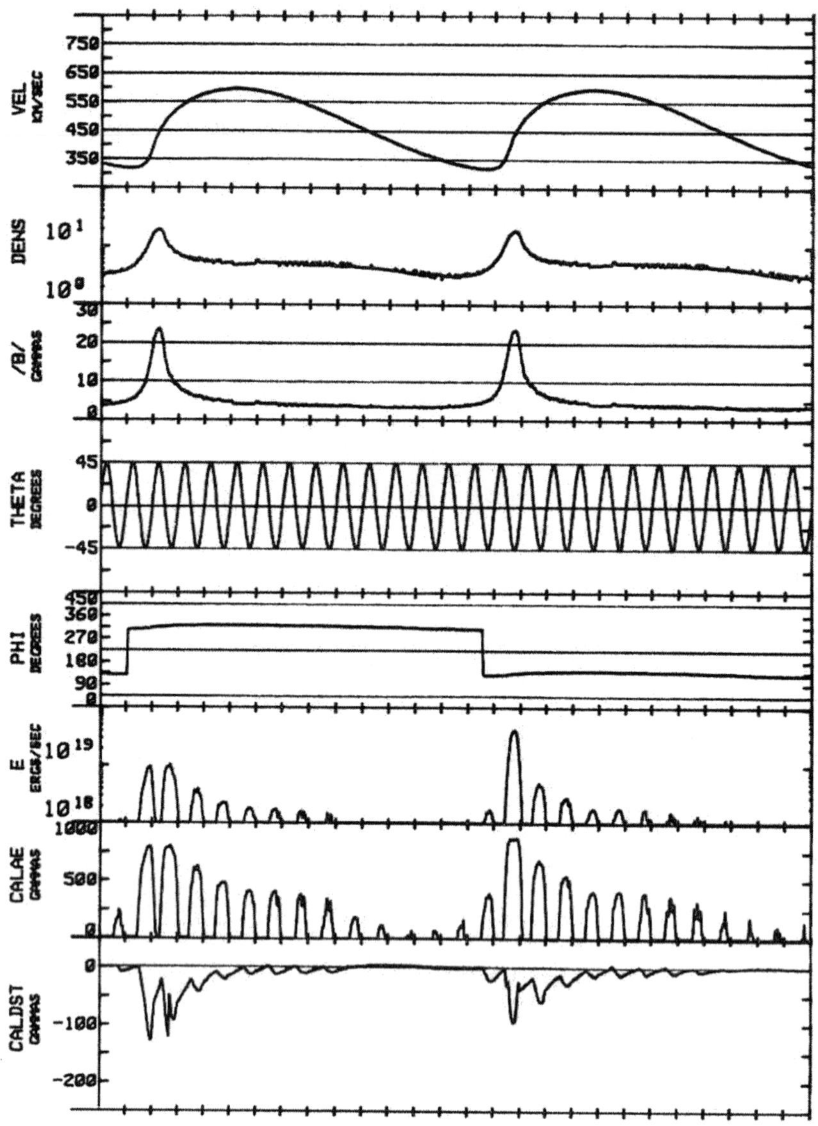

Fig. 8. Simulation of the event in Figure 7. Sinusoidal variations of the latitude angle (THETA) is superposed in representing realistic change of THETA.

Figure 9 shows graphically the adopted parameters, λ_F=180°, ϕ_F=0°, V_F=500 km/sec, σ_F=30°, and τ_F=12 hrs on the source surface for a hypothetical event. It is assumed that the event takes place on the center of the disk of the Sun at 12 UT, on December 8. Therefore, in this particular case, the center of the Sun, and the location of the solar event and the Earth, are almost on the same solar radial line at the onset of the solar event. If we consider CMEs or sudden filament disappearances, another set of parameters must be needed to describe them. Obviously, the parameterization is crucial for any modeling effort, including a MHD method.

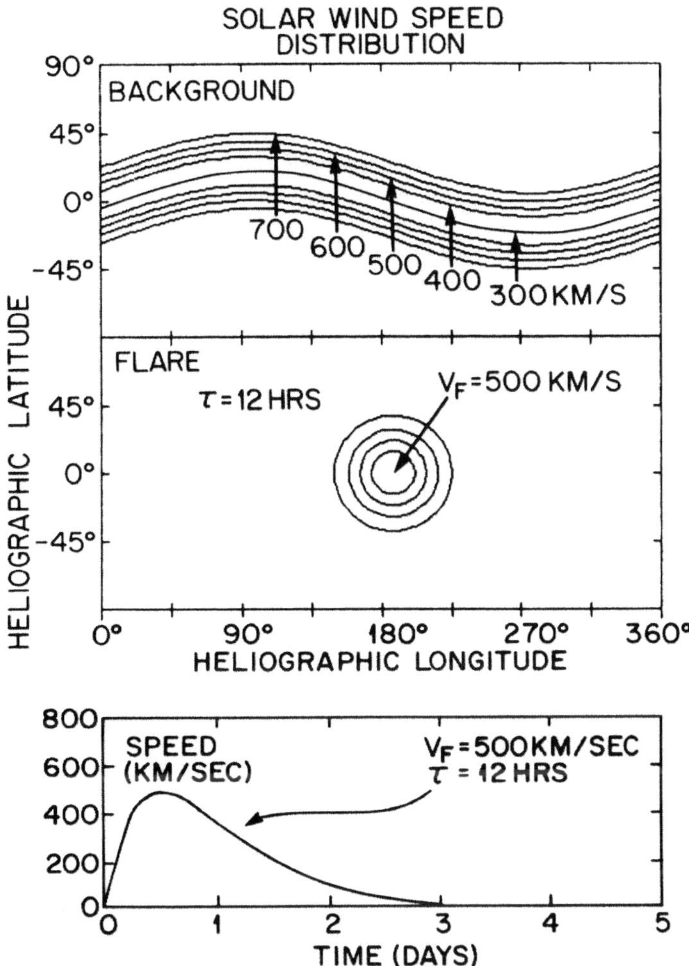

Fig. 9. From the top: The background flow speed of the solar wind on the source surface (the solar longitude-latitude map); the solar wind speed is minimum (300 km/sec) at the magnetic equator and increases toward higher latitudes (Akasofu and Fry, 1986). The middle diagram represents a solar event; a higher speed from a circular area is added to the background flow; the speed at the center of the circular area reaches $V_F = 500$ km/sec. The bottom diagram shows how the speed at the center of the circular area in the middle diagram varies in time; it reaches $V_F = 500$ km/sec 12 hours after the onset and decreases slowly (expressed by the parameter $\tau_F = 12$ hours in this particular case).

SIMULATION OF THE MARCH – APRIL 2001 EVENTS:

Figure 10 shows the locations of some of the solar events in March-April 2001 in the magnetic field map on the source surface map. The magnetic equator is given by the curve denoted by 0. The simulated interplanetary disturbances arc shown in Figure 11a. Figure 11b shows a comparison of the observed and simulated changes at the Earth (Sun et al., 2001). They showed also that the initial speed determined by metric Type II radio burst observations must be substantially reduced (30 percent in average) for most high-speed shock waves (Figure 11b). This is one of the parameterization problems.

SOUTHERN BOUNDARY OF THE EXPANDING AURORAL OVAL

The auroal oval tends to expand during major geomagnetic storms. Akasofu (1974) determined the relationship between the southern boundary of expanded oval and the *Dst* index (Figure 12). The southern boundary was determined by the location of the southernmost auroral arcs observed by all-sky cameras. It is likely that the precipitation of the auroral electrons might occur a little south of the observed arcs, but might not produce an auroral arc. Thus, one must be cautious in defining the boundary; different definitions will produce obviously different results. The observed *Dst* index is given by $\Delta H'$. The observed value of $\Delta H'$ includes the indication effect, so that the real ring current field is $\Delta H = (2/3)\Delta H'$.

SUNSPOT CYCLES AND INTENSE GEOMAGNETIC STORMS

Figure 13 shows the relationship between the sunspot cycle and the occurrence of intense geomagnetic storms. For this purpose, the occurrence of the most intense storms categorized as C9 storms are plotted with the sunspot number R for four sunspot cycles. The C9 index was devised by Bartels (see Akasofu and Chapman, 1972). As can be seen clearly in Figure 13, there is no clear tendency for most intense storms to occur at the peak of sunspot cycle. C9 storms can occur even near the minimum epoch of a solar cycle.

NEED FOR PREDICTING THE IMF POLAR ANGLE $\theta(t)$

In examining the variability of the solar wind velocity V, the IMF magnitude B, and the IMF polar angle θ, the most variable quantity is θ. In order for ε to be greater than 10^6 MW, for $V = 500-1000$ km/sec and $B \sim 10$ nT, it is necessary for θ to be greater than $90°$. This situation is generally called the southward turning of the IMF. It is likely that intense solar events produce a high value of $V \sim 500-1000$ km/sec and $B > 10$ nT. However, if θ happens to be $0°$ or very small, ε cannot reach 10^6 MW. It is thus obvious that we cannot succeed in predicting geomagnetic storms until we can find a way to predict $\theta(t)$. Thus, the prediction of $\theta(t)$ after a solar event is most crucial in predicting geomagnetic storms.

It is unfortunate that there has been little effort to predict $\theta(t)$, in spite of the fact that $\theta(t)$ is such a crucial parameter in predicting space weather. Again, most solar physicists show no interest in this effort, while magnetospheric physicists base their research on the observed $\theta(t)$ at the front of the magnetopause. It is for this reason that the problem of predicting $\theta(t)$ is left behind.

Recently, IMF changes associated with solar events are discussed in terms of magnetic clouds, magnetic flux ropes, loops, tubes, etc. (Figure 14). Perhaps some efforts are needed to standardize such terms. For example, if the structures are detached magnetically from the Sun, they may be called *clouds*, while if the structures are magnetically anchored to the Sun, they may be called *loops*.

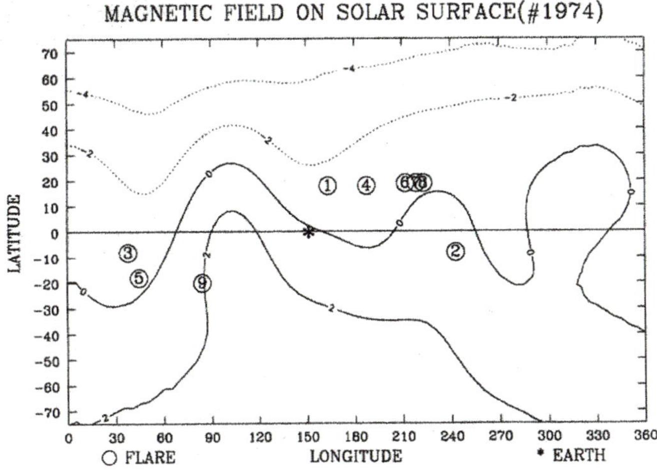

Fig. 10. Locations of some of the solar events in March-April 2001.

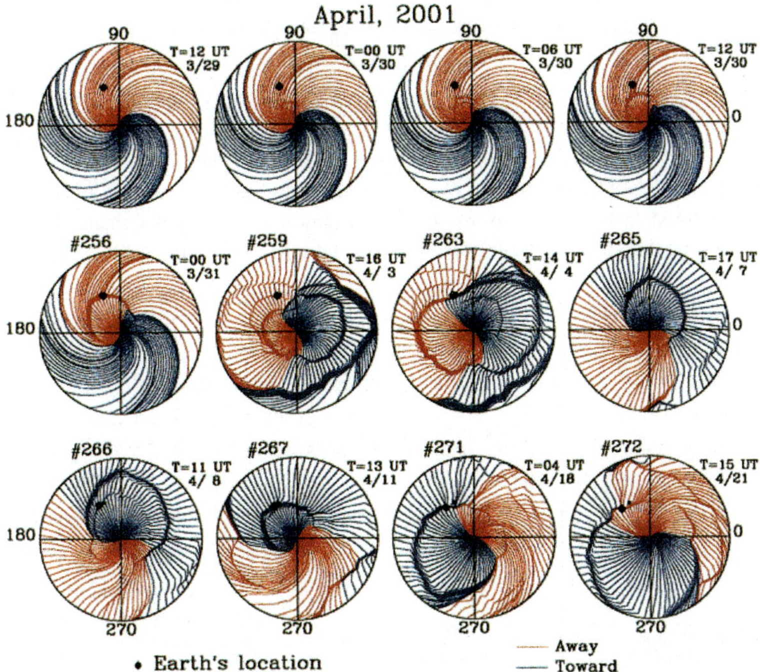

Fig. 11a. Interplanetary magnetic field lines in the ecliptic plane to a distance of 2 AU at different times. The first five diagrams show the propagation of the shock wave generated by flare/CME FFNo 256, which caused the most intense geomagnetic storm since the last decade on 31, March 2001. The other seven diagrams show the shock arrivals at the Earth for the subsequent flares/CMEs (Sun et al., 2001).

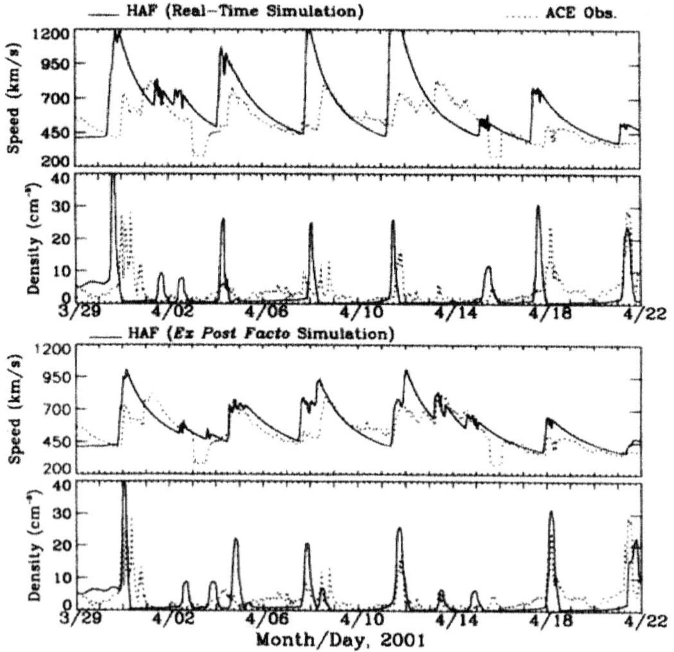

Fig. 11b. The top two diagrams show real-time simulated and observed solar wind speed and density at L1 as a function of time. The input initial speed of the shock waves is determined by the Type II radio burst observations. The lower two diagrams show the same except by using, *ex post facto*, an adjusted initial speed in simulation. The observed solar wind dropouts on 3 April and 5 April were caused by temporary ACE/SWEPAM failure (only in real time) due to contamination by prompt arrival of energetic solar flare protons. The densities at these times are also invalid (Sun et al., 2001).

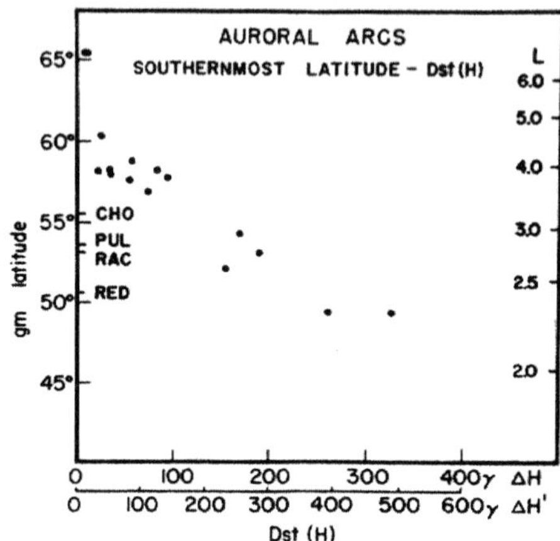

Fig. 12. Location of the southern-most auroral arc determined by all-sky cameras as a function of the *Dst* decrease.

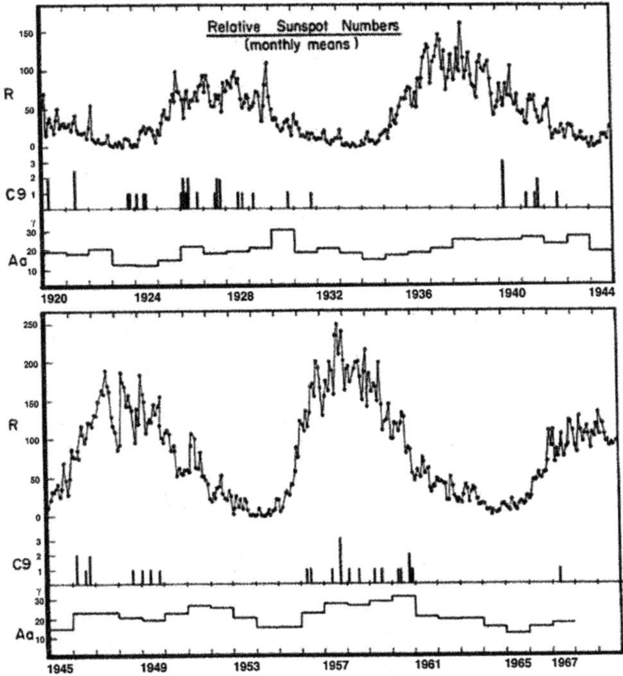

Fig. 13. The occurrence of the most intense storms (C9 storm) for several sunspot cycles.

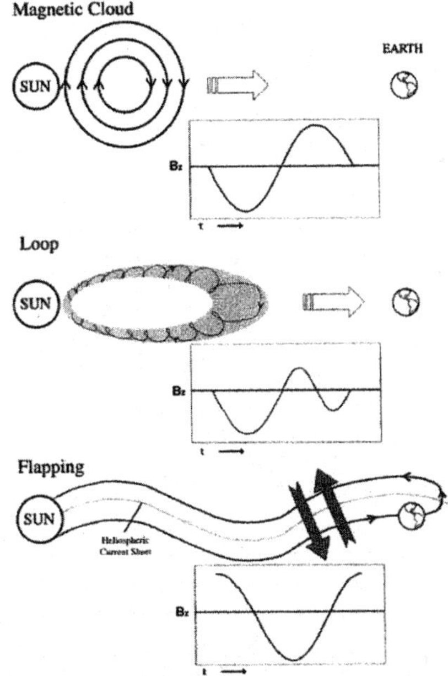

Fig. 14. Schematic diagrams illustrating three different ideas as to how IMF changes during geomagnetic storms.

Three important issues in this regard are how these structures are related to:

i. Coronal Mass Ejections (CMEs),
ii. whether the expanding CMEs constitute the so-called *driver* gas which is responsible for generating the shock wave, and
iii. if it would be at all possible to predict changes of the IMF polar angle θ as a function of time on the basis of solar observations.

Tsurutani et al. (1988) examined several possibilities for the causes of θ changes. Most sunspot pairs align in the east-west direction. However, in some exceptional cases, the pairs tend to align in the north-south direction. It was their finding that the orientation of the sunspot field is unrelated to that of the IMF at the Earth's location. Therefore, the magnetic structure within an enhanced solar wind is not a simple expansion of magnetic loops anchored in the vicinity of active spots. Figure 15 shows such an example. We were supposed to observe a strong southward oriented IMF after August 9, 1982, but no such event occurred.

It is known that there occur often several very sharp changes (in time) of the azimuth angle of the IMF, from 0° to 180° or 180° to 0°, during a major geomagnetic storm, suggesting sometimes a large-scale movement of the heliospheric current sheet, so that the earth's position with respect to the current sheet (above, before the passage of the shock) can change (below, after the passage). Some of the changes of θ may be related to flapping of the current sheet, because the field lines associated with the current sheet are supposed to be parallel to it. Figure 16 shows the flapping of the heliospheric current sheet inferred from the changes of the IMF angles. Further, the passage of the shock wave can increase the magnitude of the IMF, so that ε can be increased if $\theta > 90°$.

HIGH-SPEED STREAMS FROM CORONAL HOLES

It is now well-established that recurrent geomagnetic storms are caused by a high-speed stream from a coronal hole. It is also known that coronal holes and thus recurrent geomagnetic storms tend to develop during the declining period of sunspot cycles. One way to examine this feature is to examine the C9 geomagnetic index. Figures 17a and 17b show the C9 index between 1957 and 1966 and also between 1969 and 1975. Unfortunately, such useful figures are long forgotten. Both sunspot numbers R and the C9 index are shown with numbers with different bold types, so that a simple examination of the figures can shown many interesting trends.

First of all, it can easily be seen that recurrent geomagnetic storms and thus persistent coronal holes tend to appear a few years after the maximum phase of a sunspot cycle. The resulting high-speed streams rotate with the Sun and sweep by the Earth; it takes about one week to pass by the Earth per each solar rotation. As a result, the Earth tends to be immersed in the high-speed stream twice, each time for one week, in each solar rotation. The resulting auroral activity occurs twice (each lasting for 7-10 days) during each solar rotation, with a period of 27 days; see Figure 7. Although this fact has been well known for a long time, it appears that it has recently been forgotten. Thus, as a result, many solar physicists and auroral physicists claim that the sunspot maximum years coincide with the auroral peak year. This is not the case in high latitudes (Figure 18). It is unfortunate that auroral activity caused by the solar wind from coronal holes has to be rediscovered every eleven years.

Large equatorial coronal holes tend to be an extension of the polar coronal holes (Figure 5). We have simulated high-speed streams from equatorial coronal holes in Figure 6. The two streams in one solar rotation tend to occur because the sinusoidal magnetic equator is a representation of the extension of two coronal holes, one from the northern and the other from the southern polar region. Indeed, a typical situation, two coronal holes occur that are separated by 180° in longitude. As a result, two high-speed

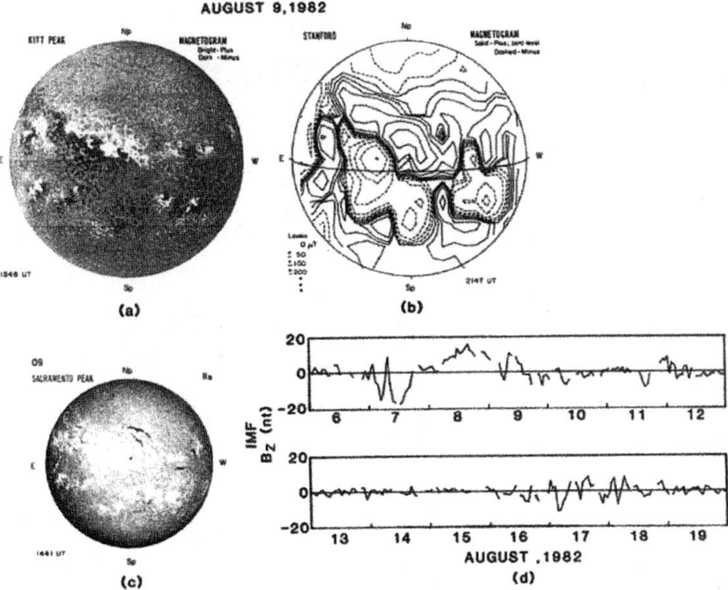

Fig. 15. A very large-scale magnetic structure in which the field orientation was such that it could produce a large southward oriented field. However, there was no clear orientation of IMF at the Earth.

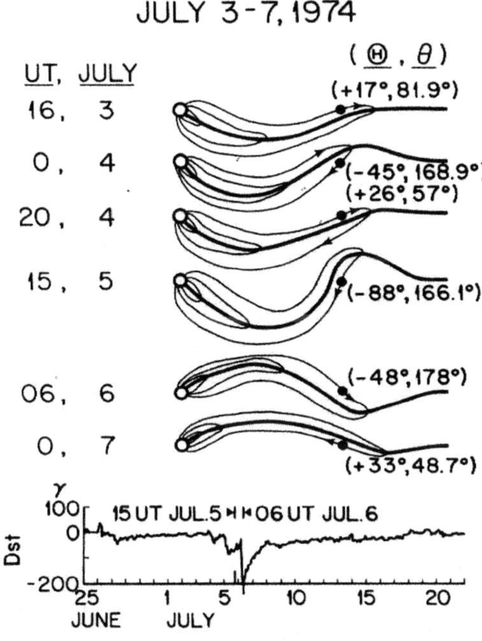

Fig. 16. Flapping of the heliospheric current sheet, which is inferred from sharp changes of the IMF azimuth angle.

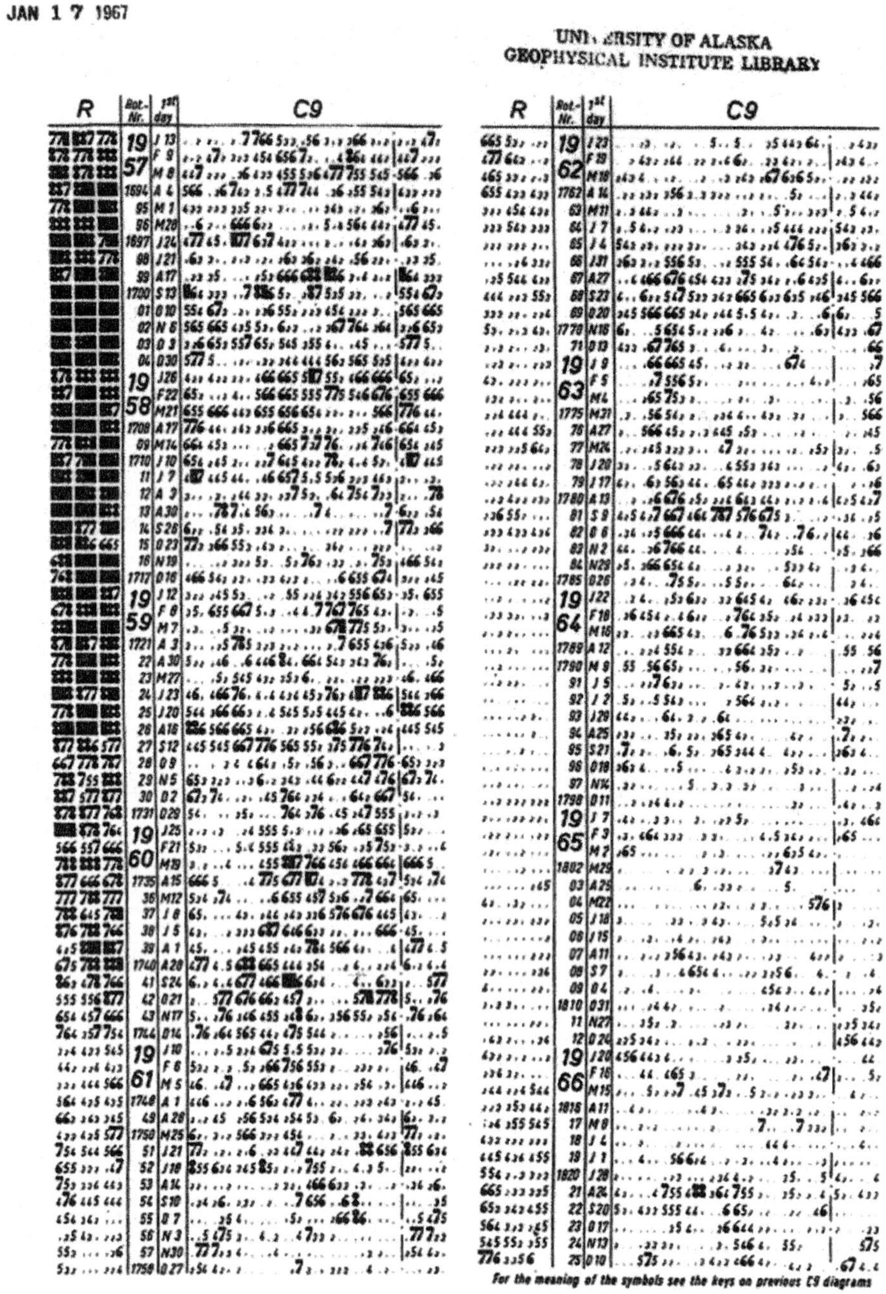

R	Rot.-Nr.	1st day	C9

R	Rot.-Nr.	1st day	C9

For the meaning of the symbols see the keys on previous C9 diagrams

Fig. 17a. Sunspot number R and the C9 geomagnetic index (1957-1966).

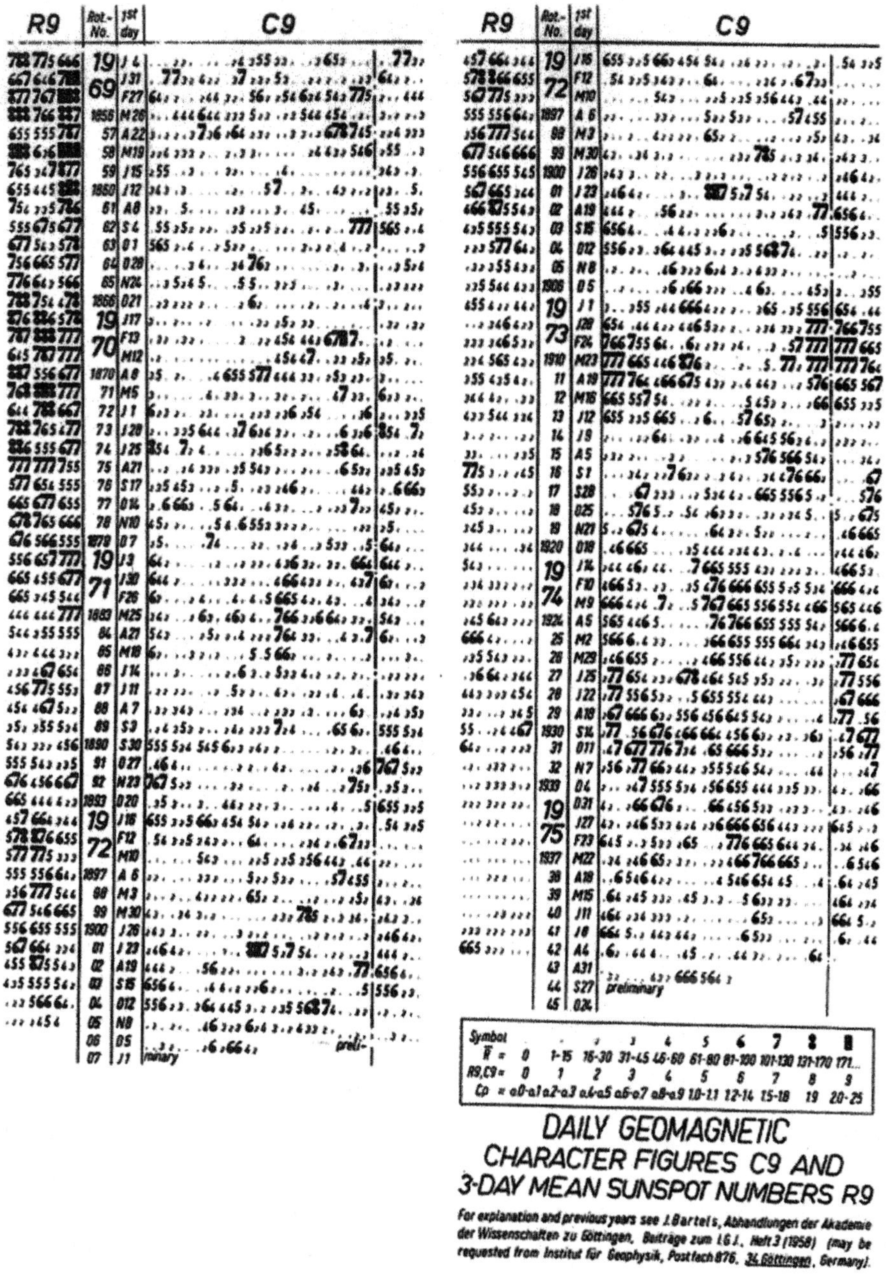

Symbol	.	₂	₃	4	5	6	7	8	9
$\bar{R}$ = 0	1-15	16-30	31-45	46-60	61-80	81-100	101-130	131-170	171...
R9,C9 = 0	1	2	3	4	5	6	7	8	9
Cp = a0-a1	a2-a3	a4-a5	a6-a7	a8-a9	1.0-1.1	12-14	15-18	19	20-25

DAILY GEOMAGNETIC CHARACTER FIGURES C9 AND 3-DAY MEAN SUNSPOT NUMBERS R9

For explanation and previous years see J. Bartels, Abhandlungen der Akademie der Wissenschaften zu Göttingen, Beiträge zum I.G.J., Heft 3 (1958) (may be requested from Institut für Geophysik, Postfach 876, 34 Göttingen, Germany).

Fig. 17b. Sunspot number R and the C9 geomagnetic index (1969-1975).

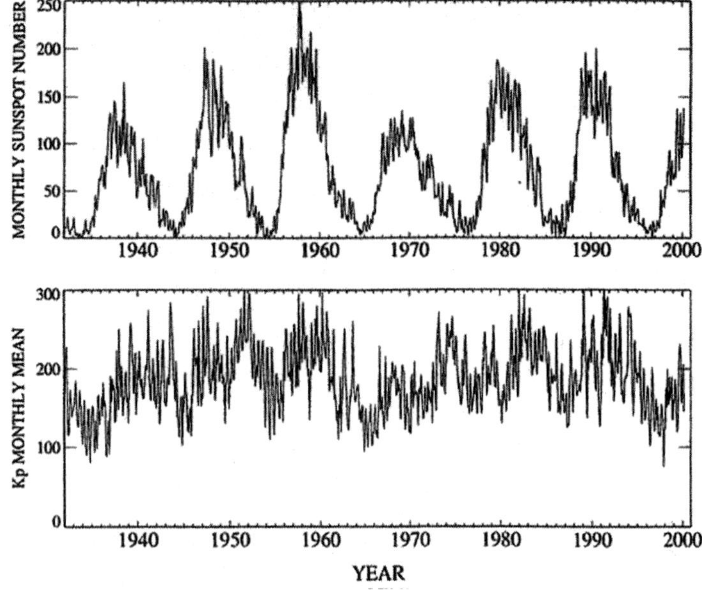

Fig. 18. Relationship between sunspot variations and the Kp index (1930-1999).

streams appear, causing two peaks of speed of the solar wind during one solar rotation; see the C9 index in 193 and 1964 in Figure 15a and the C9 index in 1973 and 1974 in Figure 15b.

ACKNOWLEDGMENTS

I would like to thank the organizing committee for inviting me to give the keynote presentation. I would like to thank also Dr. Ling-Hsiao Lyu for her conscientious editing effort of my paper.

REFERENCES

Akasofu, S.-I., The latitudinal shift of the auroral belt, *J. Atmosph. Terr. Physics,* **26**, 1, 167 (1974).

Akasofu, S.-I., Predicting geomagnetic storms as a space weather project, *Space Weather*, Geophysical Monograph, American Geophysical Union, Washington, D.C. (2001).

Akasofu, S.-I., and S. Chapman, *Solar-Terrestrial Physics*, Oxford Univ. Press, Oxford (1972).

Akasofu, S.-I., and C. D. Fry, A first generation numerical geomagnetic storm prediction scheme, *Planet. Space Sci.*, **34**, 77 (1986).

Hakamada, K., and S.-I. Akasofu, Simulation of three-dimensional solar wind disturbances and resulting geomagnetic storms, *Space Sci. Rev.*, **31**, 3 (1982).

Sun, W., M. Dryer, C. D. Fry, C. S. Deehr, Z. Smith, S.-I. Akasofu, M. D. Kartaleval, and K. G. Grigokov, Evaluation of solar Type-II radio burst estimates of initial solar wind shock speed using a kinematic model of the solar wind on April 2001 solar event swarm, *Geophys. Res. Lett.*, (in press), 2001.

Tsurutani, B. T., W. D. Gonzalez, F. Tang, S.-I. Akasofu, and E. J. Smith, Origin of interplanetary southward magnetic fields responsible for major magnetic storms near solar maximum (1978-79), *J. Geophys. Res.*, **93**, 8519 (1988).

Solar Observations and Modeling Session

DESCRIPTIONS OF CORONAL STREAMER STRUCTURES DURING THE RISING PHASE OF CYCLE 23

M. D. Andrews[1] and S. T. Wu[2]

[1]*Computational Physics, Inc., Code 7665A, Naval Research Laboratory, Washington, DC 20375, USA;*
[2]*Center for Space Plasma and Aeronomic Research and Department of Mechanical and Aerospace Engineering, The University of Alabama in Huntsville, Huntsville, Alabama 35899, USA*

ABSTRACT

Observations by the LASCO C3 telescope on SOHO have been used to illustrate the variations of coronal structures during the rising phase of Cycle 23. Most coronal streamers observed at solar minimum are seen near the equator and often exhibit a symmetric configuration. As the solar maximum approaches, the coronal streamers occur at mid to high latitude. The symmetric configuration no longer appears. We use numerical magnetohydrodynamic (MHD) models to deduce the magnetic field topology, velocity field, and plasma parameters for these observed configurations.

INTRODUCTION

Observations of the corona during total solar eclipses have been made for centuries. The invention of the coronagraph by Lyot in the 1920s allowed additional observations from terrestrial observatories. Our impression was that the corona underwent a slow evolution in its appearance over the 11-year solar activity cycle. With the recent development of space-borne white-light coronagraphs (Tousey, 1973) in the 1970s, we recognized that the corona is a very dynamic region. The launch of the Large Angle Spectroscopic Coronagraph (LASCO) on-board the Solar and Heliospheric Observatory (SOHO) gave us a whole-sun view of the extended corona from 1 $R_\odot$ to 30 $R_\odot$ ($R_\odot$ = solar radius). These instruments give us a unique opportunity to observe the variation of the corona during the rising phase of this (23rd) solar cycle.

The most spectacular manifestation of coronal dynamical activity is the coronal mass ejection (CME) that often originates from coronal streamers and occurs over a wide range of time and space scales. These transient phenomena consist of the expulsion of a few 10^{16}g of plasma and magnetic flux from the sun into interplanetary space with speeds that can be is excess of 2000km s^{-1}.

In this paper, we will discuss the large-scale structure of the corona as observed by LASCO/C3 during the rising phase of solar cycle 23. It is this large-scale structure that will control the properties of CMEs. The relationship between CMEs and the large-scale structure of the corona is an area of active research (Wu, *et al.*, 2000a) that will not be addressed in this paper. This work is limited to showing how the large-scale structure has changed as the maximum of solar activity is approached.

A magnetohydrodynamic (MHD) model can be used to reproduce these observed structures. Results of this modeling will be presented to show that the solar minimum corona corresponds to a simple global dipolar field. The structures at solar maximum require multi-polar magnetic field configurations.

LASCO C3 OBSERVATIONS: 1996-2000

The LASCO telescopes (Brueckner *et al.*, 1995) were launched on the SOHO satellite in December 1995. The C3 telescope images the region of approximately 3.7 to 32 $R_\odot$ with a resolution of ~ 113" and a pixel size of 56". All of the images used in this study were taken using the clear filter that has a band-pass of approximately 400-850nm. LASCO began taking a regular, synoptic series of images in May 1996. These images are continuously available with only minor interruptions with the exception of the period beginning in June 1998 when the SOHO mission was interrupted.

The progression of solar activity as represented by sunspot number is shown in Figure 1. The minimum of solar cycle 22 occurred in 1996. The start of the new cycle is assumed to have been May 1996 (Hathaway, *et al.* 1999). Solar activity has continued to increase through most or all of 2000. In this study, we consider data collected with the LASCO C3 telescope between May 1996 and May 2000 i.e., the rising phase of solar cycle 23.

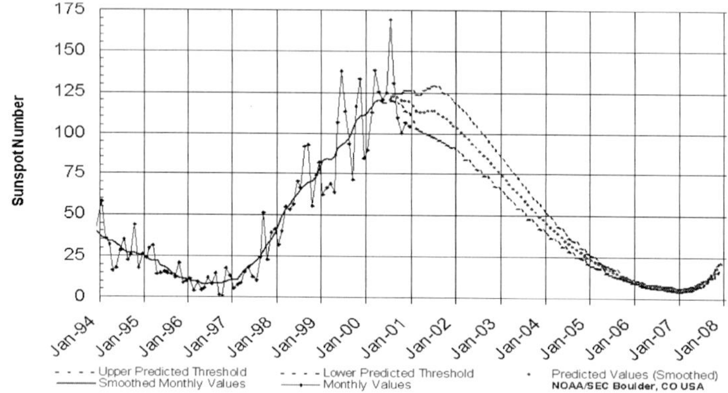

Fig. 1. Solar activity as measured by sunspot number. Shown is the measured monthly sunspot number and smoothed monthly values from January 1994 through December 2000. Also shown are three predictions of sunspot number through the year 2006. Solar cycle 23 began in 1996 and will peak in late 2000 or early 2001. Plot provided by Space Environment Center, Boulder, CO, National Oceanic and Atmospheric Administration (NOAA), US Dept. of Commerce using data collected by The International Solar Environment Service (ISES),

Data Processing

The C3 observations have been processed to generate weekly averages of the coronal brightness. The use of a one-week average was a compromise. An averaging period of one week is long enough that CMEs are not usually visible. Only the large-scale structures are seen rather than transient events. The cadence of one image per week is high enough that the effects of solar rotation can be seen with 4 images per rotation.

Each weekly image is the difference of the average brightness minus a background image. The calculation of the average and background images is described in Wu *et al.* (1997). The average brightness is the arithmetic mean of the data values for each pixel averaged over seven days running from Sunday through Saturday. Average images were calculated only if there were at least 4 days with good data coverage. Only one image per hour was used in calculating the averages. This study covered the period May 19,1996 through 27 May 2000, a period of 210 weeks. Due to gaps in the data coverage, only 176 weekly averages could be calculated.

The background image is the minimum value of the coronal brightness for each pixel over a six-week period centered on the time of the average. The background image subtraction removes the contribution of stay and scattered photospheric light and the F-corona (dust scattering), all of which are assumed to be constant in time. There will also be a contribution from the smallest observed value of the K-corona (electron scattering). The contribution of this component of the corona should be small, especially at times of significant solar activity. However, the level of the K-corona in the background image, while assumed small, is not known.

The weekly averages have been combined to generate a data animation, movie. This movie is available via anonymous ftp from ares.nrl.navy.mil, directory /pub/lasco/andrews, file c3_week.mpg. The difference images have a large dynamic range that was reduced by applying a radial-filter. This filter is calculated by transforming each difference image into a radius-angle matrix. The average is taken over all angles to obtain the mean brightness as a function of radius. Each value in the difference image is then divided by the average brightness appropriate for the radius of that pixel. This resulting image has a small dynamic range that can easily be displayed using 8-bit images. The radial filter applied to each image is calculated from that image. While this method works well to show the large-scale structure of the corona, it does not show changes in the level of the images with time.

Figure 2 shows 12 weekly averages extracted from the above referenced movie. The typical structure of the corona at solar minimum is shown in the first two images labeled 07Aug96 and 27Nov96. The coronal structure was seen as low-latitude streamers with an "arrow-head" configuration. This structure was quite stable as illustrated by the similarity of these two images that are separated by four solar rotations. An extended area of weak, magnetic activity appeared in May 1996 and persisted into 1997, and is the probable cause of this shape. This area was located about 20° S. of the equator and extended over about 40° of solar longitude.

During 1997, the coronal structure was more variable. The image labeled 21May97 shows a very simple symmetric corona where the only structures are bright, narrow streamers located directly above the solar equator. The image labeled 01Oct97 shows a slightly more complicated, but still simple, symmetric configuration.

Throughout 1996 to early December 1997, the only observed coronal structures were well-defined streamers seen only at latitude below approximately 30°. Beginning in December 1997, the observed coronal structures begin to move toward higher latitude. The images labeled 04Feb98 and 01Apr98 are separated by two rotations. The streamer above the east limb is seen at significantly higher latitude in the second image. The structure of the corona changes rapidly during early 1998. The image labeled 17Jun98 is the last weekly averaged calculated prior to the SOHO mission interruption. In this frame, streamers are seen at 45° N and 46° S. The images labeled 09Dec98 and 16Dec98 are the first two weekly averages that could be calculated after the SOHO recovery. A persistent streamer is seen at 63° N. During this period, streamers are usually observed at latitudes above 50°. The images labeled 12May99 and 02Feb00 show a feature that is commonly observed from the middle of 1999 through 2000. During this period, the corona above the solar equator can be significantly less bright than the corona above the poles. This is particularly clear in the frame labeled 02Feb00 above the west limb. The final image,

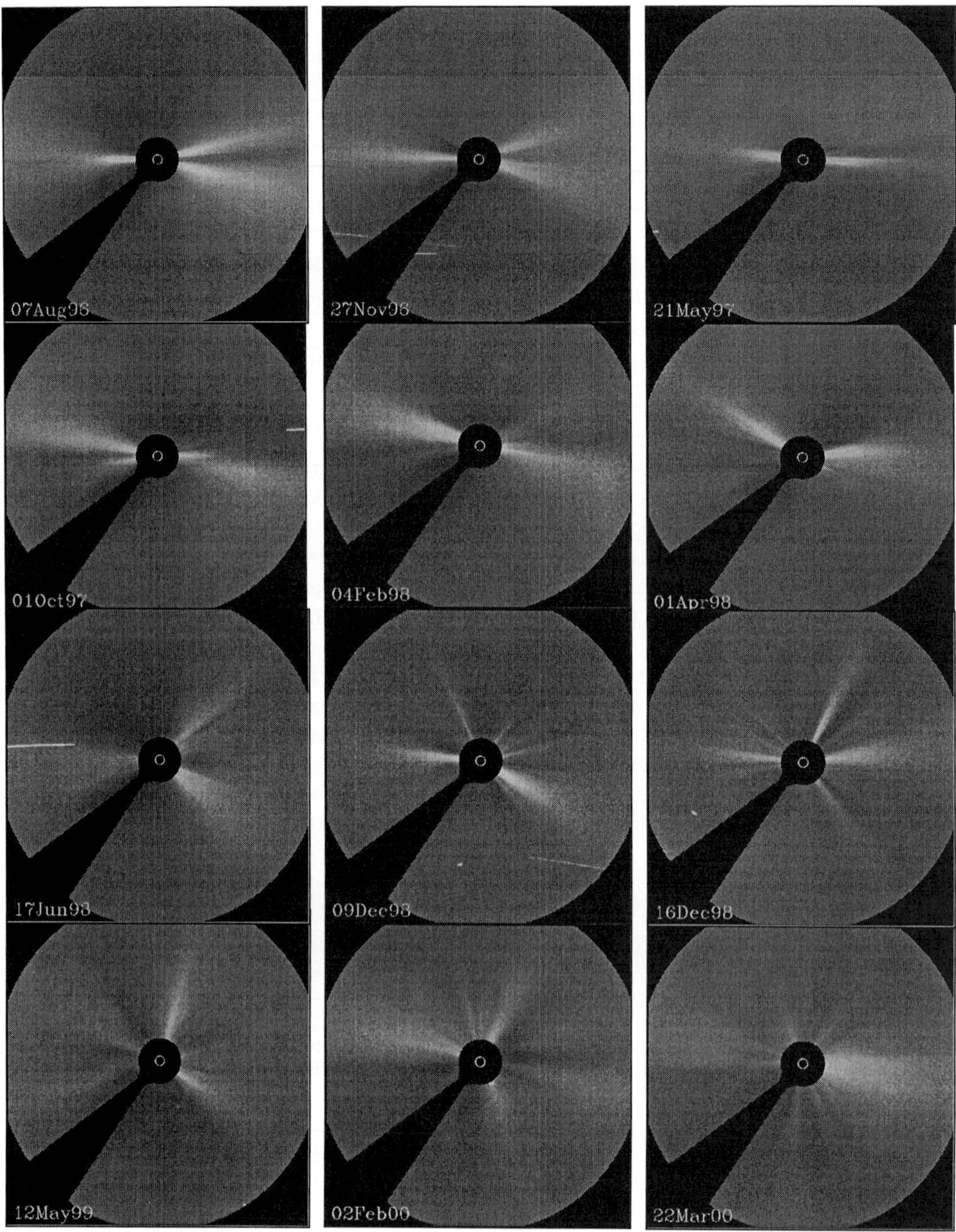

Fig. 2. LASCO C3 weekly average images. The twelve images shown illustrate the transition of coronal structure through the rising phase of solar cycle 23 (from left to right – 8/7/96; 11/27/98; 5/21/97; 10/1/97; 2/4/98; 4/1/98; 6/17/98; 12/9/98; 12/16/98; 5/12/99; 2/2/00; 3/22/00) Each image is the average of 7 days of LASCO C3 data with a background image subtracted and a radial filter applied (see text). Coronal structures are seen only at low latitude through December 1997. As activity increases, coronal structures are seen at higher latitude. The images from 2000 show coronal structures at very high latitude. The dark lane to lower-left corner is caused by blocking of the images due to a post that supports the C3 occulting disk.

labeled 22Mar00, shows the corona with no clearly defined structures. There is a suggestion of multiple, faint streamers at both extreme southern and northern latitudes. A wide region of emission is seen above the west limb that is probably the superposition of multiple, faint structures.

MHD INTERPRETATIONS OF THE OBSERVED CORONAL STUCTURES

Two distinct features of the coronal structure are immediately recognized; (i) coronal streamers are observed only near the equator at solar minimum and (ii) bright coronal structures appear everywhere as solar maximum approaches. These features are generally known, however, this is the first time that a data set of four-year space observation of the corona is presented as shown in Figure 2.

In general, we understand that these bright coronal streamers are formed due to the interactions of plasma flow and magnetic field. The enhanced brightness indicates plasma trapped by the closed configuration of the magnetic field. The theory of magnetohydrodynamics (MHD) is the proper theory to be used for the interpretation of these observed coronal features (Andrews, *et al.* 1999; Plunkett, *et al.* 2000; Linker and Mikic 2000; Wu *et al.* 1997, 1999)

We have used a MHD model to simulate the observed corona during the periods of solar activity minimum and maximum. The equations for this MHD model consists of mass, momentum, and energy conservation. In addition, the induction equation as derived from Maxwell's equation is included to describe the nonlinear interaction between the plasma flow and magnetic field. These governing equations are given by Wu *et al.* (1995). The equations are solved in a computational domain extending from the Sun (1 $R_\odot$) to 15 $R_\odot$. For the solar minimum, the corona is symmetric and it is assumed that meridional flow is zero at the pole and equator so that the calculation need only be done from the pole to the equator. The computation is done using 64 grid-points in the radial direction and 22 grid-points in the meridional direction. The radial grid size slowly increases with radius ($r_{i=1} = r_{i-1}$ $(1+\Delta\theta)$) to assure numerical accuracy. The meridian grid is divided so that points lie equidistant on either side of $\theta = 0$ and $\theta = 90°$, at $\theta = -2.25°, 2.25°, 6.75°, \ldots, 87.75°, 92.25°$.

The algorithm adopted here is the Full-Implicit-Continuous-Eulerian (FICE) scheme given by Hu and Wu (1984). For time-step, a second-order accurate forward difference scheme is used, with the step size being of the same order as given by the Courant condition (Roach, 1998) because the magnetic field is calculated explicitly. At the lower boundary, the flow is subsonic and sub-Alfvenic so that two of the six independent variables are calculated using compatability relations (Wu and Wang, 1987; Wu *et al.* 2000b). At the outer boundary, the flow is supersonic and super-Alfvenic. In this case, all variables at the boundary can be calculated by simple linear extrapolation from the first (or first two) grid points inside the boundary (Wu and Wang, 1987).

We have chosen a global dipole to represent the coronal magnetic field for the period of solar minimum. A complex multi-polar magnetic field configuration is used for the period of solar activity maximum. We used these two specific magnetic field configurations together with the flow field of a polytropic, hydrodynamic Parker solution substituted into the set of an ideal MHD governing equations. Because these given conditions are neither self-consistent nor stable solutions to the set of time-dependent ideal MHD equations, we allowed them to evolve in time according to the governing equations until their numerical solution was stable and of sufficient accuracy (Wang *et al.* 1993). This allowed us to determine the physically interesting aspects of the solution after the relaxation had proceeded long enough for an acceptably close approximation to the steady state to be reached. These solutions for the representations of the corona at solar activity minimum and maximum are described as follows.

Figure 3 shows the polarization brightness (PB) for the solutions of the initially global dipolar magnetic field (Wang *et al.* 1998). The model result should be compared the difference image of May 21, 1997 shown in Figure 2. This specific numerical solution resembles the corona at solar activity minimum very well. The corresponding radial velocity and plasma density distributions are shown in Figure 4. The results shown in Figure 4 clearly indicate the bi-modal solar wind characteristics; low speed wind near the equatorial region and high-speed wind at high latitudes (i.e. coronal hole).

Fig. 3. The magnetohydrodynamic (MHD) simulated polarization brightness (PB) by using an initially global dipolar magnetic field for the case of the corona at solar minimum. This polarization brightness is computed from the density resulting from MHD simulation according to the theory of Billing (1962) (reproduced from Wang *et al.* 1998).

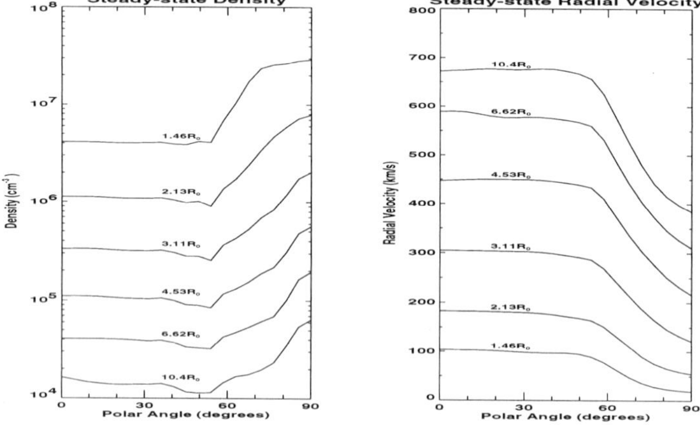

Fig. 4. The MHD simulated radial velocity and density distributions of the corona at solar minimum of Fig. 3 which shows the high speed solar wind at the high latitude and low speed solar wind at dquatorial streamer region (reproduced from Wang *et al.* 1998).

For the maximum of solar activity, we have chosen a multi-pole field as the initial condition and extended the computation to the whole sun, e.g., 360°. The other numerical procedures are identical to those used for the case for the sun at minimum activity. The numerical solution of the polarization brightness for this initially global, multi-polar magnetic field is shown in Figure 5. This result resembles the observations for February 22, 2000 as shown in Figure 2. The corresponding radial velocity and plasma density distributions are presented in Figure 6. These results show that the solar wind has little variation as a function of latitude as expected from the general characteristics of the solar wind at solar maximum. It is also interesting to note from Figure 6 that the dip of density corresponds to the peak of solar wind velocity which agrees with the images shown in Figure 5. The bright streamers are high density low velocity region.

Fig. 5. The MHD Simulated polarization brightness (PB) by using an initially global multi-polar magnetic field for the case of corona at solar maximum.

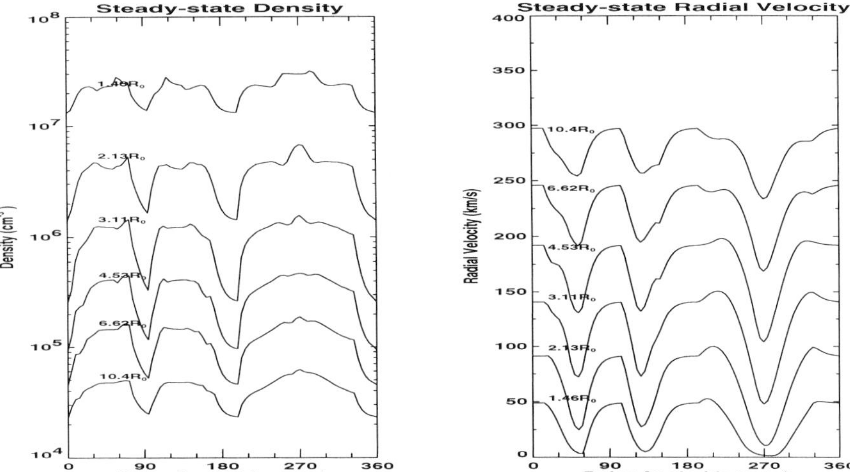

Fig. 6. The MHD simulated radial velocity and density distribution of the corona at solar maximum where 0° refers to the north pole.

DISCUSSION

We have presented the LASCO C3 observations of large-scale coronal structure during the rising phase of solar activity cycle 23: May 1996 through May 2000. The changes in coronal structures are clearly shown in Figure 2. During the solar activity minimum, coronal structures are limited to well-defined streamers that appear only at low latitude. These structures are relatively stable and can be tracked over multiple solar rotations. As solar activity increases, the coronal structures move to higher latitude. Coronal structures appear at very high latitudes as solar activity approaches the maximum in 2000.

At the minimum of solar activity, sunspots appear only at low latitude and active regions are weak or not present. The shape of the corona is primarily determined by the large-scale dipole magnetic field of the sun. The observation of May 21, 1997 can be reproduced using a MHD model that contains only the dipole field. The images which do not show this simple symmetry can probably be explained as due to local heating of the corona in the areas of weak activity present even as the minimum of solar activity.

Solar activity increases during the rising phase of the solar cycle. Sunspots and active regions appear almost everywhere on the Sun's surface. Consequently, LASCO observes coronal streamers at high latitudes. These streamers can exhibit complex structures.

These features can be reproduced using a MHD model with a complex, multi-pole magnetic field. While the complex details of the coronal structures seen at solar maximum are due to details of the distribution of active regions, the overall structure can be modeled using MHD.

ACKNOWLEDGMENT

MDA is supported at NRL under NASA contract S-86760-E. STW is supported by the National Science Foundation (ATM 0070385) and the Air Force Office of Scientific Research (F49620-00-0-0204). SOHO is a mission of international cooperation between NASA and ESA.

REFERENCES

Andrews, M. D., A.-H. Wang, S. T. Wu, Observations and Modeling of an Explosive Coronal Mass Ejection as Observed by LASCO, *Solar Phys.*, 187, 427 (1999).

Brueckner, G. E., R. A. Howard, M. J. Koomen, C. M. Korendyke, D. J. Michels, *et al.*, The large Angle Spectroscopic Coronagraph (LASCO), *Solar Phys.*, 162, 357 (1995).

Hathaway, D. H., R. M. Wilson, and E. J. Reichman, A Synthesis of Solar Cycle Prediction Techniques, *J. Geophys. Res.*, 104, A10, 22375 (1999).

Hu, Y. Q. and S. T. Wu, A Full-Implicit-Continuous-Eulerian (FICE) Scheme for Multidimensional Transient Magnetohydrodynamic (MHD) Flows, *J. Computational Phys.*, 55, No. 1, 33 (1984).

Linker, J. A. and Z. Mikic, Disruption of a Helmet Streamer by Photospheric Shear, *Astrophys. J.*, 438, L45 (1995).

Plunkett, S. P. A. Vourlidas, S. Simberova, M. Karlicky, P. Kotrc, *et al.*, Simultaneous SOHO and Ground-Based Observations of a Large Eruptive Phenomena and Coronal Mass Ejection, *Solar Phys.*, 194, 321 (2000).

Roach, P. J., Fundamentals of Computational Fluid Dynamics, P. Hermosa Publishers, P. O. Box 9110, Albuquerque, New Mexico, USA (1998).

Tousey, R., The Solar Corona, *Adv. Space Res.*, 13, 713 (1973).

Wang, A. H., S. T. Wu, S. T. Suess, and G. Poletto, A Two-dimensional MHD Global Coronal Model: Steady State Streamers, *Solar Phys.,* 147, 55 (1993).

Wang, A. H., S. T. Wu, S. T. Suess, and G. Poletto, Global Model of the Corona with Heat and Momentum Addition, *J. Geophys. Res.*, 103 (A2), 1913 (1998).

Wu, S. T. and J. F. Wang, Numerical Tests of a Modified Full-Implicit-Continous-Eulerian (FICE) Scheme with Projected Characteristic Boundary Conditions for MHD Flow, *Comp. Methods Appl. Mech. Engr.* 64, 267 (1987).

Wu, S. T., W. P. Guo and J. F. Wang, Dynamical Evolution of a Coronal Streamer-Bubble System: I. A Self-Consistent Planar Magnetohydrodynamic Simulation, *Solar Phys.*, 157, 325 (1995).

Wu, S. T., W. P. Guo, M. D. Andrews, G. E. Brueckner, R. A. Howard, et al,, MHD Interpretation of LASCO Observations of a Coronal Mass Ejection as a Disconnected Magnetic Structure, *Solar Phys.,* 175, 719 (1997).

Wu, S. T., W. P. Guo, D. J. Michels, and L. F. Burlaga, MHD Description of the Dynamical Relationships between a Flux-Rope, Streamer, Coronal Mass Ejection (CME) and Magnetic Cloud: An Analysis of the 1997 January Sun-Earth Connection Event, *J. Geophys. Res.*, 104 (A7), 14789 (1999).

Wu, S. T., W. P. Guo, S. P. Plunkett, B. Schmieder and G. M. Simnett, Coronal Mass Ejections (CMEs) Initiation: Models and Observations, *J. Atmospheric and Solar-Terrestrial Physics,* 61, 16 (2000a).

Wu, S. T., A. H. Wang, S. P. Plunkett, and D. J. Michels, Evolution of Global Scale Coronal Magnetic Field Due to Magnetic Reconnection: The Formation of the Observed Blob Motion in the Coronal Streamer Belt, *Astrophys. J.*, 545, 1101 (2000b).

TAIWAN OSCILLATION NETWORK: PROBING THE SOLAR INTERIOR

Dean-Yi Chou and the TON Team[1]

Physics Department, Tsing Hua University, Hsinchu, 30043, Taiwan, R.O.C.

ABSTRACT

In this review paper, we describe the present status of the project of the Taiwan Oscillation Network (TON) and discuss the recent scientific results using the TON data, especially the results from a new method: acoustic imaging. The TON is a ground-based network to measure solar intensity oscillations for the study of the solar interior. Four telescopes have been installed in Tenerife (Spain), Big Bear (USA), Huairou (PRC), and Tashkent (Uzbekistan). The TON telescopes take K-line full-disk solar images of diameter 1000 pixels at a rate of one image per minute. The TON high-spatial-resolution data are specially suitable for the study of local properties of the Sun. Recently we developed a new method, acoustic imaging, to construct the acoustic signals inside the Sun with the acoustic signals measured at the solar surface. From the constructed signals, we can form intensity map and phase-shift map of an active region at various depths. The direct link between the these maps and the subsurface magnetic field suffers from the poor vertical resolution of acoustic imaging. An inversion method is developed to invert the measured phase-shift distribution to estimate the source distribution based on the ray approximation. We also discuss the application of acoustic imaging to image the active regions on the far-side of the Sun that may be used for the solar activity forecast.

THE PROJECT OF TAIWAN OSCILLATION NETWORK

The Taiwan Oscillation Network (TON) is a ground-based network measuring K-line intensity oscillation for the study of the internal structure of the Sun. The TON project has been funded by the National Research Council of ROC since July of 1991. The plan of the TON project is to install several telescopes at appropriate longitudes around the world to continuously measure the solar oscillations. So far four telescopes have been installed. The first telescope was installed at the Teide Observatory, Canary Islands, Spain in August of 1993. The second and third telescopes were installed at the Huairou Solar Observing Station near Beijing and the Big Bear Solar Observatory, California, USA in 1994. The fourth telescope was installed in Tashkent, Uzbekistan in July of 1996.

The TON is designed to obtain informations on high-degree solar p-mode oscillations, along with intermediate-degree modes. A discussion of the TON project and its instrument is given by Chou et al. (1995). Here we give a brief description. The TON telescope system uses a 3.5-inch Maksutov-type telescope. The annual average diameter of the Sun is set 1000 pixels. A K-line filter, centered at 3934Å, of FWHM = 10Å and a prefilter of FWHM = 100Å are placed near the focal plane. The measured amplitude of intensity oscillation is about 2.5%. A 16-bit water-cooled CCD is used to take images. The image size is 1080 by 1080 pixels. The exposure time is set to 800-1500 ms, depending on the solar brightness. The photon noise, about 0.2%, is greater than the thermal noise of the CCD and circuit. The image data are recorded by two 8-mm Exabyte tape drives.

[1]The TON Team includes: Ming-Tsung Sun (Department of Mechanical Engineering, Chang-Gung University, Kwei-San, Taiwan); Antonio Jimenez (Instituto Astrofisica de Canarias, Observatorio del Teide, Tenerife, Spain); Guoxiang Ai and Honqi Zhang (Huairou Solar Observing Station, Beijing Observatory, Beijing); Philip Goode and William Marquette (Big Bear Solar Observatory, New Jersey Institute of Technology, Newark, NJ 07102, U.S.A.); Shuhrat Ehgamberdiev and Oleg Ladenkov (Ulugh Beg Astronomical Institute, Tashkent, Uzbekistan).

The TON full-disk images have a spatial sampling window of 1.8 arcseconds per pixel, and they can provide information of modes up to $l \approx 1000$. The TON high-resolution data is specially suitable to study local helioseismology. In this paper, we discuss a new method, acoustic imaging, first developed by the TON group to study the local property of the solar interior, such as active regions.

A NEW METHOD: ACOUSTIC IMAGING

The conventional analysis in helioseismology is the mode decomposition. A time series of intensity or Doppler images is decomposed into eigenmodes. For example, in the spherical coordinates, the eigenmodes are characterized by radial order n, spherical harmonic degree l, azimuthal order m, and frequency ω. Many mode properties can be obtained from the mode decomposition, such as (1) the dispersion relation: the relation between ω and (n, l, m), (2) power spectra: the power distribution among modes, and (3) the line profile of each mode. These mode properties can be used to infer the internal structure of the Sun. The mode decomposition can be applied to global modes or local modes. For the global modes, the mode properties provide the global information on the solar interior. For the local modes, it provides the information below the solar surface in a local area, such as an active region.

Duvall et al. (1993) first apply time-distance analysis, commonly used in earth seismology, to the solar acoustic waves. Time-distance analysis is performed directly in the spacetime domain to obtain a relation between wave travel time and travel distance, time-distance relation. The time-distance relation is determined by the physical conditions of the region through which the waves pass. Thus time-distance analysis is a useful tool to study the local structure.

Recently the TON group developed a new technique, acoustic imaging, based on the time-distance, to construct the acoustic signal at a point on the solar surface or in the solar interior at a target time with the acoustic signals measured at the solar surface. Its development was motivated by the success of imaging underwater objects with ambient acoustic noise in the ocean (Buckingham et al., 1992; Chang et al., 1997). A resonant solar p-mode is trapped and multiply reflected in a cavity between the surface and a layer in the solar interior. The acoustic signal emanating from a point at the surface propagates downward to the bottom of the cavity and back to the surface at a different horizontal distance from the original point. Different p-modes have different paths and arrive at the surface with different travel times and different distances from the original point. The modes with the same angular phase velocity ω/l have approximately the same ray path and form a wave packet, where ω is the mode frequency and l is the spherical harmonic degree. The relation between travel time and travel distance of the wave packet can be measured in the time-distance analysis (Duvall et al., 1993).

The acoustic signals measured at the surface can be coherently added, based on the time-distance relation, to construct the signal at a target point and a target time (Chou et al., 1998, 1999; Chou 2000)

$$\Psi_{\text{out,in}}(t) = \sum_{\tau=\tau_1}^{\tau_2} W(\tau, \theta) \cdot \bar{\Psi}(\theta, t \pm \tau) \, , \tag{1}$$

where $\Psi_{\text{out}}(t)$ and $\Psi_{\text{in}}(t)$ are the constructed signals at the target point at time t, $\bar{\Psi}(\theta, t \pm \tau)$ is the azimuthal-averaged signal measured at the angular distance θ from the target point at time $t \pm \tau$, where τ and θ satisfy the time-distance relation, and $W(\tau, \theta)$ is the weighting function. The positive sign corresponds to Ψ_{out} which is constructed with the waves propagating outward from the target point. The negative sign corresponds to Ψ_{in} which is constructed with the waves propagating inward toward the target point. The constructed signals Ψ_{out} and Ψ_{in} contains different information.

The coherent sum defined in equation (1) is called the computational acoustic lens since it plays a role as a lens in optics. The summation variable τ in equation (1) is evenly spaced in the interval (τ_1, τ_2). The range of τ corresponds to an annular region in space, which is called the aperture of the acoustic lens. Since each point on the time-distance curve corresponds to a wavepacket formed by the modes with the same

ω/l, each aperture corresponds to a range of ω/l. Using different apertures corresponds to using different ranges of modes to construct signals. Usually data is filtered with a Gaussian in frequency. The central frequency is defined as the frequency of the acoustic lens. For a fixed frequency, each aperture (annulus) is associated with a range of l. Thus an acoustic lens is characterized by ω, l range, and focal depth. The time-distance relation used in equation (1) is not limited to the one-bounce time-distance relation. It can be more than one bounces. The property of signals constructed with multiple bounces is discussed in Chou et al. (1998; 1999), Braun et al. (1998), and Chou and Duvall (2000). For a target point on the surface, one can use the measured time-distance relation. For a target point located below the surface, one has to use the time-distance relation computed with a standard solar model and the ray approximation (D'Silva and Duvall, 1995; Chang et al., 1997). The degree of success of the reconstruction can be tested by the correlation between the constructed time series and the observed time series for a point on the surface. With 512-minute TON data, the peak of the correlation function between the observed time series and the constructed time series (Ψ_{out} or Ψ_{in}) is about 0.47 for an aperture (annulus) of $2-17°$ (Chou et al., 1998). If Ψ_{out} and Ψ_{in} are combined after phase matching, the peak of the correlation function between the observed time series and the combined constructed time series is about 0.68. The correlation between the observed time series and constructed time series increases with the aperture size (Chou et al., 1998).

The constructed signal contains information on intensity and phase. The intensity at the target point can be computed by summing $|\Psi_{\text{out,in}}(t)|^2$ over time. The first acoustic intensity images of the solar interior were constructed with the TON data (Chang et al. 1997). One can also compute the absorption map with $(P_{\text{in}} - P_{\text{out}})/P_{\text{in}}$, where $P_{\text{in,out}}$ are ingoing and outgoing intensity, respectively. For example, the power map, constructed outgoing map and constructed ingoing map of NOAA 7978 are shown in Figure 1. The acoustic power is the local rms value of the amplitude of acoustic oscillations. The darker area in the acoustic power maps corresponds to the lower acoustic power, which correlates well with magnetic fields or plages. The constructed outgoing intensity consists of the dissipation and emission at the target point and the dissipation along the wave path from the target point to the observing point. The local suppression and enhancement of p-modes are not included in the constructed intensity, while they appear in the local acoustic power (Chou et al. 1999). The ingoing intensity map shows no signature of magnetic region as expected because the ingoing waves in the surrounding area have not been affected by the local properties of the target point before they reach the target point (Chang et al. 1997; Chou et al. 1999). The slightly darker features at the location of the active region are probably caused by the lower observed signals in magnetic regions inside the annular region (aperture), which are used to construct the signal at the target point.

The phase of constructed signals can be studied by the cross-correlation function between $\Psi_{\text{out}}(t)$ and $\Psi_{\text{in}}(t)$ (Chen et al. 1998). It is noted that $\Psi_{\text{out}}(t)$ and $\Psi_{\text{in}}(t)$ have to be constructed with the same waves; otherwise, there will be no correlation between $\Psi_{\text{out}}(t)$ and $\Psi_{\text{in}}(t)$. The cross-correlation function in magnetic regions is different from that in the quiet Sun. The phase and envelope peak of the cross-correlation function are defined as τ_{ph} and τ_{en}, respectively. Both τ_{ph} and τ_{en} are smaller in the sunspot than the quiet Sun, implying a larger sound speed in the sunspot. The changes in τ_{ph} and τ_{en} relative to the quiet Sun are defined as the phase shift $\delta\tau_{ph}$ and the envelope shift $\delta\tau_{en}$, respectively. It is noted that $\delta\tau_{ph}$ is different from $\delta\tau_{en}$ in the sunspot. $\delta\tau_{ph}$ and $\delta\tau_{en}$ have different origins (Chou and Duvall 2000; Chou et al. 2000). $\delta\tau_{en}$ in magnetic regions is caused by the change in the travel time of the wave packet, which could be due to the change in group velocity along the wave path or the change in wave path. $\delta\tau_{ph}$ is the change in phase change accumulated along the wave path. The cause of $\delta\tau_{ph}$ is more complicated than that of $\delta\tau_{en}$. Besides the changes in phase velocity and wave path, the phase changes at the boundaries of mode cavity and flux tubes also contribute to $\delta\tau_{ph}$. The presence of magnetic fields modifies the physical conditions of the plasma that results in a change in the travel time of the wave packet and the phase change accumulated along the wave path. Thus measured $\delta\tau_{ph}$ together with $\delta\tau_{en}$ in magnetic regions provide information on the structure and physical conditions in the magnetic regions.

Two-dimensional phase-shift maps at different focal depths were first given in Chen et al. (1998) using the TON data. Since the error in determining τ_{en} is larger than τ_{ph}, it needs a longer time series to determine

τ_{en} with a reasonable S/N. The envelope-shift maps were first given by Chou et al. from a 2000-minute time series taken with the SOHO/MDI instrument (Chou et al. 2000). Examples of phase-shift map and envelope-shift map of NOAA 7978 are shown in Figure 1. Figure 1 shows that the phase shift measurement is more sensitive to the magnetic fields than the acoustic intensity measurement. The features in the phase-shift maps are surprisingly similar to those in the observed power maps. The phase-shift maps also correlate well with K-line images even for weak plages in the quiet Sun (Chen et al. 1998). The features in the envelope-shift maps also correlate with acoustic power and magnetic fields, but with a larger fluctuation. Both $\delta\tau_{ph}$ and $\delta\tau_{en}$ in a sunspot increase with frequency. The frequency dependences of $\delta\tau_{ph}$ and $\delta\tau_{en}$ together provides useful information on the structure of the sunspot (Chou et al. 2000).

INVERSION PROBLEMS IN ACOUSTIC IMAGING

Although the goal of acoustic imaging is to construct the signal at the target point, the constructed signal contains not only the signal at the target point, but also the signals of other points. In other words, the spatial resolution of acoustic lens of acoustic imaging is finite. It is important to know the spatial resolution of acoustic different lenses so that the intensity maps and phase-shift maps of active regions can be correctly interpreted to learn the subsurface structure of active regions. There are two different effects involved in determining the spatial resolution of measured intensity maps and phase-shift maps. One is the wave property. Another is the non-locality effect: the waves passing through the target point also pass through the regions other than the target point. The nonlocal effect exists even in the ray approximation.

The constructed signals at the target point, $\Psi_{\rm out}(t)$ and $\Psi_{\rm in}(t)$, contain information integrated along the ray paths (Chou et al. 1999). That is the constructed signals contain not only the local effect at the target point, but also the nonlocal effect. The contribution of nonlocal effect depends on the density of the ray distribution in the ray approximation. The phase shift (or change in travel time) at the target point $\vec{r}$, $\delta\tau(\vec{r})$, can be expressed as (Chou and Sun 2000b)

$$\delta\tau_{ph}(\vec{r}) = \int K(\vec{r},\vec{r}')S(\vec{r}')\, d^3r' \tag{2}$$

where the source $S \equiv \delta c/c$ is the fraction of change in wave speed. The kernel $K(\vec{r},\vec{r}')$ is proportional to the density of the ray distribution. The effect of flow on phase shifts can be eliminated by averaging the phase shifts measured with the rays of opposite directions (Chen et al. 1998). $\delta\tau_{ph}$ discussed here is such an average. Each acoustic lens has one kernel. The kernel can be computed with a standard solar model based on the ray theory (Chou and Sun 2000a). Although the kernel $K(\vec{r},\vec{r}')$ has a maximum at the target point $\vec{r}$, its value is not negligibly small in the region near the target point. The distribution of the kernel determines the spatial resolution of the constructed phase-shift images.

Since the distribution of $\delta\tau_{ph}$ in equation (2) can be measured in acoustic imaging, with the kernels one can do both forward problems and inverse problems. In forward problems, $\delta\tau_{ph}$ is computed from the given model source with equation (2). Forward computations with model sources can provide information on the spatial resolution of various acoustic lenses. In the forward study of an active region, the given source is varied such that the computed $\delta\tau_{ph}$ is close to the measured $\delta\tau_{ph}$. The forward study of the active region NOAA 7981 has been discussed in Chou and Sun (2000a). One can also invert the measured $\delta\tau_{ph}$ to estimate the source distribution S in equation (2). We have used an inversion method, regularized least-squares fit, to invert equation (2) (Sun and Chou 2000). The inversion method has been applied to study an active region, NOAA 7981 (Chou and Sun 2000a). Figure 2 shows the result of inversion. The phase-shift maps of NOAA 7981 measured with the TON data at various depths are shown in the right column in Figure 2. The 2-d source maps from inversion at the same depths are shown in the left column. The source from inversion has a smaller vertical extent than the measured $\delta\tau_{ph}$. The dark fuzzy features corresponding to the plages in the measured phase-shift map diminish at depths of 4.4 and 8.8 Mm, and disappear at a depth of 17.6 Mm in Figure 2. In the future study, we will improve the S/N of measured $\delta\tau_{ph}$ and increase the number of $\delta\tau_{ph}$ in the vertical direction to improve the quality of the estimated source.

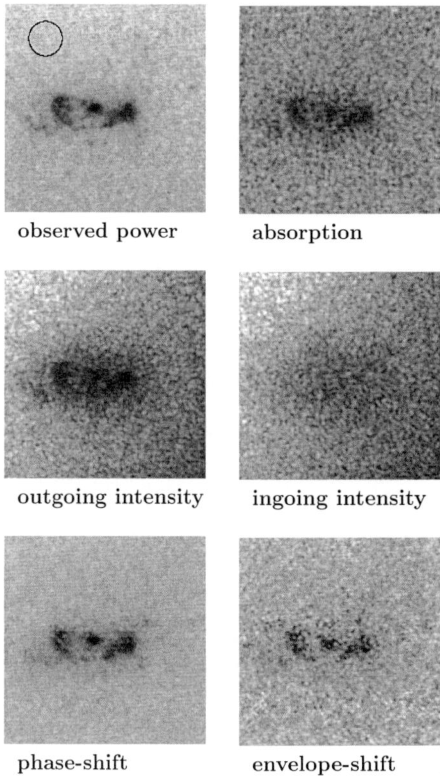

Figure 1: Observed power map, absorption map, constructed outgoing map, constructed ingoing map, phase-shift map, and envelope-shift map of NOAA 7978 using a 2000-minute time series taken with the SOHO/MDI instrument (from Chou 2000). The data is filtered with a Gaussian filter of FWHM = 1 mHz centered at 3.5 mHz. The size of aperture used is $2-6^\circ$ (a two-degree circle is shown in the upper left map). The dimension of each map is 24.0° in longitude and 24.7° in latitude. The largest phase shift and envelope shift in the central sunspot are 1.5 minutes and 5.3 minutes, respectively.

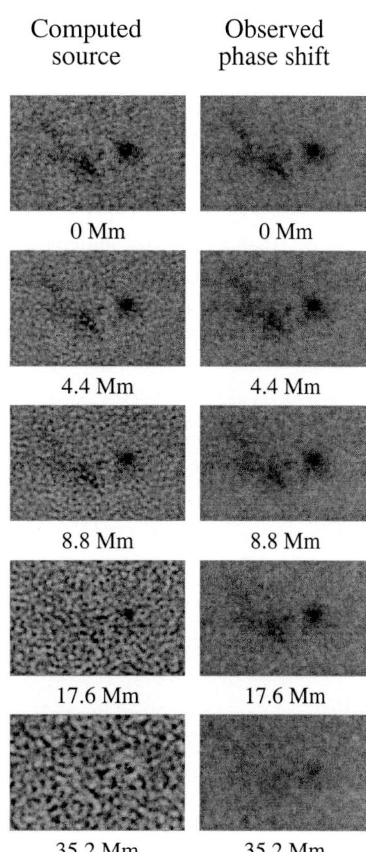

Figure 2: Measured phase-shift maps of NOAA 7981 at various depths (right column) using the TON data and 2-D source maps from inversion (left column) (from Chou and Sun 2000a). The grey scales are chosen such that the maximum values of source map and the phase-shift map at the surface have the same tone. The grey scale of the phase-shift maps is from -2 minutes to 2 minutes. The vertical extent of the estimated (computed) source is smaller than that of the measured phase shifts. The noise in estimated source increases with depth because the value of kernel decreases with depth.

APPLICATIONS TO SPACE WEATHER PREDICTIONS

The acoustic imaging technique can be used to image active regions at the far-side of the Sun. If the perturbation of acoustic waves caused by an active region at the far-side of the Sun can reach the near-side of the Sun and be detected, one can apply the technique of the acoustic imaging to image the active regions at the far-side of the Sun. We have used the TON data to construct the intensity and phase-shift maps of NOAA 7981, a medium-size active region, when it was at the far-side of the Sun; but none shows a detected signal above the noise level (Chou et al. 1998). Lindsey and Braun (2000) applied the technique of acoustic imaging (they call helioseismic holography) with the SOHO/MDI data to construct phase-shift map of the active region NOAA 8194, a very large active region, and saw its phase-shift features. Although the spatial resolution of their phase-shift map for the active region at the far-side of the Sun is rather low because only low-l modes can survive to reach the near-side of the Sun, this exciting result sheds light on the forecast of solar activities at the far-side of the Sun and on the possibility of the earlier space weather prediction.

ACKNOWLEDGMENTS

DYC and the TON project are supported by NSC of ROC under grants NSC-89-2112-M-007-038. We thank all members of the TON Team for their efforts to keep the TON project working. We are grateful to the GONG Data Team for providing the software package GRASP.

REFERENCES

Braun, D. C., Lindsey, C., Fan, Y., and Fagan, M., Seismic Holography of Solar Activity, Astrophys. J. 502, 968 (1998).

Buckingham, M. J., Berkhout, B. V., and Glegg, S. A., Imaging the Ocean with Ambient Noise, Nature 356, 327 (1992).

Chang, H.-K., Chou, D.-Y., LaBonte, B., and the TON Team, Ambient Acoustic Imaging in Helioseismolog, Nature 389, 825 (1997).

Chen, H.-R., Chou, D.-Y., Chang, H.-K., Sun, M.-T., Yeh, S.-J, LaBonte, B., and the TON Team, Probing the Subsurface Structure of Active Regions with the Phase Information in Acoustic Imaging, Astrophys. J. 501, L139 (1998).

Chou, D.-Y., Sun, M.-T., Huang, T.-Y. et al., Taiwan Oscillation Network, Solar Phys. 160, 237 (1995).

Chou, D.-Y., Chang, H.-K., Chen, H.-R., LaBonte, B., Sun, M.-T., Yeh, S.-J., and the TON Team, Results of Acoustic Imaging with the TON data, in Proc. SOHO6/GONG98 Workshop on Structure and Dynamics of the Interior of the Sun and Sun-like Stars, ed. S. Korzennik and A. Wilson (ESA SP-418; Noordwijk:ESA), 597 (1998).

Chou, D.-Y., Chang, H.-K., Sun, M.-T., LaBonte, B., Chen, H.-R., Yeh, S.-J., and the TON Team, Acoustic Imaging in Helioseismology, Astrophys. J. 514, 979, (1999).

Chou, D.-Y., Acoustic Imaging of Solar Active Regions, Solar Phys. 192, 241 (2000).

Chou, D.-Y., Sun, M.-T., and Chang, H.-K., A Study of Sunspots with Phase Time and Travel Time of p-Mode Waves in Acoustic Imaging, Astrophys. J. 532, 622 (2000).

Chou, D.-Y. and Duvall, T. L. Jr., Phase Time and Envelope Time in Time-Distance Analysis and Acoustic Imaging, Astrophys. J. 533, 568 (2000).

Chou, D.-Y. and Sun, M.-T., Phase Shifts of Acoustic Imaging and the Subsurface Structure of Active Region, in Proc. SOHO10/GONG2000 Workshop on Helio- and Astero-Seismology at the Dawn of the Millennium (ESA SP-464; Noordwijk:ESA), 157 (2000a).

Chou, D.-Y. and Sun, M.-T., Kernels of Acoustic Imaging, in Proc. SOHO10/GONG2000 Workshop on Helio- and Astero-Seismology at the Dawn of the Millennium (ESA SP-464; Noordwijk:ESA), 195 (2000b).

D'Silva, S. and Duvall, T. L. Jr., Time-Distance Helioseismology in the Vicinity of Sunspots, Astrophys. J. 438, 454 (1995).

Duvall, T. L. Jr., Jefferies, S. M., Harvey, J. W., and Pomoerantz, M. A., Time-Distance Helioseismology, Nature 362 430 (1993).

Lindsey, C. and Braun, D. C., Seismic Imnages of the Far Side of the Sun, Science, 287, 1799 (200).

Sun, M.-T. and Chou, D.-Y., The Inverse Problem in Acoustic Imaging, in Proc. SOHO10/GONG2000 Workshop on Helio- and Astero-Seismology at the Dawn of the Millennium (ESA SP-464; Noordwijk:ESA), 251 (2000).

SPACE WEATHER STUDY USING COMBINED CORONAGRAPHIC AND *IN SITU* OBSERVATIONS

N. Gopalswamy

Center for Solar Physics and Space Weather, The Catholic University of America, Washington DC 20064, and NASA/GSFC, Greenbelt, MD 20771, USA

ABSTRACT

Coronal mass ejections (CMEs) play an important role in space weather studies because of their ability to cause severe geoeffects, such as magnetic storms. Shocks driven by CMEs may also accelerate solar energetic particles. Prediction of the arrival of these CMEs is therefore of crucial importance for space weather applications. After a brief review of the prediction models currently available, a description of an empirical model to predict the 1 AU arrival CMEs is provided. This model was developed using two-point measurements: (i) the initial speeds and onset times of Earth-directed CMEs obtained by white-light coronagraphs, and (ii) the corresponding interplanetary CME speeds and onset times at 1 AU obtained *in situ*. The measurements yield an empirical relationship between the interplanetary acceleration faced by the CMEs and their initial speeds, which forms the basis of the model. Use of archival data from spacecraft in quadrature is shown to refine the acceleration versus initial speed relationship, and hence the prediction model. A brief discussion on obtaining the 1-AU speed of CMEs from their initial speeds is provided. Possible improvements to the prediction model are also suggested.

INTRODUCTION

Space Weather conditions are primarily driven by the activity on the Sun, directly related to the evolution of open and closed magnetic fields. The open magnetic fields carry high speed solar wind while solar eruptions occur in closed magnetic field regions, such as active regions, filament regions or a combination thereof. Apart from electromagnetic radiation, which reaches Earth in minutes, solar eruptions result in three propagating entities that may affect the near-Earth space environment: coronal mass ejections (CMEs) (observed in the solar wind as magnetized plasma ejecta), solar energetic particles (SEPs), and interplanetary (IP) shocks. These entities are not independent: Fast mode shocks are driven by solar ejecta and the shocks in turn accelerate SEPs. A class of IP shocks have been observed without obvious drivers behind them (Schwenn, 1996) but we now know that these shocks are driven by CMEs (Gopalswamy et al. 2001a) traveling perpendicular to the Sun-Earth line. Short-lived SEPs can also result from solar flares without accompanying mass ejections (see, e.g., Reames 1996 for a review). For space weather applications, one should be able to predict quantitative information, such as the time of arrival. It is well known that the SEPs typically arrive within an hour after their injection near the Sun and that it is very difficult to predict their onset. More important are the particles that arrive along with the IP shock, commonly referred to as "energetic storm particles" or ESPs. These "shock enhancements" could be up to two orders of magnitude larger in flux and hence pose a hazard to astronauts and space-based technological systems. Thus, prediction of the arrival of IP shocks in the vicinity of Earth is a crucial element in space weather research. Since there is a definite relationship between CMEs and IP shocks (see, e.g., Sheeley et al., 1985), prediction of one of them should be often sufficient. In this paper, we review some of the current efforts in predicting the arrival of CMEs at 1 AU based on remote sensing while the CMEs are still near the Sun.

CURRENT MODELS

It was recognized long ago that IP shocks are one of the earliest signatures of solar disturbances and have been extensively studied from the point of view of geoeffects (Chao and Lepping, 1974; Russell et al., 1983; Marsden et al., 1987; Lindsay et al., 1994). Although it was recognized early on that most of the IP shocks were associated with white-light CMEs (Sheeley et al., 1985), the initial attempts to predict the arrival of IP shocks were not based on CMEs: The Shock Time Of Arrival (STOA) model (see, e.g., Smart and Shea 1985) and the Interplanetary Shock Propagation Model (ISPM, Smith and Dryer, 1990). Both of these models use observations of metric type II bursts as the primary source of input. Metric type II bursts indicate shocks in the inner corona. The shock speed is estimated from the drift rate of type II bursts assuming a density model for the corona. Propagation of these coronal shocks through the IP medium is studied in order to predict their arrival time and strength at 1 AU.

In the STOA model, a flare explosion drives a shock which propagates initially at a constant speed, followed by a deceleration of the blast wave. The shock propagates quasi-spherically through a radially-variable solar wind, centered at the flare site. The shock is assumed to be perpendicular and driven for the duration of the flare. It is also assumed that the events are sufficiently far apart that there is no shock interaction in the IP medium. The ISPM uses the same inputs as the STOA model, but assumes an upper cut-off of two hours for the flare duration.

There are a few basic problems with these models. First, observations do not seem to support the predictions of these models. It was pointed out by Gopalswamy et al. (1998a) that there was very little correspondence between coronal shocks inferred from metric type II bursts and the IP shocks detected *in situ* over a period of 18 months. In a more comprehensive study, Gopalswamy et al. (2001a) compared 137 metric type II bursts and 49 IP events (ejecta and IP shocks) that occurred during November 1994 to June 1998. They looked for 1 AU counterparts of a subset of 44 coronal shocks inferred from on-disk metric type II bursts and solar counterparts of IP events. The reason for using on-disk type II bursts is that these are the events that are likely to have near-Earth signatures. Imposing the constraint that near-Sun and near-Earth manifestations should have corresponding signatures within 5 days, they found that (i) most (93 %) of the metric type II bursts did not have IP signatures and (ii) most (80 %) of the IP events (IP ejecta and shocks) did not have metric counterparts. Kadinsky-Cade et al. (1998) and Quigley and Kadinsky-Cade (2000) have accumulated a huge data base of metric type II bursts and flares for testing the STOA model and ISPM. Their initial finding was that the "results are disappointing". Second, these models have an inner computational boundary at 18 solar radii ($R_\odot$). Between the solar surface and 18 $R_\odot$, the corona changes its structure rapidly and the radial profiles of physical quantities are not uniform. One of the most important parameters that determines shock propagation is the speed profile of the fast mode. This speed starts at a value as small as 200 km s^{-1} at the coronal base and attains a peak value of > 500 km s^{-1} around 3 $R_\odot$. This radial profile of the fast mode speed may act as a filter because shocks with speeds less than this peak fast mode speed may not propagate past this hump (see Mann et al., 1999; Gopalswamy et al., 2001a). Furthermore, slow but accelerating CMEs with no metric radio burst signature can also produce IP shocks at distances beyond the speed hump. Finally, the assumption of perpendicular propagation is not consistent with all the CME-driven IP shocks. Recently, Berdichevsky et al. (2000) found that only a little more than half of the IP shocks are quasi-perpendicular; the rest is oblique or quasi-parallel.

The poor correlation between metric type II bursts and IP shocks does not support the numerical models (STOA model and ISPM). The small fraction of metric type II bursts that did have IP association invariably involved large-scale CMEs. Coupled with the fact that most of the IP shocks are associated with large-scale CMEs (Sheeley et al., 1985) one can conclude that CME is the primary near-Sun activity that significantly disturbs the solar wind in the vicinity of Earth (see also Gosling, 1993). Therefore, CME-based space weather prediction methods are likely to be more realistic. It must be pointed out that this paper deals only with the arrival of CMEs at 1 AU and it is straight forward to predict the arrival of IP shocks based on the known relationship between the shock and its driving CME.

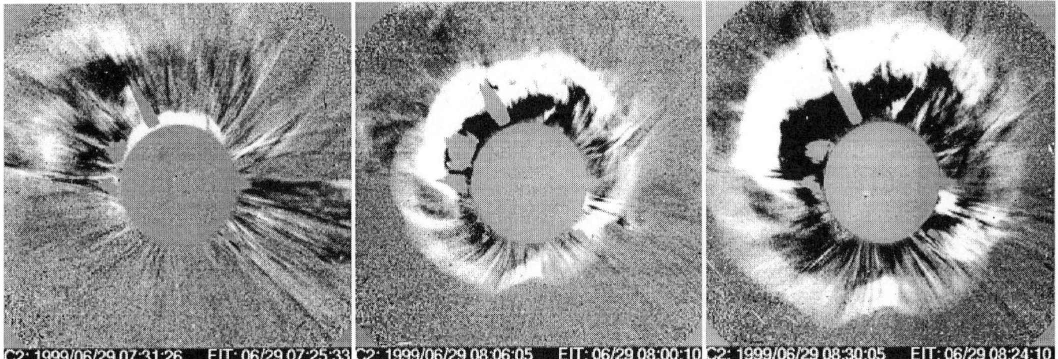

Figure 1: A near-perfect halo CME heading in the anti-Earthward direction shown in three SOHO/LASCO C2 difference images. The CME can be seen appearing above the occulting disk in the first image at 07:31 UT. In the second image at 08:06 UT, the CME has completely surrounded the occulting disk (the circular disk). SOHO/EIT difference images are superposed on SOHO/LASCO images. No changes can be seen on the disk, except for a tiny change seen in the last image at 08:30 UT by which time the CME has moved close to the edge of the LASCO field of view and hence unrelated. The CME was moving with a sky plane speed of about 630 km s^{-1}. The eruption must have occurred on the backside of the Sun.

HALO CMEs AND GEOEFFECTS

The term "halo CME" refers to a transient enhancement in the white light corona that appears to completely surround the occulting disk of a coronagraph (Howard et al., 1982). Because of the unfavorable conditions presented by halo CMEs to be detected by Thomson-scattered photospheric light (see Michels et al., 1997 for a detailed discussion), many of the halo CMEs are not very obvious. White light observations cannot show whether a halo CME is moving in the earthward or anti-Earthward direction. We need observations from inner coronal imagers such as SOHO/EIT or Yohkoh/SXT to identify the source of the eruption. Ground-based observations in Hα or microwaves can also tell us something about disk activities associated with CMEs (Gopalswamy, 1999). An anti-Earthward halo CME would not have a disk signature. Fig. 1 shows a spectacular halo event that occurred on June 29, 1999 which did not have a disk signature ("backside event"). This event was recorded by the Large Angle and Spectrometric Coronagraph (LASCO) onboard SOHO. The images are shown in running difference so the dark regions represent the location of the CME in the previous frame. Fig. 2 shows the famous "Bastille day" event (July 14, 2000). This was a frontside halo as evidenced from the EUV eruption near the disk center. These are two examples of a large number of halo CME events routinely observed by the SOHO coronagraphs. These observations have put halo CMEs in the spotlight and they are being explored as harbingers of geomagnetic storms.

Brueckner et al. (1998) examined the correlation between a set of eight halo CMEs and found that the geomagnetic storms followed the halo CMEs after about 80 hours. As these authors pointed out, the arrival time may not apply to fast events and to those occurring during the rising phase of the current solar cycle. Watari and Watanabe (1999) selected 52 geomagnetic storms with Dst < -50 nT during the same period as Brueckner et al. (1998) and identified the solar sources of these storms using Yohkoh soft X-ray data. They found that about half of the geomagnetic storms were associated with interplanetary CMEs (ICMEs) while the other half were associated with high speed streams. They also found that about 75 % of magnetic clouds were associated with geomagnetic storms. Webb et al. (2000) considered a set of seven halo CMEs and their terrestrial consequences and found that all of them were associated with magnetic clouds and geomagnetic storms. Gopalswamy et al. (2000a) studied a larger sample – a set of 23 interplanetary ejecta detected *in situ* by the Wind spacecraft – and identified the associated white-light events. They eliminated limb and backside events using optical, X-ray and EUV images and established that most of the IP ejecta

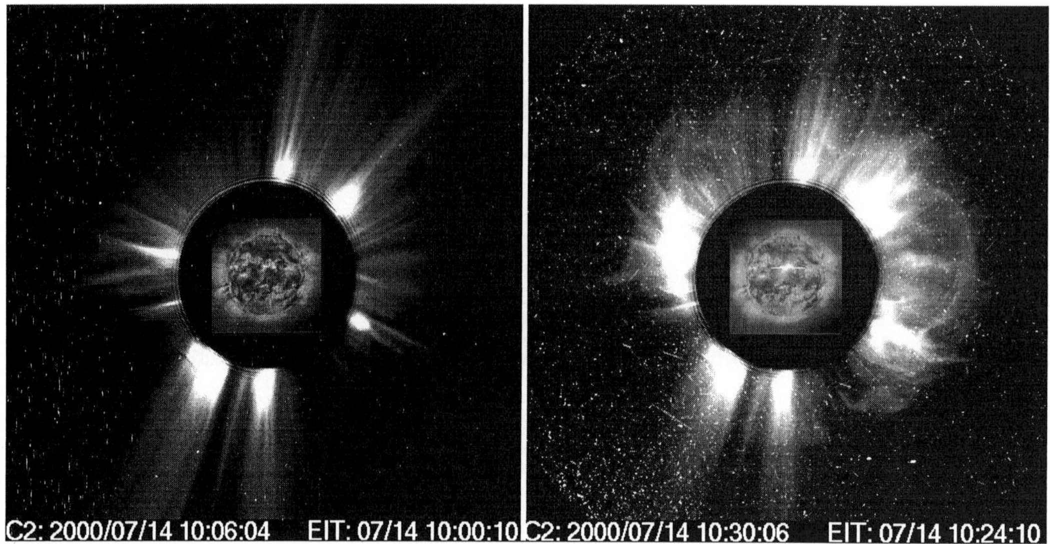

Figure 2: An Earth-directed CME that occurred on July 14, 2000 (known as the 'Bastille Day event'). EIT images (at 10:00 and 10:24 UT) are superposed on the SOHO/LASCO C2 images (at 10:06 and 10:30 UT). A bright eruption can be seen in the EIT image at 10:24 UT which is associated with the halo CME seen at 10:30 UT surrounding the occulting disk. The "snow storm" background in the right panel is due to SEPs hitting the SOHO detectors.

were associated with solar eruptions that occurred near the disk center (average latitude of 17° and average longitudinal distance of 27°). Recently, St Cyr et al. (2000) studied the annual and cumulative statistics for Kp index in comparison with halo CME events observed by SOHO. They considered 21 geomagnetic storms (Kp indices ≥ 6) that occurred during a 25 month period (January 1996 to June 1998, similar to the period of Gopalswamy et al., 2001a) and found that 15/21 (71 %) of the storms were associated with frontside halo CME events. Thus, frontside halo CMEs account for a major fraction of geomagnetic storms and need to be studied for prediction purposes. In the following, we discuss how we can predict the arrival of CMEs at 1 AU based on remote sensing of frontside halo CMEs.

AN EMPIRICAL CME ARRIVAL MODEL

Based on a set of IP ejecta detected *in situ* by Wind and the corresponding earthward CMEs detected by SOHO, Gopalswamy et al. (2000a) developed an empirical model to predict the arrival of CMEs at 1 AU. The IP ejecta had speeds in the range 350 - 650 km s^{-1} as measured *in situ*, compared with the corresponding white light CMEs, which had speeds in the range 150 - 1050 km s^{-1}. Gopalswamy et al. (2000a) set out to quantify the striking observation that the ICME-speed distribution was much narrower than the CME-speed distribution. They postulated that a CME undergoes an effective acceleration as it propagates through the IP medium and arrives at 1 AU with a different speed, assuming that they were observing the same CME at two different instances, without considering the internal structure of the CMEs. Although the exact relation between CMEs and their interplanetary counter parts (ICMEs) are not fully understood, one may expect that the spatial structure of a CME observed near the Sun is preserved as it propagates through the IP medium to produce the temporal structure observed *in situ*. For example, the ordering of substructures near the Sun (shock, frontal structure, cavity and prominence core) and at 1 AU (shock, sheath, IP ejecta and pressure pulse) may be preserved at least in some cases (Gopalswamy et al., 1998b).

Theoretically, one has to consider the resultant of the propelling and damping forces that act on the CME in order to study the dynamics of the CME through the IP medium (see, e.g., Chen, 1997). Observationally, we have only two measurements of the CME, one near the Sun performed remotely (using white light coronagraphs, such as SOHO/LASCO) and the other locally (using *in situ* plasma and magnetic field detectors, such as those onboard Wind). The effective acceleration is an average quantity, because CMEs near the Sun may show acceleration, constant speeds or even deceleration (Gopalswamy et al., 2001c). The onset time near the Sun for CMEs and the onset time for the ICMEs at 1 AU are known, so one can get the transit time, τ, for the CMEs. The respective coronagraphic and *in situ* measurements give the final and initial speeds of the CME at the two spatial points. The difference between the initial and final speeds, $\delta v = v - u$, when divided by τ gives the effective acceleration ($a = \delta v/\tau$) for each of the CMEs.

It was found that the effective acceleration and the CME initial speeds were highly correlated (correlation coefficient = 0.98). A straight-line fit to the data points yielded an empirical relation between the acceleration (a) and initial speed (u) of the CME: a = 1.41 - 0.0035u (a and u are in units if (m s^{-2}) and km s^{-1}, respectively). This relation was then used in the kinematic relation, $S = ut + 1/2\ at^2$ (where S = Sun-Earth distance (1 AU)) to predict the *arrival time, t,* of CMEs at 1 AU. The only input parameter in this model is the initial CME speed and thus provides a simple means of advance warning of solar disturbances arriving in the vicinity of Earth. Of course, we need the background information, such as disk signatures to confirm that the halo CMEs are frontside events and their location to be close to the central meridian. The all-important initial speed of the CME needs to be measured accurately to get a reliable arrival time prediction.

An Earth-directed halo CME appears to spread in the sky plane and what we measure is this spreading speed. This may or may not be the true speed of the CME. The CMEs are detected in the photospheric light Thomson-scattered by coronal material. Material in the plane of the sky is best observed by this process, so the measured speed of halo CMEs are subject to projection effects. Gopalswamy et al. (2000b) found a definite correlation between the CME speeds and the central meridian distance, with the fastest events coming from the limb. Therefore, as pointed out by Gopalswamy et al. (2000a), the empirical CME arrival model has the problem of projection effects. In order to overcome this projection effect, one needs to have stereoscopic observation, to be available in the future from the STEREO mission. However, there were some archival observations free from projection effects, which we describe below (see also Gopalswamy et al., 2001c for more details).

IMPROVING THE CME ARRIVAL MODEL

Sheeley et al. (1985) studied a large number of IP shocks detected *in situ* by the Helios-1 spacecraft and were able to identify a corresponding white light CME near the Sun using the Solwind coronagraph on board the P78-1 spacecraft (Doschek, 1983). A large fraction of these IP shocks were followed by pistons, the IP manifestations of the white light CMEs. For all these events, P78-1 was located along the Sun-Earth line while the Helios-1 spacecraft was located above east or west limb of the Sun. Thus two spacecraft were observing the same event in quadrature so that the remote sensing and local sensing corresponded to the same part (nose) of the CME. The effective acceleration derived from these observations is devoid of projection effects. Although Helios-1 was not in the vicinity of Earth, it was possible to choose events for which Helios-1 was at a distance more than 0.7 AU, similar to the Sun-Earth distance. Lindsay et al. (1999) expanded the list of such events by including data from Pioneer Venus Orbiter (PVO) which was in quadrature with either P78-1 or the Solar Maximum Mission (SMM). They found a weak correlation between the ICME and CME speeds ($v = 0.25u + 360$ km s^{-1}). This was one of the earliest attempts to link the near-Sun events to the corresponding ones in the IP medium. Gopalswamy et al (2001c) revised the list of Lindsay et al. (1999) eliminating uncertain events and came up with a set of 19 CME-ICME pairs observed by the Solwind coronagraph (remote sensing) and by PVO or Helios-1 (local sensing). When the analysis was repeated as in the case of SOHO/Wind events, the empirical relation between the effective acceleration (a) and initial speed (u) maintained the same functional form (a = 1.765 - 0.00429u), thus confirming the original method of Gopalswamy et al. (2000a).

The effective acceleration obtained from the SOHO/Wind and Helios1-PVO/P78-1 pairs is plotted in Fig. 3 (a) which shows that there is a significant deviation at low speeds. The solution of the kinematic equation with the new acceleration yielded a prediction curve as shown in Fig. 3 (b), very similar to the original curve in Gopalswamy et al. (2000a). The arrival time of low-initial speed SOHO CMEs is longer than what is seen in this curve because SOHO/LASCO underestimates the speeds of the slow CMEs and hence the model overestimates the arrival time (see Gopalswamy et al., 2000a). The prediction curve provides a simple way to estimate the arrival time of CMEs at 1 AU, once the initial speed of the Earth-directed CME is known from coronagraphic observations. Typical travel times for slow (300 km s^{-1}) and fast (1000 km s^{-1}) CMEs are shown by dashed lines as 4.4 and 2.3 days, respectively. If the IP medium had no influence, the corresponding travel times would be 5.7 and 1.7 days. The archival data and the SOHO/Wind data correspond to different phases of the solar cycle, yet the results are quite consistent. The archival data had events anywhere between 0.7 and 1 AU and the corresponding uncertainty introduced in the acceleration was ignored.

Since the ram pressure change in the solar wind impinging on the magnetosphere is important in determining the extent of geomagnetic disturbances, it is instructive to predict the final speed (v) of CMEs based on their initial speed (u). This was first attempted by Lindsay et al. (1999), who arrived at an empirical relation, $v = 0.25u + 360$ km s^{-1} as a straight line fit to the scatter plot of CME and ICME speeds. The final speed based can also be obtained using the acceleration model in the kinematic relation, $v^2 = u^2 + 2aS$ as shown in Fig. 3 (c) along with the straight line obtained by Lindsay et al. (1999). The thin and thick parabolic curves correspond to final speeds at $S = 0.7$ and $S = 1$ AU, respectively. Although the linear and parabolic curves deviate from each other at low and high speeds, both are clearly far different from the zero-acceleration case (dotted curve).

FUTURE PROSPECTS

The simple empirical model discussed above should be improved in a number of ways. The following effects need to be considered:
1) When we compared the CME speeds near the Sun and at 1 AU, we did not explicitly consider the speed of the background solar wind. If we assume that the drag acting on the CMEs is similar to the aerodynamic drag (Cargill et al., 1995), it would depend on the ambient density and flow speed. In the equatorial plane, the solar wind does not pick up for several solar radii, and the ambient density is high, resulting in a large drag force. Thus one would expect a significant drag initially. In those regions where the solar wind speed has attained a steady value, the drag would be less for a given CME speed. If a CME is launched into a region of high solar wind speed, then the drag is expected to be smaller and the CME would arrive earlier at 1 AU.
2) A similar situation arises when CMEs are launched in quick succession from the same source region. The resulting drag would depend on the speed and density of the post-CME flow of the preceding CME.
3) CMEs can interact and get deflected from the Sun-Earth line or can merge with one another (CME cannibalism – Gopalswamy et al., 2001b). The CME cannibalism is more important for solar maximum conditions because of the enhanced CME occurrence rate. The net result is that the remote sensing and local sensing would yield different counts for CMEs. One has to give careful consideration to effects like these so that false alarms can be minimized.
4) It is necessary to understand the acceleration profiles of CMEs near the Sun. There are examples of acceleration, deceleration as well as constant speed near the Sun. Current remote sensing provides data out to 30 $R_\odot$ from the Sun while local sensing is done close to 1 AU. When data points become available for other distances, we will be able to obtain a better profile for the acceleration. The acceleration profile should also consider the possibility that the IP acceleration is not constant throughout the IP medium. For example an accelerating slow CME might stop accelerating when it attains the speed of the background solar wind. The interplanetary scintillation (IPS) technique can be used to extend the height-time history

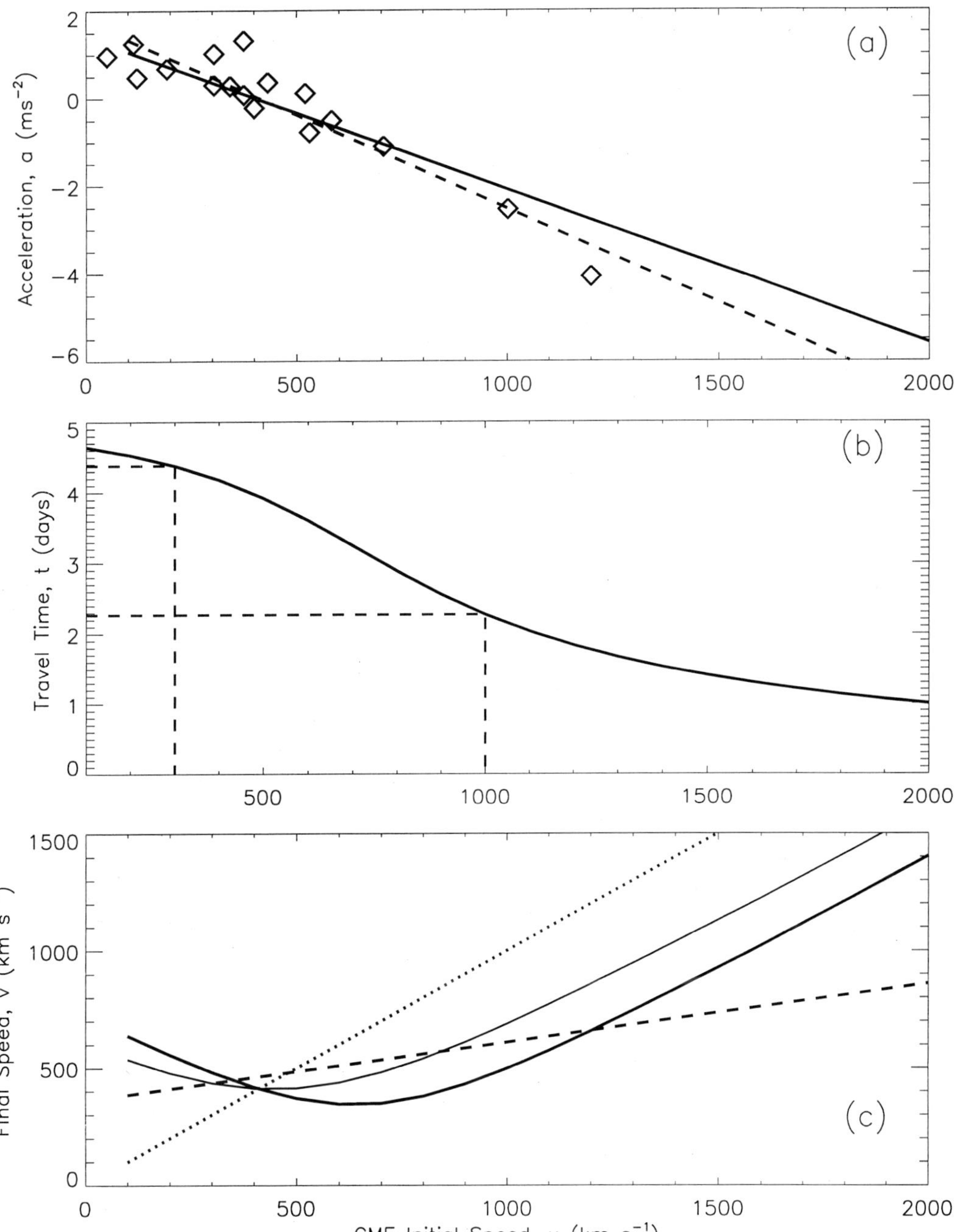

Figure 3: (a) Best-fit lines to the acceleration vs. initial speed plots for the SOHO/Wind (solid line) and Helios1-PVO/P78-1 CMEs (dashed line, fit to the data points). (b) Prediction curve using accelerations derived from Helios1-PVO/P78-1 data. Note that a 200 km s^{-1} CME would arrive in 4.5 days while a 1000 km s^{-1} CME would take just over two days to arrive at 1 AU (see the dashed lines). (c) Final speeds of CMEs predicted from initial speeds based on no acceleration (dotted line), Lindsay et al. (1999 – dashed line) and Gopalswamy et al. (2001a – thick curve for $S = 1$ AU and thin curve for $S = 0.7$ AU).

of CMEs over $\sim$ 1 AU using multiple radio sources, although it is not clear as to what part of the CME (or the CME-driven shock) is sensed by this technique.
5) Since almost all the interplanetary transient shocks are driven by CMEs, we can extend the CME arrival time to predict the arrival of IP shocks. This can be done based on the fact that there is a definite relation between the stand-off distance of the shock and the properties of the CME piston. Such a shock prediction scheme will also be useful for a better comparison with other shock arrival models.

SUMMARY

Attempts to predict the influence of large-scale Earth directed CMEs are in the beginning stages, but seem to be headed in the right direction. The most important parameter for this purpose is the initial CME speed. However, this is the most difficult parameter to measure for halo CMEs because of the nature of coronagraphic observations. Using a spacecraft located on the Sun-Earth line (to identify disk events) and another one at right angles to the Sun-Earth line (to measure the nose-speed of CMEs) seems to be the simplest way to measure the CME initial speed accurately. If ground based observations are able to identify the disk signatures, one spacecraft at right angles to the Sun-Earth line should be able to do the job. As for the effective acceleration, a two-point measurement is clearly inadequate. Measurements at several points between Sun and Earth are required to arrive at a realistic acceleration profile of CMEs. Ground based IPS observations may be used to provide a rough estimate if CMEs and radio sources are chosen carefully. A better understanding of how various substructures of the CME observed near the Sun evolve into the substructures of ICME would also help to obtain better acceleration profiles. Shock arrival times can be derived from the CME arrival times based on statistical results or theoretical considerations. The earlier models such as STOA and ISPM can also be modified by replacing the metric type II burst aspects with CME data and by moving the computational boundary closer to the Sun.

ACKNOWLEDGEMENTS

This research was supported by NASA, NSF and AFOSR. The author thanks Seiji Yashiro for help with the figures and Thomas Moran for comments on the manuscript. SOHO is a project of international cooperation between ESA and NASA.

REFERENCES

Berdichevsky, D., A. Szabo, R. P. Lepping, A. Vinas, and F. Marini, Interplanetary fast shocks and associated drivers observed through the 23rd solar minimum by Wind over its first 2.5 years, *J. Geophys. Res.*, 105, 27,289, (2000).

Brueckner, G. E., J. P. Delaboudiniere, R. A. Howard, S. E. Paswaters, O. C. St. Cyr, et al., Geomagnetic storms caused by coronal mass ejections (CMEs): March 1996 through June 1997, *Geophys. Res. Lett.*, 25, 3019, (1998).

Cargill, P. J., J. Chen, D. S. Spicer, and S. T. Zalesak, Geometry of interplanetary magnetic clouds, *Geophys. Res. Lett.*, 22, 647, (1995).

Chen, J., Coronal Mass Ejections: Causes and Consequences, A theoretical Review, in *Coronal Mass Ejections*, ed. N. Crooker, J. Joslyn, and J. Feynman, AGU monograph 99, p. 65, (1997)

Chao, J. K. and R. P. Lepping, A correlative study of ssc's interplanetary shocks, and solar activity, *J. Geophys. Res.*, 79, 1799, (1974).

Doschek, G. A., Solar instruments on the P78-1 spacecraft, *Solar Phys.*, 86, 9, (1983).

Gopalswamy, N., X-ray and Microwave signatures of Coronal Mass Ejections, in *Solar Physics with Radio Observations*, ed. T. Bastian, N. Gopalswamy and K. Shibasaki, *NRO Report* 479, Nobeyama Radio Observatory, p. 141, (1999).

Gopalswamy, N., M. L. Kaiser, R. P. Lepping, S. W. Kahler, K. Ogilvie, et al., Origin of coronal and interplanetary shocks - A new look with WIND spacecraft data, *J. Geophys. Res.*, 103, 307, (1998a).

Gopalswamy, N., Y. Hanaoka, T. Kosugi, R. P. Lepping, J. T. Steinberg, et al., *Geophys. Res. Lett.*,

25, 2485, (1998b).
Gopalswamy, N., A. Lara, R. P. Lepping, M. L. Kaiser, et al., Interplanetary Acceleration of Coronal Mass Ejections, *Geophys. Res. Lett.*, 27, 145, (2000a).
Gopalswamy, N., M. L. Kaiser, B. J. Thompson, L. Burlaga, A. Szabo, et al., Radio-rich Solar Eruptive Events, *Geophys. Res. Lett.*, 27, 1427, (2000b).
Gopalswamy, N., A. Lara, M. L. Kaiser, and J.-L. Bougeret, Near-Sun and near-Earth manifestations of solar eruptions, *J. Geophys. Res.*, 106, 25,261, (2001a).
Gopalswamy, N., S. Yashiro, M. L. Kaiser, R. A. Howard and J.-L. Bougeret, Radio Signatures of Coronal Mass Ejection Interaction: Coronal Mass Ejection Cannibalism?, *Astrophys. J.*, 548, L91, (2001b).
Gopalswamy, N., A. Lara, S. Yashiro, M. L. Kaiser and R. A. Howard, Predicting the 1-AU Arrival Times of Coronal Mass Ejections, *J. Geophys. Res.*, in press, (2001c).
Gosling, J. T., The solar flare myth, *J. Geophys. Res.*, 98, 18,937, (1993).
Howard, R. A., D. J. Michels, N. R. Sheeley, Jr., and M. J. Koomen, The observations of a coronal transient directed at Earth, *Astrophys. J.*, 90, 8173, (1982).
Kadinsky-Cade, K., S. Quigley, and G. Ginet, Validation of interplanetary shock propagation models, EOS TRANSACTIONS, 79(45), F712, (1998).
Lindsay, G. M., C. T. Russell, J. G. Luhmann, and P. Gazis, *J. Geophys. Res.*, 99, 11, (1994)
Lindsay, G. M., J. G. Luhmann, C. T. Russell, and J. T. Gosling, Relationships between coronal mass ejection speeds from coronagraph images and interplanetary characteristics of associated interplanetary coronal mass ejections, *J. Geophys. Res.*, 104, 12,515, (1999).
Mann, G., H. Aurass, A. Klassen, C. Estel, and B. J. Thompson, Coronal transient waves and coronal shock waves, in *Plasma Dynamics and Diagnostics in the Solar Transition Region and Corona*, ed. J.-C. Vial and B. Kaldeich-Schurmann, *ESA SP-446*, 477, (1999).
Marsden, R. G., T. R. Sanderson, C. Tranquille, K.-P. Wenzel, and E. J. Smith, ISEE 3 observations of low-energy proton bidirectional events and their relation to isolated interplanetary magnetic structures, *J. Geophys. Res.*, 92, 11,009, (1987).
Michels, D. J., R. A. Howard, M. J. Koomen, S. P. Plunkett, G. E. Brueckner, et al., Visibility of Earth-Directed Coronal Mass Ejections, in *Proc, Fifth SOHO Workshop*, Eur. Space Agency Publ., *ESA SP* 404, 567, (1997).
Quigley, S. and K. Kadinsky-Cade, Final Validation Results and Databases for two Solar-event Initiated Interplanetary Shock Propagation Models, in First S-RAMP Conference, Paper S1-P21, (2000).
Reames, D. V., Energetic Particles from Solar Flares and Coronal Mass Ejections, in *High Energy Solar Flares*, ed. R. Ramaty, N. Mandzhavidze, and X.-M. Hua, *AIP Conf. Proc.* 374, 35, (1996).
Russell, C. T., M. M. Mellott, E. J. Smith, and J. H. King, Multiple spacecraft observations of interplanetary shocks Four spacecraft determination of shock normals, *J. Geophys. Res.*, 88, 4739, (1983).
Schwenn, R., An Essay on Terminology, Myths and Known Facts: Solar Transient - Flare - CME - Driver Gas - Piston - BDE - Magnetic Cloud - Shock Wave - Geomagnetic Storm, *Astrophys. Space Sci.*, 243, 187, (1996).
Sheeley, N. R., R. A. Howard, M. J. Koomen, D. J. Michels, R. Schwenn, et al., *Astrophys. J.*, 484, 472, (1985).
Smart, D. F. and M. A. Shea, A simplified model for timing the arrival of solar flare-initiated shocks, *J. Geophys. Res.*, 90, 183, (1985).
Smith, Z. and M. Dryer, MHD study of temporal and spatial evolution of simulated interplanetary shocks in the ecliptic plane within 1 AU, *Solar Phys.*, 129, 387, (1990).
St. Cyr, O. C., R. A. Howard, N. R. Sheeley, Jr., S. P. Plunkett, Properties of coronal mass ejections: SOHO LASCO observations from January 1996 to June 1998 et al., *J. Geophys. Res.*, 105, A8, 18169, (2000).
Watari, S. and T. Watanabe, Interplanetary Disturbances Around the Solar Minimum of Cycle 22 in *Solar Wind Nine*, ed. S. R. Habbal, R. Esser, J. V. Hollweg, and P. A. Isenberg, p.733, (1999).
Webb, D. F., E. W. Cliver, N. U. Crooker, O. C. St. Cyr, and B. J. Thompson, Relationship of halo coronal mass ejections, magnetic clouds, and magnetic storms, *J. Geophys. Res.*, 7491, (2000).

THE SECCHI SOLAR PLASMA IMAGER FOR STEREO

D. J. Michels

Universities Space Research Association at the
Naval Research Laboratory, Code 7660M, Washington, DC 20375, USA

ABSTRACT

The Sun-Earth Connection Coronal and Heliospheric Investigation (SECCHI) consists of two identical telescope clusters, each having five telescopes. These telescope clusters will be mounted on the two spacecraft of the NASA STEREO Mission. They will be aimed so as to observe the Sun and the entire path from Sun to Earth. The two spacecraft will be in heliocentric orbit at approximately 1 AU, with one moving out ahead of the Earth in its annual path around the Sun; the other trailing behind the Earth. The spacecraft will be called STEREO-A (Ahead) and STEREO-B (Behind!) Over the course of a mission that is designed to last at least two years, each spacecraft will drift slowly away from the Sun-Earth line at a rate of about 22 degrees per year so that at the end of two years they will be able to see, for the first time, the plasma cloud of a coronal mass ejection (CME) from view angles separated by about 90 degrees. This will enable modeling of the three-dimensional structure of such a plasma cloud, direct measurement of its frontal velocity and volumetric expansion and, for CMEs heading toward Earth, detection and measurement of their earthward progress with greatly improved precision compared to previous coronagraphic experiments for which viewing has always been restricted to the Sun-Earth line of sight.

THE STEREO MISSION

The orbital configuration chosen to achieve the goals of NASA's STEREO Mission is shown in *figure 1*. Except for minor details, the two spacecraft that constitute the mission are identical. In addition to the SECCHI telescope cluster, which provides the optical imaging capabilities, each payload also includes an interplanetary radio burst tracker, SWAVES (STEREO Waves), and two experiments for measuring fields and particles in the solar wind. IMPACT, the investigation for *In-situ* Measurement of Particles And CME Transients, measures the three-dimensional velocity distribution of solar energetic particles and the local vector magnetic field at each spacecraft. PLASTIC, the PLAsma and Supra-Thermal Ion and Composition experiment, will measure the mass and charge composition of heavy ions, and properties of the plasmas comprising CMEs as they reach distances of 1 AU from the Sun. The SWAVES experiment actually provides both remote sensing and *in-situ* measurements, as it observes CME-associated type II and type III radio wave bursts, recording intensity, source direction and angular size over a range of frequencies appropriate for sources from about one solar radius (R_o) above the photosphere, out to 1 AU and beyond.

SECCHI

The SECCHI acronym characterizing the investigation is a felicitous one, as it corresponds to the name of the nineteenth century Italian astronomer, Pietro Angelo Secchi (1818-1878). In addition to pioneering the spectral classification of stars, Secchi was the first to achieve a photographic record of the corona, during the total solar eclipse of 18 July 1860, and first to study the chromospheric flash spectrum, at the eclipse of 22 December 1870. (Pohle, 1904).

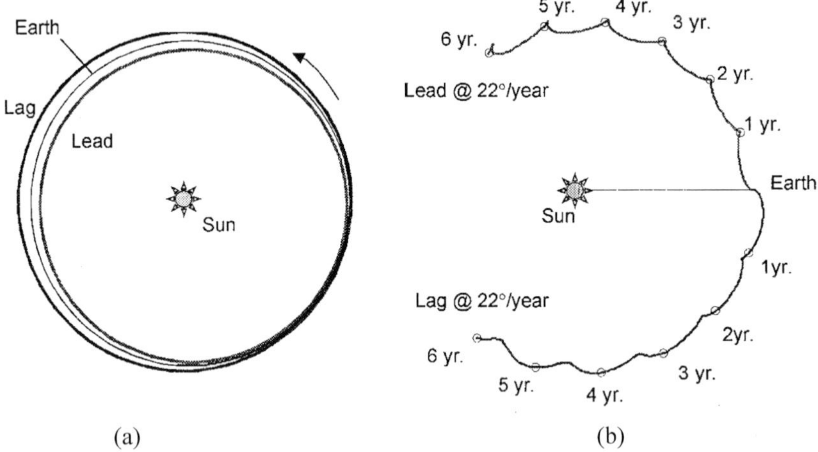

Fig. 1. Stereo Mission configuration. (a) One spacecraft is placed in an orbit having aphelion slightly less than 1 AU, giving it an orbital period under one year. Aphelion for the second spacecraft exceeds 1 AU; thus its orbital speed is slower than Earth's. (b) Orbital positions for the first six years plotted in a coordinate system in which the Sun-Earth line is fixed. Parameters are adjusted so that each platform separates from the Sun-Earth line at a rate of 22 degrees per year. A secondary goal is to minimize eccentricity of the heliocentric orbits, in order to minimize variations of the apparent solar diameter.

Scientific issues that the SECCHI investigation will pursue are focused principally on coronal mass ejections and their propagation through interplanetary space. Central questions to be asked include:

- What is the three-dimensional structure of coronal mass ejections?
- What are the origins and consequences of CMEs?
    - What magnetic configurations of the corona lead to a CME?
    - How is the requisite energy storage achieved and how is it released?
    - What initiates a CME?
- How do CMEs interact with the heliosphere?
    - What accelerates CMEs?
    - Where and how are energetic particles accelerated in the heliosphere?
    - Where and why do the fast CMEs decelerate to solar wind speeds?
- How do CMEs cause space weather disturbances?

<u>SECCHI Heliospheric Imager</u>. The configuration of the mission favors observation of CMEs headed in the general direction of Earth and the two STEREO spacecraft. To make the required measurements, each of the SECCHI instrument sets consists of an array of five telescopes; each of these specialized to view CME plasmas in a different regime as the CME ejecta progress from Sun to Earth. The telescope array

includes an extreme ultraviolet telescope, EUVI, that images the emission corona over the solar disk and out to 1.7 R_o. There are two white light coronagraphs: COR1, an internally occulted instrument with field of view extending from 1.25 to 4.0 R_o, and COR2, an externally occulted version, whose field of view extends from about 2 to 15 R_o. To track CMEs headed in the general direction of Earth (and of *in-situ* sensors aboard the STEREO spacecraft), over the great span of distances from 15 R_o to 1 AU (215 R_o), a new kind of telescope, has been developed. Actually consisting of two cameras within a single two-stage baffle system, these are collectively called the Heliospheric Imager (HI-1 and HI-2). This telescope system differs from the coronagraphs in that the camera fields of view are not centered on the Sun. They are aimed well off to one side in such a manner that, together with the coronagraphs, they cover the entire path from Sun to Earth. An optical plan of the HI instrument is shown in *figure 2*.

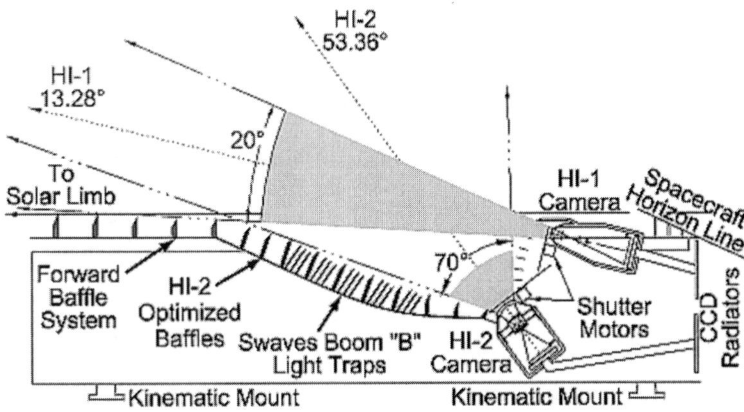

Fig. 2. Heliospheric Imager camera system, HI-1 and HI-2. Note that the solar vector is to the left; both camera objectives are well shielded from direct solar light. Residual diffracted light level at the camera apertures is a function of angle below the line to the solar limb. HI-1 is 3.28 degrees below that line; HI-2, having more stringent light rejection requirements, is buried deep behind the secondary baffle system, coming no closer than 18.36 degrees.

A summary of the unusual instrumental fields of view and of the stringent observing requirements for the SECCHI is presented in *figure 3*. The upper portion of the figure shows the fields of view of the Sun-centered COR1 and COR2, and the offset fields of HI-1 and HI-2. The Sun is at the origin, at left, and the horizontal axis gives solar elongation angle in degrees. Because of the great range of spatial scales, the fields of the Sun-centered coronagraphs, out to 15 R_o, or 4 deg., are barely discernible in the figure. The view is from the lagging spacecraft, looking west, in the plane of the ecliptic. Ecliptic north is up. At the onset of the mission, Earth will be at the extreme right, just outside the HI-2 field. As the mission progresses, with increasing Earth-Sun-spacecraft angle, the telescope axis turns to keep the coronagraphs centered on the Sun. As Earth grows more distant, and of less concern as a source of stray light, its image moves into the HI-2 field, reaching the 60 deg. point after 2.7 years. A small auxiliary occulting system may be required to avoid saturating the HI-2 CCD with bright earthshine. The lower portion of *figure 3* summarizes graphically the observing constraints and requirements for the coronagraphs and the HI. Brightness of the natural background corona, K+F, normalized to mean solar photospheric brightness, is plotted as a function of elongation angle, ε. Coronal brightness falls off as about $r^{-2.25}$ to $r^{-2.47}$, or by about 10^{-4} at $\varepsilon = 90°$ (Koutchmy and Lamy, 1985, Leinert, *et al.,* 1981). The estimated CME signal, typically 1% of the background corona, also falls at about r^{-2} to r^{-3}, or by about 10^{-4} at $\varepsilon = 90°$. (Jackson, et al., 1996). Thus CME signal detection is background noise limited at all elongation angles, regardless of how good the instrument performance may be. In addition, earthshine directly onto the HI camera lenses may be of some concern in the early phases of the mission when the spacecraft is still close to Earth. Also shown in the diagram are the one-sigma detection threshold levels for COR1, COR2, HI-1 and HI-2. The former two are based on experience with the SOHO LASCO coronagraph. For reference, the instrumental stray light of the LASCO C3 coronagraph was verified on orbit to be 10^{-12} of the solar brightness.

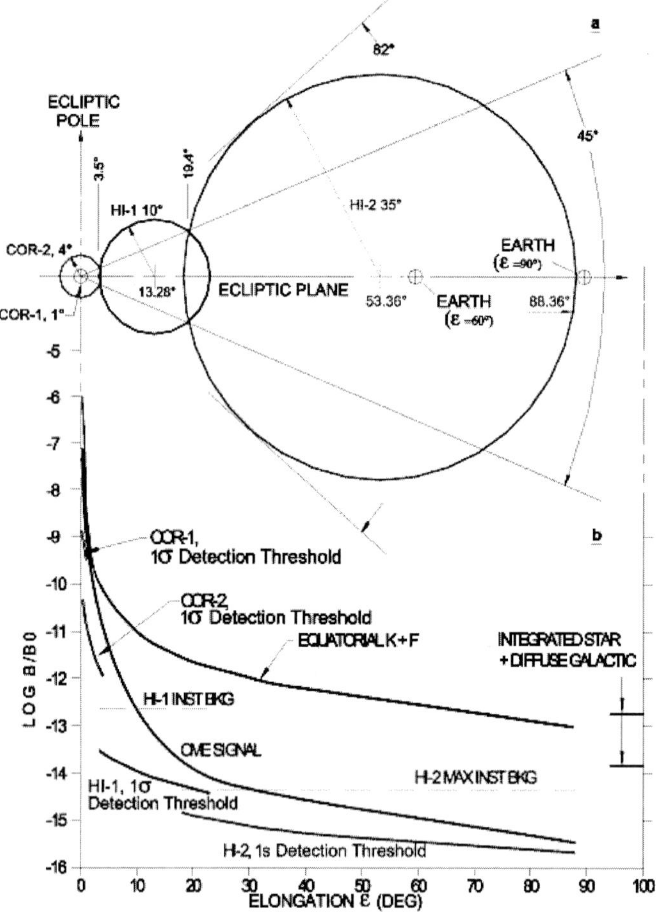

Fig. 3. SECCHI fields of view and anticipated signal and background levels in units of the solar photospheric brightness, B_o.

Extreme UltraViolet Imager. The EUVI full disk imager will observe the chromosphere and low corona in four emission lines between 17.1 and 30.4 nm. The field of view, centered on the solar disk, extends to 1.7 Ro. The instrument is similar in many respects to the EIT telescope on SOHO (Delaboudiniere, et al., 1995), but has greater resolution and sensitivity. EUV radiation enters the system through a thin-film aluminum filter, of 150 nm thickness, supported on a fine mesh. The filter blocks most longer-wavelength UV, visible, and infrared radiation, and prevents solar heat from entering the system. During launch the filter is protected by an aperture door. Optically, the system is a Ritchey-Chretien reflector.

Both primary and secondary mirrors are divided into four quadrants, each primary/secondary quadrant pair being coated with a different multilayer interference stack that provides high reflectivity over a very

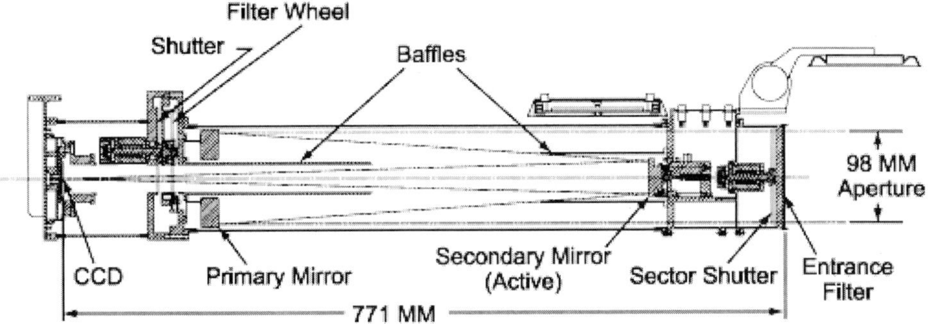

Fig. 4. Schematic drawing of the EUVI telescope. Sunlight enters from the right.

narrow passband. A sector wheel allows only one quadrant of the mirrors to be illuminated at any given time. Thus, a single optical system provides alternate images to the detector. The UV passbands are selected to correspond with regions of the solar emission spectrum that arise from narrow temperature ranges in the solar atmosphere. The CCD detector is a backside-illuminated, thinned 2048 x 2048 silicon device with EUV quantum efficiency in excess of 70%. The wavelength bands selected by the four mirror quadrants are: 30.4 nm (He II), 17.1 nm (Fe IX), 19.5 nm (Fe XII), and either 21.1 nm (Fe XIV) or 28.4 nm (Fe XV).

<u>The SECCHI White Light Coronagraphs</u>. The internally-occulted COR1 is a refractive system of 36 mm aperture, with superpolished objective. It will measure brightness and polarization brightness (pB) in the corona from 1.25 to 4.0 R_o. The spectral band covered in the final focal plane image is 650 to 750 nm, but sunlight rejection is achieved over the entire range from 300 to 1100 nm. COR1 has direct links in heritage to the Mark-3 and Mark-4 ground-based coronagraphs at Mauna Loa Observatory in Hawaii and represents an evolutionary development from those instruments. The polarization brightness measurement technique used by COR1 and COR2 will be that used by the Spartan-201 coronagraph.

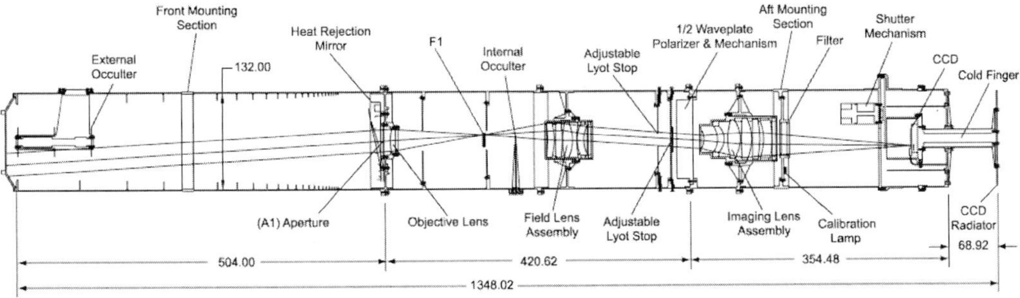

Fig. 5. COR2 optical system. The instrument borrows heavily from the SOHO LASCO C3 and C2 coronagraphs (Brueckner et al., 1995). The superpolished objective lens is currently planned to be a spaced doublet. The entrance aperture is at the left, with the aperture door shown in the closed position.

Both COR1 and COR2 will use 2048 x 2048 CCDs with capability for 2 x 2 on-chip binning of the pixels to give a 1024 x 1024 image for higher-speed readout and reduced noise. The two instruments bear many similarities with, of course, the very fundamental difference of external versus internal occulting, appropriate for the very different coronal regions each will investigate. An optical schematic of the COR2 is shown in *figure 5*.

COR2 is based on developments from the OSO-7, SOLWIND, and LASCO family of externally occulted satellite coronagraphs. Its field of view extends from 2 to 15 solar radii. It is a very fast (*f*/6) system, and can complete a full three-exposure pB measurement sequence within 22 seconds, allowing polarization brightness measurement of a rapidly moving CME within the time it takes for a CME feature to cross a single 14 arc-second pixel.

ACKNOWLEDGEMENTS

The author wishes to thank Dr. R. A. Howard, SECCHI Principal Investigator, and the entire SECCHI Consortium team, for access to design information regarding the SECCHI experiment. The STEREO Mission is funded by NASA's Office of Space Science under the Solar Terrestrial Probes (STP) Program, as part of its long term investigation theme in Sun-Earth Connections.

REFERENCES

Brueckner, G. E., R. A. Howard, M. J. Koomen, C. M. Korendyke, D. J. Michels, J. D. Moses, D. G. Socker, K. P. Dere, P. L. Lamy, A. Llebaria, M. V. Bout, R. Schwenn, G. M. Simnett, D. K. Bedford, and C. J. Eyles , The Large Angle Spectrometric Coronagraph (LASCO), *Solar Physics*, **162**, 357-402 (1995).

Delaboudinere, J. P., G. E. Artzner, J Brunaud, A. H. Gabriel, J. F. Hochedez, F. Millier, X. Y. Song, B. Au, K. P. Dere, R. A. Howard, R. Kreplin, D. J. Michels, J. D. Moses, J. M. Defise, C. Jamar, P. Rochus, J. P. Chauvineau, J. P. Marioge, R. C. Catura, J. R. Lemen, L. Shing, R. A. Stern, J. B. Gurman, W. M. Neupert, A. Maucherat, F. Clette, P. Cugnon, and E. L. Van Dessel, EIT: The Extreme-Ultraviolet Imaging Telescope for the SOHO Mission, *Solar Physics*, **162**: 291-312 (1995).

Jackson, B. V., A. Buffington, P. L. Hick, S. W. Kahler, R. C. Altrock, R. E. Gold, and D. F.Webb, 'The Solar Mass Ejection Imager,' *Solar Wind Eight*, Winterhalter, D., Gosling, J.T., Habal, S.R., Kurth, W.S., and Neugebauer, M., eds., *AIP Conf. Proc.* **382**, 536-539 (1996).

Koutchmy, S. and Ph. L. Lamy, The F-Corona and The Circum-Solar Dust: Evidences and Properties, *Proceedings of the Eighty-Fifth Colloquium: IAU* **85**, 63 (1985).

Leinert, C., I. Richter, E. Pitz, and B. Planck, The Zodiacal Light From 1.0 to 0.3 A.U. as Observed by the Helios Space Probes, *Astronomy and Astrophysics*, **103**, 177 (1981).

Pohle, J., *P. Angelo Secchi, Ein Lebens- u. Kulturbild aus dem 19 Jahrhundert*, 2nd ed., Cologne (1904).

TOMOGRAPHIC ANALYSIS OF SOLAR WIND STRUCTURE USING INTERPLANETARY SCINTILLATION

M. Kojima[1], K. Fujiki[1], M. Tokumaru[1], T. Ohmi[1], Y. Shimizu[1], A. Yokobe[1], B.V. Jackson[2], and P.L. Hick[2]

[1]*Solar-Terrestrial Environment Laboratory, Nagoya University, Toyokawa, Japan*
[2]*Center for Astrophysics and Space Sciences, University of California at San Diego, La Jolla, CA 92093-0424 U.S.A.*

ABSTRACT

For space weather research it is important to know the quiet solar wind structure existing as background for transient interplanetary phenomena. Once we know the quiet background structure, transient phenomena are easily recognized as soon as they appear in interplanetary space. The background structure is also important to understand how interplanetary disturbances propagate in it and interact with it. Interplanetary scintillation (IPS) observations have several advantages over in situ spacecraft observations for solar wind studies. Although IPS measurements are biased by the effects of line-of-sight integration, they provide global information about the solar wind. To obtain unbiased solar wind parameters we have developed a computer assisted tomography (CAT) technique. The CAT analysis can retrieve not only unbiased solar wind parameters but also provide high spatial resolution. We introduce the IPS CATanalysis and discuss its reliability by comparing with Ulysses observations. We also introduce the real-time space weather forecast project currently carried out by UCSD/CASS and STELab under a US-Japan cooperative project.

INTRODUCTION

Ulysses found in its first rapid latitudinal scan that the high-latitude solar wind had a speed in range of 700-800 km/s and that there was a small but noticeable gradual increase of the solar wind speed towards higher latitudes (Woch et al., 1997; Neugebauer et al., 1998). Ulysses also observed higher velocities at northern high latitudes than at southern high latitude (Goldstein et al., 1996; McComas et al., 2000). From in situ (Woch et al., 1997) and IPS (Kakinuma, 1977; Coles, 1996) observations, it is well known that the solar wind has a bimodal structure in the solar minimum phase, that is, a low-speed region at low latitudes is separated from high-speed regions at high latitudes with a sharp velocity gradient in between. We compare these 3-D solar wind structures with those obtained from the IPS CAT analysis to confirm the validity of the IPS CAT method.

INTERPLANETARY SCINTILLATION

From the time of analyses using the IPS phenomenon by Hewish et al. (1964), these studies have been used to probe the solar wind plasma (velocity and density turbulence) remotely from the ground. Although biased by a line-of-sight integration, these analyses have several advantages as a remote sensing tool: (1) They can measure dynamics and structure of the solar wind in three dimensions, including regions near the Sun and at high latitudes where *in situ* measurements are difficult; (2) They provide continuous, long-term coverage of the solar wind over the whole solar cycle; and (3) They can measure vast expanses of interplanetary space in a few solar rotations period by using a large number of IPS radio sources.

The Solar-Terrestrial Environment Laboratory (STELab) has been carrying out the IPS observations over two solar cycles with a facility fully dedicated to IPS. The facility consists of four antennas operated at a frequency of 327 MHz, which enables us to observe the solar wind at distances of 0.1–1 AU. On a campaign basis, several facilities, MERLIN, EISCAT (Moran et al., 2000), Ooty (Manoharan et al., 2000), and STELab (Tokumaru et al., 2000), are now making collaborative observations at different longitudinal locations and different radio frequencies around the world.

IPS TOMOGRAPHY ANALYSIS

Although IPS is a useful means to measure the global structure of the solar wind, the IPS measurements are biased by line-of-sight integration through the 3-D structured solar wind. The bias leads to large underestimation of the velocity of the fast solar wind from a polar coronal hole (Kojima and Kakinuma, 1990). Several attempts have been made to remove the bias from the IPS observations. Grall et al. (1996) determined the best fit solar wind distribution along a line of sight by model-fitting the shape of a cross-correlation function. Kakinuma (1977) optimized a latitudinal speed distribution using a model-fitting method. Asai et al. (1998), Jackson et al. (1998), and Kojima et al. (1998) have recently developed a method which uses a computer assisted tomography (CAT) technique to obtain unbiased longitudinal and latitudinal distributions of speed and electron density fluctuations. The CAT analysis can retrieve not only unbiased solar wind parameters but also has high spatial resolution as shown in the next section.

The IPS CAT analysis reconstructs three dimensional views of the interplanetary medium by combining multi-direction information from a large number of IPS radio sources. IPS observation system of STELab can measure more than 30 IPS sources per day covering various elongation angles, and thus a thousand lines of sight are available in one solar rotation period. Using a thousand of lines of sight, both solar rotation and solar wind outward motion give perspective views of three dimensional solar wind structure.

SOLAR WIND STRUCTURE DERIVED FROM IPS CAT ANALYSIS

We made synoptic maps of the velocity distribution in the Carrington coordinates from IPS data obtained at STELab during years of 1994 to 1997. Each map was derived from observations covering several Carrington rotations so that there are many numbers of lines of sight. In these reconstructed velocity maps, we analyze velocity asymmetry between the northern and the southern hemispheres, latitudinal velocity gradient, and bimodal structure and compare them with the Ulysses observations.

North-South asymmetry

Averaged velocities in latitudes of 70°–80° were compared between two hemispheres in Figure 1a. The velocity difference observed by Ulysses in 1994–1995 is also shown for comparison; Ulysses observed 24 km/s difference at a latitude of 40° and 13 km/s at 80° (Goldstein et a., 1996). Ulysses sampled a velocity along its trajectory in latitude and longitude, while our analyses were made by averaging velocities along all longitudes in the latitude range of 70°–80°. Although this may produce some differences between IPS and Ulysses results, the NS-asymmetries in 1994 which are close to zero agree very well with each other. The N-S asymmetry observed by IPS was in the same sense as Ulysses observed, that is, the northern hemisphere has higher velocities than the southern hemisphere. This implies that the solar wind asymmetry measured by Ulysses had a solar origin. We find that this asymmetry was far greater in the years 1995-1996.

BIMODAL STRUCTURE AND VELOCITY GRADIENT

We derived a velocity map from the IPS data obtained during April-June 1995, and sampled the velocity from the map along the projected trajectory of Ulysses on the source surface (2.5 Rs). This period is close to the period when Ulysses made a rapid latitudinal scan from the south pole to the north poles of the Sun (from September 1994 to July 1995). Sampled velocities are plotted together with those from Ulysses

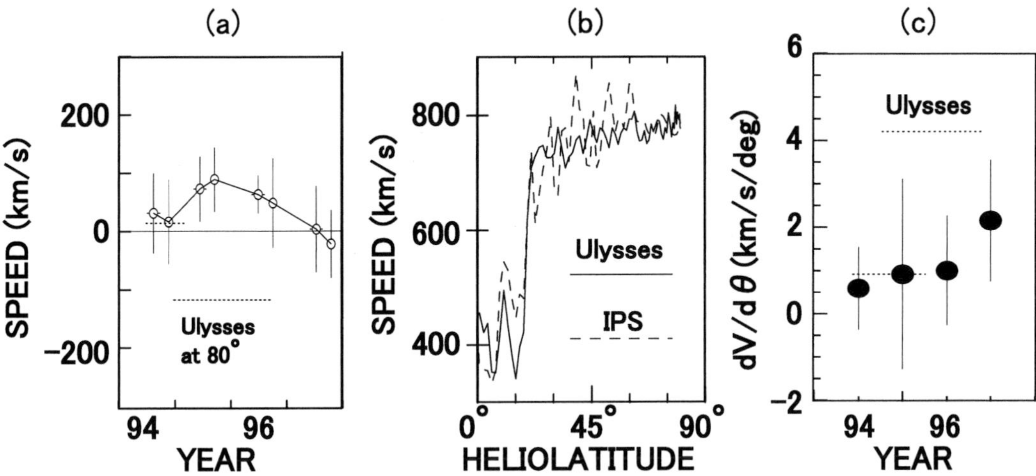

Figure 1: (a) Velocity difference between northern and southern high latitudes. Ulysses observations at 80° (Goldstein et a., 1996) are shown by a dotted line. (b) Latitudinal plot of the speeds from the Ulysses observations and the IPS CAT analysis. (c) Latitudinal gradient of the high-speed wind. Ulysses observations (McComas et al., 2000) are shown by a dotted line.

in Figure 1b. Both plots show fairly good agreement in the velocity values of low- and high-speed winds, recurrent high-speed winds at low latitudes, and the sharp velocity gradient between low- and high-speed regions.

The gradual velocity increase at high latitudes, also observed by Ulysses (Woch et al., 1997; McComas et al., 2000), is easily recognized. We analyzed velocity gradients at high latitudes for years of 1994-1997 and show them in Figure 1c. They agree with the velocity gradient, 0.95 km $s^{-1}deg^{-1}$, observed by Ulysses (McComas et al., 2000). It will be interesting to see whether the gradient increases as solar activity increases.

APPLICATION TO SPACE WEATHER

In the previous section we showed that the IPS CAT analysis can retrieve the recurrent solar wind structure. Using this CAT analysis method, we derive quiet solar wind structure (corotating structure) and predict the solar wind which will be observed at the earth. This is the forecast of the solar wind existing as background structure. This daily forecast is carried out at
UCSD/CASS (http://casswww.ucsd.edu/solar/forecast/index.html) and
STELab (http://stesun5.stelab.nagoya-u.ac.jp/allsky_gmap/).

Once we know the background structure, it is easy to find transient interplanetary disturbances. We derive a synoptic velocity map on the source surface using the IPS CAT analysis, and the map is expanded into interplanetary space to make 3-D solar wind model. IPS observations for hypothetical radio sources across the sky are simulated in this solar wind, and an IPS sky map is produced as shown in Figure 2 for July 15, 2000. This Figure shows a "fisheye" velocity map of the sky as observed from Earth. The Sun is centered in the day side map, the center of the night side map is the anti-sun direction, and the ecliptic plane is in the horizontal axis. This map shows a prediction of the IPS speed observations in the sky plane for the

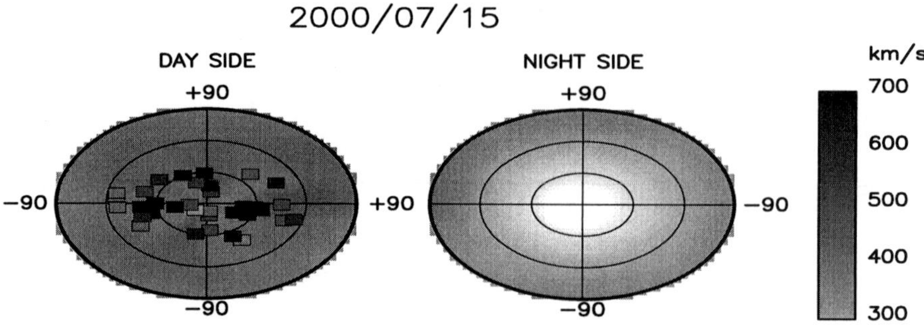

Figure 2: A fisheye velocity map of the sky as observed from Earth predicted for July 15, 2000. Squares in the map are velocities observed on July 15.

corotating solar wind. On this fisheye map, velocities observed on July 15th are shown with squares. If the velocity in the square is the same as the background, it means corotating structure is observed. At $\pm 30^\circ$ elongations near the ecliptic plane, high-speed winds were observed. Since they were not expected from the corotating solar wind, we can classify these as transient phenomena.

CONCLUSION

We analyzed IPS data obtained at STELab during years of 1994–1997 using the tomography method. We compared several structures in the map (N-S asymmetry, bimodal structure and latitudinal gradient) to Ulysses' in situ observations and confirmed that the IPS CAT can deconvolve those structures from line-of-sight integrated measurements.

We are using the CAT analysis to predict corotating solar wind structures and watching transient interplanetary phenomena. This project is carried out by the groups at the UCSD and the STELab under the US-Japan Cooperative Research Project. Daily observed IPS data are transferred fully automatically from the STELab to UCSD in near-real-time; the data are analyzed at UCSD, and the forecast is then broadcast on the web. We currently are planning a project to build a larger IPS antenna named "Heliospheric Imager", which will enable us to increase the number of radio sources observed on a daily basis significantly. This will enable the CAT analyis of events over much shorter time periods.

ACKNOWLEDGMETNS

We would like to acknowledge engineering support for the IPS observations from Y. Ishida, K. Maruyama and N. Yoshimi. This work was partially supported by the Scientific Research Fund of Japan Society for the Promotion of Science (JSPS) (grant 12440130) and by the funds for the Cooperative Research under the Japan-U.S. Cooperative Science Program of the JSPS and the NSF. B. Jackson and P. Hick were supported at UCSD under NSF grants ATM-9819947 and INT-9815377.

REFERENCES

Asai, K., M. Kojima, M. Tokumaru, A. Yokobe, B. V. Jackson, P. Hick, and P. K. Manoharan, Heliospheric tomography using interplanetary scintillation observations, 3, Correlation between speed and electron density fluctuations in the solar wind, *J. Geophys. Res.*, 103, 1991–2001, 1998.

Coles, W.A., A Bimodal Model of the Solar Wind Speed, *Astrophys. and Space Sci.*, *243*, 87-96, 1996.

Goldstein, B. E., M. Neugebauer, J.L. Phillips, S. Bame, J.T. Gosling, D. McComas, Y.-M. Wang, N.R. Sheeley, and S.T. Suess, Ulysses plasma parameters: latitudinal, radial, and temporal variations, *Astron.*

Astrophys., *316*, 296–303, 1996.

Grall, R.R., W.A. Coles, M.T. Klinglesmith, A.R. Breen, P.J.S. Williams, J.Markkanen, and R. Esser, Rapid acceleration of the polar solar wind, *Nature*, *379*, 429–432, 1996.

Hewish, A., P.F. Scott, and D. Wills, Interplanetary scintillation of small diameter radio sources, *Nature*, *203*, 1214–1217, 1964.

Jackson, B. V., P. L. Hick, M. Kojima, and A. Yokobe, Heliospheric tomography using interplanetary scintillation observations, 1, Combined Nagoya and Cambridge data, *J. Geophys. Res.*, *103*, 12049–12067, 1998.

Kakinuma, T, Observations of interplanetary scintillation: Solar wind velocity measurements, in *Study of Traveling Interplanetary Phenomena*, edited by M. A. Shea, D. F. Smart, and S. T. Wu, 101–118, D. Reidel, Norwell, Mass., 1977.

Kojima, M., and T. Kakinuma, Solar cycle dependence of global distribution of solar wind speed, *Space Sci. Rev.*, *53*, 173–222, 1990.

Kojima, M., M. Tokumaru, H. Watanabe, A. Yokobe, K. Asai, B. V. Jackson, and P. L. Hick, Heliospheric tomography using interplanetary scintillation observations, 2, Latitude and heliocentric distance dependence of solar wind structure at 0.1-1 AU, *J. Geophys. Res.*, *103*, 1981–1989, 1998.

Manoharan, P.K., M. Kojima, N. Gopalswamy, T. Kondo, and Z. Smith, Radial evolution and turbulence characteristics of a coronal mass ejection, *Astrophys. J.*, *530*, 1061–1070, 2000.

McComas, D. J., B. L. Barraclough, H. O. Funsten, J. T. Gosling, E. Santiago- Muñoz, R. M. Skoug, B. E. Goldstein, M. Neugebauer, P. Riley, and A. Balogh, Solar wind observation over Ulysses' first full polar orbit, *J. Geophys. Res.*, *105*, 10419–10422, 2000.

Moran, P.J., A.R. Breen, A. Canals, J. Markkanen, J Padmanabhan, M. Tokumaru, and P.J.S. Williams, Observations of interplanetary scintillation during the 1998 whole sun month: a comparison between EISCAT, ORT and Nagoya data, *Annales Geophysicae*, June 2000 (in press).

Neugebauer, M., R.J. Forsyth, A.B. Galvin, K.L. Harvey, J.T. Hoeksema, A.J. Lazarus, R.P. Lepping, J.A. Linker, Z. Mikic, J.T. Steiberg, R. von Steiger, Y.-M. Wang, and R.F. Wimmer-Schweingruber, Spatial structure of the solar wind and comparisons with solar data and models, *J. Geophys. Res.*, *103*, 14587–14599, 1998.

Tokumaru, M., M. kojima, K. Fujiki, and A. Yokobe, Three-dimensional propagation of interplanetary disturbances detected with radio scintillation measurements at 327 MHz, *J. Geophys. Res.*, *105*, 10435–10453, 2000.

Woch, J., W. I. Axford, U. Mall, B. Wilken, S. Livi, J. Geiss, G. Gloeckler, and R. J. Forsyth, SWICS/Ulysses observations: The three-dimensional structure of the heliosphere in the declining/minimum phase of the solar cycle, *Geophys. Res. Lett*, *24*, 2885–2888, 1997.

POLAR PLUMES IN CORONAL EXPANSION

Wing-Huen Ip

Institutes of Astronomy and Space Science, National Central University Chung-Li, 320 Taiwan

ABSTRACT

Several major structures in the solar corona might share similar formation mechanisms. A case in point is the magnetic field reconnection effect which could provide the basic driving force behind the production of spicules and polar plumes. A more realistic consideration of the three-dimensional reconnection process leads to the idea that the solar spicules and coronal plumes should be characterized by twisted flux loops carrying strong current flows. This provides the linkage between the solar corona to the large-scale heliosphere. Even though the polar plumes do not supply the majority of the solar wind, they could play a key role in guiding the low-frequency Alfven waves to large coronal heights. The high level of perpendicular temperatures of the soalr wind protons and minor ions – as probably observed by SOHO/UVCS – might be maintained in this way.

INTRODUCTION

In the context of space weather research, the acceleration and heating of the solar wind is one of the key issues. The recent observations by the UVCS experiment on SOHO suggested that in the solar corona wihtin 2 to 4 $R_{\odot}$, the temperatures of protons and oxygen (O^{5+}) ions are much hotter than predicted by thermal wind models (e.g. Kohl *et al.*, 1998; Kohl *et al.*, 1999; Cranmer *et al.*, 1999a, 1999b) Furthermore, according to these authors the high perpendicular ion temperatures and strong temperature anisotropies all pointed to the presence of resonant ion cyclotron wave interaction advocated by Axford and McKenzie (1992). Another potentially important observational result is that the maintenance of the large temperature anisotropies ($T_{per}/T_{par} \sim$ 10-100) in competition against adiabatic expansion effect would probably require an extended source region of such ion cyclotron waves. If this preliminary result remains valid after more detailed data analysis, a certain modification might have to be given to the Axford-McKenzie mechanism of high-frequency wave generation via nano-flares on the solar surface. This would be necessary because the dissipation length scale of the high-frequency ion cyclotron waves is relatively short. For wave heating above 1.5 $R_{\odot}$, MHD turbulent cascade of low-frequency Alfven waves (Isenberg and Hollweg, 1983) and other mechanisms would have to be invoked. For example, the recent suggestion by Lee and Wu (2000) that shock waves generated by nano-flares could play an important role in perpendicular heating of protons and minor ions is an interesting possibility. Anothre point of view to be pursued here is to identify a mechanism by which Alfven waves (or energy flux) generated at the photosphere could be channeled to larger coronal heights. As will be discussed in the following, polar plumes are of particular interest in this respect. The organization of the present paper is to produce a brief overview and comparison of the spicules and plumes in Section 2. This is followed by an examination (in Section 3) of possible electrodynamical coupling between the polar plumes and the high speed solar wind via electric current systems in the extended coronal region.

SPICULES AND PLUMES: ARE THEY NON-IDENTICAL TWINS?

Figure 1 shows three images of the same region of a coronal hole (CH) taken at the same time but in three different wavelengths by the SOHO/SUMER instrument (Wilhelm, 2000). The lower image shows the chromospheric network at CI (124.94 nm) which are organized by magnetic field. Very complicated magnetic loop systems with mixed polarity exist at the network boundaries. Most of the fine-scaled structures find their origin there. The middle image taken at O V (62.97 nm) shows an abundance of linear spicules plus the underlying chromospheric network.

Wilhelm (2000) made a careful study of the morphology of these spicules and came to the conclusion that they must have formed via merging of the network loop system. While the correlation of the spicule formation site with the chromospheric network boundaries remains the same as the model originally proposed by Uchida (1969). More complex versions of the merging and reconnection of the magnetic field lines have been envisioned. In Figure 2, several examples are given for the coalescence of two loop systems plus the merging of a dipolar loop with a unipolar flux tube.

One problem with the above scenarios has to do with the fact that they all assumed that the magnetic field merging process should be a two-dimensional process. In reality, reconnection is most likely to proceed in a three-dimensional context. A well known example is the flux transfer events detected at Earth's magnetopause (Russell and Elphic, 1979; Lee and Fu, 1985). The interconnection of two sheets of magnetic field lines with non-zero inclination will lead to the production of helical structure in the flux rope (see Figure 3). If the scheme previously proposed by several authors is valid (e.g. Wilhelm, 2000; Uchida, 1969; Pikel'ner, 1971; Blake and Sturrock, 1985), it is difficult to avoid the conclusion that many of the spicules observed are twisted flux ropes and hence current flows. From this point of view, such twisting motion might have been indicated by the report of detection of transverse motions up to a speed of 19 km/s relative to the spicule propagation direction (Nikolskii and Paltova, 1971). In the study of Wilhelm (2000), strong red and blue shifts (+/- 30 km/s) along a single spicule can be found. Could this be related to the unwinding of the twisted flux ropes? Note that Kudoh and Shibata (1999) produced an alternative model in which the spicules are genearated by the upward propagation of Alfven waves from the photosphere to the transition region. Sharp changes in the velocity component perpendicular to the magnetic field direction could also be created.

The upper diagram in Figure 1 shows the formation of ray-like structures in Mg X (624.94 A). These polar plumes have a higher density (by a factor of 4-6) compared with the interplume region cover about 10% of the CH (Saito, 1965). For a plume velocity of $V_p \sim 50$ km/s and an interplume velocity of $V_i \sim 299$ km/s, the mass flux of the polar plumes is only 10% of the total solar wind flux. The polar plumes are therefore not the major source of the high-speed solar wind as far as the mass flux is concerned. However, as we will discuss in the following, the polar plumes could play a very important role in the heating and acceleration of the high-speed solar wind.

Figure 1. A portion of the north polar cap (~390"x200") of the Sun on August 31, 1996, taken in three emission lines (top to bottom: Mg X at 62.494 nm O V at 62.973 nm and C I at 124.941 nm) with different formation temperatures from 1.1×10^6 K (top picture) to 3×10^4 K (bottom picture). From Wilhelm (2000) with copy right permission of Astronomy and Astrophysics.

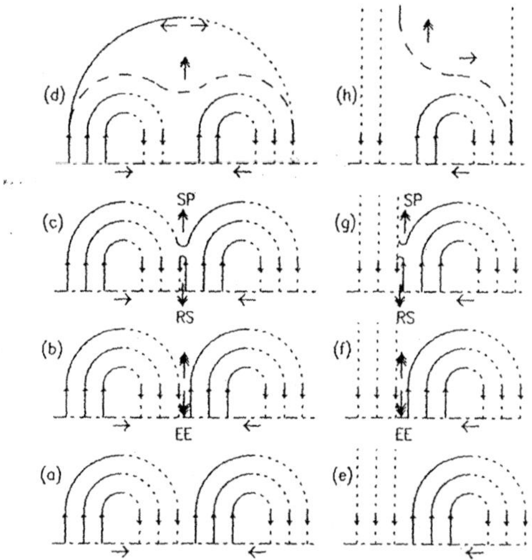

Figure 2. A summary of two different possible merging processes of magnetic field lines leading to the formation of spicules. In this representation, EE stands for explosive events produced by magnetic field reconnection; SP stands for upward lifting material, e.g., spicules; and RS stands for downward moving material in redshift. The lift-hands side figures (a-d) represent the merging g two network loop system. The sight-hand side figures (e-h) represent the merging of a loop system with an open field line region. A generalization of the planar reconnection in two dimension to a three-dimensional situation will lead to the production of flux loops containing helical field lines. From Wilhelm (2000) with copy right permission of Astronomy and Astrophysics.

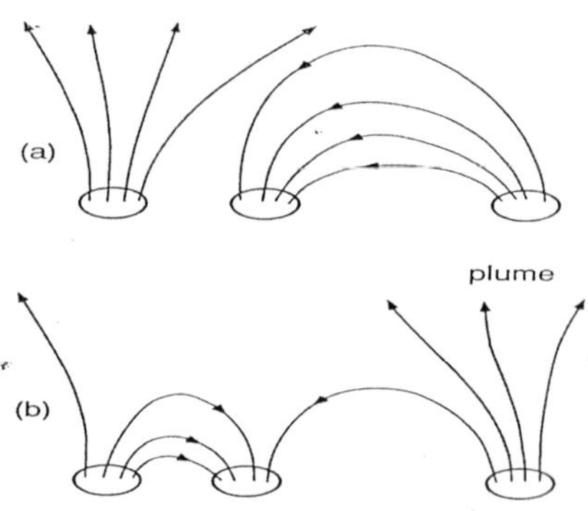

Figure 3. A schematic view of the formation of helical magnetic field structures by reconnection of two flux loops with mutual inclination. From Wang and Sheeley (1995).

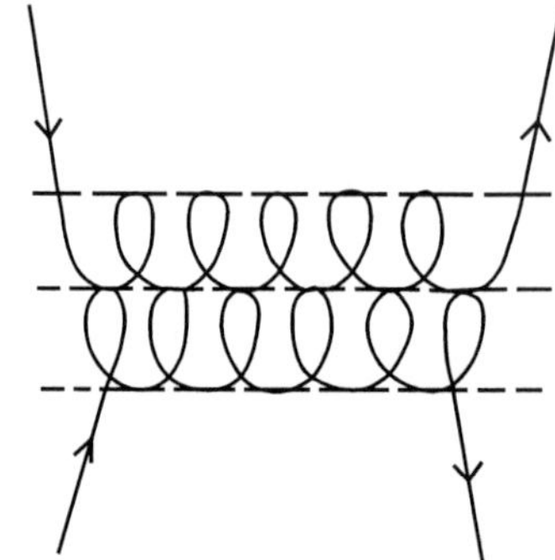

Figure 4. A schematic view of coronal plume formation via three-dimensional magnetic field reconnection on the solar surface.

PLUMES AS DRIVING FORCE ?

According to Wang and Sheeley (1995) the formation of the polar plumes can be understood in terms of reconnection of bipolar magnetic structures with a mono-field structure (see Figure 4). This scenario is very similar to the one proposed for the generation of the spicules - and for the nano-flares for this matter. Once this happens, the chromospheric material initially contained in the bipolar loop system will be injected into the open field region. Since the magnetic field lines in the loop system should be tightly wound, the released flux tube could carry this signature of helical configuration. This feature has recently been reported by Zhukov *et al.* (2000) based on their study of the white light images obtained by the SOHO/LASCO C2 coronagraph. These authors found that the helical structure along a plume would be associated with an electric current of $I_p \sim 10^9$ Amperes in total.

If the opening angle of a CH is 20°, its total surface area will be $A_{CH} \sim 2.8x10^{11}$ km^2. Given the coverage by the polar plumes to be only 10%, the corresponding surface area occupied by the coroal plumes will be $A_p \sim 2.8x10^9$ km^2. Thus, with an average radius of $R_p \sim 1.6x10^4$ km (Zhukov *et al.*, 2000), the number of plumes would be approximately $N_p \sim 36$. The total current flow carried by the polar plumes would consequently be of the order of $I_{CH} \sim N_p I_p \sim 3.6x10^{10}$ A. This current is more than enough to close the global heliospheric current system postulated by Alfven (1981) with a total strength of $I_A \sim 3x10^9$ A. In fact, Alfven (1981) had also suggested that the filamentary structures of the polar plumes could be the result of field-aligned currents. This conjecture has two major implications. First, the solar corona must be immersed in a system of field-aligned electric currents forging the linkage of the solar corona to the large-scaled solar wind (see Figure 5). Second, the plume structures collimated by the current could produce a distributed source of ion cyclotron waves to accelerate the solar wind. The related new observation of interest has to do with the possible detection of compressive slow-mode Alfven waves propagating along polar plumes up to 1.3 $R_\odot$ (Deforest and Gurman, 1998). These SOHO/EIT observations produced evidence of quasi-periodic intensity fluctuations with periods of 10-12 minutes. The outward speed of these structures was estimated to be ~ 75-150 km/s. This effect suggests that the plumes may serve as a waveguide for low-frequency slow waves generated at the footpoints of the flux tubes (e.g. Ofman *et al.*, 1999; Ofman *et al.*, 2000). According to DeForest and Gurman (1998), the energy flux carried by these waves was of the order of a few 10^5 ergs/cm^2/s which is comparable to the energy input rate required to heat the solar corona. Plumes can be found to a radial distance of 5-10 $R_\odot$ (e.g. Zhukov et al., 2000; Fisher and Guhathakurt, 1995). Could this waveguide effect contribute to the distributed heating mechanism in the extended solar corona as indicated by SOHO/UVCS observations (e.g. Kohl *et al.*, 1998; Kohl *et al.*, 1999; Cranmer *et al.*, 1999a, 1999b)?

Interpretation of the UVCS data is not straightforward (Giordano and Antonucci, 2000). It is possible that there is no need for a distributed heating mechanism and the frequency-scanning resonant ion cyclotron wave heating effect near the coronal base is sufficient to account for the heating and acceleration of the high-speed solar wind (Axford and McKenzie, 1992). It is nevertheless important to reflect on the wealth of new information from SOHO and to see what they all mean.

DISCUSSION

How far away in the solar wind will the polar plumes be able to maintain their identity? The SOHO/LASCO measurements by Zhukov *et al.*(2000) suggested that the polar plumes could be detected to radial distance r ~ 20 $R_\odot$. With much foresight, Parker (1964) investigated the MHD instability of counter-streaming solar wind flux tubes. He estimated that there will be onset of Kelvin-Helmholtz shear instability (KHI) at r ~ 0.1 AU (20 $R_\odot$). Suess (1998) and Prahi *et al.* (1999) derived a smaller value of the critical distance (r ~ 10 $R_\odot$). Plasma measurements by the Helios spacecraft at perihelion of about 0.3 AU (~ 60 $R_\odot$) did not indicate any obvious signs of systematic density variations with an enhancement factor of 3-4. Intermixing of the slow-moving plume structures with the bulk flow must therefore have taken place between 10 and 60 $R_\odot$. Besides KHI, other physical effects could also play a role in smoothing the velocity and density differences between the plumes and inter-plume lanes. For example, electrostatic ion clcyltron (EIC) waves might be generated by velocity shear at the plume boundary (Shukla and Stenflo, 1999). The EIC waves so generated could in turn lead to a certain amount of solar wind ion heating as suggested for the polar wind acceleration by Lysak *et al.* (1980). Indeed, when the Solar Probe goes through the inner corona at its closest approach to the Sun, the coronal hole region might appear very similar to the polar region of Earth's ionosphere which is filled with wave activity and current flows.

ACKNOWLEDGMENT

This work was partially supported by NSC grant NSC 89-2111-M-008-017.

REFERENCES

Alfven, H., *Cosmic Plasma*, D. Reidel Publishing Company, Dordrecht Holland (1981).

Axford, W. I. and J. F. McKenzie, The origin of high speed solar wind streams, in *Proc. Solar Wind Seven*, edited by E. Marsch and R. Schwenn, pp. 1-5, Pergamon (1992).

Blake, M. L. and P.A. Sturrock, Spicules and surges, *Astrophys. J.*, **290**, 359 (1985).

Cranmer, S. R., J. L. Kohl, G. Noci, E. Antonucci, G. Tondello, *et al.*, An empirical model of a polar coronal hole at a solar minimum, *Astrophys. J.*, **511**, 481 (1999a).

Cranmer, S. R., G. B. Field, and J. L. Kohl, Spectroscopic constraints on models of ion cyclotron resonance heating in the polar solar corona and high-speed solar wind, *Astrophys. J.*, **518**, 937 (1999b).

DeForest, C. E. and J. B. Gurman, Observation of quasi-periodic compressive waves in solar polar plumes, *Astrophys. J.*, **501**, L217 (1998).

Fisher, R. and M. Guhathakurt, Physical properties of polar coronal rays and holes as observed with the SPARTAN 201-01 coronagraph, *Astrophys. J.*, **447**, L139 (1995).

Giordano, S. and E. Antonucci, Identification of the coronal sources of the fast solar wind, *Astrophys.J.*, **531**, L79 (2000).

Isenberg, P. and J. V. Hollweg, On the preferential acceleration and heating of solar wind heavy ions, *J. Geophys. Res.*, **88**, 3923 (1983).

Kohl, J. L., G. Noci, E. Antonucci, G. Tondello, M. C. E. Huber, *et al.*, UVCS/SOHO empirical determinations of anisotropic velocity distributions in the solar corona, *Astrophys. J.*, **501**, L127 (1998).

Kohl, J. L., R. Esser, S. R. Cranmer, S. Fineschi, L. D. Gardner, *et al.*, EUV spectral line profiles in polar coronal holes from 1.3 to 3.0 R , *Astrophys. J.*, **510**, L59 (1999a).

Kudoh, T. and K. Shibata, Alfven wave model of spicules and coronal heating, *Astrophys. J.*, **514**, 493 (1999).

Lee, L. C. and B. H. Wu, Heating and acceleration of protons and minor ions by fast shocks in the solar corona, *Astrophys. J.*, **535**, 1014 (2000).

Lee, L. C. and Z. F. Fu, A theory of magnetic flux transfer at the Earth's magnetopause, *Geophys. Res. Lett.*, **12**, 105 (1985).

Lysak, R. L., M. K. Hudson, and M. Temerin, Ion heating by strong electrostatic ion cyclotron turbulence, *J. Geophys. Res.*, **85**, 678 (1980).

Nikolskii, G. M. and A. G. Paltova, Motions of Hα-spicules along the solar limb, *Solar Phys.*, **18**, 403 (1971).

Ofman, L., V. M. Nakariakov and C. E. DeForest, Slow magnetosonic waves in coronal plumes, *Astrophys. J.*, **514**, 441 (1999).

Ofman, L., V. M. Nakariakov and N. Sehgal, Dissipation of slow magnetosonic waves in coronal plumes, *Astrophys. J.*, **533**, 1071 (2000).

Parhi, S., S. T. Suess, and M. Sulkanen, Can Kelvin-Helmholtz instabilities of jet-like structures and plumes cause solar wind fluctuations at 1 AU? *J. Geophys. Res.*, **104,** 14781 (1999).

Parker, E. N., Dynamical properties of stellar coronas and stellar winds. III. The dynamics of coronal streamers, *Astrophys. J.*, **139**, 690 (1964).

Pikel'ner, S. B., Nature of the fine structure of the middle chromosphere, *Solar Phys.*, **20**, 286 (1971).

Russell, C. T. and R. C. Elphic, ISEE observations of flux transfer events at the dayside magnetopause, *Geophys. Res. Lett.*, **6**, 33 (1979).

Saito, K., Polar rays of the solar corona. II, *Publ. Astron. Soc. Japan*, **17**, 1 (1965).

Shukla, P. K. and L. Stenflo, Velocity-gradient-driven electrostatic ion-cyclotron drift waves and associated ion acceleration in the auroral ionosphere, *Plasma Phys. Rep.*, **25**, 355 (1999).

Suess, S. T., Models of plumes: their flow, their geometric spreading, and their mixing with interplume flow, in "*Solar Jets and Coronal Plumes*", ESA SP-421 (1998).

Uchida, Y., A mechanism for the acceleration of solar chromospheric spicules, *Publ. Astron. Soc. Japan*, **21**, 128 (1969).

Wang Y. M. and N. R. Sheeley, Jr., Coronal plumes and their relationship to network activity, *Astrophys. J.*, **452**, 457 (1995).

Wilhelm, K., Solar spicules and macrospicules observed by SUMER, *Astron. Astrophys.*, **360**, 351 (2000).

Zhukov, A. N., I. S. Veselovsky, S. Koutchmy and A. Llebaria, Helical magnetic structure of white light polar plumes, in *Recent Insights into the Physics of the Sun and Heliosphere – Highlights from SOHO and Other Space Missions*, ASP Conference Series, Vol 200, in press (2000).

SOLAR CORONAL HEATING AND WEAK FAST SHOCKS

L. C. Lee and B. H. Wu

Department of Physics, National Cheng Kung University, Tainan, 701 Taiwan

ABSTRACT

It is suggested that protons and minor ions in the solar corona can be heated and accelerated by fast shocks. Weak fast shocks with an Alfvén Mach number $M_A < 1.5$ can be generated by the interaction between network magnetic fields and emerging intranetwork fields which leads to magnetic reconnection and microflares. The nearly nondeflection of ion motion across the shock ramp leads to a large perpendicular thermal velocity ($v_{th\perp}$), which is an increasing function of the mass/charge ratio. The heating by subcritical shocks with $M_A \approx 1.3$ ($1.1 \leq M_A \leq 1.5$) leads to a large temperature anisotropy with $T_\perp/T_\parallel \approx 50$ for O^{5+} ions and a mild anisotropy with $T_\perp/T_\parallel \approx 1.2$ for protons. The large perpendicular thermal velocity of He^{++} and O^{5+} ions can be partly converted to the radial outflow velocity (u) in the divergent coronal field lines. The heating and acceleration by shocks with $M_A \approx 1.3$ can lead to $u(O^{5+}) \approx v_{th\perp}(O^{5+}) \approx 460$ km/s for O^{5+} ions, $u(He^{++}) \approx v_{th\perp}(He^{++}) \approx 360$ km/s for He^{++} ions, and $u(H^+) \approx v_{th\perp}(H^+) \approx 240$ km/s for protons at $r = 3-4\,R_\odot$. These results can explain recent SOHO observations of the heating and acceleration of protons and minor ions in the solar corona. It is suggested that an observational search for weak fast shocks in the coronal holes is important.

INTRODUCTION

Recent spectroscopic measurements of protons and minor ions from the Ultraviolet Coronagraph Spectrometer (UVCS) instruments onboard the SOHO provided some very stringent constraints on coronal heating processes (Kohl *et al.*, 1998; Cranmer *et al.*, 1999). Three important observed features are listed below. (1) The protons (H^+) are mildly anisotropic with $T_\perp/T_\parallel \approx 1-1.5$ above the heights of $2-3\,R_\odot$, while the O^{5+} ions are strongly anisotropic with $T_\perp/T_\parallel \approx 10-100$. (2) At $r = 3\,R_\odot$, the O^{5+} ions have a perpendicular temperature $T_\perp \sim 2\times10^8$ °K corresponding to a thermal speed ~ 450 km/s, while the protons have $T_\perp \sim 3\times10^6$ °K and thermal speed ~ 225 km/s. (3) At $r = 3.5\,R_\odot$, the outflow speed (~ 450 km/s) of O^{5+} ions is nearly twice the proton outflow speed (~ 250 km/s).

The conventional coronal heating mechanisms, such as dissipation of acoustic waves, heating by current sheets and resonant absorption of Alfvén waves (see Parker, 1991; Zirker, 1993; Narain and Ulmschneider, 1996 for recent reviews), cannot explain the observed properties in the lower solar corona. Low-frequency (0.001 – 0.1 Hz) Alfvén waves have been demonstrated to transport a significant outward momentum to the solar wind particles via a mean wave pressure (e.g., Belcher, 1971; Ofman and Davila, 1997). However, these low-frequency Alfvén waves are not likely to accelerate the heavier ions (helium and oxygen) to flow speeds faster than that of protons or heat the heavier ions preferentially in the perpendicular direction ($T_\perp >> T_\parallel$) at the lower solar corona.

Ion-cyclotron waves ($10-10^4$ Hz) have been suggested to be responsible for the observed ion temperature anisotropies ($T_\perp > T_\parallel$), and more than mass propotional ion temperature (e.g., Tu and Marsch, 1997; Hollweg, 1999). However, ion-cyclotron waves with frequencies ≥ 10 Hz have not yet been observed in the solar wind or corona. Furthermore, the fluctuating power of ion cyclotron waves is much too low to heat and accelerate the solar wind.

Recently, Lee and Wu (2000) proposed a new coronal heating mechanism, in which the protons and minor ions are heated and accelerated by weak fast magnetosonic shocks. In this paper, we briefly present the model and discuss some consequences of the mechanism.

ION HEATING BY FAST SHOCKS

The large-scale magnetic fields in solar coronal holes are open and connected to the interplanetary magnetic field. These magnetic fields are rooted in the network which corresponds to the boundaries of supergranulation. The upwelling motions in the center of supergranules carry upward the magnetic fields, which are transported by diverging flows towards supergranular cell boundaries. The newly emerging intranetwork fields have a closed bipolar structure. Magnetic reconnection may take place between the network fields and the emerging intranetwork fields. A fast shock may be generated from the eruption region with nearly same speed in all directions as shown in Figure 1. Quasi-parallel shocks ($Q_\parallel$) are formed along field lines near the reconnection region, while quasi-perpendicular ($Q_\perp$) shocks are formed along field lines away from the eruption region.

Fig. 1. Magnetic reconnection between the emerging magnetic flux and the network open fields, which leads to a high-speed plasma flow and the generation of a shock propagating outward from the reconnection region.

We now demonstrate how the ions can be heated across the ramp of a fast shock. Consider a simple model for the magnetic field and particle motion across the shock ramp as shown in Figure 2. The shock normal is along the z-axis and the magnetic field lies in the x-z plane. The angle between the upstream (downstream) magnetic field $\boldsymbol{B}_1$ ($\boldsymbol{B}_2$) and the shock normal is θ_1 (θ_2), and $\delta \equiv \theta_2 - \theta_1$. The incident ion has a normal velocity V_1. In the de Hoffman-Teller frame, this ion is streaming along the upstream field line ($\boldsymbol{B}_1$) with $V_S = V_1/\cos\theta_1$. Assume the ion velocity immediately behind the shock is still parallel to $\boldsymbol{B}_1$ with a magnitude V_S. The component of the ion velocity in the direction perpendicular to $\boldsymbol{B}_2$ is $V_{g2} = V_S \sin\delta = V_1 \sin(\theta_2 - \theta_1)/\cos\theta_1$, which provides a large gyration speed for the downstream ion.

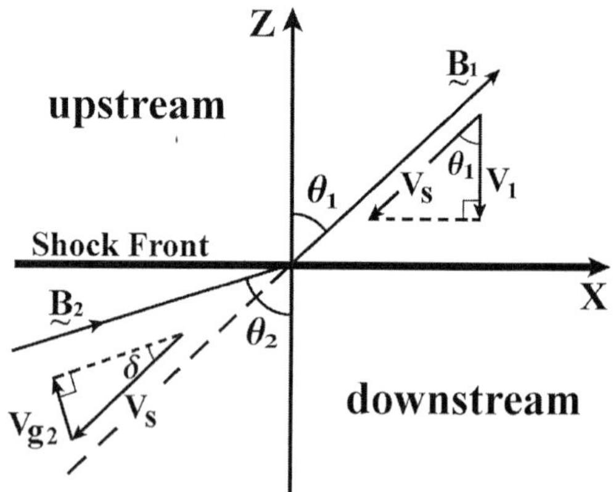

Fig. 2. The shock front is located in the plane $z = 0$. $\boldsymbol{B}_1$ and $\boldsymbol{B}_2$ are respectively the upstream and downstream magnetic fields; V_1 is the upstream normal velocity; V_S is the streaming velocity along $\boldsymbol{B}_1$ in the de Hoffman-Teller frame; and V_{g2} is the downstream ion gyration velocity.

In the realistic case, in which there exists a jump of electric potential $\Delta\phi$ at the shock ramp to slow down the normal component of the ion velocity. Let m_p be the proton mass, m_i be the ion mass, Qe be the ion charge, $M \equiv m_i/m_p$ be ion/proton mass ratio, $\alpha \equiv e\Delta\phi/\varepsilon_0$, and $\varepsilon_0 = m_p V_1^2/2$ be the incident proton energy. The ion gyration velocity immediately downstream of the shock ramp can be written as (Lee et al., 1986; Lee and Wu, 2000)

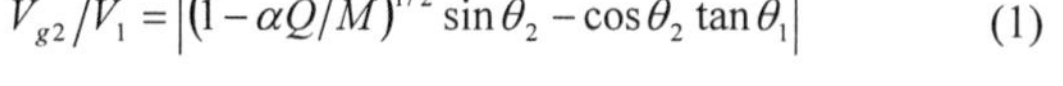

$$V_{g2}/V_1 = \left|(1-\alpha Q/M)^{1/2}\sin\theta_2 - \cos\theta_2\tan\theta_1\right| \quad (1)$$

For the same shock with fixed θ_1, θ_2 and α, the downstream gyration speed V_{g2} in Eq. (1) is an increasing function of the mass/charge ratio (M/Q).

RESULTS AND DISCUSSIONS

We also study the heating of protons and minor ions across fast shocks by one-dimensional hybrid simulations, in which the protons are treated as particles and electrons as a massless fluid. The heavier ions with mass/charge ratio

M/Q = 2, 3, 16/5 (for O^{5+}), 4, 5, 6, 8 are treated as test particles which do not affect the charge density and currents in the system.

Figure 3 shows the perpendicular thermal velocity $v_{th\perp}$ and the temperature anisotropy $T_\perp/T_\parallel$ as a function of mass/charge ratio (M/Q). The solid squares are the simulated values for the $Q_\perp$ shock with $M_A = 2$, $\theta_1 = 80°$ and $\beta_1 = 0.01$. The solid circles are for the $Q_\parallel$ shock with $M_A = 1.5$, $\theta_1 = 15°$ and $\beta_1 = 0.01$. In both cases, $v_{th\perp}$ and $T_\perp/T_\parallel$ increase with M/Q. For the $Q_\perp$ shock, we also plot the predicted values of V_{g2}/V_1 (solid line) obtained from Eq. (1) based on $\alpha \sim 0.75$, $B_{t2}/B_{t1} \sim 2.2$, and $\theta_2 \sim 86°$. The predicted values of V_{g2} agree well with the simulated values of $v_{th\perp}$. Equation (1) further predicts that as $M/Q \to \infty$, the V_{g2} (or $v_{th\perp}$) would approach the asymptotic value $(1 - B_{t1}/B_{t2}) = 0.54$. For the $Q_\parallel$ shock, the predicted value of V_{g2}/V_1 obtained from Eq. (1) with $\alpha \sim 0.9$, $\theta_1 = 15°$ and $\theta_2 \sim 54°$ ($B_{t2}/B_{t1} \sim 5.1$) are plotted as a solid line. The predicted V_{g2} is smaller than the simulated value of $v_{th\perp}$ for M/Q = 1. Note that we obtain $\beta \sim 0.006$ for B = 0.3 Gauss, $n = 5\times10^4$ cm^{-3}, $T = 2.3\times10^6$ °K at $r = 3R_\odot$ (Kohl *et al.*, 1998; Cranmer *et al.*, 1999; Lee and Wu, 2000).

Due to the conservation of magnetic moment ($\mu = T_\perp/B$), the ion gyration velocity is partly converted to the outflow velocity. As ions move outward along the diverging field lines, they are continuously heated mainly in the perpendicular direction and convert part of the perpendicular thermal energy ($kT_\perp$) to the outflow energy and gravitational potential energy.

The ion heating depends on the upstream Alfvén Mach number M_A. We have simulated proton and oxygen heating as a function of Alfvén Mach number, M_A, for shocks with $\beta_1 = 0.01$ and 0.001, and $\theta_1 = 80°$, 60°, 40°, and 15°. It is found that weak fast shocks with $M_A \approx 1.3$ ($1.1 \le M_A \le 1.5$) can lead to (a) outflow speed $u(O^{5+}) \approx$ perpendicular thermal speed $v_{th\perp}(O^{5+}) \approx 0.2\,V_A \approx$ 480km/s for O^{5+} ions, (b) $u(He^{++}) \approx v_{th\perp}(He^{++}) \approx 0.15\,V_A \approx$ 360km/s for He^{++} ions, and (c) $u(H^+) \approx v_{th\perp}(H^+) \approx 0.1\,V_A \approx$ 240km/s for protons. Here we have chosen the average Alfvén speed $V_A \approx$ 2400km/s in the region with $r \approx 1-3R_\odot$. The reaults are summarized in Table 1. The shock heating also leads to a large temperature anistropy with $T_\perp/T_\parallel \approx 50$ for O^{5+} ions and a mild anistropy with $T_\perp/T_\parallel \approx 1.2$ for protons. The results are consistent with recent SOHO observations at $r \approx 3R_\odot$ in the polar coronal holes.

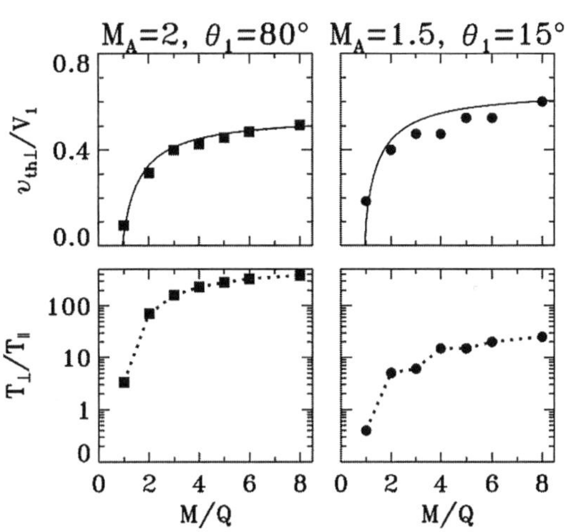

Fig. 3. The perpendicular thermal velocity $v_{th\perp}$ and the temperature anisotropy $T_\perp/T_\parallel$ are plotted as a function of the mass/charge ratio (M/Q) for two shocks, while the solid line are theoretical curves.

Table 1. Ion Heating and Outflow in the Polar Corona Holes

Species	H^+	He^{++}	O^{5+}
u	$0.1\,V_A \approx 240$ km/s	$0.15\,V_A \approx 360$ km/s	$0.2\,V_A \approx 480$ km/s
$v_{th\perp}$	$0.1\,V_A \approx 240$ km/s	$0.15\,V_A \approx 360$ km/s	$0.2\,V_A \approx 480$ km/s
$T_\perp$	3×10^6 K	3×10^7 K	2×10^8 K

* Alfvén speed $V_A \approx$ 2400km/s in the polar corona.

* If $V_A \approx$ 1200km/s in the equatorial corona, the speeds u and $v_{th\perp}$ are reduced by half and the temperature are reduced to one quarter of listed values.

Recent UVCS observations (Kohl, 2000, presented in UVCS/SOHO 2000 Science Meeting held in Northeast Harbor, Maine) also found that in the equatorial coronal holes, the heating and acceleration of protons at $r \le 3.5\,R_\odot$

are not as effective as in the polar coronal holes. The observed proton thermal speed and outflow speed are ~ 120km/s. This result can be explained by the shock heating mechanism if the Alfvén speed in the equatorial corona is ~ 1200km/s or half the Alfvén speed in the polar corona, which leads to $u(H^+) \approx v_{th\perp}(H^+) \approx 0.1\, V_A \approx$ 120km/s. The reduction of Alfvén speed can be realized if the proton density in the equatorial corona is four times larger than that in the polar corona.

Finally, we suggest that a search for the weak ($M_A \approx 1.3$) fast magnetosonic shocks in the coronal holes is very important for the identification of coronal heating mechanism. However, these weak shocks may not be readily seen or identified as the strong fast shocks associated with large flares or coronal mass ejections (CMEs). The differences between the Alfvén speed (or proton density) in the polar and equatorial coronal holes can also be searched observationally.

ACKNOWLEDGEMENTS

This work is supported by a grant from the National Science Council (NSC 90-2111-M-006-001) in Taiwan.

REFERENCES

Belcher, J. W., Alfvénic wave pressures and the solar wind, *Astrophys. J.*, 168, 509, 1971.

Cranmer, S. R., G. B. Field, and J. L. Kohl, An empirical model of a polar coronal hole at solar minimum, *Astrophys. J.*, 511, 481 (1999).

Hollweg, J. V., Potential wells, the cyclotron resonance, and ion heating in coronal holes, *J. Geophys. Res.*, 104, 505 (1999).

Kohl, J. L., et al., UVCS/SOHO empirical determinations of anisotropic velocity distributions in the solar corona, *Astrophys. J.*, 501, L127 (1998).

Lee, L. C. and B. H. Wu, Heating and acceleration of protons and minor ions by fast shocks in the solar corona, *Astrophys. J.*, 535, 1014 (2000).

Lee, L. C., Wu, C. S., and X. W. Hu, Increase of ion kinetic temperature across a collisionless shock: I. A new mechanism, *Geophys. Res. Lett.*, 13, 209 (1986).

Narian, U. and P. Ulmschneider, Chromospheric and coronal heating mechanisms. 2., *Space Sci. Rev.*, 75, 453 (1996).

Ofman, L. and J. M. Davila, Solar-wind acceleration by solitary waves in coronal holes, *Astrophys. J.*, 476, 357 (1997).

Parker, E. N., Heating solar coronal holes, *Astrophys. J.*, 372, 719 (1991).

Tu, C. -Y. and E. Marsch, Two-fluid model for heating of the solar corona and acceleration of the solar wind by high-frequency Alfvén waves, *Solar Physics*, 171, 363 (1997).

Zirker, J. B., Coronal heating, *Solar Physics*, 148, 43 (1993).

AN ALGORITHM OF CALCULATION FOR THE MOTION OF LOOPLIKE CORONAL MASS EJECTIONS

Tyan Yeh

Institute of Space Science, National Central University, Chung-Li, Taiwan

ABSTRACT

Coronal mass ejections, which manifest as magnetic clouds in interplanetary space, are now commonly accepted as the main cause of interplanetary disturbances that may trigger the earth's space weather storms. Model computations of the motion of coronal mass ejections may provide quantitative predictions for the arrivals of interplanetary disturbances at the earth. In this paper we present an algorithm of calculation for the motion of a looplike coronal mass ejection which is represented by a flux rope moving through the coronal-interplanetary medium. The moving flux rope is characterized with several parameters for its position, size, velocity, magnetic field, mass density, and temperature. These parameters evolve temporally as the flux rope is accelerated by the hydromagnetic buoyancy force exerted by the surrounding medium in overcoming the gravitational force exerted by the sun. We devise a set of ordinary differential equations that describe the temporal evolution of the parameters. Thus, the initial position and conditions together with the ambient conditions encountered by the coronal mass ejection will determine how close the latter will pass to the earth.

INTRODUCTION

Observations on the temporal changes of the solar corona by orbiting coronagraphs have revealed that coronal mass ejections are common daily occurrences (MacQueen et al., 1974; House et al.,1981; Howard et al, 1982; Howard et al., 1997). Their passages in the solar atmosphere and subsequent manifestations as magnetic clouds in interplanetary space may cause traveling disturbances to reach the Earth. The arrivals of such interplanetary disturbances are now commonly accepted as the main cause of triggering the earth's space weather storms. By virtue of the high electrical conductivity of the solar plasma, coronal mass ejections carry the magnetic field frozen into them. The accompanying magnetic field makes the dynamics of coronal mass ejections a complicated magnetohydrodynamic process.

There is a kind of so-called looplike coronal mass ejections that are well distinguished from their surrounding medium of the solar wind. Each looplike coronal mass ejection is a huge flux rope that has two footpoints anchored to the solar surface. The outward motion of the ejected mass stretches the axis of the flux rope and enlarges the cross sections of the flux rope. Thus, kinematically the velocity of a mass element in a moving flux rope comprises a translational velocity common with the axis and an expansional velocity relative to the axis. Dynamically, the translational motion is driven by a force density that overcomes the gravitational force exerted by the sun and the expansional motion is driven by a force density that is directed away from the axis. Based on this perception a theory of hydromagnetic buoyancy force for looplike coronal mass ejections has been formulated by Yeh (1982, 1983, 1985).

The flux rope that represents a looplike coronal mass ejection is regarded as a separate body, not magnetically interconnected to the exterior, immersed in the vast medium of the solar wind. The thermal and magnetic pressures exerted by the surrrounding medium constitute the hydromagnetic buoyancy force. The internal stress that is caused by the ambient stress is mainly in the form of thermal pressure. The gradient force associated with the internal thermal pressure is a distributive manifestation of the hydromagnetic buoyancy force. The Lorentz force of the cospatial current and magnetic field inside a flux rope is associated with the internal magnetic stress that is caused by internal currents alone. The former gradient force of thermal pressure has a major part that is essentially uniform in each cross section of the flux rope, much like the gravitational force exerted by the sun. The latter Lorentz force as well as the remaining part of the gradient force is essentially in the radial direction relative to the axis. Accordingly, the translational motion away from the sun is to be driven by the hydromagnetic buoyancy force. The Lorentz force drives the expansional motion relative to the axis. The partition of motion into translation and expansion allows the dynamics of a looplike coronal mass ejection to be described by the evolution of several parameters that characterize the flux rope.

Previously the formulated theory is applied to a straight flux rope (Yeh, 1995). In this paper we shall apply the theory to a curved flux rope. The characterizing parameters vary spatially along the curved axis. Their temporal evolution is determined by a set of ordinary differential equations. In a sense the solution obtained is a similarity solution of the partial differential equations of magnetohydrodynamics. Their determination requires specifications of the initial conditions of the coronal mass ejection and certain local property of the coronal-interplanetary medium through which the coronal mass ejection moves.

LOOPLIKE CORONAL MASS EJECTIONS

A flux rope that represents a looplike coronal mass ejection evolves temporally in accordance with the magnetohydrodynamic equations. The two main equations are the equation of motion

$$\rho_O \frac{d}{dt}\mathbf{u}_O = -\nabla p_O + \mathbf{J}_O \times \mathbf{B}_O + \rho_O \mathbf{g} \tag{1}$$

and the equation of induction

$$\frac{d}{dt}\mathbf{B}_O = -\mathbf{B}_O(\nabla \cdot \mathbf{u}_O) + (\mathbf{B}_O \cdot \nabla)\mathbf{u}_O \tag{2}$$

for the flow velocity $\mathbf{u}_O$ and the divergence-free magnetic field $\mathbf{B}_O$. The other two equations are the equation of continuity

$$\frac{d}{dt}\rho_O = -\rho_O (\nabla \cdot \mathbf{u}_O) \tag{3}$$

and the equation of energy

$$\rho_O C_v T_O(\rho_O^{C_p/C_v}/p_O) \frac{d}{dt}(p_O/\rho_O^{C_p/C_v}) = \nabla \cdot (\kappa \nabla T_O) \tag{4}$$

for the mass density ρ_O and the thermal pressure p_O. The latter equation expresses the increase of entropy due to heat conduction (κ being the thermal conductivity, C_v and C_p being the specific heats under constant volume or pressure). The current density $\mathbf{J}_O=\mu_0^{-1}\nabla\times\mathbf{B}_O$ is determined by the curl of the magnetic field (μ_0 being the magnetic permeability). The temperature $T_O=p_O/K\rho_O$ is determined by the thermal pressure and the mass dendity (K being the gas constant for the proton-electron solar plasma). We use subscript O to indicate quantities in the immersed body of the flux rope.

As to the surrounding medium, the thermal pressure p_I and the magnetic field $\mathbf{B}_I$ are related to the mass density ρ_I, the acceleration $\mathbf{a}_I=(\partial/\partial t+\mathbf{u}_I \cdot \nabla)\mathbf{u}_I$, the magnetic tensile force density $\tau_I=\mu_0^{-1}(\mathbf{B}_I \cdot \nabla)\mathbf{B}_I$, and the solar gravity $\mathbf{g}$ by

$$-\nabla(p_I + \tfrac{1}{2}\mu_0^{-1}\mathbf{B}_I^2) = -\rho_I \mathbf{g} + \rho_I \mathbf{a}_I - \tau_I. \tag{5}$$

We use subscript I to indicate quantities pre-existing at the location where the flux rope intrudes, hence the local values of p_I, $\mathbf{B}_I$, and $\nabla(p_I+\frac{1}{2}\mu_0^{-1}\mathbf{B}_I^2)$ are regarded quasi-constant.

The set of partial differential equations (1)-(4) can be reduced, in useful context and good approximation, to a set of ordinary equations when the flow velocity is partitioned as

$$\mathbf{u}_O(t;\ \ell, q, \phi) \cong \mathbf{u}_0 + \mathbf{1}_q V \frac{q}{Q} \tag{6}$$

and the magnetic field has the distribution

$$\mathbf{B}_O(t;\ \ell, q, \phi) \cong \mathbf{1}_l B_0 \frac{(1-q^2/Q^2)^{1/2}}{1-(q/R)\cos(\phi-\Phi)} + \mathbf{1}_\phi\, \mu_0 J_0 Q \frac{\frac{1}{2}q/Q}{1-(q/R)\cos(\phi-\Phi)} - \mathbf{1}_q \frac{dB_0}{d\ell} \frac{\frac{1}{2}q(1-q^2/Q^2)^{1/2}}{[1-(q/R)\cos(\phi-\Phi)]^2}\ . \tag{7}$$

The prescribed magnetic field is exactly divergence-free if the radius-of-curvature R is infinite. Otherwise, an inessential modification is needed to the radial component. Moreover, the thermal pressure is assumed to have the distribution

$$p_O(t;\ \ell, q, \phi) \approx p_0\left(1 - \frac{q^2}{Q^2}\right) + \left[-\frac{\mathbf{I}_O\times\mathbf{B}_I}{\pi Q^2} + \nabla_\perp\left(p_I + \frac{1}{2}\mu_0^{-1}\mathbf{B}_I^2\right)\right] \cdot \mathbf{1}_q q + \left(p_I + \frac{1}{2}\mu_0^{-1}\mathbf{B}_I^2\right)\frac{q^2}{Q^2} \tag{8}$$

and the mass density

$$\rho_O(t;\ \ell, q, \phi) \approx \rho_0(t;\ \ell) \tag{9}$$

is assumed to have negligible variation in each cross section of the flux rope.

We have used tubal coordinates (ℓ, q, ϕ) aligned with the curved axis of the flux rope, which has radius-of-curvature R(t; ℓ) and torsion τ(t; ℓ). The arc-length ℓ is measured along the axis from a conveniently chosen mass element and the perpendicular distance q is measured from the axis. To make the coordinate system orthogonal, with the differential metric

$$d\mathbf{x} = \mathbf{1}_l\left[1 - \frac{q}{R}\cos(\phi-\Phi)\right] dl + \mathbf{1}_q dq + \mathbf{1}_\phi\, q\, d\phi, \tag{10}$$

the azimuthal angle ϕ is measured from a certain azimuthal surface so that the center of curvature has the ϕ-value of $\Phi(t;\ \ell) \equiv \int_0^{\ell} \tau(t;\ \acute{\ell})\, d\acute{\ell}$ (Yeh, 1986).

The parameters $\mathbf{u}_0$ (which signifies the translational velocity common to all mass elements in the same cross section), V (which signifies the expansional speed of the periphery), Q (which signifies the cross-sectional radius of the circular periphery), B_0 (which signifies the axial value of the magnetic field), J_0 (which signifies the axial value of the current density), p_0 (which signifies the axial value of the thermal pressure), and ρ_0 (which signifies the axial value of the mass density) are all functions of ℓ in addition to the time t. We use subscript 0 to indicate quantities on the axis of the flux rope.

The current density is

$$\mathbf{J}_O(t;\ \ell, q, \phi) \cong \mathbf{1}_l J_0 \frac{1}{1-(q/R)\cos(\phi-\Phi)} + \mathbf{1}_\phi \frac{\mu_0^{-1} B_0}{Q} \frac{(q/Q)(1-q^2/Q^2)^{-1/2}}{1-(q/R)\cos(\phi-\Phi} - \mathbf{1}_q \frac{dJ_0}{d\ell} \frac{\frac{1}{2}q}{[1-(q/R)\cos(\phi-\Phi)]^2}. \tag{11}$$

Thus, the flux rope has a longitudinal current $\mathbf{I}_O \equiv \mathbf{1}_l I_O$ with a magnitude

$$I_O(t;\ \ell) = \int_0^Q \int_0^{2\pi} (\mathbf{1}_l \cdot \mathbf{J}_O)\, q\, dq\, d\phi \approx J_0 \pi Q^2. \tag{12}$$

GEOMETRY OF A FLUX ROPE

We assume that the periphery of the flux rope has circular cross-sections at all times with the radii varying along the axis and also varying in time. Moreover, we assume that as the flux rope moves, its axis and its periphery remain to be composed of the same mass elements. So do each of its cross-sectional concentric circles and each of its cross-sectional radial lines. Accordingly, each mass element of the flux rope can be specified by its Lagrangian coordinates (λ, μ, ν), with $\lambda \equiv \int_0^{\ell} \pi Q^2 \rho_0 d\ell$ specifying a cross-sectional plane perpendicular to the axis, $\mu \equiv q/Q$ specifying a radial distance from the axis, and $\nu \equiv \phi - \Phi$ specifying a radial line.

The motion of a mass element of a flux rope can be described by the temporal change of its heliocentric position

$$\mathbf{x}_O(t;\lambda,\mu,\nu) \equiv \mathbf{x}_0(t;\lambda) + \mathbf{1}_q Q(t;\lambda)\,\mu. \tag{13}$$

The heliocentric vector $\mathbf{x}_0$ is from the solar center to an axial mass element which has the same λ-coordinate as the mass element under consideration. The radial vector $\mathbf{1}_q Q\mu$ is from the axis perpendicularly to the mass element, with the unit vector $\mathbf{1}_q$ dependent on λ, μ, and ν. At any instant of time, the axis of the flux rope is described by $\mathbf{x}_0$ as a function of ℓ, which is related to λ by

$$d\lambda = \rho_0 \pi Q^2 dl. \tag{14}$$

Through this mapping between λ and ℓ all the parameters are functions of t and λ. Their dependences on λ are prescribed at initial time. Their changes in t are to be determined. Now the unit tangent vector is

$$\mathbf{1}_l(t;\lambda) \equiv \frac{d\mathbf{x}_0}{d\ell}. \tag{15}$$

The normal vector $\mathbf{1}_c$ toward the center of curvature is in the direction of $d\mathbf{1}_l/dl$ and the radius-of-curvature R is the reciprocal of the magnitude of $d\mathbf{1}_l/dl$. Namely,

$$\mathbf{1}_c \frac{1}{R} = \frac{d\mathbf{1}_\ell}{d\ell}. \tag{16}$$

The binormal vector $\mathbf{1}_b$ perpendicular to the osculatory plane is $\mathbf{1}_l \times \mathbf{1}_c$. The torsion τ can be calculated from the Frenet-Serret formula $\mathbf{1}_c \tau = d\mathbf{1}_b/dl$.

The vectorial differential area of the peripheral surface of the flux rope, given by $\mathbf{1}_\phi Q\, d\phi \times \{\mathbf{1}_l[1-(Q/R)\cos(\phi-\Phi)]dl + \mathbf{1}_q dQ\}$, can be written $\mathbf{1}_n\, ds\, dl$, with

$$\mathbf{1}_n ds \equiv \left\{ \mathbf{1}_q\left[1 - \frac{Q}{R}\cos(\phi-\Phi)\right] - \mathbf{1}_l \frac{dQ}{d\ell} \right\} Q\, d\phi. \tag{17}$$

The differential area of the cross-section is $dS \equiv q\, dq\, d\phi$.

Indeed, the velocity given by equation (6) is obtainable from equation (13) upon the use of the kinematic relationship $\mathbf{u}_O = d\mathbf{x}_O/dt$, namely

$$\frac{d\mathbf{x}_0}{dt} = \mathbf{u}_0, \tag{18}$$

and

$$\frac{dQ}{dt} = V \tag{19}$$

by virtue that q/Q does not change in time and $\mathbf{1}_q$ as well as $\mathbf{1}_\phi$ are assumed not changing in time. The acceleration is

$$\frac{d\mathbf{u}_O}{dt} \cong \frac{d\mathbf{u}_0}{dt} + \mathbf{1}_q \frac{dV}{dt}\frac{q}{Q}. \tag{20}$$

From equation (6) we obtain

$$\nabla \cdot \mathbf{u}_O \approx \mathbf{1}_l \cdot \frac{d\mathbf{u}_0}{d\ell} + \frac{\mathbf{1}_c \cdot \mathbf{u}_0}{R} + 2\frac{V}{Q} \tag{21}$$

and

$$(\mathbf{B}_O \cdot \nabla)\mathbf{u}_O \approx \mathbf{1}_l (\mathbf{1}_l \cdot \frac{}{d\ell})(\mathbf{1}_l \cdot \mathbf{B}_O) + \mathbf{1}_q \frac{}{Q}(\mathbf{1}_q \cdot \mathbf{B}_O) + \mathbf{1}_\phi \frac{}{Q}(\mathbf{1}_\phi \cdot \mathbf{B}_O). \tag{22}$$

LORENTZ FORCE IN A FLUX ROPE

In the absence of field line reconnection with external magnetic field, the internal magnetic field due to the internal current in the flux rope must be spatially so distributed that field lines do not cross the periphery. So must the internal current itself. Moreover, the boundary magnetic field on the periphery should be azimuthal because no peripheral longitudinal component is produced by the internal current. These requirements are satisfied by the assumed magnetic field $\mathbf{B}_O$ and current density $\mathbf{J}_O$. In fact, the field lines are torsionally helical, with pitch angles increasing from 0° at the axis to 90° at the periphery. The current is tangential on the periphery. The curvature of the axis has only a small effect of the order of Q/R. The torsion of the axis has a much smaller effect.

The flux rope has a longitudinal magnetic flux

$$\Psi_l = \int_0^Q \int_0^{2\pi} (\mathbf{1}_l \cdot \mathbf{B}_O)\, q\, dq\, d\phi = B_0 \int_0^Q (1 - q^2/Q^2)^{1/2} q \int_0^{2\pi} \frac{d\phi}{1 - (q/R)\cos\phi}\, dq \approx \frac{2}{3} B_0 \pi Q^2 \tag{23}$$

and an azimuthal flux

$$\Psi_\phi = \int dl \int_0^Q (\mathbf{1}_\phi \cdot \mathbf{B}_O)\left[1 - \frac{q}{R}\cos(\phi - \Phi)\right] dq = \int \frac{1}{4} \mu_0 J_0 Q^2 dl. \tag{24}$$

Continuities of field lines and current lines require that both the longitudinal magnetic flux and the longitudinal current be invariant along the flux rope. Hence, the field strength B_0 and the current density J_0 at the axis must vary spatially as

$$\frac{d}{d\ell} B_0 Q^2 \approx 0, \tag{25}$$

$$\frac{d}{d\ell} J_0 Q^2 \approx 0 \tag{26}$$

in connection with the spatial variation of the cross-sectional radius of the flux rope.

The helical distributions of magnetic field and current in a flux rope has the property that the Lorentz force

$$\mathbf{J}_O\times\mathbf{B}_O \cong \mathbf{1}_q\left(\frac{\mu_0^{-1}B_0^{\,2}}{Q^2} - \frac{1}{2}\mu_0 J_0^{\,2}\right)\frac{q}{[1-(q/R)\cos(\phi-\Phi)]^2}$$

$$+\ \mathbf{1}_l\left(\frac{1}{2}\mu_0^{-1}\frac{B}{Q^2}\frac{dB_0}{d\ell} - \frac{1}{4}\mu_0 J_0\frac{dJ_0}{d\ell}\right)\frac{q^2}{[1-(q/R)\cos(\phi-\Phi)]^2} + \mathbf{1}_\phi\left(\frac{1}{2}\frac{dJ_0}{d\ell}B_0 - \frac{1}{2}\frac{dB_0}{d\ell}J_0\right)\frac{q(1-q^2/Q^2)^{1/2}}{[1-(q/R)\cos(\phi-\Phi)]^3} \tag{27}$$

is zero on the axis and is in the normal direction on the periphery. In between the Lorentz force is mainly in the radial direction with a magnitude proportional to the radial distance from the axis. This Lorentz force results from interaction among different parts of the internal current only, without involving any currents outside the flux rope. The longitudinal current, which produces the azimuthal magnetic field, gives rise to a radial force toward the axis. The azimuthal current, which produces the longitudinal magnetic field, gives rise to a radial force away from the axis (Yeh, 1986). Clearly these radial forces drive motion relative to the axis. They do not provide acceleration for the motion away from the sun.

The volume integral

$$\int dl \int_0^Q \int_0^{2\pi} \mathbf{J}_O\times\mathbf{B}_O\, q\, dq\, d\phi \cong \int \mathbf{1}_l\frac{1}{2}\left(\frac{1}{2}\mu_0^{-1}B_0\frac{dB_0}{d\ell} - \frac{1}{4}\mu_0 J_0\frac{dJ_0}{d\ell}Q^2\right)\pi Q^2\, dl \tag{28}$$

has a high-order value. It will be zero if the radius is invariant along the flux rope. For a curved flux rope the integral has a residual value of the order of Q/R. Clearly, such a residual sum of Lorentz force is not capable of overcoming the solar gravity, not to mention to provide the acceleration $d\mathbf{u}_0/dt$ as a whole.

DIAMAGNETIC FORCE ON A FLUX ROPE

The force needed to overcome the solar gravity must involve interaction with the surrounding medium, including the external current that sustains the magnetic field carried by the solar wind. It turns out that the interaction between the current in the flux rope and the currents outside causes an inhomogeneity in the thermal pressure inside the flux rope. The latter inhomogeneity produces a gradient force of thermal force that has a quasi-uniform part across a cross section of the flux rope, much like the gravitational force exerted by the sun. Such a quasi-uniform force density is capable of overcoming the solar gravity and driving the common motion of all mass elements in a cross section of the flux rope.

As a flux rope moves through the magnetized plasma of the solar atmosphere and interplanetary medium, current and vorticity are induced in an interface layer between the flux rope and the surrounding medum. The induced current as well as the internal current in the flux rope produces a perturbation magnetic field to realign the external field lines to become tangential to the flux rope. Likewise, the induced vorticity deflects the streamlines of the surrounding flow to become tangential too. Because the solar plasma has very high electrical conductivity, the induced current forms a thin layer that contains insignificant amount of mass. The realignments of surrounding field lines and streamlines produce inhomogeneity in the ambient hydromagnetic stress in addition to any pre-existing inhomogeneity in the external medium.

The magnetic field outside the flux rope is due to three currents of different causes. The heliospheric current produces the magnetic field $\mathbf{B}_I(\mathbf{x})$. Its local value is denoted hereafter by

$$\mathbf{B}_I \approx B_I\left\{\mathbf{1}_l \cos\theta_I + [\mathbf{1}_q\cos(\phi-\phi_I) - \mathbf{1}_\phi\sin(\phi-\phi_I)]\sin\theta_I\right\} \tag{29}$$

and its local inhomogeneity is to be accounted for by the local value of $\nabla\frac{1}{2}\mu_0^{-1}\mathbf{B}_I^2$. The internal current in the flux rope produces a magnetic field, outside the flux rope, which is $\mathbf{1}_\phi\mu_0 I_O/2\pi q$ in approximation. The peripheral current induced in the thin interface layer produces a current-free magnetic field

$$\mathbf{B}_P(\mathbf{x}) \approx -B_I\,[\mathbf{1}_q\cos(\phi-\phi_I) + \mathbf{1}_\phi\sin(\phi-\phi_I)]\sin\theta_I\,\frac{Q^2}{q^2} \tag{30}$$

in the exterior region. Thus, the induced current has an internal image of a lineal dipole at the axis. Accordingly, the exterior magnetic field is

$$\mathbf{B}^{(ext)}(\mathbf{x}) \approx \mathbf{B}_I + \mathbf{B}_P + \mathbf{1}_\phi \frac{\mu_0 I_O}{2\pi q} \tag{31}$$

if the higher-order effect of the curvature of the flux rope and the inhomogeneity of the pre-existing magnetic field are ignored. Thus, the ambient magnetic field on the outer face of the peripheral layer is

$$\mathbf{B}_+ \approx \mathbf{1}_l B_I \cos\theta_I + \mathbf{1}_\phi [-2B_I \sin\theta_I \sin(\phi-\phi_I) + \frac{\mu_0 I_O}{2\pi Q}]. \tag{32}$$

There the perturbation magnetic field due to the peripheral current cancels the radial component of the pre-existing magnetic field due to the external current and doubles the azimuthal component, in zeroth order.

On the other hand, the induced current and the external current together produces a null magnetic field in the region occupied by the flux rope. In other words, the induced current on the interface shields the flux rope magnetically. Thus, the internal magnetic field

$$\mathbf{B}^{(int)}(\mathbf{x}) = \mathbf{B}_O \tag{33}$$

remains to be due to the internal current only. The boundary magnetic field on the inner face of the peripheral layer is

$$\mathbf{B}_- = \mathbf{1}_\phi \mu_0 J_0 Q \frac{\frac{1}{2}}{1-(Q/R)\cos(\phi-\Phi)} \approx \mathbf{1}_\phi \frac{\mu_0 I_O}{2\pi Q}, \tag{34}$$

which is quite different from the ambient magnetic field on the outer face. The difference between the ambient and boundary magnetic fields is sustained by the sheet current

$$\mathbf{i}_P = \mathbf{1}_n \times \mu_0^{-1}(\mathbf{B}_+-\mathbf{B}_-) \approx -\mu_0^{-1} B_I [\mathbf{1}_l\, 2\sin\theta_I \sin(\phi-\phi_I) + \mathbf{1}_\phi \cos\theta_I]. \tag{35}$$

In this thin layer of peripheral current the magnetic field is well represented by

$$\frac{1}{2}(\mathbf{B}_++\mathbf{B}_-) \approx \frac{1}{2}\mathbf{1}_l B_I \cos\theta_I + \mathbf{1}_\phi [-B_I \sin\theta_I \sin(\phi-\phi_I) + \frac{\mu_0 I_O}{2\pi Q}]. \tag{36}$$

Thus the magnetic force in the current layer has the surface density

$$\begin{aligned} \mathbf{i}_P \times \frac{1}{2}(\mathbf{B}_++\mathbf{B}_-) &= -\mathbf{1}_n(\frac{1}{2}\mu_0^{-1}\mathbf{B}_+^2 - \frac{1}{2}\mu_0^{-1}\mathbf{B}_-^2) \\ &\cong \mathbf{1}_q\{-\frac{1}{2}\mu_0^{-1} B_I^2[\cos^2\theta_I + 4\sin^2\theta_I \sin^2(\phi-\phi_I)] + \frac{I_O B_I \sin\theta_I}{\pi Q}\sin(\phi-\phi_I)\} \end{aligned} \tag{37}$$

which amounts to the difference of the magnetic pressures on the outer and inner faces of the current sheet. The volume integral of this magnetic force

$$\int dl \int_0^{2\pi} \mathbf{i}_P \times \frac{1}{2}(\mathbf{B}_++\mathbf{B}_-)\, Q\, d\phi \cong \int \mathbf{I}_O \times \mathbf{B}_O dl, \tag{38}$$

which is equal to the surface integral of $\frac{1}{2}\mu_0^{-1}\mathbf{B}_+^2 - \frac{1}{2}\mu_0^{-1}\mathbf{B}_-^2$ regardless of the distribution of magnetic field by virtue of the relationship between force density and stress gradient, sums up essentially to the diamagnetic force

$$\int dl \int -\mathbf{1}_n \tfrac{1}{2}\mu_0^{-1}\mathbf{B}_+^2\, ds \cong \int \mathbf{I}_O \times \mathbf{B}_O dl \tag{39}$$

exerted on the flux rope by the ambient magnetic pressure (Yeh, 1983). This diamagnetic force is to be counterbalanced by a thermal force which is sustained by a pressure difference across the peripheral layer. In turn, the matching inhomogeneity in the boundary thermal pressure incurs a gradient force of thermal pressure inside the flux rope. The quasi-uniform part of that gradient force manifests the diamagnetic force to act at various mass elements.

GRADIENT FORCE OF THERMAL PRESSURE IN A FLUX ROPE

Since the thin layer of peripheral current has a negligible mass, there the hydromagnetic force density must be zero. This is so because, with a finite acceleration, a non-zero force can only be sustained by a non-zero mass of the same order. The vanishing hydromagnetic force density has a volume integral over a pill box equal to zero. Hence the corresponding surface integral of hydromagnetic stress is zero too. Accordingly, the ambient hydromagnetic pressure on the outer face of the current layer is equal to the boundary hydromagnetic pressure on the inner face. Namely,

$$p_+ + \tfrac{1}{2}\mu_0^{-1}\mathbf{B}_+^2 = p_- + \tfrac{1}{2}\mu_0^{-1}\mathbf{B}_-^2 . \tag{40}$$

Although the sum of thermal and magnetic pressures does not change across the peripheral layer, the thermal pressure does undergo a change to compensate the change in the magnetic pressure. Since the boundary magnetic pressure is essentially homogenous circumferentially, the inhomogeneity of ambient hydromagnetic pressure is transformed spatially into a circumferential inhomogenity of thermal pressure on the boundary surface of the flux rope. The latter inhomogeneity has a significant contribution from the magnetic perturbation to the surrounding medium because the internal current carried by the flux rope causes an inhomogeneity in the ambient magnetic pressure even the pre-existing magnetic field is uniform. It follows from Eqs. (32) and (34) that

$$\tfrac{1}{2}\mu_0^{-1}\mathbf{B}_+^2 \approx \tfrac{1}{2}\mu_0^{-1}\mathbf{B}_-^2 - \frac{I_O B_I \sin\theta_I}{\pi Q}\sin(\phi-\phi_I) + \tfrac{1}{2}\mu_0^{-1}B_I^2 + \mathbf{1}_q Q \cdot \nabla_\perp \tfrac{1}{2}\mu_0^{-1}B_I^2, \tag{41}$$

with the last term added to account roughly for the pre-existing inhomogeneity in the external magnetic field at where the flux rope intrudes into, with the subscript $\perp$ indicating the transverse component. Besides the magnetic perturbation, the intrusion of the flux rope also incurs pressure perturbation. But the incurred so-called dynamic pressure is a higher-order modification to the zeroth-order force associated with the pre-existing pressure inhomogeneity in view that a uniform flow diverted around an immersed body with irrotational streamlines causes no lift force at all. Ignoring this higher-order inhomogeneity, the ambient thermal pressure

$$p_+ \approx p_I + \mathbf{1}_q Q \cdot \nabla_\perp p_I \tag{42}$$

has the inhomogeneity from the transverse component of the pre-existing gradient of the thermal pressure of the surrounding medium. Here p_I and ∇p_I denote the local values of the pre-existing pressure and its gradient at the position occupied by the axis of the intrusive flux rope. The inhomogeneity from the longitudinal component of the pre-existing pressure gradient is implicitly included in the variation of p_I along the axis. Accordingly, the inhomogeneous boundary thermal pressure is

$$p_- \approx p_I + \tfrac{1}{2}\mu_0^{-1}B_I^2 + \mathbf{1}_q Q \cdot \nabla_\perp(p_I + \tfrac{1}{2}\mu_0^{-1}B_I^2) - \frac{I_O B_I \sin\theta_I}{\pi Q}\sin(\phi-\phi_I). \tag{43}$$

Since pressure as well as stress is spatially continuous, the boundary thermal pressure is matched by the internal thermal pressure continuously. Inhomogeneity in the boundary pressure will cause corresponding inhomogeneity in the internal pressure. The thermal pressure in the flux rope varies from its boundary value

p_- to its axial value p_O at the axis of the flux rope. It is well represented by equation (8). Thus, the pressure gradient

$$-\nabla p_O \approx \frac{\mathbf{I}_O\times\mathbf{B}_I}{\pi Q^2} - \nabla_\perp(p_I+\tfrac{1}{2}\mu_0^{-1}B_I^2) - \mathbf{1}_l\frac{dp_O}{d\ell} + \mathbf{1}_q 2\frac{p_O-p_I-\frac{1}{2}\mu_0^{-1}B_I^2}{Q}\frac{q}{Q} \tag{44}$$

has a part that manifests the quasi-uniform hydromagnetic buoyancy force and another part that is proportional to the radial distance. Its volume integral

$$\int dl \int_0^Q \int_0^{2\pi} -\nabla p_O\, q\, dq\, d\phi \approx \int [\mathbf{I}_O\times\mathbf{B}_I - \nabla_\perp(p_I+\tfrac{1}{2}\mu_0^{-1}B_I^2)\pi Q^2 - \mathbf{1}_l\frac{dp_O}{d\ell}\pi Q^2]dl \tag{45}$$

shows that the part of radial force density sums up to zero essentially, just like the Lorentz force density in the flux rope.

Accordingly, the gradient force of thermal pressure consists of two parts. The quasi-uniform part is caused by the peripheral inhomogeneity. The remaining part is caused by difference between the internal and external pressures. The former is capable of overcoming the solar gravity and providing the needed acceleration for the motion to move away from the sun. The latter, together with the Lorentz force, provides the acceleration for the motion of peripheral expansion.

HYDROMAGNETIC BUOYANCY FORCE ON A FLUX ROPE

By virtue of the relationship between force density and stress gradient, the volume integral of hydromagnetic force density is equal to the surface integral of hydromagnetic stress on the bounding surfaces. Namely,

$$\iiint(-\nabla p_O+\mathbf{J}_O\times\mathbf{B}_O)dV$$
$$= \int\Big\{\int -\mathbf{1}_n(p_-+\tfrac{1}{2}\mu_0^{-1}\mathbf{B}_-^2)ds + \frac{d}{d\ell}\iint[-\mathbf{1}_l(p_O+\tfrac{1}{2}\mu_0^{-1}\mathbf{B}_O^2)+\mu_0^{-1}(\mathbf{1}_l\cdot\mathbf{B}_O)\mathbf{B}_O]dS\Big\}dl. \tag{46}$$

The integral on the lateral surface, where the magnetic field is tangential hence no magnetic tensile stress, amounts to the hydromagnetic buoyancy force

$$\int dl \int -\mathbf{1}_n(p_++\tfrac{1}{2}\mu_0^{-1}\mathbf{B}_+^2)ds \approx \int\Big\{\mathbf{I}_O\times\mathbf{B}_I + [-\nabla_\perp(p_I+\tfrac{1}{2}\mu_0^{-1}B_I^2) + \mathbf{1}_l\tfrac{1}{4}\mu_0 J_0^2 Q\frac{dQ}{d\ell}]\pi Q^2\Big\}dl \tag{47}$$

exerted by the surrounding medium (Yeh, 1985). The integral on the cross-sectional surfaces accounts for the forces mainly in the radial direction. It has a residual value

$$\int dl(d/dl)\{\mathbf{1}_l[-\tfrac{1}{2}(p_O+p_I+\tfrac{1}{2}\mu_0^{-1}B_I^2)+\tfrac{1}{2}(\tfrac{1}{2}\mu_0^{-1}B_O^2-\tfrac{1}{8}\mu_0 J_0^2Q^2)]\pi Q^2\}$$

that becomes zero when the flux rope is straight. This residual force, which is not distributed uniformly over a cross section, does not drive the translational motion.

EQUATIONS FOR PARAMETERS

Besides hydromagnetic force, each mass element is also exerted by the sun's gravitational force. The total force density may be split into a part which is quasi-uniform in a cross section and another part which is mainly in the radial direction. The former force will drive the translational motion of all mass elements in a cross section as a whole and the latter force will drive the expansional motion of each mass element relative to the axis.

Upon the substitution of equations (7), (9), (12), (20), (21), (22) and (44), equation (1) yields

$$\rho_O\frac{d}{dt}\mathbf{u}_O = \rho_O\mathbf{g} + \frac{\mathbf{I}_O\times\mathbf{B}_I}{\pi Q^2} - \nabla_\perp(p_I+\tfrac{1}{2}\mu_0^{-1}B_I^2) - \mathbf{1}_l\frac{dp_O}{d\ell}, \tag{48}$$

$$\rho_0 \frac{d}{dt}V = \frac{\mu_0^{-1}B_0^2}{Q} - \frac{1}{2}\mu_0 J_0^2 Q + 2\frac{p_0 - (p_I + \frac{1}{2}\mu_0^{-1}B_I^2)}{Q} \tag{49}$$

and equation (2) yields

$$\frac{d}{dt}B_0 = -(\frac{\mathbf{1}_c \cdot \mathbf{u}_0}{R} + 2\frac{V}{Q})B_0, \tag{50}$$

$$\frac{d}{dt}J_0 Q = -(\mathbf{1}_l \cdot \frac{d\mathbf{u}_0}{d\ell} + \frac{\mathbf{1}_c \cdot \mathbf{u}_0}{R} + \frac{V}{Q})J_0 Q. \tag{51}$$

Equation (3) yields

$$\frac{d}{dt}\rho_0 = -(\mathbf{1}_l \cdot \frac{d\mathbf{u}_0}{d\ell} + \frac{\mathbf{1}_c \cdot \mathbf{u}_0}{R} + 2\frac{V}{Q})\rho_0. \tag{52}$$

As to equatio (4), we adopt

$$\frac{d}{dt}\frac{p_0}{\rho_0^{C_p/C_v}} = 0. \tag{53}$$

We remark that for mass elements off the axis we leave the ratio $p_0/\rho_0^{C_p/C_v}$ varying in accordance with the value of p_0 determined by equation (8). This means that there are heat fluxes going through the flux rope.

CONCLUSION

To recapitulate, a looplike coronal mass ejection is represented by a flux rope which is assumed to have self-similar spatial distributions for the position, velocity, magnetic field, thermal pressure, and mass density associated to its mass elements. These spatial distributions are characterized by several parameters (viz., $\mathbf{x}_0$, Q, $\mathbf{u}_0$, V, B_0, J_0, p_0, and ρ_0) which vary along the axis of the flux rope as prescribed at the initial time. The temporal evolutions of the parameters are described by a set of ordinary differential equations and require specifications of the medium (viz., ρ_I, p_I, and $\mathbf{B}_I$) which is encounterd by the flux rope in its passage through the corona-interplanetary space.

The calculation of the temporal evolution can be effected by a numerical procedure as follows. First, the time-invariant Lagrangian coordinates are determined for each mass element on the axis of the flux rope. Then, the system of ordinary differential equations that comprise equations (18)-(19) and (48)-(53) is integrated in time steps. At each step the spatial differentiations on $\mathbf{x}_0$ and $\mathbf{u}_0$ as well as p_0 are performed numerically in order to calculate $\mathbf{1}_l$, $\mathbf{1}_c$, and R in accordance with equations (15) and (16).

ACKNOWLEDGMENTS

This research was supported by the National Science Council of Taiwan under grant NSC 89-2111-M-008-018-AP8 to the National Central University.

REFERENCES

House, L. L., W. J. Wagner. E. Hildner, C. Sawyer, and H. U. Schmidt, Studies of the corona with the solar maximum mission coronagraph/polarimeter, *Astrophys. J. Letters* , *244* , L117, (1981).

Howard, R. A., D.J. Michels , N.R. Sheeley, Jr., and M.J. Koomen, The observation of a coronal transient directed at earth, *Astrophys. J. Letters, 263*, L101, (1982).

Howard, R. A., G.E. Brueckner, O.C. St. Cyr, D.A. Biesecker, K.P. Dere, M.J. Koomen, C.M. Korendyke, P.L. Lamy, A. Llebaria, M.V. Bout, D.J. Michels, J.D. Moses, S.E. Paswaters, S.P. Plunkett, R. Schwenn, G.M. Simnett, D.G. Socker, S.J. Tappin, and D. Wang, Observation of CMEs from

SOHO/LASCO, in *Coronal Mass Ejections,* edited by N. Crooker, J. Joselyn, and J. Feynman, 17-26. American Geophysical Union, (1997).
MacQueen, R. M., J. A. Eddy, J. T. Gosling, E. Hildner, R. H. Munro, G. A. Newkirk, Jr., A. I. Poland, and C. L. Ross, The outer solar corona as observed from Skylab: preliminary results, *Astrophys. J. Letters*, *187*, L85, (1974).
Yeh, T., A magnetohydrodynamic theory of coronal loop transient, *Solar Pyhs., 78*, 287, (1982).
Yeh, T., Diamagnetic force on a flux tube, *Astrophys. J., 264*, 603, (1983).
Yeh, T., Hydromagnetic buoyancy force in the solar atmosphere, *Solar Phys., 95*, 83, (1985).
Yeh, T., Magnetic structure of a flux rope, *Astrophys. J., 305*, 884, (1986).
Yeh, T., A dynamical model of magnetic clouds, *Astrophys. J., 305*, 884, (1995).

Interplanetary Observations and Modeling Session

UPSTREAM SHOCKS AND INTERPLANETARY MAGNETIC CLOUD SPEED AND EXPANSION: SUN, WIND, AND EARTH OBSERVATIONS

R. P. Lepping[1], D. Berdichevsky[1,2], A. Szabo[1], A. J. Lazarus[3], and B. J. Thompson[1]

[1] *Laboratory for Extraterrestrial Physics, NASA's Goddard Space Flight Center, Greenbelt, MD 20771*
[2] *ITSS, Raytheon Corporation, Lanham, MD 20706*
[3] *Center for Space Research, MIT, Cambridge, MA 02139*

ABSTRACT

Identifications have been made of the probable sources on the Sun (from SOHO data), where possible, of the magnetic clouds observed by WIND near Earth over the span of about 4 years (1995-1998), i.e., mainly during the quiet solar phase. The timing of the probable source transients at the Sun allow the estimations of transit times from the Sun to Earth for a set of 20 magnetic clouds with the result that acceleration effects in the interplanetary medium appear to be small. For a larger set of magnetic clouds (N = 27) it is found that the relative speed of the front part of a cloud at 1 AU (its ramming speed) correlates better than cloud expansion with the existence of an upstream interplanetary shock. Hence, cloud expansion is not likely the primary agent responsible for clouds driving shocks. Approximately 1/2 of the magnetic clouds (i.e., 14 cases) were accompanied by upstream shocks, and for an additional 1/4 of the cases pressure pulses ahead of the magnetic clouds were present. It is shown that these upstream shocks (or pressure pulses) are well correlated with SSC's at Earth. We also examine the effects of the engulfing of the magnetosphere by the shocked solar wind-magnetic cloud complex, by comparing the integrated solar wind Poynting flux input ($\Sigma\ \varepsilon\Delta t$ = energy, where ε is the Akasofu ε–function) with the associated minimum Dst, and we develop a formal relationship between them.

INTRODUCTION

This study deals primarily with the expansion and "ramming" speeds of magnetic clouds seen by the WIND spacecraft and the relationship they have with upstream interplanetary shocks. Specifically examined are the relative speeds of the shocks, upstream solar wind, average (or center point-) plasma of the clouds, as well as the gradient of speed throughout the clouds. We also compare these magnetic cloud speeds with their average transit speeds from Sun-to-Earth from identified solar source regions, and finally we briefly consider the consequences of the interaction of these magnetic clouds and their upstream shocks (when present) with the magnetosphere. The intense southward magnetic fields within magnetic clouds, as within or around various other large interplanetary structures [e.g., *Tsurutani and Gonzalez*, 1997], usually interact strongly with the Earth's magnetosphere. We will briefly investigate some features of this interaction.

A magnetic cloud is a structure in the interplanetary medium having relatively high magnetic field intensity, an observed smoothly changing field direction as the observer passes thorough the cloud, and lower than average proton temperature [*Burlaga*, 1995]. On average, clouds are 1/4 AU in diameter at 1 AU and often have the field structure of magnetic flux ropes. Their density profiles are usually unique (and therefore not predictable), so consideration of density cannot be used to identify them. However, their densities, on average, are comparable to those upstream of the cloud [*Lepping and Berdichevsky*, 2000]. Typically magnetic clouds have very low proton plasma beta, where a value of 0.05 is not uncommon at 1 AU. These structures are like huge magnetospheres moving out from the Sun [e.g., *Marubashi*, 1997; *Burlaga et al.*, 1990], often at speeds slightly faster than the speed of the solar wind ahead of them, at least at 1 AU. Some magnetic clouds, which often act kinematically like coronal mass ejections (CME's) [e.g., *Gosling*, 1990, 1997; *Lepping and Berdichevsky*, 2000], have been shown to have an effective acceleration (for slow CME's) or deceleration (for fast CME's) tending toward measured solar wind speeds at 1 AU (ranging 320 to 650 km/s) from those speeds estimated at the Sun (ranging from 124 km/s to 1056 km/s) [*Gopalswamy et al.*, 2000]. We examine this finding by comparing the timing of measured magnetic clouds at 1 AU to the apparent solar source timing, in order to obtain average transit speeds, which are then appropriately compared to the measured cloud speeds.

Interplanetary shocks can arise from a variety of sources [see e.g., *Borrini et al.*, 1982]. They can be driven by various forms of solar ejecta (magnetic clouds being one form), by faster streams impinging on slower solar wind, or from a "blast" of energy at the Sun with no further energy input, although the latter seem to be uncommon [see, e.g., *Berdichevsky et al.*, 2000; *Lindsay et al.*, 1994]. We are concerned here with magnetic-cloud-driven shock waves only. We examine the local relationship at 1 AU of the speeds of many magnetic clouds to their upstream solar wind speeds, in order to see how the speed differences relate to the existence of driven interplanetary shocks. In the case of magnetic cloud-driven shocks we make the analogy with the fast solar wind impinging on the Earth's magnetosphere (which is the driver in that case) causing an upstream bow shock, much like the upstream shock of the driving cloud is caused by the cloud. But the Earth's bow shock is apparently always present (or almost always so), but not all magnetic clouds drive interplanetary shocks.

Figure 1 shows magnetic field magnitude $|B|$, most probable thermal speed V_{Th}, and bulk speed $|V|$ for each of three magnetic clouds observed by the WIND spacecraft, to exemplify some speed-gradient conditions within magnetic clouds, in relation to the characteristics of the regions upstream of the magnetic clouds; see Appendix A for definitions of quantities used. Example A shows a rather typical speed profile indicating a uniform drop in speed from front to back; in this case it decreases by over 100 km/s. In many cases the cloud must be expanding in a direction perpendicular to its axis [see, e.g., *Osherovich et al.*, 1993]. In case B, by contrast, we see a rather flat (unusual) profile in speed. But in both cases a moderately strong upstream shock existed, as seen in all three physical parameters. In case C there is a complex drop in speed across the cloud and a relatively fast cloud front (of over 550 km/s), but no upstream shock existed. To understand general features of the magnetic cloud-shock relationship mainly for the quiet solar epoch, we analyze a large set of WIND magnetic clouds, stressing their front interfaces and degrees of expansion.

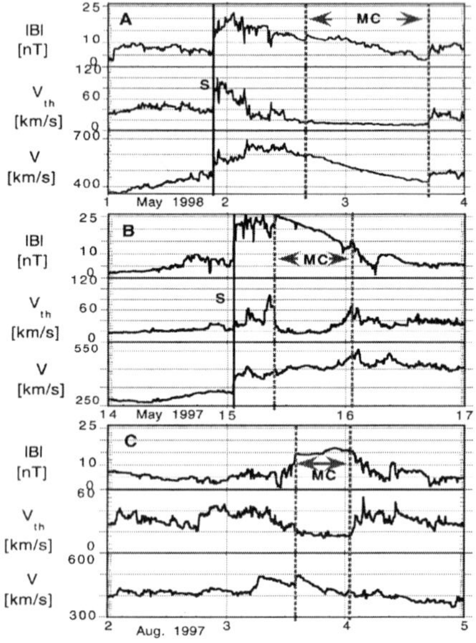

Fig. 1. Examples of magnetic field ($|B|$), most probable thermal speed (V_{Th}), and bulk speed (V) profiles for three magnetic clouds revealing markedly contrasting types: (A) the May 1998 case shows an almost uniform drop in V across the cloud (3rd panel down), (B) the May 1997 case shows an almost flat speed profile, and (C) the August 1997 case shows a two-step drop in V across the cloud and no shock. The first two clouds had upstream shock waves (indicated by the solid vertical lines marked "S"). The dashed vertical lines indicate the boundaries of the magnetic clouds (MC).

SOLAR SOURCES OF THE MAGNETIC CLOUDS AND SUN-EARTH TRANSIT SPEEDS

This study is based on a limited set (27) of WIND magnetic clouds, selected on the basis of the quality of the cloud modeling of the candidates. It arose from an initial set of 32 clouds that were observed from early 1995 until late August of 1998. The start times, durations, assumed solar source (based on SOHO/EIT and LASCO images) of the set of 27 are given in Table 1. Five cases were dropped from the original list for this study, because they did not satisfy selection criteria that depended on the magnetic cloud modeling fit parameter, χ^2, and/or symmetry considerations, as described by *Lepping and Berdichevsky* [2000]. The model for the magnetic field of the cloud's field is described by *Lepping et al.* [1990] and the means of obtaining χ^2 is described there. The magnetic field and solar wind plasma observations used in this study are from the WIND MFI [*Lepping et al.*, 1995] and SWE instruments [*Ogilvie et al.*, 1995], respectively.

Table 1. Start Times, Durations, Transit Times from the Sun, and Sources of 27 WIND Magnetic Clouds

Case #	Year	Month	Day	Hour	Δt * [Hr]	ΔT_T + [Hr]	Solar Source Month day, hour	Solar Source Position(Long, Lat)
1	95	Feb	8	03	20			
2	95	Mar	4	11	18			
5	95	Aug	22	22	22			
6	95	Oct	18	19	30			
8	96	May	27	15	41	172.1	May 21, 17	(-27° to -15° E, 0° to 28° N)
9	96	Jul	1	17	17	168.7	Jun 25, 00	(-60° to -10° E, 37° to 49° N)
10	96	Aug	7	13	22	159.0	Aug 1, 16	(-4° to 4° E-W, -9° to 9° N-S)
11	96	Dec	24	03	32	120.1	Dec 19, 16	(0° to 19° W, -6° to -20° S)
12	97	Jan	10	05	22	96.45	Jan 6, 14.5	(1° W, -24° S)
14	97	Apr	21	15	41	148.8	Apr 16, 14.5	(4°, 23°S)
15	97	May	15	09	17	83.0	May 12, 05	(5° to 12°W, 17° to 26° N)
16	97	Jun	8	22	28	121.1	Jun 4, 00	(?)
17	97	Jul	15	06	20	?		
18	97	Aug	3	14	12	110.2	Jul 30, 05.7	(NW)
19	97	Sep	18	00	61	?		
20	97	Sep	21	22	26	115.7	Sep 17, 15	(-82°E to 47° W, 21° to 66° N)
21	97	Oct	1	16	32	101.6	Sep 27, 23	(-13° E to 30° W, 14° to 39° N)
22	97	Oct	10	23	26	105.0	Oct 6, 14.5	(-59° to -11° E, -32°to -60°S)
23	97	Nov	7	05	32	71.74	Nov 4, 05.7	(20° W, -18°S)
24	97	Nov	22	14	29	82.0	Nov19, 17	(-29° to 0° E, 15° to 35° N)
25	98	Jan	7	03	31	115.1	Jan 2, 23.5	(?: Multiple Locations)
27	98	Feb	4	04	43	82.8	Feb 1, 16.3	(?: 37° to 52° W, 2° to 21° N)
28	98	Mar	4	14	41	131.8	Feb 28, 08	(4° to 37° W, 21° to 22° N)
29	98	May	2	12	30	82.2	Apr 29, 16.3	(Very Complex Active Region)
30	98	Jun	2	10.5	5.4	112.0	May 29, 10.3	(8° to 22° W, -8° to 23° S)
31	98	Jun	24	14	27	92.2	Jun 21, 04.7	(Near 20° W, 20° N)
32	98	Aug	20	10	34	?		

* Δt is the duration of the magnetic cloud.
\+ ΔT_T is the magnetic cloud transit time from the Sun to the center of the observed cloud.

The transit times (ΔT_T's) from the Sun to the measured clouds, from which average transit speeds ($V_{\Delta T}$'s) are determined, are presented in Table 1. Our interest is in comparing the transit speeds with the locally measured cloud speeds. [However, there is a slight improvement in agreement, in most cases, when $V_{X\text{-}GSE}$ (instead of local cloud central *speed*, V_C) and transit speed ($V_{\Delta T}$) are compared, where $V_{X\text{-}GSE}$ is the *absolute value* of the x-component of the cloud's central velocity in GSE coordinates, i.e., $|V_{X\text{-}GSE}|$; see Table 2 and Appendix A. Hence, we use $V_{X\text{-}GSE}$ instead of V_C for the comparison.] Figure 2 shows ΔV [= ($V_{\Delta T}$ - $V_{X\text{-}GSE}$)] vs. time over an interval of slightly more than 2 years for which we have SOHO data, giving 20 cases. The overall average transit time for these cases was 113 hrs corresponding to an overall average transit speed of 393 km/s. This speed is somewhat lower than the long-term average solar wind speed of 420 km/s (but not lower than the average upstream solar wind speed, 363 km/s, encountered by these clouds. In all cases, the distance from Sun to the center of the magnetic cloud was assumed to

Table 2. Magnetic Cloud and Shock Speeds and Speed Comparisons

Case No.	Yr	$<V_U>$ km/s	V_F km/s	V_C km/s	V_F/V_C "EXP"	$V_F/<V_U>$ "RAM"	$V_{\Delta T}$ + km/s	$V_{S,R}$ # km/s	ΔV_C* km/s	$\Delta V_{S,F}$ ++ km/s	$\Delta V_{S,U}$ ## km/s	SSC
1	95	371	440	410	1.073	1.185	?		30			
2	95	424	440	457	0.963	1.038	?	474	-17 ?	34	50	?[b]
5	95	318	375	350	1.071	1.179	?	381	25	6	63	Yes
6	95	288	425	405	1.049	1.476	?	408	20	- 17	120	Yes
8	96	411	412	349	1.181	1.002	242		63			
9	96	321	370	345	1.072	1.153	247		25		PP	Yes
10	96	336	355	335	1.060	1.057	262		20			
11	96	413	390	340	1.147	0.944	347		50			
12	97	372	465	440	1.057	1.250	432	439	25	- 26	67	Yes
14	97	368	390	338	1.154	1.060	280		52		PP	
15	97	327	430	470	0.915	1.315	502	422	- 40 ?	- 8	95	Yes
16	97	360	380	350	1.086	1.056	344		30		PP(?)	Yes
17	97	303	350	350	1.000	1.155	?		0 ?		PP	Yes
18	97	412	500	397	1.259	1.214	378		103		PP	Yes
19	97	330	367	295	1.244	1.112	?		72			
20	97	330	460	400	1.150	1.394	360	438	60	- 22	108	Yes
21	97	410	490	450	1.089	1.195	410	485	40	- 5	75	Yes
22	97	409	431	400	1.078	1.054	397	475	31	44	66	Yes
23	97	335	460	440	1.045	1.373	583	512	20	52	177	Yes
24	97	349	510	490	1.041	1.461	508	524	20	14	175	Yes
25	98	302	415	395	1.051	1.374	362	420	20	5	118	Yes
27	98	400	360	320	1.125	0.900	503		40		PP	
28	98	351	390	340	1.147	1.111	316	420	50	30	69	Yes
29	98	465	635	508	1.250	1.366	507	620	127	-15	155	Yes
30	98	409	420	405	1.037	1.027	372		15			
31	98	405	525	465	1.129	1.296	452		60			
32	98	280	335	305	1.098	1.196	?	340	30	5	60	Yes
Averages:		363	427	391			393	430	43.5[a]	6.9	99.9	

\+ $V_{\Delta T}$ is the average transit speed from the Sun to the center of the cloud, i.e., 1 AU/ΔT_T; see Table 1.

\# $V_{S,R}$ is the radial component (to Sun) of the local upstream shock velocity in GSE coords, but where positive (+) is outward from the Sun.

* ΔV_C is defined as $(V_F - V_C)$, a measure of speed gradient in the cloud.

++ $\Delta V_{S,F}$ is defined as $(V_{S,R} - V_F)$.

\## $\Delta V_{S,U}$ is defined as $(V_{S,R} - <V_U>)$, the radial speed of the shock in the solar wind frame of reference. All cases with a numerical value represent the existence of an upstream shock. Wherever "PP" is listed an upstream ram pressure pulse was seen instead of a shock.

a The average of ΔV_C's ignores odd cases nos. 2, 15, and 17.

b Why is there no SSC?

be exactly 1 AU taken to be 1.499 x 10^8 km. In many cases the spacecraft at the observing time was not near Earth; it can be about 1 hour transit time upstream of Earth when it is as far as the L1 point. At most then, when the WIND spacecraft is that far upstream of Earth, our assumption of a 1 AU transit distance will give an error of - 3.2 km/s in estimated transit speed which we consider negligible for this study. Figure 2 shows that, within ±35 km/s, most cases show good agreement between $V_{\Delta T}$ and $V_{X\text{-}GSE}$. There are five suspect cases lying outside this band that may not have the proper source identification. Excluding these cases there appears to be clear independence of ΔV (= $V_{\Delta T}$ - $V_{X\text{-}GSE}$) on the transit time (ΔT_T; see Table 1), as seen in Table 2 [i.e., see V_C ($\approx V_{X\text{-}GSE}$), and $V_{\Delta T}$ in cols. 5 and 8, respectively]. The agreement between $V_{\Delta T}$ and $V_{X\text{-}GSE}$ for most cases suggests that either (1) there are apparently some but *not many cases* in which significant (de-)acceleration takes place over 1 AU or (2) in general these large structures tend to move (in bulk) with approximately constant velocity from near the Sun to Earth. [But, cases of CME (de-) acceleration close to the Sun are inferred from coronograph observations. [e.g., *Gosling*, 1997]. The question of interplanetary (de-) acceleration is complex, and we considered it still to be an open one.

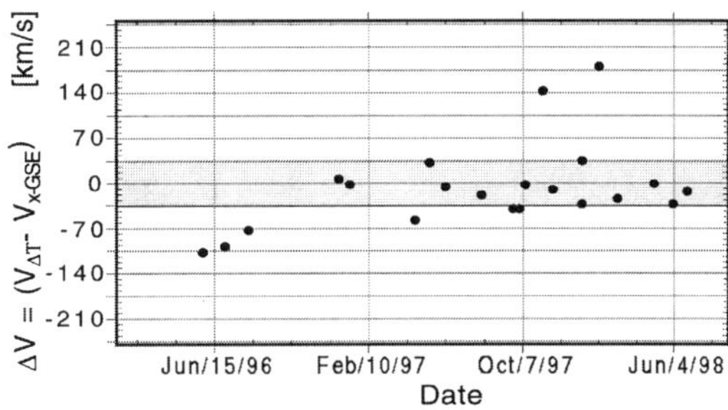

Fig. 2. Sun-to-cloud transit speed of magnetic clouds compared to their central speed. Specifically ΔV is ($V_{\Delta T}$ - $V_{X\text{-}GSE}$); see text.

CLOUD AND SHOCK SPEED RELATIONSHIPS: RAMMING AND EXPANSION

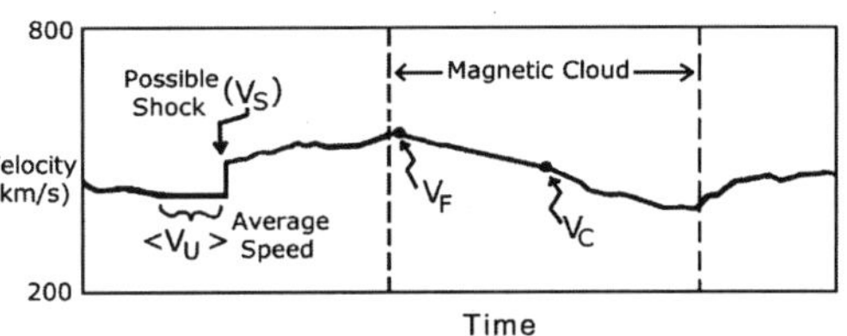

Fig. 3. A sketch of a speed profile of a typical magnetic cloud with a (possible) upstream interplanetary shock shown to define key quantities: V_S (shock velocity in the spacecraft frame of reference), $<V_U>$ (average upstream speed of the solar wind), V_F (speed of the plasma just inside the "front" of the cloud), and V_C (speed of the plasma at the center of the cloud).

Figure 3 shows a sketch of a speed profile of a shock/cloud complex and the relative positions within the complex where speed values are determined; see Appendix A. We define $<V_U>$ as the average solar wind speed just *upstream* of the cloud or just in front of an interplanetary shock (when a shock was present) and V_F as the observed speed of the solar wind (proton) plasma just within the *front* boundary of the magnetic cloud. V_U is usually averaged over from 2 to 4 hours, as necessary. V_C is the *central* speed of the cloud (which is usually very close to the average speed of the cloud). We define RAM as $V_F/<V_U>$. When the front of the cloud is moving faster than the immediate upstream solar wind RAM is greater than 1.0, and we say that the cloud is *ram*ming into the upstream solar wind. When the front of the cloud is moving faster than the center speed of the cloud (which is almost the same as the average speed of the cloud), we say that the cloud is *exp*anding [EXP (= V_F/V_C)]. Columns 4 through 8 of Table 2 present the relevant speeds and ratios (RAM and EXP) for the 27 magnetic clouds considered here. The table also indicates whether an upstream shock exists or not, by cols. 9, 11, and 12, where the values related to the radial shock speed ($V_{S,R}$) are given when a shock existed; also notice that col. 12 explicitly tells which clouds had shocks or pressure pulses. It is interesting that the frequency of occurrence of clouds, with or without shocks, is fairly uniformly distributed in time; the related average period between cloud appearances was about 53 days.

Figure 4 shows RAM vs. EXP for all 27 events split according to the same two sub-categories of whether an upstream shock was present (closed circle events - 14 cases) or not (open circle events - 13 cases). First, we notice that more than ½ of the total cases have upstream shocks. And generally the "no shock" cases exist for almost the same range of EXP as the shock cases with the two shock-cases of EXP *lower* than 1.0 being exceptions. These two exceptions are obviously due to unusual speed profiles; see, for example, case B of Figure 1. Also there were two no-shock cases with EXP around 1.25, along with the highest RAM shock case. Hence, cloud expansion into the upstream region is apparently not primarily responsible for the existence of these shocks, although we cannot ignore the possibility that it still plays some role. Considering all 27 cases together RAM and EXP are not significantly correlated; the linear correlation coefficient (C. C.) was - 0.12. Low correlations are found for both subsets separately as well. RAM values go only from 0.90 to 1.50. The upper part of the diagram, at higher RAM, contains most of the shock cases. Specifically, of the 10 clouds having largest RAM values 8 possess shocks. The lower part of Figure 4 is mostly devoid of shock cases. There was overlap of cases in the region of RAM from 1.0 to 1.3. Obviously the RAM factor is related to the existence of upstream shocks, and (depending on specific upstream conditions) it appears that the relatively faster ramming magnetic clouds drive shocks.

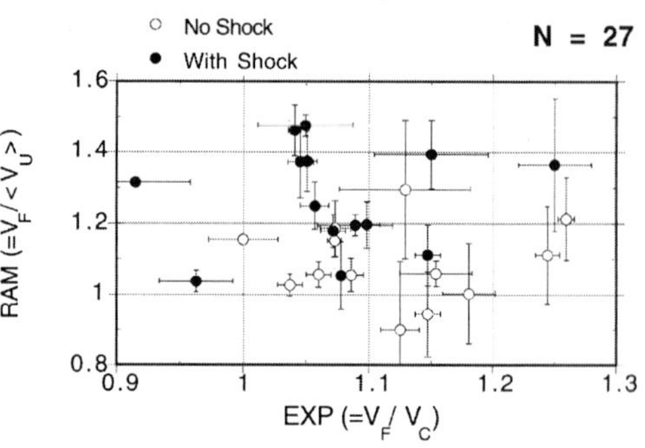

Fig. 4. RAM vs. cloud EXP for the shock (N = 14) and non-shock (N = 13) cases separately.

We now consider two quantities in Table 2 concerning upstream shock speeds. The first of these is $\Delta V_{S,F}$ in col. 11 [defined as ($V_{S,R}$ - V_F), where $V_{S,R}$ is the x-component of a shock velocity in GSE coordinates (col. 9), but taken positive *away* from the Sun]. Note that $\Delta V_{S,F}$ is the difference between the absolute value of the x-component of the shock speed ($V_{S,R}$) and the speed of the front of the cloud; it is a measure of the expansion rate of the intervening sheath. The average of $\Delta V_{S,F}$ over the 14 events is 6.9 km/s, which is consistent with 0.0, considering the large spread of the $\Delta V_{S,F}$ values ($\pm$ 25 km/s, measured by $\pm$1.0 rms) and the significant errors (less well know) expected for both V_F and $V_{S,R}$, especially the latter. Hence, the radial shock speeds are, on average, very close to the speeds of the fronts of the clouds, as might be expected, if the clouds are playing a significant role in driving the shocks. The second quantity of interest is $\Delta V_{S,U}$, in col. 12 of Table 2 [defined as ($V_{S,R}$ - $<V_U>$)]. This is the difference between the average upstream solar wind speed and $V_{S,R}$. $V_{S,R}$ itself is independent of RAM (with a correlation coef. of - 0.11), probably because of the dominance of the average upstream solar wind speed, which is a principal component of $V_{S,R}$. However, if we carefully examine $\Delta V_{S,U}$ (i.e., shock speed in the upstream solar wind frame of reference), we see that it strongly depends on RAM. Figure 5 shows the log of $\Delta V_{S,U}$ vs. RAM with a straight line overplot. The correlation coefficient of the original plot is 0.87 and the straight line gives a "prediction" that, on average:

$$\Delta V_{S,U} = 3.7 \times 10^{1.1 RAM} \quad [\text{km/s}] \tag{1}$$

When the RAM factor is 1.00 (no ramming), $\Delta V_{S,U}$ is predicted to be 47 km/s, which is fairly typical fast mode MHD wave speed in the upstream regions. The RAM is 1.27 on average for the 14 cases that have upstream shocks. We see that the magnetic clouds without upstream shocks generally have small RAM's as Table 2 shows. As the table shows, only seven cases have neither an upstream shock or a pressure pulse (PP), i.e., case #'s 1, 8, 10, 11, 19, 30, and 31. Case # 31 is an exception (large RAM) which is probably due to unusual upstream conditions. The average RAM for the other 6 cases is 1.04. The PP's are presumably caused by the ramming clouds also and may have been (or will be) shocks as upstream conditions allow.

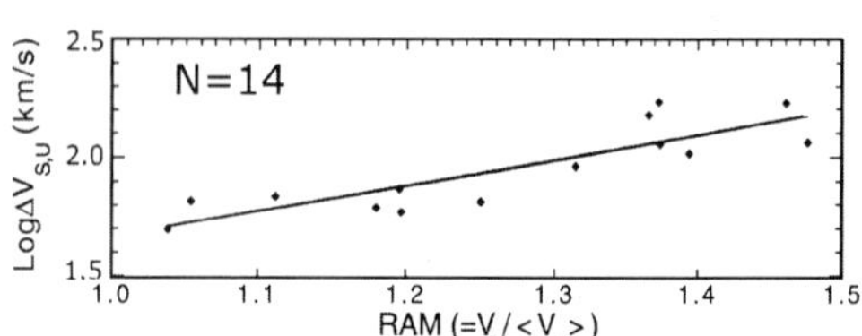

Fig. 5. Log $\Delta V_{S,U}$ vs. RAM for the 15 shock cases, where $\Delta V_{S,U}$ = ($V_{S,R}$ - $<V_U>$), the speed of the shock relative to the average upstream solar wind speed. The points represent the actual data. The straight line fit to these data had a good correlation coefficient of 0.87.

RELATED EFFECTS AT EARTH

We examine here only a few types of effects at Earth of the passage of the magnetic clouds and upstream shocks (or pressure pulses, PP) by comparing an estimated solar wind energy input to the magnetosphere to minimum Dst. For each case the full cloud complex (i.e., including the upstream shock or PP and sheath) is considered, and we check to see if an SSC took place when a shock/PP occurred. Long-duration magnetospheric activity (i.e., on a scale of many hours) traditionally has been measured by Dst (e.g., see studies of the relationship of solar wind input to the magnetosphere vs. Dst [*Burton et al.*, 1975; *Klimas et al.*, 1998], especially when major events, such as magnetic storms are studied, because of the well-known effect that long-duration solar wind input (for southward IMF) has in increasing the westward moving ring current in the outer radiation belt, which is primarily responsible for changes in Dst [*Feldstein*, 1992]. For consideration of the input energy we compute the integrated Akasofu-epsilon [*Akasofu*, 1981] (i.e., $\Sigma\varepsilon\Delta t$) over the time interval from the upstream shock (or PP) to the end of the magnetic cloud. The Akasofu epsilon is

$$\varepsilon = VB^2 l_0^2 \sin^4(\psi/2), \tag{2}$$

and where V is the speed of the solar wind, B is the magnitude of the IMF, l_0^2 is an empirically determined value of the effective interaction area at the front of the magnetosphere ($l_0 = 7\ R_E$), and ψ is the clock angle of the IMF which is defined as the polar angle between the IMF as projected into the y-z plane and the z-axis in GSM coordinates. Integrating ε over the full interval, as stated, gives a total Poynting flux input energy (E) into the magnetosphere for that interval; note that $E = \Sigma\varepsilon\Delta t$, where Δt is the sample period for each. The importance of measuring the solar wind energy input in terms of an integrated ε for magnetic clouds has been discussed by *Farrugia et al.* [2000]. (See *Farrugia et al.* [1998] concerning the geoeffectiveness of the impact of three magnetic clouds on the magnetosphere in terms of integrated ε during solar minimum.) Also see *Freeman and Farrugia* [1999], who compare various coupling functions, including ε, in terms of effects in the magnetotail. The results for ε were in broad agreement with the findings of *Klimas et al.*, [1998] concerning theories of substorm generation based on deterministic chaos. Figure 6 shows a plot of log E vs. log of |min-Dst| for each magnetic cloud interval. A magnetic storm resulted from most cloud passages. The straight line least-squares fit to these data (solid curve shown in the figure) indicates that $|Dst| = (E/E_0)^{0.55}$ (where E_0 is 2.1×10^{12} joules and Dst is in nT) with a relatively good correlation coefficient of 0.80. A convenient form for this (where E is in units of 10^{15} joules) is

$$|Dst| = 29\ E^{0.55} \qquad [\text{for - Dst(min): } 10 \leftrightarrow 130\ \text{nT}], \tag{3}$$

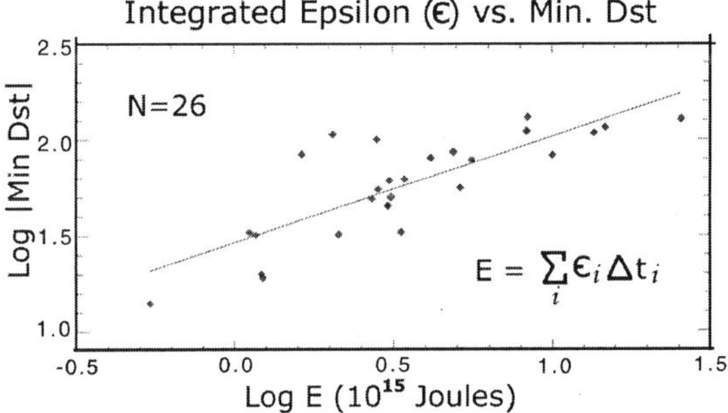

Fig. 6. Time-integrated Akasofu-epsilon (ε) vs. minimum Dst for 26 magnetic clouds. (One case is missing, because the sheath region in front of one cloud was missing due to the fact that the cloud had just departed the bow sheath when it was observed.) The points represent the actual data. The straight line fit to these data had a relatively good correlation coefficient of 0.80.

At the top of Figure 7 we show a histogram of ΔT_D's (a solar wind-magnetosphere delay time), where by ΔT_D we strictly mean the interval from the start of any distinct increase in ε at or within the shock-cloud complex to the time of minimum Dst. The distinct increase in ε may start at the upstream shock, at the start of the cloud, in the sheath between or somewhere in the cloud. The bottom of the figure shows diagrammatically how a ΔT_D is chosen. We see that ΔT_D peaks close to 10 hours and is 14 hours on average.

Finally, in the last two cols. of Table 2, we compare the existence of a SSC with that of an upstream shock (or PP) from WIND data. As expected, there is a good correspondence between the shocks (or PP's) and the occurrence of SSC's.

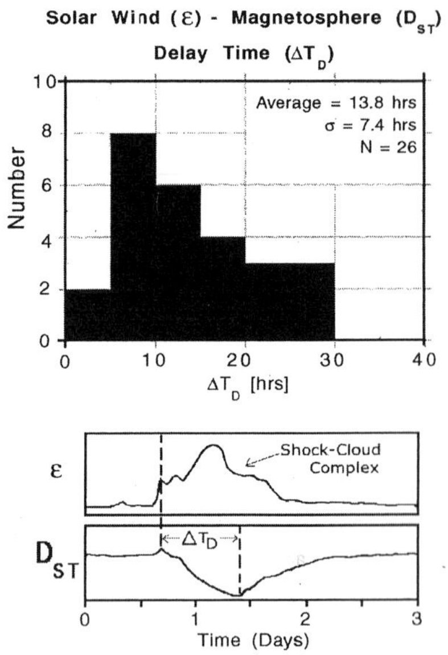

Fig. 7. (Top) The histogram of ΔT_D's, where ΔT_D is the interval from the start of the distinct increase in ε within the shock-cloud complex to the time of minimum Dst. (Bottom) A sketch giving a pictorial example of how ΔT_D is defined.

SUMMARY AND COMMENTS

We briefly present the main findings of this study:

• There is a broad variety of cloud speed profiles, but most show a slowing down indicating expansion.

• There was good agreement (> 75% of cases) between the average magnetic cloud speed (or cloud center speed, V_C) and the Sun-Earth transit speed, $V_{\Delta T}$, within ± 35 km/s. Also when averaged over 27 clouds, $<V_C>$ = 391 km/s and $<V_{\Delta T}>$ = 393 km/s for the 20 cases for which solar sources are "determined." Since this agreement may be controversial [e.g., see Gopalswamy et al., 2000] further investigation is needed, first, to increase sample size and, second, to accommodate higher speed clouds.

• In slightly more than ½ of the 27 cases studied, magnetic clouds drive upstream shocks. When upstream shocks or pressure pulses (PP's) are considered, 3/4 of the magnetic clouds have such an upstream event. Shocks appear to be more common than PP's by more than 2 to 1.

• Since $<\Delta V_{S,F}> \approx 0$, within ± 7.0 km/s for the full set of cases studied, then radial (from the Sun) shock speeds are very close, on average, to the speeds of the fronts of the magnetic clouds.

• As expected, a cloud with a high RAM factor ($\equiv V_F/<V_U>$), which on average is 1.27, is more likely to have an upstream shock. The average RAM for cases without shocks or PP's is 1.04.

• The degree of expansion apparently does not control whether an upstream shock will exist or not (at 1 AU). Since the controlling speeds are V_F = (Initial Speed + Expansion Speed) and the characteristic plasma speeds, such as, the Alfvén and sound speeds, then this is consistent with the Initial Speed >> Expansion Speed.

• There is good agreement between Log $\Delta V_{S,U}$ and RAM, where specifically we find that

$$\Delta V_{S,U} = 3.7 \times 10^{1.1\,\mathrm{RAM}}, \qquad \text{where } \Delta V_{S,U} = (V_{SR} - <V_U>) \qquad [\text{C.C.} = 0.87]$$

• There is a relatively good agreement between Log(|Min-Dst|) and Log E, where assumed solar wind energy input is $E = \Sigma\, \varepsilon\, \Delta t$, and ε is the Akasofu-epsilon (integrated Poynting flux to the magnetosphere), where we find that:

$$|\text{Dst}| = 29\, E^{0.55} \quad (\text{For - Dst(min): } 10 - 130 \text{ nT}), \qquad [\text{C.C.} = 0.80]$$

where E is in units of 10^{15} joules. It is hoped that this relationship may play a role in the prediction of the"strength" of a magnetic storm based on interplanetary parameters (e.g., V and the IMF) from a station upstream of Earth.

• A preliminary finding is that there is typically 10 hrs (on average, 14 hrs) between the first sharp increase of ε (anywhere in the shock-cloud complex) and the time Dst.

• The upstream shocks (or pressure pulses) almost always cause SSC's (i.e., > 85% of the time).

It would be interesting to use future observations of shocks upstream of clouds from an appropriate (and especially widely-spaced) cluster of interplanetary spacecraft, to see, by proper tracking, if shocks always tend to exist at all projected measuring positions and for all points in time. Or do these shocks sometimes dissipate, according to hostile upstream conditions, and temporarily disappear, only to be energetically refurbished again by the perpetually oncoming magnetic cloud (or other ejecta)? For weak shocks the latter seems likely, but this question should be investigated.

ACKNOWLEDGMENTS

We thank the WIND MFI and SWE teams and in particular Keith W. Ogilvie, the principal investigator of SWE, for all of their help and cooperation in this work. We are especially grateful to Claudia Arqueros for assistance with field data handling, Franco Mariani for magnetic field instrument calibration, and J. K. Chao for kindly agreeing to present the COSPAR talk associated with this paper.

APPENDIX A. DEFINITIONS OF QUANTITIES USED

- $|\mathbf{B}|$ is the interplanetary magnetic field magnitude.
- V_{Th} is the most probable thermal speed.
- V is the solar wind bulk speed.
- Δt is the duration of the magnetic cloud, from observed front to back boundary.
- ΔT_T is the magnetic cloud's transit time from the Sun (SOHO observations) to the center of the observed cloud.
- ΔT_D is the interval from the start of the distinct increase in ε within the shock-cloud complex to the time of minimum DST.
- $V_{\Delta T}$ is the average transit speed from the Sun to center of the cloud, i.e., 1 AU/ΔT_T.
- $v_{X\text{-}GSE}$ is the x-component of the cloud's central velocity in GSE coordinates.
- $V_{X\text{-}GSE}$ is the *absolute value* of the x-component of the cloud's central velocity in GSE coordinates, i.e., $|v_{X\text{-}GSE}|$.
- ΔV is $(V_{\Delta T} - V_{X\text{-}GSE})$.
- $<V_U>$ is the average solar wind speed just *upstream* of the cloud or just in front of an interplanetary shock (when a shock was present).
- V_F is the observed speed of the solar wind (proton) plasma just within the *front* boundary of the magnetic cloud.
- V_C is the *central* speed of the cloud which is usually very close to the *average* speed of the cloud.
- RAM is $V_F/<V_U>$, the ramming factor.
- EXP is V_F/V_C, the expansion factor.
- $V_{S,R}$ is the radial component (to Sun) of the local upstream shock velocity in GSE coords, but where positive (+) is outward from the Sun.
- ΔV_C is defined as $(V_F - V_C)$, a measure of speed "gradient" in the cloud.
- $\Delta V_{S,F}$ is defined as $(V_{S,R} - V_F)$.
- $\Delta V_{S,U}$ is defined as $(V_{S,R} - <V_U>)$, the radial speed of the shock in the solar wind frame of reference.

REFERENCES

Akasofu, S.-I, Energy coupling between the solar wind and the magnetosphere, *Space Sci. Rev., 28*, 121, 1981.

Berdichevsky, D.B., A. Szabo, R. P. Lepping, A. F. Viñas, and F. Mariani, Interplanetary fast shocks and associated drivers observed through the twenty-third solar minimum by WIND over its first 2.5 years, *J. Geophys. Res.*, Vol. 105, No. A12, 2000.

Borrini, G., J. T. Gosling, S. J. Bame, An analysis of shock wave disturbances observed at 1AU from 1971 through 1978, *J. Geophys. Res.*, 87, 4365, 1982.

Burlaga, L. F., *Interplanetary Magnetohydrodynamics*, Oxford Univ. Press, New York, 1995.

Burlaga, L. F., R. P. Lepping, and J. A. Jones, Global configuration of a magnetic cloud, *Physics of Magnetic Flux Ropes, Geophys. Monogr. Ser.*, vol. 58, edited by C. T. Russell, E. R. Priest, and L. C. Lee, p. 373, AGU, Washington, D. C., 1990.

Burton, R. K., R. L. McPherron, and C. T. Russell, An empirical relationship between interplanetary conditions and *Dst*, *J. Geophys. Res., 80*, 4204, 1975.

Farrugia et al., Geoeffectiveness of three Wind magnetic clouds, *J. Geophys. Res.*, *103*, 17,261, 1998.

Farrugia, C. J., L. F. Burlaga, V. K. Jordanova, M .P. Freeman, G. Lawrence, C. C. Cocheci, R. L. Arnoldy, J. D. Scudder, K. W. Ogilvie, and R. P. Lepping, Power to the magnetosphere: May 4, 1998, *Adv Space Res.*, in press, 2000.

Feldstein, Y. I., Modelling of the magnetic field of magnetospheric ring current as a function of interplanetary medium parameters, *Space Sci. Rev., 59*, 83, 1992.

Freeman, M. P., and C. H. Farrugia, The October 1995 magnetic cloud: A comparison of measures of the inter-substorm solar wind input, *J. Geophys. Res., 104*, 22,729, 1999.

Gopalswamy, A. Lara, R. P. Lepping, M. L. Kaiser, D. Berdichevsky, and O. C. St. Cyr, Interplanetary acceleration of coronal mass ejections, *Geophys. Res. Lett., 27,* 145, 2000.

Gosling, J. T., Coronal mass ejections and magnetic flux ropes in interplanetary space, *Physics of Magnetic Flux Ropes, Geophys. Monogr. Ser.*, vol. 58, edited by C. T. Russell, E. R. Priest, and L. C. Lee, p. 343, AGU, Washington, D. C., 1990.

Gosling, J. T., Coronal mass ejections: An overview, *Coronal Mass Ejections*, edited by N. Crooker et al., *Geophys. Monogr. Ser., vol 99*, p. 9, Washington, D. C., American Geophysical Union, 1997.

Klimas, A. J., D.Vassiliadis, and D. N. Baker, Dst index prediction using data-derived analogues of the magnetospheric dynamics, *J. Geophys. Res., 103*, 20,435, 1998.

Lepping, R. P., J. A. Jones, and L. F. Burlaga, Magnetic field structure of interplanetary magnetic clouds at 1 AU, *J. Geophys. Res., 95,* 11,957, 1990.

Lepping, R. P., et al., The WIND magnetic field investigation, The Global Geospace Mission, *Space Sci. Rev. 71*, 207, 1995.

Lepping, R. P, and D. Berdichevsky, Interplanetary magnetic clouds: sources, properties, modeling, and geomagnetic relationship, *Recent Research Developments in Geophysical Research*, issue 3, Research Signpost, Trivandrum-8, India, 77, 2000.

Lindsay, J. D., C. T. Russell, J. G. Luhmann and P. Gazis, On the sources of interplanetary shocks at 0.72 AU, *J. Geophys. Res.*, 99, 11, 1994.

Marubashi, K., Interplanetary magnetic flux ropes and solar filaments, *Coronal Mass Ejections*, edited by N. Crooker et al., *Geophys. Monogr. Ser., vol 99,* p. 147, Washington, D. C., American Geophysical Union, 1997.

Ogilvie, K. W., et al., SWE, A comprehensive plasma instrument for the WIND spacecraft, The Global Geospace Mission, *Space Sci. Rev., 71*, 55, 1995.

Osherovich, V. I., C. J. Farrugia, and L. F. Burlaga, Dynamics of aging magnetic clouds, *Adv. Space Res., 13(6),* 57, 1993.

Tsurutani, B. T., and W. D. Gonzalez, The interplanetary causes of magnetic storms: A review, in *Magnetic Storms*, edited by B. T. Tsurutani, W. D. Gonzalez, Y. Kamide, and J. K. Arballo, AGU, Washington D. C., *98*, p. 77, 1999

ELECTROMAGNETIC ELECTRON AND PROTON CYCLOTRON WAVES IN GEOSPACE: A CASSINI SNAPSHOT

B. T. Tsurutani[1], J. K. Arballo[1], X.-Y. Zhou[1], C. Galvan[1], and J. K. Chao[2]

[1]*Jet Propulsion Laboratory, California Institute of Technology, 4800 Oak Grove Drive, Pasadena, CA 91109, USA*
[2]*Institute of Space Sciences, National Central University, Chung-li, Taiwan, Republic of China*

ABSTRACT

The Cassini spacecraft flew past the Earth in a trajectory almost along the Sun-Earth line, giving a unique perspective of low frequency waves in geospace. Geotail was immediately upstream of the bow shock nose, allowing for an accurate assessment of the solar wind conditions driving geospace macroscale processes, the latter of which led to microinstabilities and electromagnetic plasma wave growth. We demonstrate the presence of nonlinear cyclotron waves in the foreshock, magnetosheath, outer magnetosphere, and tail lobe. A predominance of right-hand cyclotron waves with $f \approx 1.0$ Hz (in the spacecraft frame) are detected in the foreshock. The waves are compressional with $\Delta\vec{B} / B_0 \sim 0.25$. It is argued that these waves are propagating in the electron cyclotron (whistler) mode and that the waves are generated by low-energy electrons streaming into the upstream region with parallel kinetic energies of tens of eV. This region appears to be a pure electron foreshock. The magnetosheath waves are the largest amplitude (~15 nT peak-to-peak in a ~30 nT field) proton cyclotron waves detected in geospace. The wave amplitudes are largest near the bow shock and decrease to ~8 nT peak-to-peak near the magnetopause. No discernible mirror mode structures were detected in the magnetosheath. It is possible that a low abundance of He^{++} ions in the solar wind high-speed stream may be the cause of the ion cyclotron wave dominance, as previously predicted by theory. However, other explanations are possible as well. The outer magnetospheric proton cyclotron waves are of low intensity (~4 nT peak-to-peak), and can cause weak proton pitch angle diffusion and concomitant diffuse proton aurora. Large-scale solar wind ram pressure pulses do not appear to be the cause of the magnetospheric waves. However, small scale fluctuations are possible. Two new low-intensity wave modes are identified in the tail lobes associated with substorm events. One mode is a transverse, elliptically polarized ~7.5 s wave and the other is a purely compressional (~0.1 nT amplitude) wave, with a ~8 s quasiperiod. The latter mode is propagating obliquely (~70° to 84°) to the magnetic field.

INTRODUCTION

The Cassini spacecraft had a unique trajectory in its flyby past the Earth on 18 August 1999. Cassini had a path through the subsolar point of the Earth's foreshock and magnetosheath and then flew through the magnetosphere and down the north lobe of the magnetotail. Cassini exited the dawnside of the magnetotail at $X \approx -70\ R_e$. This flyby took a total time of ~10 hours, thus one can get a unique "snapshot" of low

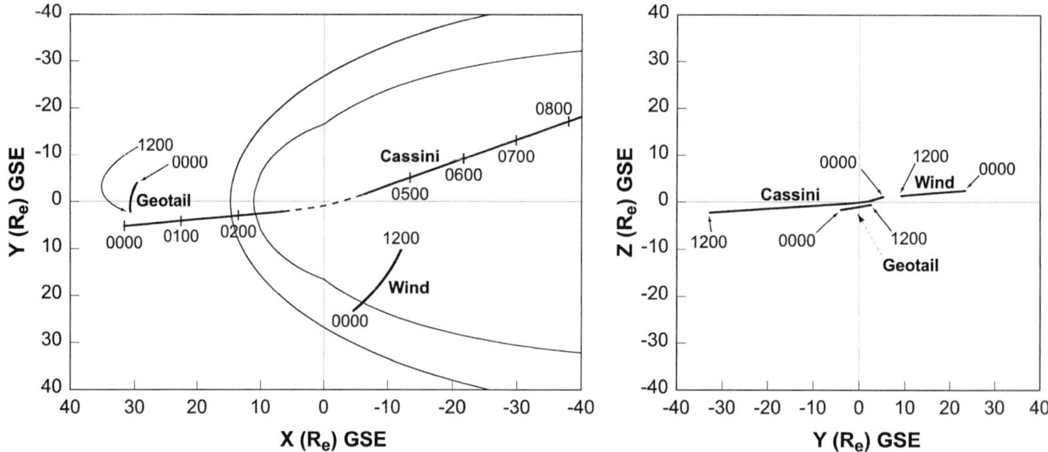

Fig. 1. Trajectories of Cassini, WIND, and Geotail during the Cassini Earth flyby, in GSE coordinates.

frequency plasma waves in the near-Earth environment. What was particularly striking was that much of this interval was filled with intense, nonlinear electromagnetic plasma waves.

The solar wind interaction with the magnetosphere/magnetotail influences all of geospace, and thus the solar wind is the ultimate source of local plasma instabilities and concomitant wave growth. It was particularly fortunate, during the Cassini flyby, to have the Geotail spacecraft positioned just upstream of the bow shock nose so that it could serve as an upstream monitor. ACE was located farther upstream of the Earth, near the L_1 libration point. Solar wind ram pressure variations lead to magnetosheath/outer magnetosphere/near-Earth magnetotail compressions and rarefractions, and the embedded interplanetary B_S magnetic fields can lead to magnetic reconnection and substorms (Dungey, 1961). Solar wind velocity and magnetic field directional variations can result in bow shock strength and shock Mach number fluctuations (Kennel *et al.*, 1985). These large scale processes in turn cause microscale processes and plasma instabilities, the topic of this paper.

The Earth's foreshock is typically dominated by low frequency (10 to 60 s periods) electromagnetic plasma waves occurring in the foreshock (Fairfield, 1969; Hoppe *et al.*, 1981). The waves are associated with energetic ion beams streaming into the upstream medium (Asbridge *et al.*, 1968; Thomsen, 1985, and references therein), and are believed to be generated by an ion-ion instability (Papadopoulos, 1985; Gary, 1991). However, it has been argued that the particle source may be either bow-shock reflected solar wind ions (Bonifazi and Moreno, 1981a,b), escaping magnetospheric ions (Baker *et al.*, 1988; Sibeck *et al.*, 1988), or possibly even energetic electrons (Smith *et al.*, 1976; Goldstein *et al.*, 1985). Foreshock/shock particle energization processes can occur through particle interaction with the waves through either Fermi (Lee, 1982; Forman and Webb, 1985) and/or gradient *B* drift shock (Armstrong *et al.*, 1985) acceleration.

The Earth's magnetosheath is typically dominated by nonoscillatory mirror mode structures (Tsurutani *et al.*, 1982; Anderson *et al.*, 1994), even though the ion cyclotron mode theoretically has higher growth rates (Price *et al.*, 1986). These structures are large-amplitude quasiperiodic oscillations in magnetic field magnitude where peak to valley ratios can be as large as 10:1. Gary *et al.* (1992), in agreement with the work of Price *et al.* (1986), explained the dominance of mirror mode structures in the magnetosheath as being due to the presence of small amounts of He^{++} in the solar wind and magnetosheath (see also Brinca

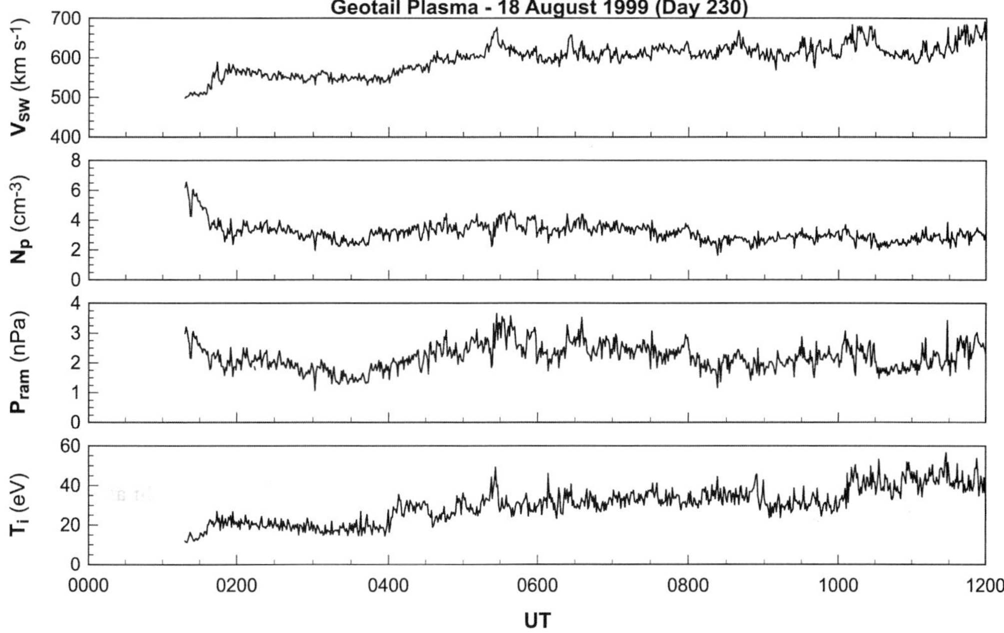

Fig. 2. The upstream solar wind during the time of the Cassini Earth-encounter and beyond. From top to bottom are the solar wind speed, density, ram pressure, and proton temperature. This is an edge of a high speed stream.

and Tsurutani, 1989). In this scenario, the He^{++} ions create a stop-band in the proton cyclotron dispersion relationship, effectively reducing the proton cyclotron wave growth rate.

Proton cyclotron waves have previously been detected in the outer magnetosphere during solar wind compression events (Anderson and Hamilton, 1993). The waves are generated by a proton temperature anisotropy ($T_{\perp}/T_{\parallel} > 1$) instability, the latter which results presumably from adiabatic compression of pre-existing energetic protons in the outer magnetosphere.

Low frequency electromagnetic waves in the tail (Tsurutani and Smith, 1984) have been ascribed as being due to generation by ion beams streaming away from the tail reconnection *X*-line (Cowley *et al.*, 1984; Tsurutani *et al.*, 1985) in the low beta plasmasheet boundary layer. This instability has been discussed in detail by Gary *et al.* (1985).

The purpose of this paper is to analyze and characterize the low frequency electromagnetic plasma waves in the four regions of geospace (foreshock, magnetosheath, outer magnetosphere and near-Earth magnetotail lobe) to determine the solar wind influence (or lack of it) on plasma waves in each region. We will demonstrate that the macroscale properties of the solar wind have significant influence on the wave modes and intensities for these various regions of geospace.

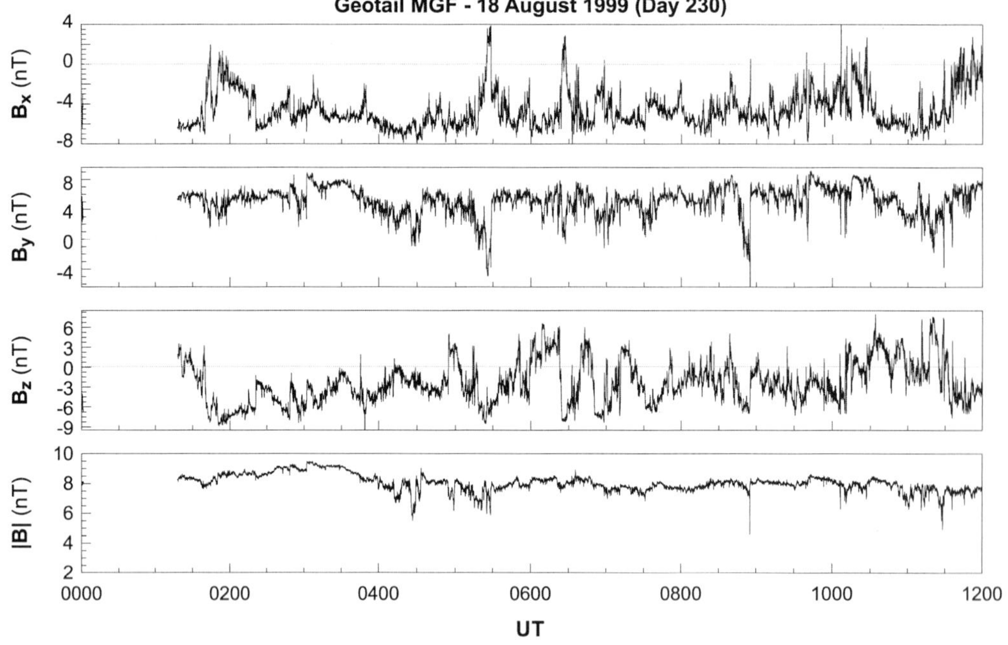

Fig. 3. The interplanetary magnetic field during the time of the Cassini Earth-encounter, in GSM coordinates.

RESULTS

Overview

The Cassini encounter trajectory is shown in Figure 1. The left-hand panel shows the trajectory projected into the GSE *X-Y* plane and the right-hand panel shows the projection into the GSE *Y-Z* plane. In this coordinate system $\hat{x}$ is in the Earth-Sun direction, $\hat{y}$ is in the plane of the ecliptic in the direction of motion opposite to the Earth's rotation about the Sun, and $\hat{z}$ forms the right-hand system. Cassini enters the magnetosheath/magnetosphere essentially along the Sun-Earth line.

The figure also shows the trajectories of two other spacecraft: Geotail (designated by GE), and WIND. Geotail is positioned ~20 R_e upstream of the magnetopause nose, an ideal position to monitor the solar wind conditions and their variations during the encounter. In the figure, WIND starts (at 0000 UT) in the solar wind, then enters the magnetosheath shortly thereafter. Not shown is the ACE spacecraft which is orbiting the L_1 libration point. ACE was at (248, –4, 24 R_e).

Solar Wind

Before going into a detailed discussion of the plasma waves, the solar wind plasma and field conditions should be discussed. Both the general state and variations of this interplanetary driver can have significant impact on the magnetosphere interaction, as we will show later.

Figure 2 shows the Geotail solar wind plasma parameters for the Cassini encounter interval (there was a spacecraft tracking gap from 0000 to ~0120 UT but this is not important for this study, as Cassini did not

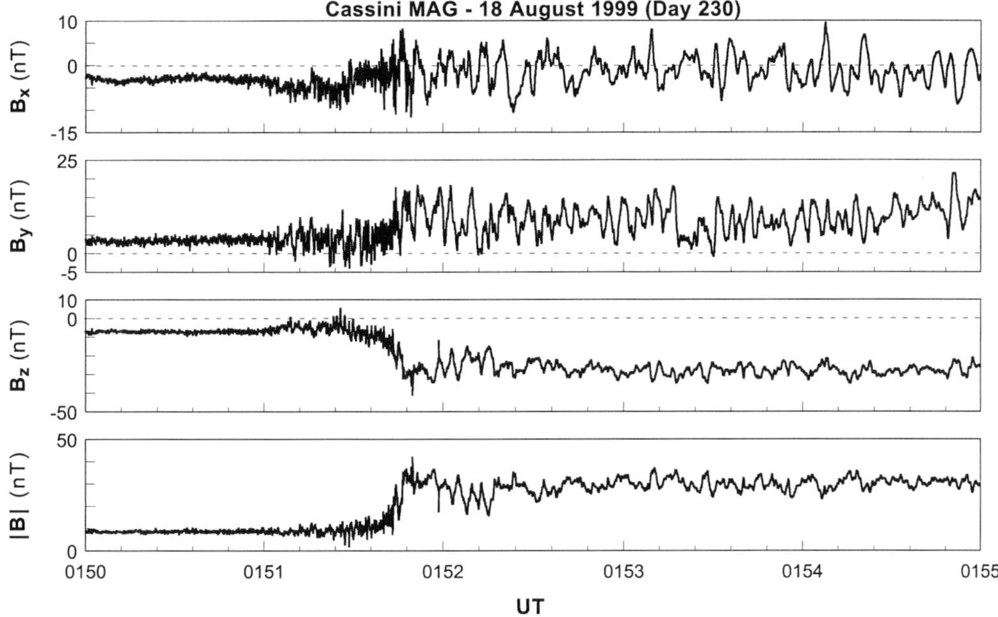

Fig. 4. The Earth's foreshock and magnetosheath waves detected by Cassini. The bow shock is located at ~0151:50 UT. The coordinate system is GSM.

enter the foreshock until much later. The solar wind speed is shown in the top panel of the figure. The speed gradually increased from ~500 km s^{-1} to ~680 km s^{-1} over the ~10 hour Cassini interval.

Coincident with this solar velocity increase was a decrease in proton density. The proton density was ~6.0 cm^{-3} at 0120 UT and decreased to ~3.0 cm^{-3} by 1200 UT. The proton thermal energy increased from ~10 eV to ~45 eV in the same time interval. The variation of the three parameters indicates that Geotail (and the Earth) were becoming engulfed in a high-speed (coronal hole) stream (Phillips *et al.*, 1994). This general scenario has been verified using the ACE plasma data. A peak velocity of ~800 km s^{-1}, minimum plasma density of ~2.0 cm^{-3}, and peak proton thermal energy of ~50 eV was attained at ~1200 UT day 231 (one day later, not shown).

The third panel in Figure 2 gives the solar wind ram pressure. During the Cassini encounter, Geotail was located ~20 R_e upstream of the Earth's magnetopause (assumed to be located at x = 10 R_e). The solar wind, with a speed of ~560 km s^{-1}, would take ~4 min. to convect from Geotail to the magnetopause (this propagation time has been taken out for the purposes of the following discussion). During the first part of the foreshock traversal, ~0151:00 to ~0151:50 UT, the solar wind ram pressure was more or less constant, with some small variations. During the first portion of the magnetosheath traversal, the ram pressure increased during the interval 0152 to ~0205 UT, and then was more-or-less constant thereafter. The ram pressure was generally decreasing during the outer magnetosphere passage (0226 to 0240 UT).

The Geotail interplanetary magnetic fields are shown in Figure 3. The components were highly fluctuating, while the field magnitude remained relatively constant. The field component often changed abruptly. Previous studies by Ulysses investigations (Tsurutani *et al.*, 1994; Smith *et al.*, 1995) have shown that such magnetic fluctuations are a characteristic of high-speed streams. The fluctuations are outwardly

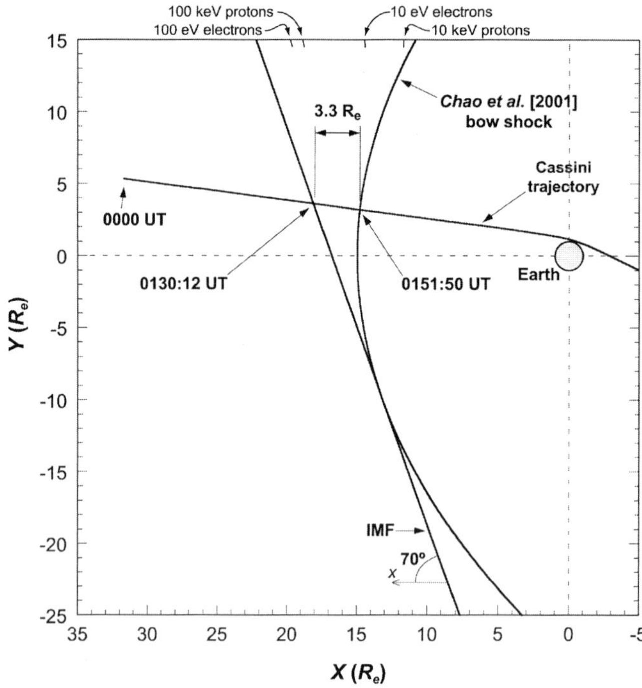

Fig. 5. The Cassini encounter trajectory. A magnetic field line with a 70° orientation with respect to the Sun-Earth line is shown. The shock shape and position has been modeled. The coordinate system is defined as having the magnetic field line lie in the *x*-*y* plane.

propagating Alfvén waves. The abrupt component directional changes (rotational discontinuities) are the phase-steepened edges of the nonlinear Alfvén waves. The interplanetary Alfvén waves have been noted to be arc-polarized and spherical in nature (Tsurutani and Ho, 1999).

The interval when Cassini was in the Earth's foreshock, magnetosheath, and outer magnetosphere (~0149 to 0238 UT) is characterized by a large negative B_z component. This is noted to be a portion of an interplanetary Alfvén wave.

Foreshock Waves

An overview of the Cassini near-Earth foreshock and magnetosheath waves is shown in Figure 4. The upstream interplanetary magnetic field (for reference, the bow shock is located at ~0151:50 UT) is (B_x, B_y, B_z) = −3, +3, −8 nT, in GSM coordinates. In this system, $\hat{x}$ is in the direction from the Earth to the Sun, $\hat{y} = \hat{x} \times \hat{M}/|\hat{x} \times \hat{M}|$, where $\hat{M}$ is the south magnetic dipole direction, and $\hat{z}$ completes the right-hand system. The interplanetary field is a positive sector field (outward from the Sun) with a strongly negative B_z component. The magnetic field has an orientation ~70° relative to the Sun-Earth line.

Both the B_y and B_z components are intensified at and behind the bow shock (~0151:50 UT). The magnitude of the magnetic field increases from ~8 nT upstream (B_1) to ~32 nT downstream (B_2). The ratio $B_2/B_1 \approx 4.0$, the maximum expected for MHD shocks (Kennel *et al.*, 1985). There is a significant shock

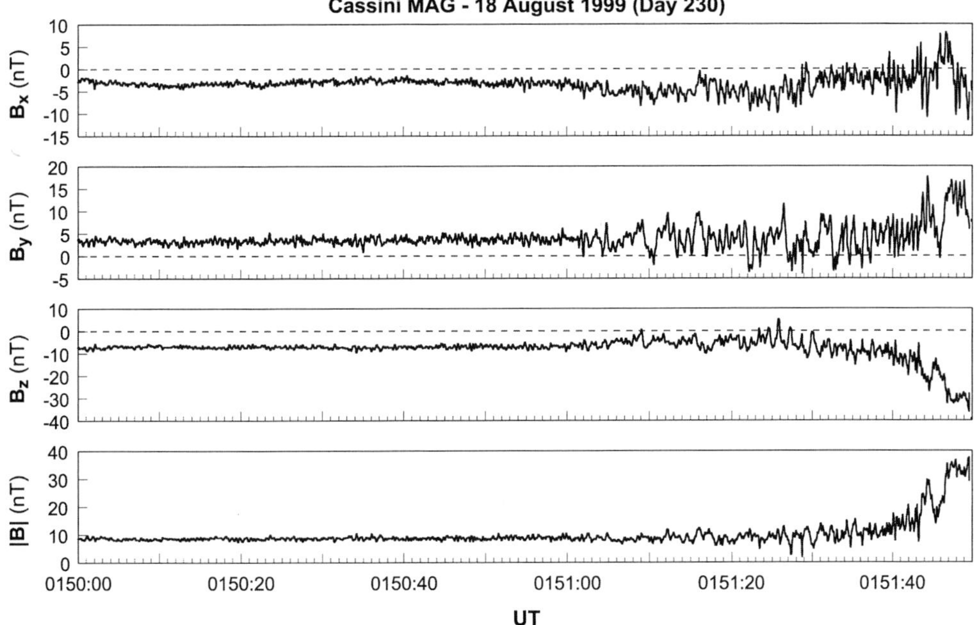

Fig. 6. The Earth's foreshock waves shown in GSM coordinates.

"overshoot", where the field reaches ~47 nT. "Overshoots" are typically associated with ion deflection at supercritical shocks (Goodrich, 1985).

At 0148 UT at Geotail (Figure 3), the solar wind speed was ~560 km s^{-1}, the proton density was 3 cm^{-3}, the proton thermal energy was ~20 eV and the magnetic field strength was 8.5 nT. The Alfvén speed $(B^2/4\pi\rho)^{1/2}$ was therefore ~107 km s^{-1}, the sound speed was $(\gamma P/\rho)^{1/2} \approx 58$ km s^{-1}, and the magnetosonic speed was 121 km s^{-1}. The ion cyclotron, electron cyclotron and electron plasma frequencies were 1.5×10^{-1}, 2.3×10^{2}, and 1.5×10^{4} Hz, respectively. From magnetic coplanetary analyses, the shock normal was determined to be $\theta_{Bn} = 63°$, oriented at (0.96, –0.29, –0.04) in GSE coordinates. The shock magnetosonic Mach number was ~4.6.

Upstream of the shock (Figure 4), foreshock waves had higher frequencies (in the spacecraft frame) than the downstream magnetosheath waves. The foreshock wave amplitudes were largest nearest the shock, and they decreased with increasing upstream distance. This amplitude falloff is consistent with the free energy for wave growth being largest near the shock and decreasing with increasing distance (consistent with a shock-reflected particle beam or magnetospheric particle escape wave generation mechanism).

The post-shock magnetosheath waves were of lower frequency, beginning directly at the shock and continuing into the downstream region to the magnetopause (not shown in this figure). In Figure 4, the waves are noted to be quasiperiodic and are exceptionally large in amplitude. Immediately behind the shock, the waves have ~15 nT peak-to-peak amplitudes in a ~30 nT field.

The solar wind parameters were used as inputs to the Chao *et al.* (2001) bow shock model. The shock standoff distance was predicted to be 14.83 R_e compared to the Cassini measured value of 14.78 R_e. Fig-

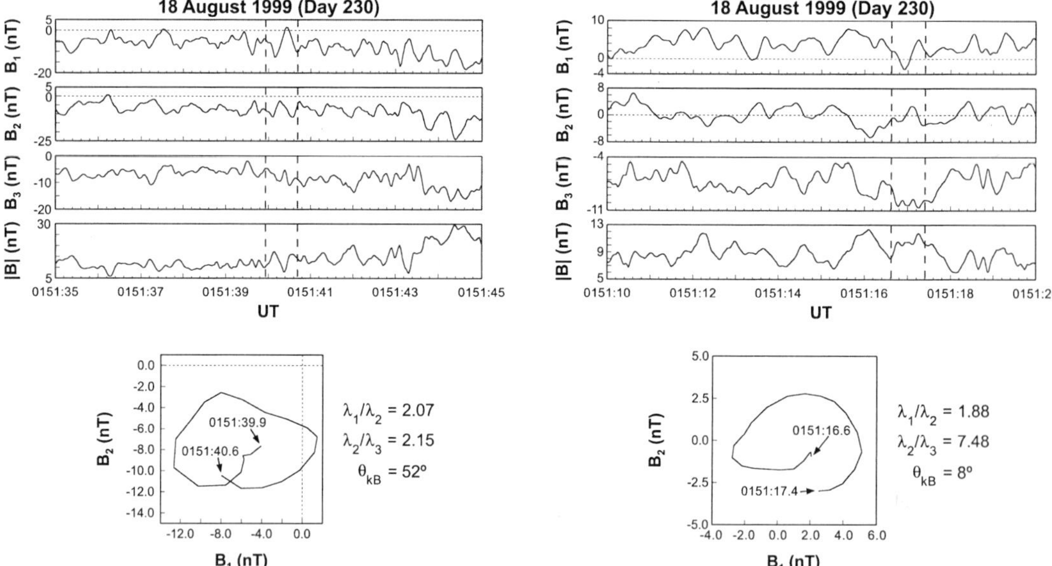

Fig. 7. Magnetosheath waves in minimum variance coordinates. The bottom panels show the hodogram of the wave cycle from 0151:39.9 to 0151:40.6 UT.

Fig. 8. Same as Figure 7. The bottom panel shows the hodogram of the wave cycle from 0151:16.6 to 0151:17.4 UT.

ure 5 is a schematic using the Chao *et al.* shock shape and the Cassini trajectory. Also indicated in the figure is the magnetic field line tangent to the shock. From measurements just prior to the Cassini bow shock crossing, it was ascertained that the field angle relative to the Sun-Earth line was ~70°. This figure assumes that the bow shock is symmetric relative to the x-axis and the magnetic field is constant in orientation. The distance between the edge of the foreshock and the shock is 3.3 R_e, taken along the Cassini trajectory. This corresponds to ~22 min time duration. Thus for steady IMF directions, Cassini should be in the electron foreshock region from ~0130 UT to the shock crossing (0151:50 UT).

Figure 6 is an overview plot of the upstream waves, shown in higher resolution. The wave amplitudes were largest near the shock, ~14 nT peak-to-peak in a ~35 nT field with decreasing amplitudes with increasing upstream distance. The waves had ~7.0 nT peak-to-peak amplitudes in an ~8 nT field at 0151:00 UT. The quasiperiod of the waves was ~0.8 s. There were waves detected upstream of the interval shown in this figure and in Figure 4. They are discussed in some detail in Tsurutani *et al.* (2001).

Figure 7 is an example of one wave cycle from 0151:39.9 to 0151:40.6 UT, shown in minimum variance coordinates. The minimum variance coordinate system is determined by diagonalizing the covariance matrix. The three eigenvectors λ_1, λ_2, and λ_3 are the directions of maximum, intermediate, and minimum variance (Smith and Tsurutani, 1976). The wave peak-to-peak amplitude is ~14.0 nT in a ~15 nT ambient field. In the hodogram it is noted that the wave was right-hand polarized in the spacecraft frame, nearly circularly polarized (λ_1/λ_2 = 2.1) and planar (λ_2/λ_3 = 2.2). Here λ_1, λ_2, and λ_3 correspond to the maximum, intermediate, and minimum eigenvalues. Electromagnetic wave $\hat{k}$ vectors are along λ_3. The angle of propagation relative to the ambient magnetic field θ_{kB} was ~52°.

Figure 8 is a wave cycle far from the shock, from 0151:16.6 to 0151:17.4 UT. The wave peak-to-peak amplitude was ~8.0 nT in a ~10.0 nT ambient field. The wave was right-hand circularly polarized

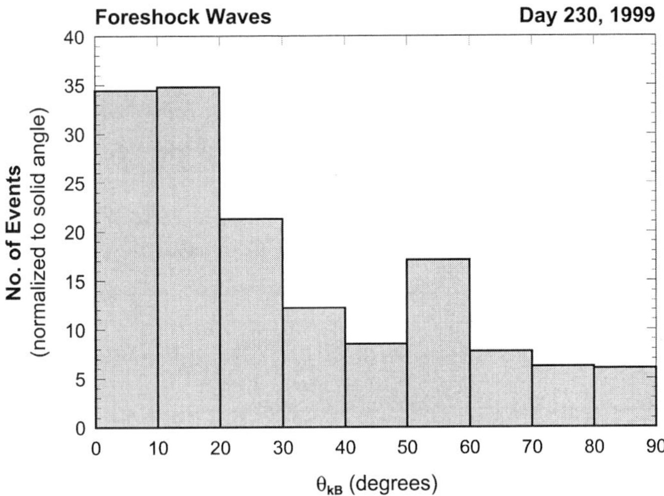

Fig. 9. A normalized histogram of the foreshock wave propagation directions.

($\lambda_1/\lambda_2 = 1.9$), planar ($\lambda_2/\lambda_3 = 7.5$) and was propagating at an angle of ~8° relative to the magnetic field. There was a ~2 nT compressional component associated with the wave.

Table 1 summarizes the wave properties for the region immediately upstream of the shock. The waves were almost all right-hand circularly to elliptically polarized in the spacecraft frame. There were only a few left-hand wave cycles identified. The wave periods range between 0.3 s and 1.3 s in the spacecraft frame, with a median of 0.8 s. Of the waves, 45% had λ_1/λ_2 values between 1.0 and 3.0, indicating that the waves were often circularly to elliptically polarized. A normalized (to spherical area) histogram of θ_{kB} is given in Figure 9. The waves were propagating at all angles with respect to $\vec{B}$, with a slight tendency to propagate along the field line direction.

Calculation of Resonant Particle Energies (Generating the Foreshock Waves)

Foreshock plasma waves are generated by energetic protons or electrons that are streaming away from the bowshock/magnetosheath into the upstream region. In this particular case, the solar wind speed was ~560 km s^{-1} and the Alfvén speed ~107 km s^{-1}. The relevant ion, electron and plasma frequencies were stated previously.

Several features of the foreshock waves were unusual. First, the vast majority of the waves were right-hand polarized in the spacecraft frame. The waves had magnetic magnitude variations of ~25%, strong compressional components. Frequencies in the spacecraft frame were ~1.25 Hz (~0.8 s periods), considerably higher than ~10^{-1} to 1.3×10^{-2} Hz usual foreshock waves (Hoppe *et al.*, 1981). Also, foreshock waves are typically left-hand polarized (in the spacecraft frame) rather than right-hand polarized as shown here.

Sentman *et al.* (1983) have shown that the most likely source of upstream whistler mode waves is ~20 eV electrons reflected off of the bow shock. They reported a lack of 1-20 keV electrons during the intervals of $f \approx 1.25$ Hz waves. In another article focusing on the foreshock waves, Tsurutani *et al.* (2001) have demonstrated that small angular changes in the magnetic field orientation lead to orders of magnitude changes in wave power. These latter results were consistent with low-energy electron generation.

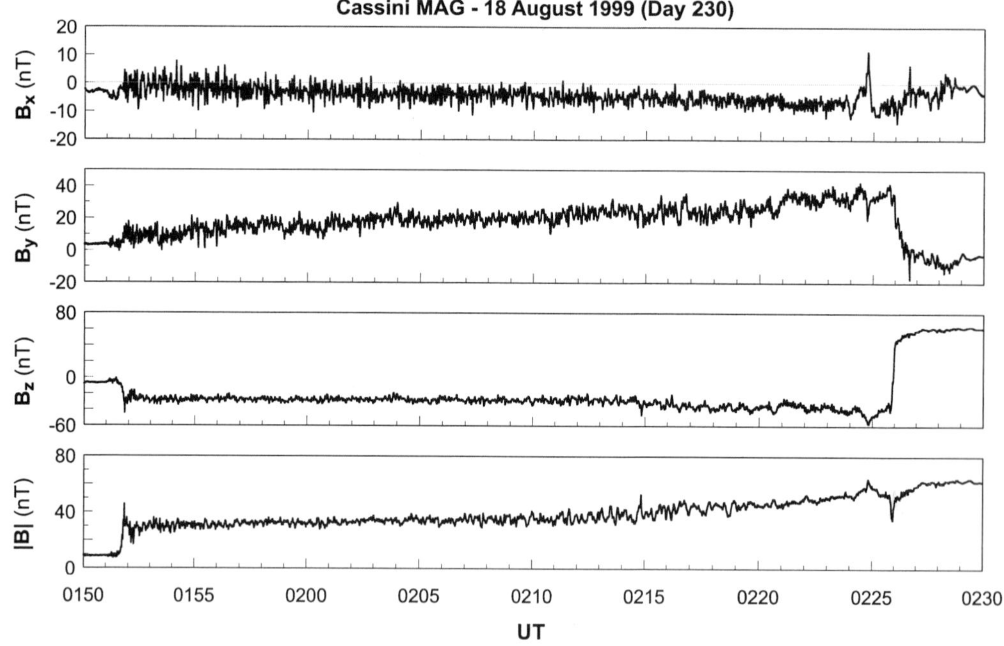

Fig. 10. The magnetosheath magnetic fields and superposed waves, in GSM coordinates.

Magnetosheath Waves

Figure 10 shows the full sheath, up to and past the magnetopause (located at ~0226 UT). The magnetic field magnitude slowly increased from the post-shock region (~32 nT as previously stated), to ~60 nT just prior to the magnetopause. B_y increased from ~10 nT to ~30 nT and B_z from ~ −30 nT to ~ −55 nT during the same time interval of time. This steady magnetic field magnitude increase with decreasing distance from the magnetopause is due to the Zwan-Wolf (1976) field draping effect, where magnetosheath magnetic fields "drape" around the magnetosphere. As the magnetosheath plasma and fields are convected towards the stagnation point (the magnetopause nose), more and more plasma gets squeezed out the ends of the flux tubes leading to higher magnetic field strengths and lower plasma densities (lower plasma betas) near the magnetopause (Tsurutani *et al.*, 1982). This general trend can be noted in the figure.

Although the wave amplitudes were large throughout the magnetosheath, they gradually decreased from the bow shock to the magnetopause. This can be noted in the B_x component, the component most transverse to the ambient field direction. Another feature of the waves is that they were generally noncompressive. The largest field magnitude variations were present at ~0215 UT with ~6 nT variations in a ~40 nT field. These field decreases were quite small compared to the 50 - 80% quasiperiodic variations typical of mirror mode structures (Tsurutani *et al.*, 1982).

A sample of the magnetosheath waves is shown in Figure 11. The waves have 5-6 s quasisinusoidal periods with peak-to-peak amplitudes of ~8-10 nT in a ~34 nT field. The waves are thus highly nonlinear (the wave amplitudes cannot be considered as a small perturbation of the ambient field). There are large field magnitude decreases associated with some of the waves. The decreases involve field directional changes, thus they do not appear to be parts of mirror mode structures. It should be noted, however, that

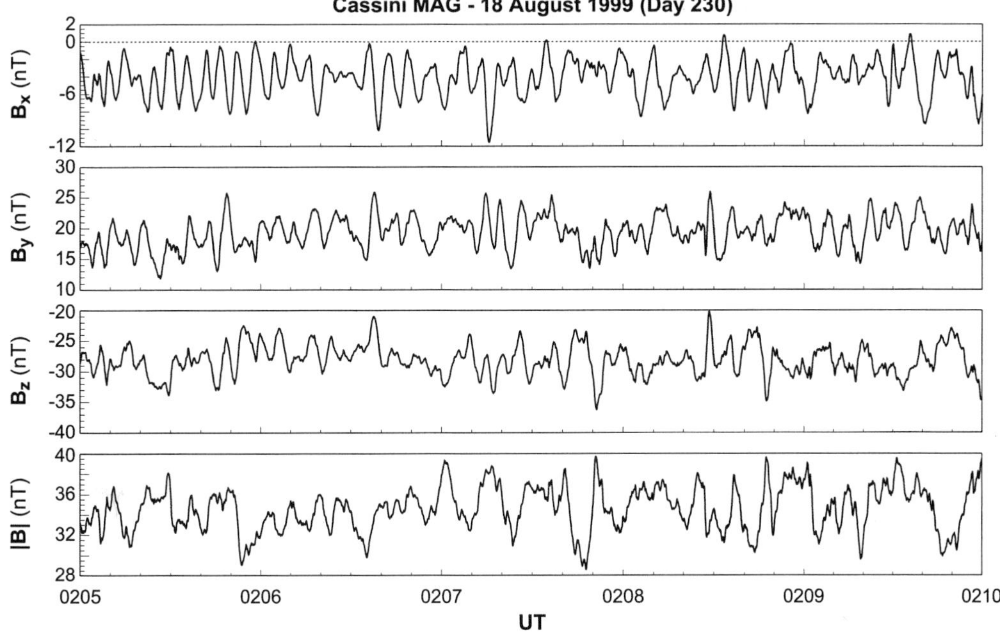

Fig. 11. Magnetosheath waves with ~5 to 6 second quasiperiods and ~8-10 nT peak-to-peak amplitudes, shown in GSM coordinates.

there is not a one-to-one association between waves and field decreases. There are more wave cycles than field decreases.

Figure 12 shows the wave-magnetic field decrease relationship in greater detail. Between the dashed vertical lines are 4 cycles of the waves. The magnetic field magnitude displays variations during two of the largest amplitude wave cycles (0157:07-0157:12 and 0157:12-0157:17 UT).

We will use two slightly different methods to determine the properties of the electromagnetic waves. We will use a minimum variance technique applied to single wave cycles and also applied to several wave cycles ("packets" - a packet is loosely defined as a number of cycles that appear to be roughly coherent) to compare and contrast the differences in the results. In some regions of space, wave properties vary from cycle-to-cycle, and thus multiple-cycle analyses could lead to a misunderstanding of the wave properties.

In Figure 13 we show the minimum variance analysis results for two consecutive wave cycles analyzed separately and also analyzed together (bottom panel). On the top is the cycle from 0157:03 - :08 UT and in the center, the cycle from 0157:08 - :15 UT. The wave cycle in the center is nearly circularly polarized in a left-hand sense, and is propagating at an angle of 6° relative to the ambient field. The wave is planar, $\lambda_2/\lambda_3 = 14.2$. The wave cycle in the center shows similar, but slightly different properties. The wave is planar ($\lambda_2/\lambda_3 = 11.2$) and is polarized in the left-hand sense. The wave is propagating at 18° relative to the ambient magnetic field and is elliptically polarized ($\lambda_1/\lambda_2 = 3.8$). The bottom panel shows the minimum variance results from the entire interval. Several differences can be noted: the multiple cycle "average" value indicates that the waves are far less planar ($\lambda_2/\lambda_3 = 3.7$), they are more circularly polarized ($\lambda_1/\lambda_2 =$

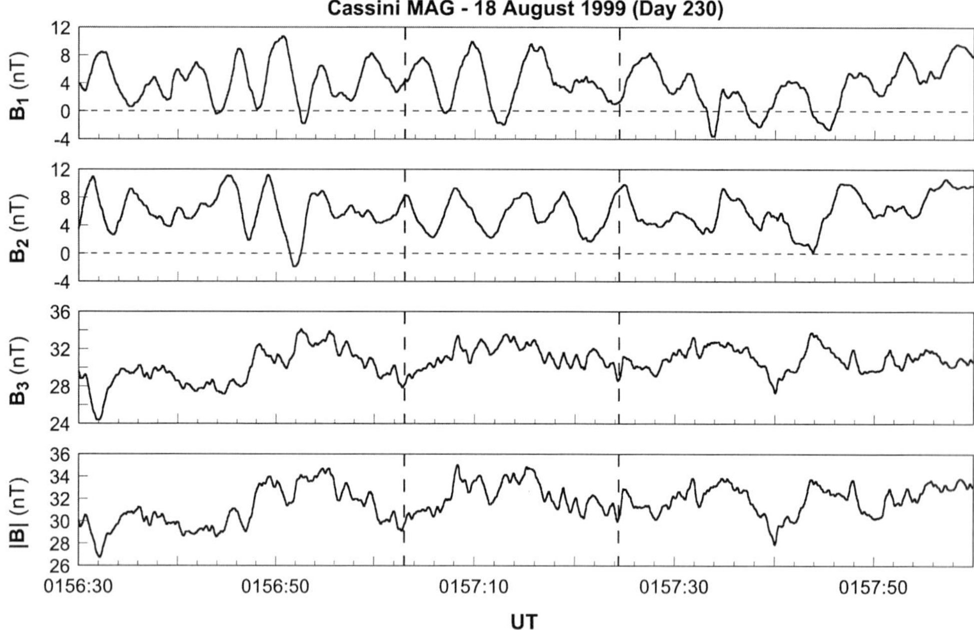

Fig. 12. Magnetosheath waves in minimum variance coordinates.

1.8), and the average propagation direction is nearly along $\vec{B}_0$ ($\theta_{kB} = 12.7°$).

Wave "packets" and individual cycles throughout the magnetosheath passage have been analyzed. A summary of the results is given in Table 2. The second and third columns are the start and stop times of the wave interval analyzed, the fourth column the average period of the waves in the packets ("packets" are denoted by the roman numerals in the first column). B is the average ambient field strength in units of nT. Almost all cycles examined were found to be left-hand polarized in the spacecraft frame (right-hand column).

There are several interesting features to note in this table. The waves vary from circular to elliptical polarizations and the angles of propagation relative to the ambient magnetic field vary from 1° to 73°. Most of the waves are propagating obliquely to the ambient magnetic field. There is a general tendency, although not a strong one, for the obliquely propagating waves to be more elliptically polarized. However, there are examples of circularly polarized waves found to be propagating at very large θ_{kB} angles. An example can be found at 0205:48.6 to 05:55 UT ($\lambda_1/\lambda_2 = 1.8$ and $\theta_{kB} = 78°$).

The properties of the entire "packets" are shown in the first row of each event. The level of polarization (λ_1/λ_2) is typically near the minimum value of the individual cycle values within the packet. The angle of propagation θ_{kB} is also near the minimum of values of the individual cycles. Individual wave cycles within a "packet" often have similar angles of propagation (but not always). On the other hand, the degree of polarization of cycles within a packet is typically highly mixed.

Figure 14 shows the magnetosheath wave frequency (f) and the proton cyclotron frequency (f_p) as a function of spacecraft location (and time). Magnetospheric waves are also included in this plot ($T > 0226$ UT)

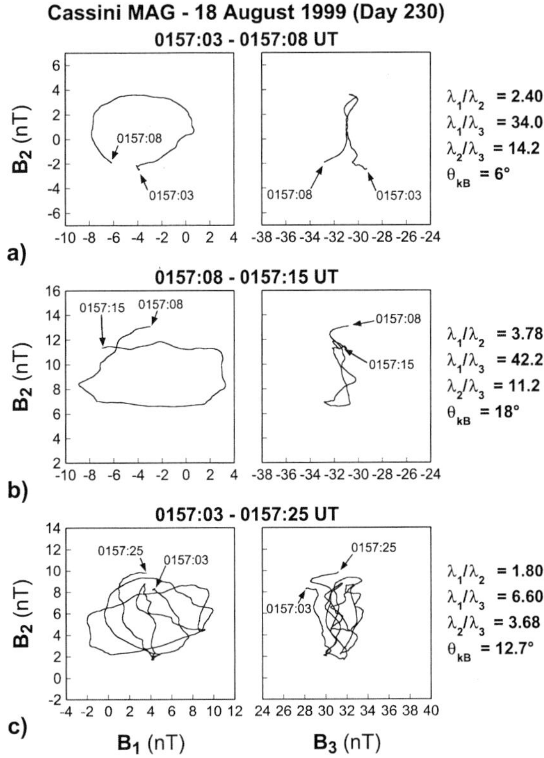

Fig. 13. Several magnetosheath wave cycles in minimum variance coordinates. The events were taken from the interval indicated by vertical dashed lines in Figure 12.

but will not be discussed for the moment. The magnetosheath waves are noted to have frequencies of ~0.3 to 0.6 f_p.

Since the magnetosheath fields were oriented essentially orthogonal to the Sun-Earth line, the Doppler shift of the waves due to convection past the spacecraft was negligible. Thus the waves are left-hand polarized in the plasma frame and were propagating in the proton (ion) cyclotron mode. We believe that these are the largest amplitude proton cyclotron waves detected in geospace to date.

From an inspection of the magnetic field data, it was found that there was little or no indication of the presence of mirror mode structures. A search for field decreases with little or no changes in direction proved to be unfruitful. Only a few intense whistler mode lion roars (often associated with the high beta portions of mirror mode structures: Tsurutani *et al.*, 1982; Zhang *et al.*, 1998; Baumjohann *et al.*, 1999) were detected in the magnetosheath (G. Hospodarsky, personal communication, 2001).

Magnetopause

The minimum variance analysis results of the magnetopause are shown in Figure 15. The analysis was performed on 32 vector-per-second fields, the highest time resolution available for the Cassini magnetometer (Dougherty *et al.*, 2001). The magnetopause current layer is quite broad, extending from 0225:44 UT to 0226:05 UT. There is a large, steady normal field component (B_3) of ~8 nT present. The large

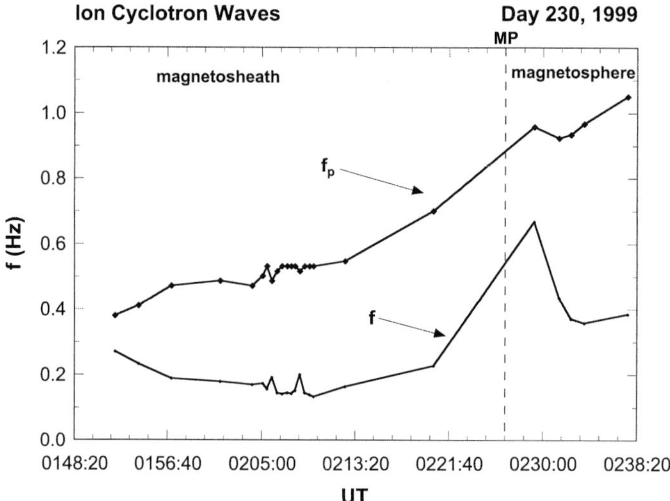

Fig. 14. Magnetosheath and magnetospheric wave frequencies (f). The proton cyclotron frequency (f_p) is also shown for comparison.

normal component indicates that this is a rotational discontinuity (Sonnerup and Ledley, 1974), consistent with the concept that magnetic reconnection is taking place at this boundary (the magnetopause).

Outer Magnetospheric Waves

An example of the outer magnetospheric waves is shown in Figure 16. The waves are quasi-sinusoidal with ~3 s periods. The maximum peak-to-peak amplitudes are ~3.0 nT in a ~61 nT field (at ~0232:30 UT). The wave amplitudes are largest in the transverse components (GSM B_y and B_x). The ambient field is primarily in B_z, as expected for magnetospheric fields. The amplitude is largest in the B_y component and is considerably smaller in the B_x component. The waves sometimes have a compressional component. At ~0232:30 UT, the magnetic magnitude (compressional) variations are on the order of ~0.5 nT.

Figure 17 shows the waves in minimum variance coordinates and in higher time resolution. It is interesting to note that the amplitudes are largest (8.5 nT peak-to-peak) where the magnetic field increased from ~61 to ~62 nT. The interval between 0232:30 and 0232:46 UT has been used to determine the minimum variance coordinates for the whole interval of the figure. The B_3 component oscillations are minimum in this region, indicating an accurate determination of the minimum variance direction. On the bottom (panel b) are the corresponding hodograms. The B_2 - B_1 hodogram indicates that the waves are highly linearly polarized (consistent with the previous description of B_y and B_x amplitudes). The individual wave cycles are shifted from each other in the hodogram because of the spacecraft motion toward the Earth into higher magnetic field strengths (we specifically did not detrend the data). The B_2 - B_3 hodogram shows that the waves are reasonably planar. The angle of propagation of the waves relative to the magnetic field was ~42°.

Figure 18 shows another example of the outer magnetospheric waves, but later in time. The amplitudes are as large as ~2.0 nT peak-to-peak in a ~69.0 nT ambient field. There is a slight magnetic field magnitude gradient where the waves were detected. The interval where the minimum variance analyses were performed is indicated by the two vertical dashed lines at 0237:24 UT and 0237:41 UT. Again there is

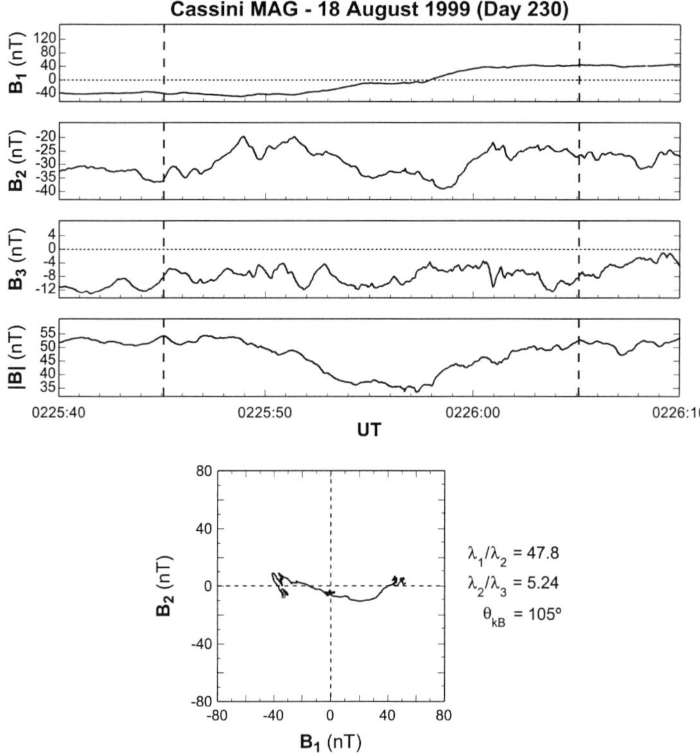

Fig. 15. The magnetic field at the magnetopause (top 4 panels), in minimum variance coordinates. The bottom panel is a hodogram of the magnetopause field during the interval indicated by the vertical dashed lines.

very little variation along B_3, indicating a "good determination" of the minimum variance direction. The magnetic magnitude variations associated with the waves are quite small, ~0.2 nT (note that the scale of the B_3 component and $|\boldsymbol{B}|$ is much finer than for the other components).

The bottom panel (b) gives the hodogram results. These waves are left-hand elliptically polarized with $\lambda_1/\lambda_2 = 2.9$. The average angle of propagation is 8.6° relative to $\vec{B}$. The waves are highly plane polarized, $\lambda_2/\lambda_3 = 73.1$.

The magnetospheric wave results are itemized in Table 3. All of the waves analyzed were found to be left-hand circularly to linearly polarized (λ_1/λ_2 up to 175) in the spacecraft frame. The θ_{kB} values range from 6° to 71°. Some packets have nearly parallel propagating waves and other packets are off-axis propagating waves. Since there is little or no plasma convection to speak of, the waves must be left-hand polarized in the plasma frame. From Figure 14, the waves have a frequency f between 0.3 and 0.5 f_p. The waves are therefore propagating in the proton cyclotron mode.

To determine the solar wind effects on the magnetosphere during this interval, we again examine the Geotail ram pressure values/variations given in Figure 2. The interval of outer magnetospheric waves occurs from 0226 UT to 0238 UT (Figures 16 to 18). As previously stated, the time delay for the solar wind to

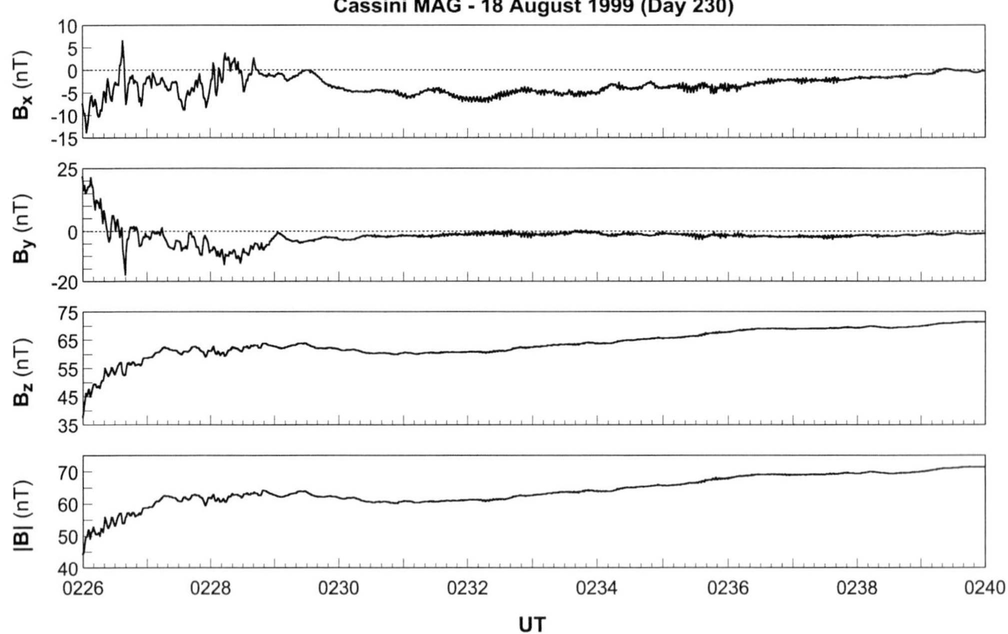

Fig. 16. Outer magnetospheric waves shown in GSM coordinates. Small amplitude waves can be noted in the B_x and B_y components from ~0230 to 0238 UT.

travel from Geotail to the magnetopause is ~4 min. Thus, the relevant solar wind plasma parcel is from ~0222 to 0234 UT.

Can one find solar wind features that lead to proton temperature anisotropies ($T_{\perp}/T_{\parallel} > 1$) that cause ion cyclotron wave growth? There are no obvious large scale solar wind ram pressure fluctuations that caused the growth of the waves such as were found by Anderson and Hamilton (1993) in their events. However, the relationship between large amplitude waves and magnetic field increases noted in Figures 17 and 18 indicates that small ram pressure fluctuations may be present. If so, there could be small scale density enhancements that would easily be missed by Geotail. Perhaps future Cluster II searches for small solar wind features and their effects will answer this question.

Particle Losses and Proton Auroras

The peak proton cyclotron wave amplitudes of ±1.7 nT in a 61 nT field occurred at ~0232 UT. The Cassini spacecraft was located at (8.7, 2.4, -0.3 R_e) GSM at this local time. The proton cyclotron frequency f_p was 0.93 Hz. The wave frequency f was ~0.35 Hz. Assuming a wave phase velocity $V_{ph} \approx 10^8$ cm^{-2} s^{-1}, and using the first-order cyclotron resonance condition (1), one obtains a resonant proton parallel energy of ~10 keV. The pitch angle diffusion rate is given by:

$$D_{\alpha\alpha} = 2\pi f_p \left(\frac{B_\omega}{B} \right)^2 \eta \tag{5}$$

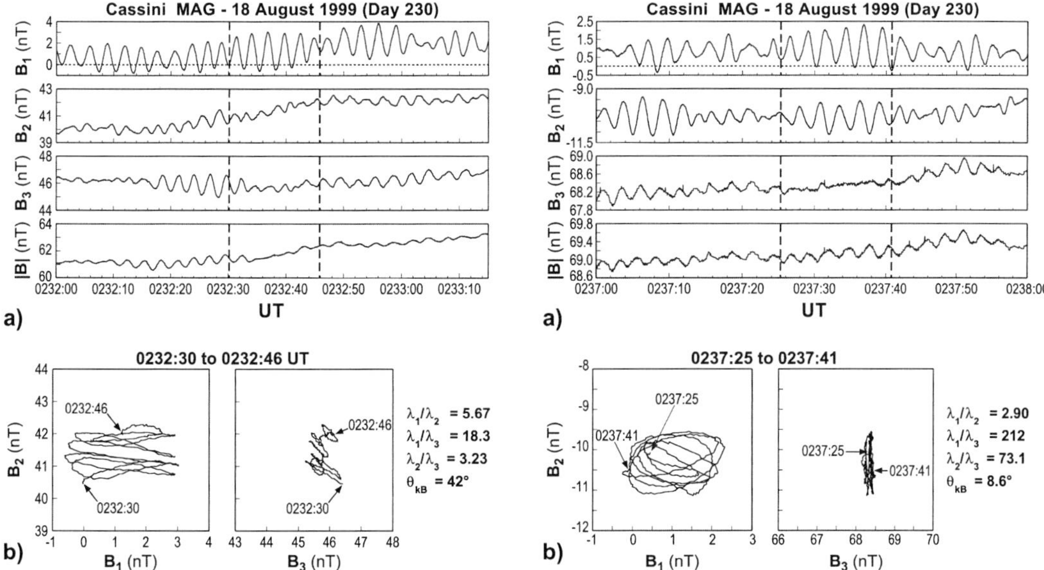

Fig. 17. Magnetospheric waves shown in minimum variance coordinates. The bottom panels are the hodograms for the 0232:30 to 0232:46 UT magnetospheric wave packet.

Fig. 18. Same as Figure 17. The bottom panels are the hodograms for the 0237:24 to 0237:41 UT wave packet.

from Kennel and Petschek (1966). With a value of $B_\omega \approx 1.7$ nT, and assuming η the fractional amount of time the particle is in resonance with the wave is ≈ 0.1, $D_{\alpha\alpha}$ is therefore $\sim 7.2 \times 10^5$ s^{-1}, or $T = 1.3 \times 10^4$ s. The bounce time T_b of a 10 keV proton is 1.2×10^2 s. Thus the protons are on weak pitch angle diffusion. The resulting proton aurora due to wave-particle interactions would be expected to be weak and diffuse.

Magnetotail Waves

The Alfvénic B_s fluctuations (shown in Figure 3) are believed to trigger magnetospheric substorms through magnetic reconnection processes. Corresponding geomagnetic activity was high throughout the Cassini pass. This is shown in Figure 19 by the *AU*, *AL*, and *AE* indices. Substorms were occurring from ~0700 to ~1200 UT, the Cassini tail passage. Large amplitude waves in the tail lobe were detected primarily near two time regions, ~0820 and 1003 UT. This is indicated in the figure.

One example of waves near the former interval is shown in Figure 20. The waves are transverse oscillations, and have ~7.5 s periods. The peak-to-peak amplitudes are ~0.3 nT in a ~16.1 nT field. The compressional components are ≤ 0.1 nT. The three cycles from 0819:43.7 to 0820:07.6 UT are elliptically polarized ($\lambda_1/\lambda_2 = 2.3$ to 3.0) and are propagating nearly along $\vec{B}$ (6° to 26°).

Figure 21 shows an example of the waves detected in the second interval. Nearly purely compressional (longitudinal) oscillations with ~7.3 s periods are present. The magnitudes are again small, ~0.1 nT. The four cycles from 1003:15.6 to 1003:44.4 UT are compressive waves. The angles of propagation for the four wave cycles are highly oblique to $\vec{B}$ (76°, 84°, 70°, and 75°). We believe that this is the first time such purely compressional waves have been detected in the magnetotail.

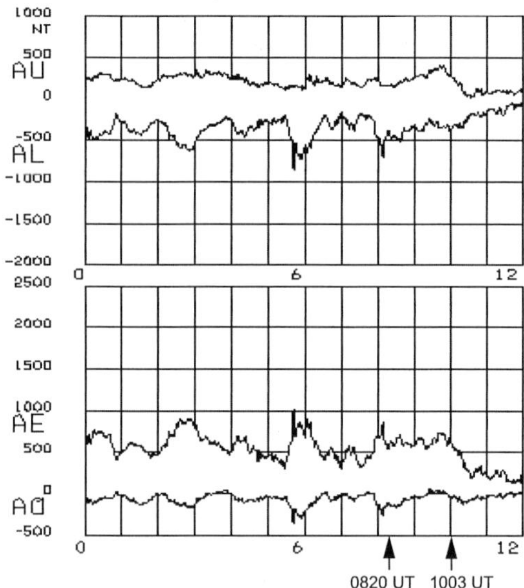

Fig. 19. Auroral zone geomagnetic indices during the Cassini flyby. Two intervals where LF waves were detected and analyzed are indicated by arrows.

The extremely thorough Cassini magnetic control program instituted by the Cassini Project has developed this very low-noise spacecraft (Mehlem and Narvaez, 1999). Without such a rigorous program it is probable that such new wave modes could not have been detected.

SUMMARY

Foreshock Waves

Foreshock whistler mode waves are detected well upstream of the bow shock. The waves have ~0.8 s quasiperiods in the plasma frame and are right-hand circularly to elliptically polarized in the spacecraft frame. The angle of propagation to $\vec{B}$ varies from 6° to 88°. The waves are compressive. It is argued that the waves are whistler mode emissions generated by electrons tens of eV in energy streaming towards the Sun. A more comprehensive description and analysis is presented in Tsurutani *et al.* (2001).

Magnetosheath Waves

The magnetosheath waves are found to be left-hand circular to elliptically polarized with $f = 0.3$ to $0.6\ f_p$. They are generated by a temperature anisotropy instability with $T_\perp/T_\parallel > 1$. The anisotropy is created by the shock heating of solar wind protons, in a direction perpendicular to the magnetic field. Behind the shock and throughout the sheath, the Zwan-Wolf (1976) draping effect will maintain a strong $T_\perp/T_\parallel > 1$ anisotropy. The magnetosheath proton cyclotron waves are the most intense ion cyclotron waves detected in geospace to date.

Magnetospheric Waves

The outer magnetosphere waves are also identified as proton cyclotron waves with frequencies between 0.2 to 0.5 f_p. These waves are often highly elliptical to linearly polarized. The source of free energy is,

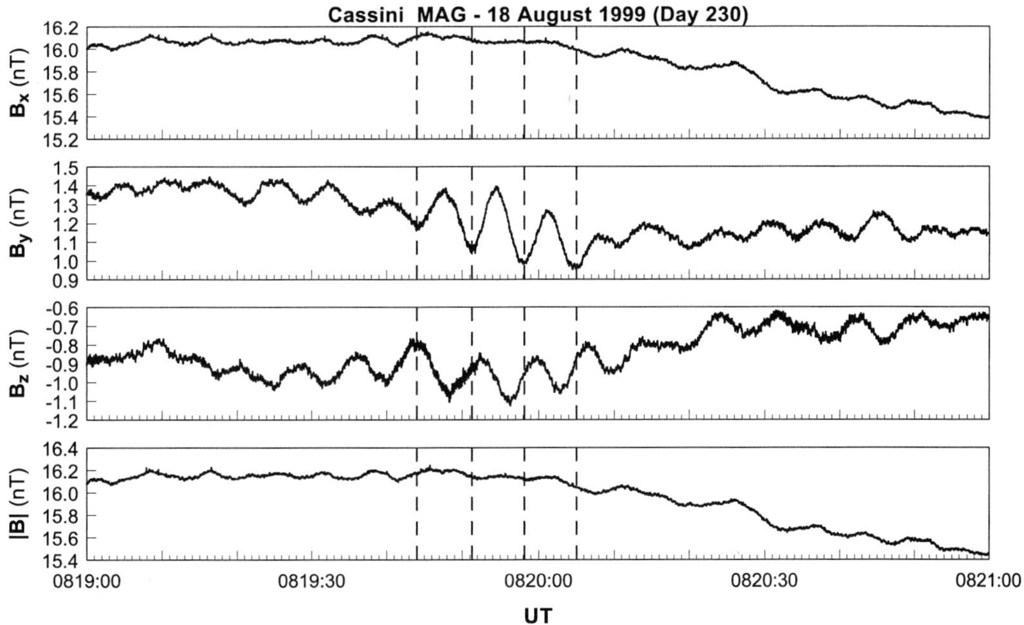

Fig. 20. Transverse ~7.5 second oscillations in the magnetotail, shown in GSM coordinates.

however, somewhat of a mystery. At the time of wave occurrence, the large scale solar wind ram pressure was steady to decreasing. Small scale solar wind fluctuations cannot be ruled out, however.

Magnetotail Waves

Two new modes were identified in the distant tail lobes, primarily due to the unusual sensitivity of the Cassini magnetometers and the spacecraft cleanliness program. A purely compressive (~0.1 nT) mode as well as a short period transverse wave mode were illustrated.

CONCLUSIONS AND DISCUSSION

In the Cassini geospace wave survey, we find that much can be done to understand microinstabilities generating LF waves using magnetic field data alone. From our results, we have been able to make a variety of predictions of plasma and energetic particle pitch angle distributions (in the foreshock, magnetosheath and outer magnetosphere). In a follow-up effort, we will examine the plasma and energetic particle and wave properties to determine if our conclusions/speculations in this paper were indeed correct or not.

This paper has shown that even though these four regions of space have been previously studied in very great detail, there is still much to learn and it is very worthwhile examining the waves in even greater detail. Cluster II will provide an opportunity for detailed studies of several of these regions.

Many surprises were found from this examination. In the upstream region, the standard picture is that ion beams stream from the shock/magnetosphere and populate this region. These beams generate foreshock waves. The waves and particles would be detected on field lines antisunward of the field tangent to the bow shock. This was shown in Figure 5. The mode typically reported in the literature (Hoppe *et al.*,

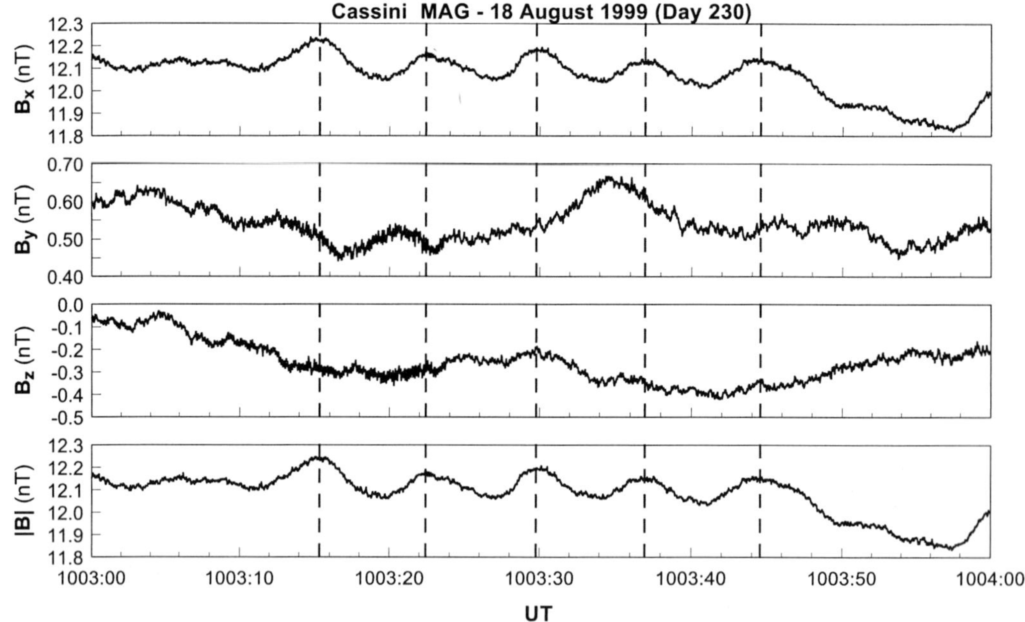

Fig. 21. Nearly purely compressional ~7.3 s waves in the magnetotail, shown in GSM coordinates.

1981; Smith *et al.*, 1983; Tsurutani *et al.*, 1993) has 10 - 60 s periods is and left-hand polarized in the spacecraft frame. These waves have been shown to be anomalously Doppler-shifted magnetosonic (right-hand) waves convected past the spacecraft.

Figure 5 shows a schematic of the electron and proton foreshocks for an interplanetary magnetic field angle of ~70° relative to the Sun-Earth line. This is the correct geometry for the present case. Indicated along the top are the particle parallel kinetic energies needed to propagate upstream against a ~560 km s^{-1} solar wind flow. It is clear that only very energetic protons can get upstream at all, while low energy electrons have easy access. Thus the "foreshock" is an almost purely electron foreshock. The access of shock accelerated/reflected electrons to the upstream region is controlled by the connection of the interplanetary magnetic field line to the bow shock (the IMF angle). Cases where the magnetic field angle connected and disconnected to the shock led to the presence of waves and then the disappearance of waves (not shown). One interesting new detail is the strong compressibility of the waves, presumably due to the large angles of $\vec{k}$ relative to the ambient magnetic field (Figure 9). The strong off-axis character has been explained by Sentman *et al.* (1983) and Wong and Smith (1994) as being due to a Landau interaction with the electrons. The full distribution of wave k values (Figure 9) may be related to the unusual electron distribution functions found in various parts of the foreshock (see Savoini and Lembege, 1994, 2001).

The magnetosheath waves are perhaps the biggest surprise. No obvious mirror mode structures were detected. Only proton cyclotron waves were present and these had the largest amplitude (of ion cyclotron waves) detected in geospace to date. The ACE He^{++}/H^{+} density ratio was ~ 2.0% for the solar wind plasma parcel that hit the bowshock at the time of the Cassini encounter. This low He^{++} density ratio (an average value of 4.4% has been found for high speed streams: McComas *et al.*, 1996) is in qualitative agreement with the Price *et al.* (1986) and Gary *et al.* (1992) scenario of a reduction in size of the stop-band, thus an increase in the ion cyclotron growth rate. However, further modeling work needs to be done to understand if this is indeed the correct explanation or not. We further note that the He^{++}/H^{+} ratio varied from <1% (at the peak of the event) to 4% (at the edges of the event) within the high-speed stream. Whether this variability is a feature of high-speed streams detected in the ecliptic plane where stream "edge effects" and fast stream-slow stream interactions are important, needs to be studied. Another possible explanation is that the presently used threshold for mirror instability is incorrect. Leubner and Schupfer (2001), examining non-thermal distributions, have indicated that the threshold can be significantly higher than previously thought. For these reasons, the variability of Jovian and Saturnian magnetosheath wave modes and intensities (also under the different plasma beta conditions) are of particular interest for future Cassini studies.

One possible scenario explaining the presence of dayside instabilities/waves is that the previously occurring substorms have led to proton injection in the midnight sector. With subsequent particle gradient and curvature drift to the dayside, the asymmetric shape of the magnetosphere (compressed dayside) leads to particle drift-shell splitting (Roederer, 1967; Schulz, 1972), where particles with ~90° pitch angle will drift to larger L than particles with near ~0° pitch. This picture will lead to "patches" in the outer dayside magnetosphere where the proton cyclotron mode would be unstable. A second scenario, working in conjunction with the first, can lead to the enhanced wave growth. If sunlight heating of the ionosphere leads to upward flow of plasma and an enhanced magnetospheric cold population, the presence of this plasma would lead to a lowering of the resonant energy threshold for instability. A similar explanation has been proposed to explain enhanced chorus (electron temperature anisotropy instability) at local dawn (Jensch, 1976, Tsurutani and Smith, 1977) through the electron temperature anisotropy instability. A third factor discussed previously is the possible presence of small scale solar wind ram pressure pulses. Ram pressure pulses will compress magnetospheric electrons and ions in the perpendicular direction, leading to $T_{\perp}/T_{\parallel} > 1$ and the loss cone instability (Zhou and Tsurutani, 1999). At any rate, the proton cyclotron waves are clearly present and will lead to a steady "drizzle" of particles into the upper ionosphere. To determine the specific mechanisms, energetic particle pitch-angle information and plasma densities will have to be examined in detail. We hope to do this in the near future.

ACKNOWLEDGMENTS

Portions of the research presented here were performed at the Jet Propulsion Laboratory, California Institute of Technology, Pasadena, under contract with the National Aeronautics and Space Administration.

REFERENCES

Anderson, B. T., and D. C. Hamilton, Electromagnetic Ion Cyclotron Waves Stimulated by Modest Magnetospheric Compressions, *J. Geophys. Res.*, **98**, 11369 (1993).

Anderson, B. J., S. A. Fuselier, S. P. Gary, and R. E. Denton, Magnetic Spectral Signatures in the Earth's Magnetosheath and Plasma Depletion Layer, *J. Geophys. Res.*, **99**, 5877 (1994).

Armstrong, T. P., M. E. Pesses, and R. B. Decker, Shock Drift Acceleration, in *Collisionless Shocks in the Heliosphere: Reviews of Current Research*, ed. B. T. Tsurutani and R. G. Stone, Geophys. Mon., **35**, pp. 271-285, Amer. Geophys. Union Press, Wash. D.C. (1985).

Asbridge, J. R., S. J. Bame, and I. B. Strong, Outward Flow of Protons from the Earth's Bow Shock, *J. Geophys. Res.*, **73**, 5777 (1968).

Baker, D. N., R. D. Belian, T. A. Fritz, P. R. Higbie, S. M. Krimigis, D. G. Sibeck, and R. D. Zwickl, Simultaneous Energetic Particle Observations at Geostationary Orbit and in the Upstream Solar Wind: Evidence for Leakage During the Magnetospheric Compression Event of November 1, 1984, *J. Geophys. Res.*, **93**, 14317 (1988).

Baumjohann, W., R. A. Treumann, E. Georgescu, G. Haerendel, K.-H. Fornacon, and U. Auster, Waveform and Packet Structure of Lion Roars, *Ann. Geophys.*, **17**, 1528 (1999).

Bonifazi, C., and G. Moreno, Reflected and Diffuse Ions Backstreaming from the Earth's Bow Shock, 1. Basic Properties, *J. Geophys. Res.*, **86**, 4397 (1981a).

Bonifazi, C., and G. Moreno, Reflected and Diffuse Ions Backstreaming from the Earth's Bow Shock, 2. Origin, *J. Geophys. Res.*, **86**, 4405 (1981b).

Brinca, A. L., and B. T. Tsurutani, Influence of Multiple Ion Species on Low-Frequency Electromagnetic Wave Instabilities, *J. Geophys. Res.*, **94**, 13565 (1989).

Chao, J. K., C. H. Lin, Y. H. Yang, X. Y. Wang, M. Kessel, C. H. Lin, S. H. Chen, and R. P. Lepping, Models for the size and shape of the Earth's magnetopause and bow shock, *J. Atmos. Sol. Terr. Phys.*, in press 2001.

Cowley, S. W. H., R. J. Hynds, I. G. Richardson, P. W. Daly, T. R. Sanderson, K.-P. Wenzel, J. A. Slavin, and B. T. Tsurutani, Energetic Ion Regimes in the Deep Geomagnetic Tail: ISEE-3, *Geophys. Res. Lett.*, **11**, 275 (1984).

Dougherty, M. K., S. Kellock, D. J. Southwood, A. Balogh, M. Barlow, T. Beek, M. W. Dunlop, R. White, E. J. Smith, L. Wigglesworth, M. Fong, R. Marquedant, B. T. Tsurutani, B. Gerlach, G. Musmann, K.-H. Glassmeier, H. Hartje, M. Rahm, I. Richter, C. T. Russell, D. Huddleston, R. C. Snare, G. Erdos, S. Szalai, F. M. Neubauer, S. W. H. Cowley, and G. L. Siscoe, The Cassini Magnetic Field Investigation, *Space Sci. Rev.*, submitted (2001).

Dungey, J. W., Interplanetary Magnetic Field and the Auroral Zones, *Phys. Res. Lett.*, **6**, 47 (1961).

Fairfield, D. H., Bow Shock Associated Waves Observed in the Far Upstream Interplanetary Medium, *J. Geophys. Res.*, **74**, 3541 (1969).

Forman, M. A., and G. M. Webb, Acceleration of Energetic Particles, in *Collisionless Shocks in the Heliosphere: A Tutorial Review*, ed. R. G. Stone and B. T. Tsurutani, Geophys. Mon., **34**, pp. 91-114, Amer. Geophys. Union Press, Wash. D.C. (1985).

Gary, S. P., C. D. Madland, and B. T. Tsurutani, Electromagnetic Ion Beam Instabilities, II., *Phys. Fluids*, **28**, 3691 (1985).

Gary, S. P., Electromagnetic Ion/Ion Instabilities and their Consequences in Space Plasmas: A Review, *Space Sci. Rev.*, **56**, 373 (1991).

Gary, S. P., The Mirror and Ion Cyclotron Anisotropy Instability, *J. Geophys. Res.*, **97**, 8519 (1992).

Goldstein, M. L., H. K. Wong, A. F. Vinas, and C. W. Smith, Large Amplitude MHD Waves Up Stream of the Jovian Bow Shock: Reinterpretation, *J. Geophys. Res.*, **90**, 302 (1985).

Goldstein, M. L., H. K. Wong, and A. Eviatar, Excitation of MHD Waves Upstream of Jupiter by Energetic Sulfur on Oxygen Ions, *J. Geophys. Res.*, **91**, 7954 (1986).

Goodrich, C. C., Numerical Simulations of Quasi-Perpendicular Collisionless Shocks, in *Collisionless Shocks in the Heliosphere: Reviews of Current Research*, ed. B. T. Tsurutani and R. G. Stone, Geophys. Mon., **35**, pp. 153-168, Amer. Geophys. Union Press, Wash. D.C. (1985).

Hoppe, M. M., C. T. Russell, L. A. Frank, T. E. Eastman, and E. W. Greenstadt, Upstream Hydromagnetic Waves and their Association with Backstreaming Ion Populations: ISEE 1 and 2 Observations, *J. Geophys. Res.*, **86**, 4471 (1981).

Jensch, V., Electron Precipitation in the Morning Sector of the Auroral Zone, *J. Geophys. Res.*, **81**, 135 (1976).

Kennel, C. F., and H. E. Petschek, Limit on stably trapped particle fluxes, *J. Geophys. Res.*, **71**, 1, 1966.

Kennel, C. F., J. P. Edmiston, and T. Hada, A Quarter Century of Collisionless Shock Research, in *Collisionless Shocks in the Heliosphere: A Tutorial Review*, ed. R. G. Stone and B. T. Tsurutani, Geophys. Mon., **34**, pp. 1-36, Amer. Geophys. Union, Wash. D.C. (1985).

Lee, M. A., Coupled Hydromagnetic Wave Excitation and Ion Acceleration Upstream of the Earth's Bow Shock, *J. Geophys. Res.*, **87**, 5063 (1982).

Leubner, M.P., and N. Schupfer, A Universal Mirror Wave-mode Threshold Condition for Non-thermal Space Plasma Environments, *Eur. Geophys. Soc. Abstracts*, 215, 2001.

McComas, D. J., G. W. Hoogeveen, J. T. Gosling, J. L. Phillips, M. Neugebauer, A. Balogh, and R. Forsyth, Ulysses Observations of Pressure-Balance Structures in the Polar Solar Wind, *Astron. Astrophys.*, **316**, 368 (1996).

Mehlem, K., and P. Narvaez, Magnetostatic Cleanliness of Radioisotopic Thermoelectric Generators (RTGs) of Cassini, IEEE, 899 (1999).

Papadopoulos, K., Microinstabilities and Anomalous Transport, in *Collisionless Shock in the Heliosphere: A Tutorial Review*, ed. R. G. Stone and B. T. Tsurutani, Geophys. Mon., **34**, pp. 59-90, Amer. Geophys. Union Press, Wash. D.C. (1985).

Phillips, J. L., A. Balogh, S. J. Bame, B. E. Goldstein, J. T. Gosling, J. T. Hoeksema, D. J. McComas, M. Neugebauer, N. R. Sheeley, Jr., and Y. M. Wang, Ulysses at 50° South: Constant Immersion in the High-Speed Solar Wind, *Geophys. Res. Lett.*, **21**, 1105 (1994).

Price, C. P., D. W. Swift, and L. C. Lee, Numerical Simulation of Nonoscillatory Mirror Waves at the Earth's Magnetosheath, *J. Geophys. Res.*, **91**, 101 (1986).

Roederer, J. G., On the Adiabatic Motion of Energetic Particles in a Model Magnetosphere, *J. Geophys. Res.*, **72**, 981 (1967).

Savoini, P., and B. Lembége, Electron Dynamics in Two- and One-dimensional Oblique Supercritical Collisionless Magnetosonic Shocks, *J. Geophys. Res.*, **99**, 6609, 1994.

Savoini, P., and B. Lembége, Two-dimensional Simulations of a Curved Shock: Self-consistent Formation of the Electron Foreshock, to appear in *J. Geophys. Res.*, 2001.

Schulz, M., Drift-Shell Splitting at Arbitrary Pitch Angle, *J. Geophys. Res.*, **77**, 624 (1972).

Sentman, D. D., M. F. Thomsen, S. P. Gary, W. C. Feldman, and M. M. Hoppe, The oblique whistler instability in the Earth's foreshock, *J. Geophys. Res.*, **88**, 2048, 1983.

Sibeck, D. G., R. W. McEntire, S. M. Krimigis, and D. N. Baker, The Magnetosphere as a Sufficient Source for Upstream Ions on November 1, 1984, *J. Geophys. Res.*, **93**, 14328 (1988).

Smith, E. J., and B. T. Tsurutani, Magnetosheath Lion Roars, *J. Geophys. Res.*, **81**, 2261 (1976).

Smith, E. J., B. T. Tsurutani, D. L. Chenette, T. F. Conlan, and J. A. Simpson, Jovian Electron Bursts, Correlation with the Interplanetary Field Direction and Hydromagnetic Waves, *J. Geophys. Res.*, **81**, 65 (1976).

Smith, E. J., A. Balogh, M. Neugebauer, and D. McComas, Ulysses Observations of Alfvén Waves in the Southern and Northern Hemispheres, *Geophys. Res. Lett.*, **22**, 3381 (1995).

Smith, C. W., M. L. Goldstein, and W. H. Mathaeus, Turbulence Analysis of the Jovian Upstream Wave Phenomena, *J. Geophys. Res.*, **88**, 5581 (1983).

Sonnerup, B. U. O., and B. G. Ledley, Magnetopause Rotational Forms, *J. Geophys. Res.*, **79**, 4309 (1974).

Stix, T. M., *The Theory of Plasma Waves*, McGraw Hill, New York (1962).

Thomsen, M. F., Upstream Suprathermal Ions, in *Collisionless Shocks in the Heliosphere: Reviews of Current Research*, ed. B. T. Tsurutani and R. G. Stone, Geophys. Mon., **35**, pp. 253-270, Amer. Geophys. Union Press, Wash D.C. (1985).

Tsurutani, B. T., and C. M. Ho, A Review of Discontinuities and Alfvén Waves in Interplanetary Space: Ulysses Results, *Rev. Geophys.*, **37**, 517 (1999).

Tsurutani, B. T., and G. S. Lakhina, Some Basic Concepts of Wave-Particle Interactions in Collisionless Plasmas, *Rev. Geophys.*, **35**, 491 (1997).

Tsurutani, B. T., and E. J. Smith, Two Types of Magnetospheric ELF Chorus and their Substorm Dependences, *J. Geophys,. Res.*, **82**, 5112 (1977).

Tsurutani, B. T., and E. J. Smith, Magnetosonic Waves Adjacent to the Plasma Sheet in the Distant Magnetotail: ISEE-3, *Geophys. Res. Lett.*, **11**, 331 (1984).

Tsurutani, B. T., E. J. Smith, R. R. Anderson, K. W. Ogilvie, J. D. Scudder, D. N. Baker, and S. J. Bame, Lion Roars and Nonoscillatory Drift Mirror Waves in the Magnetosheath, *J. Geophys. Res.*, **87**, 6060 (1982).

Tsurutani, B. T., I. G. Richardson, R. M. Thorne, W. Butler, E. J. Smith, S. W. H. Cowley, S. P. Gary, S.-I. Akasofu, and R. D. Zwickl, Observations of the Right-Hand Resonant Ion Beam Instability in the Distant Plasma Sheet Boundary Layer, *J. Geophys. Res.*, **90**, 12159 (1985).

Tsurutani, B. T., D. J. Southwood, E. J. Smith, and A. Balogh, A Survey of Low Frequency Waves at Jupiter: The Ulysses Encounter, *J. Geophys. Res.*, **98**, 21203 (1993).

Tsurutani, B. T., C. M. Ho, E. J. Smith, M. Neugebauer, B. E. Goldstein, J. S. Mok, J. K. Arballo, A. Balogh, D. J. Southwood, and W. C. Feldman, The Relationship Between Interplanetary Discontinuities and Alfvén Waves: Ulysses Observations, *Geophys. Res. Lett.*, **21**, 2267 (1994).

Tsurutani, B. T., E. J. Smith, M. E. Burton, J. K. Arballo, C. Galvan, X.-Y. Zhou, D. J. Southwood, M. K. Dougherty, K.-H. Glassmeier, F. M. Neubauer, and J. K. Chao, Oblique "1-Hz" whistler mode waves in an electron foreshock: The Cassini near-Earth encounter, *J. Geophys. Res.*, in press 2001.

Wong, H. K., and C. W. Smith, Electron beam excitation of upstream waves in the whistler mode frequency range, *J. Geophys. Res.*, **99**, 13,373, 1994.

Zwan, B. J., and R. A. Wolf, Depletion of the Solar Wind Plasma Near a Planetary Boundary, *J. Geophys. Res.*, **81**, 1636 (1976).

Zhang, Y., H. Matsumoto, and H. Kojima, Lion Roars in the Magnetosheath: Geotail Observations, *J. Geophys. Res.*, **103**, 4615 (1998).

Zhou, X.-Y., and B. T. Tsurutani, Rapid intensification and propagation of the dayside aurora: Large-scale interplanetary pressure pulses (fast shocks), *Geophys. Res. Lett.*, **26**, 1097 (1999).

Table 1. Foreshock Waves. Day 230, 1999.

	Start Time	Stop Time	T	$\|\boldsymbol{B}\|$	λ_1/λ_2	θ_{KB}	Polarization
1	0151:00.2	0151:00.6	0.4	8	17.5	32	RH
2	0151:00.7	0151:00.9	0.2	8	1.9	69	RH
3	0151:01.9	0151:02.5	0.6	9	1.4	25	RH
4	0151:02.5	0151:03.7	1.2	9	1.3	44	RH
5	0151:03.7	0151:05.0	1.3	9	1.9	21	RH
6	0151:05.0	0151:05.6	0.6	9	3.7	35	RH
7	0151:05.6	0151:06.4	0.8	8	3.0	18	RH
8	0151:06.7	0151:07.3	0.6	8	1.7	60	LH
9	0151:07.3	0151:07.8	0.5	8	3.7	22	RH
10	0151:07.8	0151:08.2	0.4	8	7.3	11	RH
11	0151:08.2	0151:08.7	0.5	8	2.0	72	RH
12	0151:08.8	0151:09.3	0.5	9	4.3	28	RH
13	0151:09.3	0151:10.0	0.7	9	5.1	32	RH
14	0151:10.2	0151:10.7	0.5	8	5.0	79	LH
15	0151:10.8	0151:11.5	0.7	8	8.5	15	RH
16	0151:11.6	0151:12.0	0.4	8	3.0	47	RH
17	0151:12.0	0151:12.5	0.5	10	2.1	55	RH
18	0151:12.5	0151:13.2	0.7	9	3.3	39	RH
19	0151:13.2	0151:13.8	0.6	9	2.7	6	RH
20	0151:13.8	0151:14.6	0.8	8	3.5	86	RH
21	0151:14.7	0151:15.2	0.5	9	7.3	62	linear
22	0151:15.3	0151:16.5	1.2	10	4.6	46	RH
23	0151:16.6	0151:17.4	0.8	10	1.9	8	RH
24	0151:17.5	0151:18.7	1.2	8	2.9	15	RH
25	0151:18.7	0151:19.5	0.8	8	2.1	27	RH
26	0151:19.5	0151:20.3	0.8	8	3.1	17	RH
27	0151:20.3	0151:21.3	1.0	9	4.5	50	RH
28	0151:21.2	0151:21.9	0.7	10	2.6	25	RH
29	0151:22.0	0151:23.0	1.0	10	2.8	9	RH
30	0151:23.0	0151:23.5	0.5	8	4.4	67	RH
31	0151:23.6	0151:24.2	0.6	8	3.1	14	RH
32	0151:24.2	0151:25.1	0.9	9	3.0	18	RH
33	0151:25.1	0151:26.4	1.3	10	7.5	50	linear
34	0151:26.5	0151:27.4	0.9	8	2.7	37	RH
35	0151:27.4	0151:27.8	0.4	7	5.6	72	LH
36	0151:27.9	0151:28.8	0.9	8	1.4	85	LH
37	0151:28.8	0151:29.3	0.5	8	4.7	43	RH
38	0151:29.2	0151:30.5	1.3	15	10.3	55	RH
39	0151:30.7	0151:31.5	0.8	15	5.1	59	LH
40	0151:31.5	0151:32.3	0.8	15	2.7	50	RH
41	0151:32.4	0151:33.1	0.7	15	4.5	64	RH
42	0151:33.1	0151:33.8	0.7	15	7.4	59	linear
43	0151:33.8	0151:34.3	0.5	15	2.2	27	RH
44	0151:34.3	0151:34.9	0.6	15	2.1	64	RH
45	0151:34.9	0151:35.3	0.4	15	1.5	50	RH
46	0151:35.3	0151:36.1	0.8	15	8.2	51	RH
47	0151:36.1	0151:36.6	0.5	15	3.3	56	RH
48	0151:36.8	0151:37.1	0.3	15	7.8	44	RH
49	0151:37.2	0151:37.8	0.6	15	8.2	26	RH
50	0151:37.9	0151:38.6	0.7	15	1.3	59	RH
51	0151:38.6	0151:39.2	0.6	15	3.8	66	linear
52	0151:39.5	0151:39.8	0.3	15	2.4	20	RH

53	0151:39.9	0151:40.6	0.7	15	2.0	52	RH
54	0151:40.7	0151:41.4	0.7	15	1.8	38	RH
55	0151:41.5	0151:41.9	0.4	15	12.2	80	linear
56	0151:41.9	0151:42.5	0.6	15	3.3	88	RH
57	0151:42.5	0151:42.8	0.3	15	1.7	36	RH
58	0151:42.8	0151:43.2	0.4	15	1.6	77	RH
59	0151:43.2	0151:44.1	0.9	15	2.7	87	linear
60	0151:44.1	0151:44.8	0.7	25	8.4	76	RH
61	0151:44.8	0151:45.3	0.5	20	2.2	77	RH
62	0151:45.4	0151:45.9	0.5	20	2.7	52	LH
63	0151:46.0	0151:46.7	0.7	20	3.0	46	RH
64	0151:46.8	0151:47.1	0.3	30	4.3	53	RH
65	0151:47.4	0151:48.1	0.7	35	6.6	15	RH
66	0151:48.2	0151:48.6	0.4	35	8.2	19	RH
67	0151:49.2	0151:49.7	0.5	35	192.4	81	-

Table 2. Magnetosheath Waves. Day 230, 1999.

	Start Time	Stop Time	T	$\|\boldsymbol{B}\|$	λ_1/λ_2	θ_{KB}	Polarization
1	0152:02.6	0152:17.4	3.7	25	1.8	72	-
	0152:02.6	0152:06.6	-	-	16.9	71	linear
	0152:06.6	0152:09.8	-	-	3.4	73	-
	0152:09.8	0152:12.9	-	-	4.1	52	LH
	0152:12.9	0152:17.4	-	-	1.4	70	LH
2	0154:07.8	0154:20.7	4.3	27	1.9	32	LH
	0154:07.8	0154:12.2	-	-	11.6	56	LH
	0154:12.2	0154:16.1	-	-	1.5	31	LH
	0154:16.2	0154:20.7	-	-	5.1	50	LH
3	0157:03.1	0157:24.4	5.3	31	2.0	15	LH
	0157:03.1	0157:07.7	-	-	2.6	6	LH
	0157:07.8	0157:12.9	-	-	4.0	13	LH
	0157:13.0	0157:18.8	-	-	5.1	18	LH
	0157:18.9	0157:24.4	-	-	11.1	31	LH elliptical
4	0201:22.9	0201:45.4	5.6	32	3.2	24	LH
	0201:22.9	0201:28.7	-	-	4.5	3	LH
	0201:28.7	0201:34.4	-	-	4.0	49	LH
	0201:34.4	0201:39.3	-	-	10.8	79	LH
	0201:39.4	0201:45.4	-	-	3.0	16	LH
5	0204:10.3	0204.33.8	5.9	31	1.3	11	LH
	0204:10.3	0204:16.5	-	-	7.2	11	LH
	0204:16.5	0204:21.3	-	-	3.5	37	LH
	0204:21.6	0204:27.0	-	-	5.6	28	LH
	0204:27.0	0204:33.8	-	-	2.2	9	LH
6	0205:06.2	0205:29.2	5.8	33	2.8	32	-
	0205:06.2	0205:10.9	-	-	2.0	45	LH
	0205:10.9	0205:15.5	-	-	5.3	42	LH
	0205:15.5	0205:21.3	-	-	3.0	75	RH
	0205:21.6	0205:29.2	-	-	1.4	10	RH
7	0205:29.4	0205:55.0	6.4	35	2.1	6	LH
	0205:29.4	0205:36.1	-	-	1.7	28	LH
	0205:36.1	0205:41.9	-	-	2.1	56	LH
	0205:41:9	0205:48.6	-	-	4.9	12	LH
	0205:48.6	0205:55.0	-	-	1.8	78	LH
8	0205:55.0	0206:21.2	5.2	32	1.9	15	LH
	0205:55.0	0206:00.0	-	-	3.5	14	LH
	0206:00.0	0206:06.3	-	-	2.4	20	LH
	0206:06.3	0206:11.1	-	-	8.5	32	LH
	0206:11.1	0206:16.4	-	-	9.9	81	linear
	0206:16.4	0206:21.2	-	-	2.0	31	LH
9	0206:21.2	0206:48.7	6.9	34	1.8	12	LH
	0206:21.2	0206:27.8	-	-	2.3	84	LH
	0206:35.4	0206:42.5	-	-	1.7	19	LH
	0206:42.5	0206:48.7	-	-	1.6	9	LH
10	0206:48.7	0207:17.0	7.1	35	1.2	30	LH
	0206:48.7	0206:56.6	-	-	2.3	24	linear
	0206:56.6	0207:04.4	-	-	2.4	50	LH
	0207:04.4	0207:11.6	-	-	1.4	11	LH/linear
	0207:11.6	0207:17.0	-	-	1.8	13	LH
11	0207:17.0	0207:37.6	6.9	35	3.3	15	LH
	0207:17.0	0207:22.8	-	-	4.1	63	LH
	0207:22.8	0207:29.8	-	-	3.8	80	LH

	0207:29.8	0207:37.6	-	-	1.3	7	LH
12	0207:37.6	0207:58.5	7.0	35	2.7	90	-
	0207:37.6	0207:44.2	-	-	2.0	50	LH
	0207:44.2	0207:50.7	-	-	2.1	77	linear
	0207:50.7	0207:58.5	-	-	6.2	69	linear
13	0207:58.5	0208:24.7	6.6	35	1.1	26	LH
	0207:58.5	0208:03.7	-	-	7.0	48	LH
	0208:03.7	0208:10.2	-	-	1.9	9	linear
	0208:10.2	0208:18.0	-	-	2.2	34	LH
	0208:18.0	0208:24.7	-	-	2.7	79	LH
14	0208:24.7	0208:49.7	5.0	34	2.3	52	LH
	0208:24.7	0208:28.9	-	-	5.2	69	LH
	0208:28.9	0208:33.1	-	-	4.7	27	LH
	0208:33.1	0208:39.8	-	-	1.7	44	LH
	0208:39.8	0208:44.1	-	-	4.9	60	linear
	0208:44.2	0208:49.7	-	-	11.5	43	LH/elliptical
15	0208:49.7	0209:17.5	6.9	35	1.7	52	-
	0208:49.7	0208:53.6	-	-	2.6	60	linear
	0208:53.6	0209:05.3	-	-	1.3	64	LH
	0209:05.6	0209:11.8	-	-	4.5	61	LH
	0209:11.8	0209:17.5	-	-	3.4	78	RH
16	0209:17.5	0209:39.0	7.2	35	2.1	63	-
	0209:17.5	0209:24.7	-	-	3.1	65	RH
	0209:25.3	0209:30.0	-	-	2.9	31	LH
	0209:30.0	0209:39.0	-	-	3.9	32	LH
17	0209:39.0	0210:01.5	7.5	35	2.7	50	LH
	0209:39.0	0209:47.6	-	-	2.9	42	-
	0209:47.6	0209:56.5	-	-	3.8	81	LH
	0209:56.5	0210:01.5	-	-	1.4	40	LH
18	0212:27.1	0212:45.5	6.1	36	2.7	11	LH
	0212:27.1	0212:32.7	-	-	8.9	35	LH
	0212:32.7	0212:37.9	-	-	6.1	66	RH
	0212:37.9	0212:45.5	-	-	1.6	35	LH
19	0220:18.3	0220:31.5	4.4	46	5.0	2	LH
	0220:18.3	0220:22.2	-	-	1.8	14	LH
	0220:22.2	0220:25.7	-	-	2.7	9	LH
	0220:25.7	0220:31.5	-	-	33.0	1	LH

Table 3. Magnetospheric Waves. Day 230, 1999.

	Start Time	Stop Time	T	$\|\boldsymbol{B}\|$	λ_1/λ_2	θ_{KB}	Polarization
1	0229:15.2	0229:18.3	1.5	63	1.9	29	LH
	0229:15.2	0229:16.7	-	-	3.4	32	LH
	0229:16.7	0229:18.3	-	-	1.6	26	LH
2	0231:27.1	0231:33.9	2.3	60.7	4.0	7	LH
	0231:27.1	0231:29.3	-	-	1.5	13	LH
	0231:29.3	0231:31.6	-	-	27.1	18	LH
	0231:31.6	0231:33.9	-	-	8.3	7	LH
3	0232:29.9	0232:46.1	2.7	61.4	5.5	41	linear
	0232:29.9	0232:32.7	-	-	20.2	23	LH elliptical
	0232:32.7	0232:35.4	-	-	42.9	17	LH elliptical
	0232:35.5	0232:38.1	-	-	174.8	62	linear
	0232:38.2	0232:41.1	-	-	106.4	37	linear
	0232:41.1	0232:43.6	-	-	24.7	71	LH elliptical
	0232:43.6	0232:46.1	-	-	21.4	20	linear
4	0233:40.0	0233:51.1	2.8	63.6	1.6	21	LH
	0233:40.0	0233:42.7	-	-	5.2	18	LH
	0233:42.8	0233:45.2	-	-	2.7	31	LH
	0233:45.3	0233:48.0	-	-	2.5	23	LH
	0233:48.1	0233:51.1	-	-	2.4	23	LH
5	0237:30.5	0237:43.4	2.6	69.1	2.9	9	LH
	0237:30.5	0237:33.3	-	-	2.5	6	LH
	0237:33.5	0237:35.8	-	-	4.0	10	LH
	0237:35.8	0237:38.1	-	-	2.8	10	LH
	0237:38.1	0237:40.8	-	-	9.5	13	LH
	0237:40.8	0237:43.4	-	-	10.0	17	linear

MODELS FOR THE SIZE AND SHAPE OF THE EARTH'S MAGNETOPAUSE AND BOW SHOCK

J. K. Chao[1], D. J. Wu [1,2], C.-H. Lin[1,6], Y.-H. Yang[1], X.Y. Wang [1,3], M. Kessel [4], S.H. Chen [4], and R.P. Lepping [5]

[1] *Institute of Space Science, National Central University, Chung-li, Taiwan 32001*
[2] *Purple Mountain Observatory, Academia Sinica, Nanjing 210008*
[3] *Center for Space Science and Applied Research, Academia Sinica, Beijing 100080*
[4] *NSSDC, GSFC, NASA, Greenbelt, MD 20771*
[5] *Planetary Magnetosphere Branch, Laboratory for Extraterrestrial Physics GSFC/NASA, Greenbelt, MD 20771*
[6] *Department of Electrical Engineering, Nan-Jeon Institute of Technology, Yen-Shui, Taiwan 737*

ABSTRACT

New models for the size and shape of the Earth's magnetopause and bow shock are derived, based on a criterion for selecting the crossing events and their corresponding up-stream solar wind parameters. In this study, we emphasize the importance of accurate interplanetary parameters for predicting the size and shape of the magnetopause and bow shock. The time lag of the solar wind between the solar wind monitor and the location of crossings is carefully considered, ensuring more reliable up-stream solar wind parameters. With this database new functional forms for the magnetopause and bow shock surfaces are deduced. In this paper, we briefly present the preliminary results. For a given up-stream solar wind dynamic pressure D_p, an IMF north–south component B_z, a solar wind β and a magnetosonic Mach number M_{ms}, the parameters that describe the magnetopause and bow shock surfaces r_0 and α can be expressed in terms of a set of coefficients determined with a multi-parameter fitting. Applications of these models to extreme solar wind conditions are demonstrated. For convenience, we have assumed that r_0, B_z and D_p retain their units, except in equations where they are normalized by 1 R_E(Earth radius), 1 nT and 1 nPa, respectively.

INTRODUCTION

The magnetopause (MP) and bow shock (BS) play an important role in protecting the Earth from the harmful effects of the solar wind. Generally speaking, subsolar points of the MP and BS are located at about 10 and 15 R_E, respectively, from the Earth under normal solar wind conditions. *Chapman and Ferraro* [1931] first suggested the existence of an MP boundary. Subsequently, many MP models have been proposed [*Fairfield*, 1971; *Holzer and Slavin*, 1978; *Sibeck et al.*, 1991; *Petrinec et al.*, 1991; *Petrinec and Russell*, 1993, 1996; *Roelof and Sibeck*, 1993; *Shue at al.*, 1997, 1998; *Kuznetsov and Suvorova*, 1998; *Kawano et al.*, 1999]. Most MP models use either a general equation of an ellipsoid with two parameters (the eccentricity and standoff distance) or a general quadratic equation. *Cairns et al.* [1995] discussed the limitations of using an elliptic equation. Recently, *Shue et al.* [1997] presented a new function to fit the size and shape of the MP:

$$r = r_0\left(\frac{2}{1+\cos\theta}\right)^{\alpha} \qquad (1)$$

where r_0 and α are the standoff distance and the level of tail flaring, respectively, and (r, θ) are polar coordinates in the ecliptic plane with the origin at the Earth's center and the axis along the Sun–Earth line. The model is derived from crossings near the ecliptic plane. Thus, the model may not be suitable for use with high-latitude crossings. For high-latitude crossings, the model of *Boardsen et al.* [2000] can be applied. In the present study, we use the same functional form as *Shue et al.* [1997]. However, the database differs from that of their model by using a more accurate time shift of the solar wind propagating from IMP 8 (or ISEE 3) to the Earth. So that the database can be used for extreme solar wind conditions, the MP crossings by geosynchronous satellites are included. We will discuss the possible bias due to the satellite's particular orbits and how it can be minimized in our model.

The study of the large-scale size and shape of the Earth's BS has a long history [see for example *Beard*, 1960, 1962; *Midgley and Davis*, 1962; *Spreiter et al.*, 1966; *Dryer and Heckman*, 1967; *Gosling et al.*, 1967; *Russell et al.*, 1968; *Behannon*, 1968; *Binsack and Vasyliunas*, 1968; *Egidi et al.*, 1970; *Fairfield*, 1971; *Formisano et al.*, 1971, 1973; *Formisano et al*, 1973; *Spreiter and Rizzi*, 1974; *Villante*, 1976; *Russell*, 1985; *Formisano*, 1979; *Slavin and Holzer*, 1981; *Slavin et al.*, 1984; *Greenstadt et al.*, 1990; *Farris et al.*, 1991; *Nemecek and Safrankova*, 1991; *Cairns and Grabbe*, 1994; *Farris and Russell*, 1994; *Cairns and Lyon*,1995; *Cairns et al.*, 1995, 1996; *Peredo et al.*, 1995; *Lepidi et al.*, 1996; *Russell and Petrinec*, 1996; *Slavin et al.*, 1996; *Bennett et al.*, 1997, and references therein]. On the theoretical side, the studies are mainly based on numerical simulations using gas dynamic and hydromagnetic approaches [*Spreiter et al.*, 1966; *Dryer and Heckmann*, 1967; *Spreiter and Rizzi*, 1974; *Cairns and Lyon*, 1995; *Song et al.*, 1999]. Much attention has been given to the fitting and empirical modeling of the BS boundaries. The purpose of those works was to test and refine theoretical models of the solar wind flow about the magnetosphere. This boundary is readily identifiable and the observed position and shape can be used to validate theoretical predictions. In addition, the development of empirical models can meet operational needs and the contributions of such models may improve our understanding of the solar wind/magnetosphere coupling which is important for space weather studies. For our BS model we use a similar functional form to that used by *Shue et al.* [1997] for the MP (i.e. the same form as Eq. 1). The parameters r_0 and α that describe the BS surface are functions of B_z, D_p, β and M_{ms}, determined with a multi-parameter fitting. In the following sections we describe our derived models of the MP and BS.

A MAGNETOPAUSE MODEL

We have improved the database of MP crossings used by *Shue et al.* [1997, 1998] based on the following considerations. In the old database, the time shift of solar wind propagating from IMP 8 (or ISEE 3) to the Earth is assumed to be constant for simplicity (10 minutes for IMP 8 and 50 minutes for ISEE 3). However, the speed of the solar wind can change substantially in a 10-minute (or 50-minute) interval and hence the transient time can vary by a large amount. This transient effect appears to be very important especially for those MP crossings that occur near the Earth. Since these crossings are associated almost exclusively with sudden changes in the solar wind speed and density, using only a default time delay would lead to solar wind conditions different from the actual enhancements. Therefore, we use the actual satellite positions and the measured solar wind speed to estimate a more accurate time shift for the database. In the flank region of the magnetosphere, the BS is so weak that it cannot be detected easily. Thus, the data from IMP 8 could have been taken incorrectly as the solar wind values when the satellite was in fact in the magnetosheath. Such a situation is eliminated in the new database.

Finally, 552 crossings have been selected. The ranges of the solar wind parameters are $-15 < B_z < 15$ nT and $0.4 < D_p < 9$ nPa. All crossings have been corrected for aberration in the GSM coordinate system. Our model has the same functional form as that of *Shue et al.* [1997], but the way in which r_0 and α depend on B_z and D_p is different. The predicted errors in our model are approximately 10% less than those in the *Shue et al.* [1998] model. In order to distinguish magnetopause responses under normal and extreme solar-wind conditions, the dependence of r_0 is derived separately. Those observed crossings with $r > 6.7\ R_E$ and $r \le 6.7\ R_E$ are used to derive a relationship for normal and extreme solar-wind conditions, respectively. Then these relationships are used as the models for normal and extreme solar-wind conditions with $r_0 \ge 7.0$ R_E and $r_0 \le 6.4\ R_E$, respectively. The reason for choosing $r_0 \ge 7.0\ R_E$ as the boundary of normal solar wind conditions is somewhat arbitrary. Since all three models under consideration can predict the magnetopause crossings very well for $r_0 \ge 7.0\ R_E$, we use one model to describe this region. Because the r_0 vs D_p dependence is very different in the regions $r > 6.7$ R_E and $r \le 6.7\ R_E$, we choose $r_0 \le 6.4\ R_E$ as the boundary for extreme solar wind conditions. This choice is obtained through an iterative procedure, described in the latter part of this section. A continuous and smooth variation of r_0 is required between 6.4 and 7.0 R_E. It is found that if we assume $\ln(r_0)$ vs D_p is linear in the region $6.7 < r_0 < 7.0\ R_E$, but r_0 vs D_p is linear in the region $6.4 < r_0 \le 6.7\ R_E$, then we have a continuous and smooth transition such that both r_0 and dr_0/dD_p are continuous functions of D_p, while dr_0/dD_p is always negative from normal to extreme conditions. The total errors from the best fit with interpolation are found to remain the same as that without interpolation. Then, our model is obtained as follows:

$$r0 = \begin{cases} (a_1)(D_p)^{-1/a_4} & for \quad B_z \ge 0 \\ (a_1 + a_2 B_z)(D_p)^{-1/a_4} & for \quad -8 \le B_z < 0 \\ (a_1 + 8a_3 - 8a_2 + a_3 B_z)(D_p)^{-1/a_4} & for \quad B_z < -8 \end{cases} \tag{2}$$

$$\alpha = (a_5 + a_6 B_z)(1 + a_7 D_p)$$

In deriving our model, the following three assumptions have been made. First, the functional forms in terms of D_p and B_z for normal and extreme conditions are assumed to be the same, except that the coefficients are different. Second, the subsolar distances r_0 for $B_z > 0$ depend only on D_p with a power index equal to $-1/a_4$, where a_4 can differ between normal and extreme conditions. Third, the coefficients a_1, a_5, a_6 and a_7 do not change between normal and extreme conditions. With these assumptions, we derive the coefficients a_1 = 11.646, a_2 = 0.216, a_3 = 0.122, a_4 = 6.215, a_5 = 0.578, a_6 = - 0.009 and a_7 = 0.012 for the normal solar wind conditions (i.e. $r_0 \ge 7.0\ R_E$). The standard deviation (SD) from the best fit for the normal solar wind conditions is 1.14 R_E. It is also found that the coefficients for extreme conditions (i.e. $r_0 < 6.4\ R_E$) are $a_1 = 11.646$, $a_2 = 0.169$, $a_3 = 0.158$, $a_4 = 6.800$, $a_5 = 0.578$, $a_6 = -0.009$ and $a_7 = 0.012$, with SD = 0.55 R_E from the best-fit procedure. For $B_z > 0$, the function for region $6.7 < r_0 < 7.0\ R_E$ is $\ln(r_0) - \ln(7) = C_1(D_p - C_2)$, where $C_1 = -0.003$ and $C_2 = 23.7$. The function for $6.4 < r_0 < 6.7\ R_E$ is $r_0 - 6.4 = C_3(D_p - C_4)$, where $C_3 = -0.015$ and $C_4 = 58.6$ calculated from the first equation of Eq. (2). As for $B_z < 0$, the coefficients C_1, C_2, C_3 and C_4 can also be calculated from the other equations in Eq. (2) which have a B_z dependence.

In deriving the model for the extreme solar wind conditions, the MP crossings observed by geosynchronous satellites are taken into consideration. The possible bias due to the satellite's particular orbits may be minimized in deriving a model for $r_0 = 6.6\ R_E$ if we use the following procedure. Figure 1 shows the predicted contour plots of $r_0 = 6.6\ R_E$ for various MP models. The models of *Shue et al.* [1998] and *Petrinec and Russell* [1996] are shown as dashed and dotted lines, respectively. The solid line between the dashed and dotted lines is the prediction of $r_0 = 6.6\ R_E$ from our model for the normal solar wind condition. The crosses in the figure are the interplanetary D_p and B_z values associated with those crossings that were observed by GOES satellites, implying that these MP have already moved inside the geosynchronous orbit. By inspecting the positions of these three contours with respect to the crosses, it is apparent that the predicted $r_0 = 6.6\ R_E$ contours are slightly outside the boundaries of the crosses. Ideally, the contours should be as close as possible to the outer edge of the crosses.

Intuitively, a better model may be obtained if the contour $r_0 = 6.6\ R_E$ is moved toward the boundary of these crosses. So that the model does not over-predict and a false alarm can be minimized, we need the contour of $r_0 = 6.6\ R_E$ to be well inside the outer boundary of the crosses. Once the location of $r_0 = 6.6\ R_E$ is determined, we anticipate that our model will cover the region $r_0 > 6.4\ R_E$. In deriving the model for the extreme condition (i.e. $r_0 < 6.7\ R_E$), we first require that the contour $r_0 = 6.6\ R_E$ is as close as possible to the boundary of the crosses. Secondly, we require a smooth transition from the normal to extreme conditions of r_0 in the region (7.0, 6.4) in which a linear $\ln(r_0)$ vs D_p relationship is assumed for the interval (7.0, 6.7), but a linear r_0 vs D_p applies in interval (6.7, 6.4). This procedure requires iteration.

We have a model covering the ranges of both the normal and extreme solar wind conditions. Since there were no crossings available for $r_0 < 6.4\ R_E$ in the derivation of our model, the extreme solar wind condition cannot be applied for regions much less than $r_0 = 6.6\ R_E$. Hence, our prediction is not reliable in the region $r_0 < 6.4\ R_E$. The contour $r_0 = 6.6\ R_E$ thus obtained is shown in Figure 1 as the solid line on the far right-hand side. The prediction of our model is now quite different from that of the other two models. However, under normal solar wind conditions, all three models predict very similar values of r_0. Note that no change of the a coefficients for α is assumed from the normal to extreme conditions

The contours of r_0 and α for the normal and extreme solar wind conditions are given in Figures 2(a) and 2(b). Under extreme solar wind conditions, *Yang et al.* [2001] have shown that our model predicts more accurate geosynchronous MP crossings, as given in Table 1, where the three parameters are defined: probability of prediction (PoP), probability of detection (PoD) and false alarm rate (FAR). These parameters quantify the forecasting capability of a model. Higher PoP and PoD values with a lower FAR imply a better forecasting model. It is evident that our model gives a better prediction.

Table 1. Comparison of prediction capabilities for three models from 1986-1992(unit: %)

Model	PoP	PoD	FAR
Yang et al. [2001]	79	96	69
Shue et al. [1998]	67	93	78
Petrinec and Russell [1996]	45	97	85

The models of Petrinic and Russell (1996) and Chao et al. (1997) predict that a larger southward value of B_z can push the magnetopause further closer to the Earth regardless how large the value of the negative B_z is. On the other hand, the models of *Shue et al.* [1998] and *Kuznetsov and Suvorova* [1998] suggest that the value of the negative Bz has a limit. In addition, the sheath encounters shown in Figure 6 (pluses) of *Yang et al.* [2001] seem to suggest that the negative B_z cannot further reduce the magnetopause distance when B_z become less than a certain negative value. We describe the B_z influence as 'saturated' [*Dmitriev et al.*, 2001]. From an event study, we also found that the saturation can be approximated by $B_z = (-12 - D_p)$ if we have $B_z < (-12 - D_p)$. The parameter α in Eq. (1) controls the level of tail

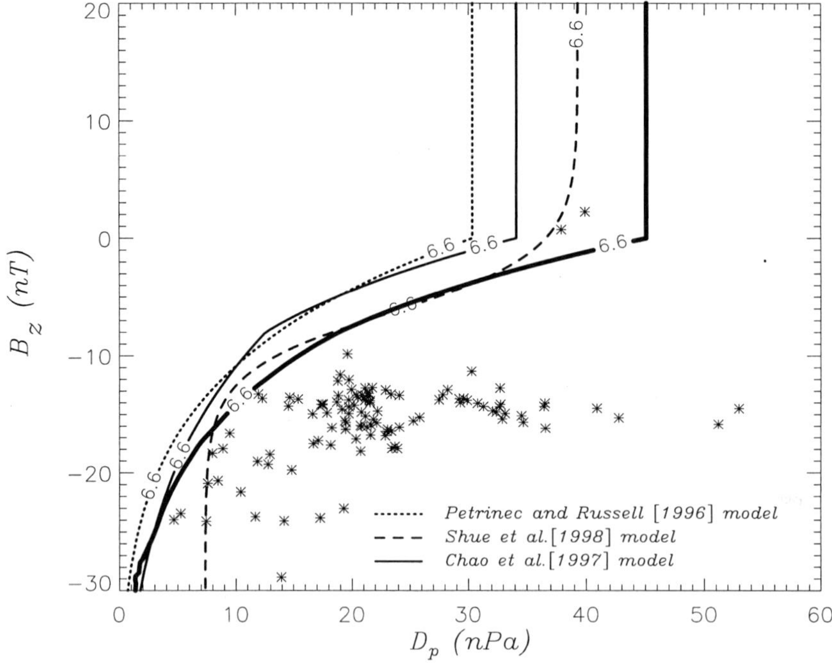

Fig. 1. Predicted contour plots of r_0 = 6.6 Re for various magnetopause models. Models of Shue et al. [1998] and Petrinec and Russell [1996] are shown as dashed and dotted lines, respectively. The solid line between the dashed and dotted lines is the extrapolated values for r_0 = 6.6 Re from our model in the normal solar wind condition. The other solid line is from our present model. The cross-signs show the D_p and B_z values when GOES's are inside the magnetosheath.

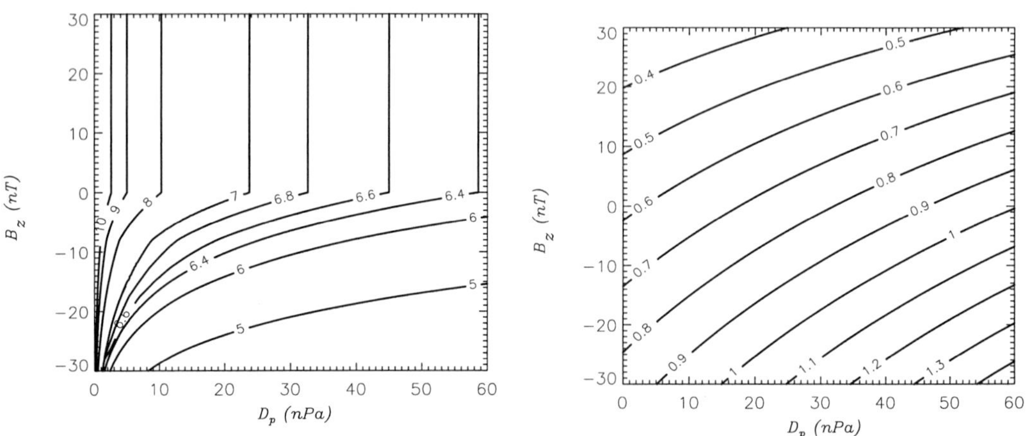

Fig. 2. Left: Contour plots for r_0 as a function of D_p and B_z for both normal and extreme solar wind conditions(Chao, 1997). Right: Contour plots for α as a function of D_p and B_z for our new model (Chao, 1997).

flaring of the magnetopause. However, the contour plots of α in Figure 2 are derived from crossings under normal solar wind conditions that may not be reasonable for large D_p. Therefore, we suggest that the dependence of α on D_p should gradually diminish when D_p approaches a large value. Thus, we assume that the coefficient $a_7 = a_7{}^*\exp(-D_p/10)$. Eventually, these two effects will be considered in detail in future studies.

A BOW SHOCK MODEL

To develop an empirical model for the size and shape of the Earth's bow shock, the selection of the BS crossings and their associated upstream parameters are important. The solar wind parameters are not constant during the periods of the BS crossing. A sharp change in solar wind parameters often causes the crossings. The hourly average values of the solar wind are included in the BS database in some earlier work, but they may not be correct. The actual values responsible for the BS crossings can be quite different from the hourly average values. When there are sharp changes in the solar wind parameters, the BS will move from a previous equilibrium to a new one. Since the satellite's location is taken as the BS equilibrium position when the satellite passes by, the real equilibrium position of the BS is not properly obtained.

To derive a BS database, we have used high-resolution data from the ISTP-Key parameters of WIND and GEOTAIL. These BS crossings in quasi-steady conditions are chosen for our BS database. In one of the following two cases, with upstream solar wind conditions, we have selected the two BS crossings:

(1) Only one BS crossing is observed and occurs slowly.
(2) In the case of multiple crossings, only the middle one is selected to represent and weighted all crossings.

The crossings that are apparently transient events caused by sharp changes of solar wind parameters have not been selected for our database. To demonstrate the selection of quasi-steady BS and their corresponding upstream solar wind parameters, we present two examples. In Figure 3, the solar wind parameters observed by WIND (solid curves) and the magnetic fields observed by GEOTAIL (dashed curves) during 0700–1400UT, April 11, 1996 are shown. At this time, WIND is located at ~80 R_E upstream from the GEOTAIL and observed solar wind velocity at ~425 km/sec. The times of GEOTAIL have been properly shifted by an amount equal to solar wind travel-times from WIND to GEOTAIL.

From the top to bottom panels of Figure 3, the magnitudes of the magnetic field B_t and its two components B_y and B_z, together with D_p and M_{ms}, are shown. By comparing the solid and the dashed curves, it is possible to identify GEOTAIL's repeated crossings of the BS. We select two crossings observed under steady-state solar wind conditions in this figure. During these two periods, the solar wind parameters were very stable, as can be seen in the figure. Therefore we suggest that the observed BS is at equilibrium for the corresponding solar wind conditions.

In our second example, multiple crossings observed by GEOTAIL are shown in Figure 4. Multiple crossings occurring with quasi-steady upstream conditions indicate that the BS is close to a steady state. We take the middle crossing as our selected event. At this time, WIND is 231.7 R_E upstream of the Earth observed at a solar wind velocity of ~640 km/sec. Thus, the solar wind traveling time is approximately 38 minutes. We take this to be the time shift for GEOTAIL as shown by the dashed curves in Figure 4. Note that the first BS crossing at about 16:45 UT cannot be used for our database because the fast crossing by GEOTAIL indicates that the BS has moved out too far from the location of GEOTAIL. With this constraint, only 265 such events were selected for the period 1995–97 from GEOTAIL using WIND as the solar wind monitor. With proper weight added to the events, we finally obtained an equivalent of 1438 crossings for our fitting to derive a BS model.

The size and shape of the BS surface in aberrated polar coordinates with the origin at the Earth's center is described by a nonlinear function similar to Eq. (1) for the MP:

$$r = r_0\left(\frac{1+\varepsilon}{1+\varepsilon\cos\theta}\right)^{\alpha} \qquad (3)$$

where r is the radial distance from the Earth to the predicted BS along the Earth–satellite line, r_0 is the standoff distance of the BS, θ is the cone angle from the x-axis, ε is a parameter similar to the eccentricity, and α is the level of tail flaring.

In order that our BS model is consistent with the results of the distant crossings (*Bennett et al.*, 1997), the parameter ε has been included in Eq. (3). The parameters r_0 and α are both functions of B_z, D_p, β and M_{ms}. The BS with aberration corrected is treated to be axi-symmetric. The effect of the y-component of solar wind velocity V_y is also considered in the aberration. The BS model is obtained by using a multi-parameter fitting scheme. The optimization is carried out by a chi-squared minimization of the difference between model predictions and observations. The best-fit results for the parameters r_0, α and ε can be expressed as the following two sets of equations, depending on the sign of B_z (*Chao et al.*, 1999).

$$
\begin{aligned}
r_0 &= a_1(1 + a_2 B_z)(1 + a_9 \beta)(1 + a_4 \frac{(a_8 - 1)M_{ms}^2 + 2}{(a_8 + 1)M_{ms}^2})D_p^{-1/a_{11}} \quad \text{for } B_z \geq 0 \\
\alpha &= a_5(1 + a_{13} B_z)(1 + a_7 D_p)[1 + a_{10}\ln(1 + \beta))](1 + a_{14} M_{ms}) \\
r_0 &= a_1(1 + a_3 B_z)(1 + a_9 \beta)(1 + a_4 \frac{(a_8 - 1)M_{ms}^2 + 2}{(a_8 + 1)M_{ms}^2})D_p^{-1/a_{11}} \quad \text{for } B_z < 0 \\
\alpha &= a_5(1 + a_6 B_z)(1 + a_7 D_p)[1 + a_{10}\ln(1 + \beta)](1 + a_{14} M_{ms})
\end{aligned}
\tag{4}
$$

where $\varepsilon = a_{12}$ and the coefficients are shown below:

$a_1 = 11.1266$	$a_2 = 0.0010$	$a_3 = -0.0005$	$a_4 = 2.5966$	$a_5 = 0.8182$
$a_6 = -0.0170$	$a_7 = -0.0122$	$a_8 = 1.3007$	$a_9 = -0.0049$	$a_{10} = -0.0328$
$a_{11} = 6.047$	$a_{12} = 1.029$	$a_{13} = 0.0231$	$a_{14} = -0.002$	

The errors from the best-fit procedure give a standard deviation of 1.2 R_E. The functional dependences of r_0 and α on the parameters D_p, B_z, β and M_{ms} are given in Figure 5.

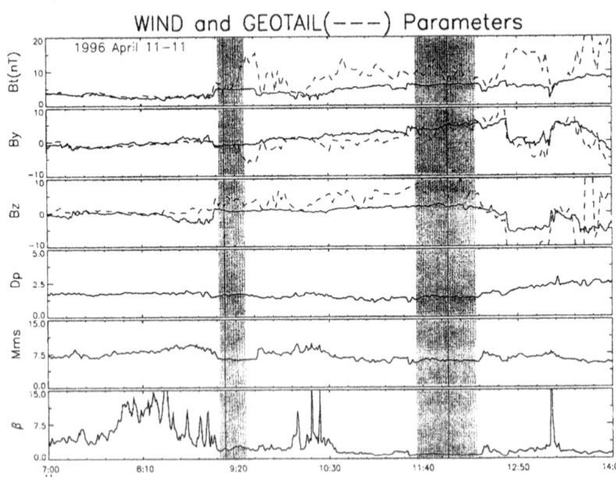

Fig. 3. Two bow shock crossings (shaded area) observed by the GEOTAIL satellite (dashed lines) at (− 7.0, 27.5, − 2.9) in GSE coordinates on April 11, 1996. The satellite passes through the shock gradually indicating that the shock is in an equilibrium position, whereas the upstream parameters (solid lines) observed by WIND during the same periods show little change. From top to bottom, the magnetic field and its two components in GSE coordinates, the solar wind dynamic pressure D_p, the magneto-sonic Mach number M_{ms}, and the ratio of thermal to magnetic pressure β are shown.

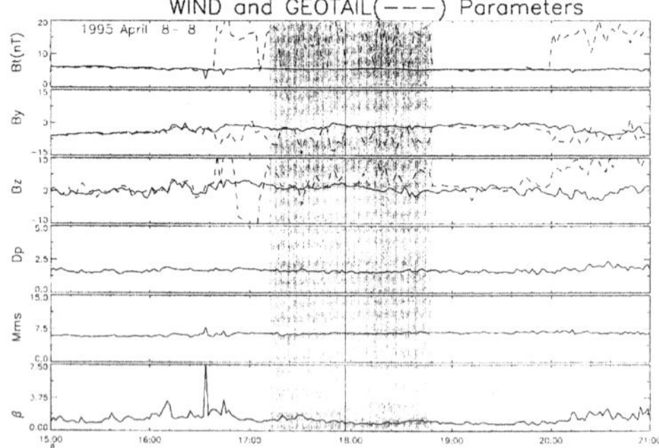

Fig. 4. Multiple bow shock crossings (shaded area) observed by GEOTAIL (dashed lines) at (0., 28., -3.5) in GSE coordinates on April 8, 1995. The satellite GEOTAIL observed multiple bow shock crossings whereas the upstream parameters (solid lines) observed by WIND remain unchanged during this period indicating the bow shock is in equilibrium position. The same parameters as in Figure 1 are shown from the top to the bottom.

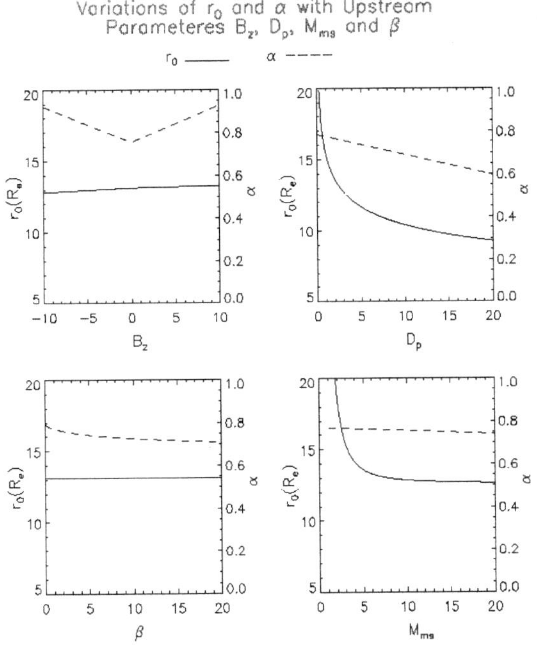

Fig. 5. Plots of r_0 (solid lines) and α (dashed lines) as functions of D_p, B_z, M_{ms} and β for the normal solar wind conditions, where $B_z = -0.35$ nT, $D_p = 2.48$ nPa, $M_{ms} = 6.96$ and $\beta = 2.08$.

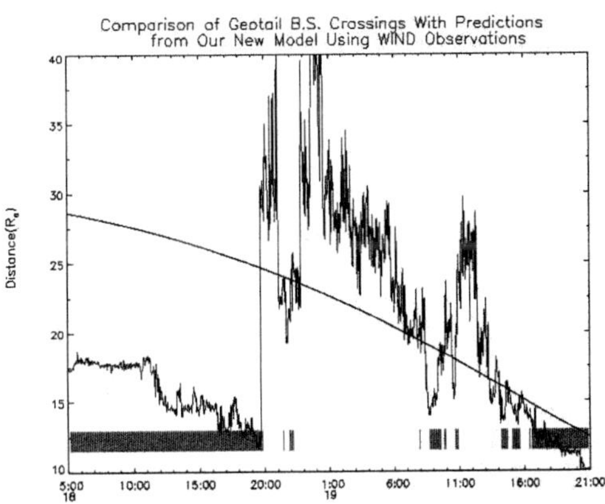

Fig. 6. Predicted distances r of the bow shock (fluctuated line) and the GEOTAIL's orbit (smooth line) from the Earth's center (in units of R_E). The horizontal bars in the lower part of the figure indicate the actual positions of GEOTAIL in interplanetary space.

To demonstrate the capability of the model for predicting the position and shape of the Earth's BS, we give the following example. During the period October 18–20, 1995, an interplanetary magnetic cloud (IMC) was observed to pass the Earth. WIND recorded its interplanetary characteristics at 175 R_E upstream of the Earth's BS, and about 45 minutes later GEOTAIL, being near the nominal location of the dawn-side of the BS, detected multiple BS crossings. Using the data from WIND, we predicted the changes in position and shape of the BS from our semi-empirical BS model, caused by the interaction of the IMC with the magnetosphere. Figure 6 shows a comparison of the BS motion predicted by the model with GEOTAIL observations of the actual BS crossings. Both the distance of GEOTAIL from the Earth's center (i.e. corresponding to GEOTAIL's trajectory) and the distance of the BS from the Earth's center along the Earth-GEOTAIL direction, as predicted from our semi-empirical model at a time resolution of 1.5 minutes, are shown as a function of time. The transient time from WIND to GEOTAIL has been correctly shifted. Our prediction of the BS crossings is shown as the intersection points of the predicted BS distances (solid lines) and the trajectory of GEOTAIL. The horizontal bars in the lower part of the figure indicate the actual positions (or periods) of GEOTAIL being in interplanetary space; otherwise, it is in the magnetosheath. Thus, the positions when GEOTAIL changes from one region to the other represent the observed crossings by GEOTAIL. It is apparent from this figure that our predictions agree very well with observation.

DISCUSSION AND SUMMARY

In our derivation of the models for both the MP and BS, we have assumed that the subsolar distances are proportional to D_p to a power of $-1/a$, as in Eqs (2) and (4) where in theory $a = 6.0$ for an ideal dipole magnetosphere. The empirically determined a values for both the MP and BS are close to 6.0, indicating our assumed functional form of D_p is close to the ideal dipole result. For the case of the MP, r_0 and α are functions only of D_p and B_z. Since the observation of a southward B_z related to inward motion of the MP has long been recognized, magnetic reconnection may play a role in the physics of this inward motion. It is possible that the $V_x B_z$ term is more important than B_z for controlling this inward motion. However, our preliminary study shows that using $V_x B_z$ instead of B_z increases the error in the fitting. Thus, we have continued to use B_z as a control parameter in this study. According to Eq. (2), the subsolar position of the MP is independent of B_z when it is northward and more strongly depends on the southward B_z (i.e. a_3 is not much smaller than 1). Since most of the crossings of the MP are on the dayside, the parameter α controlling the flaring of the MP for the tail part may not be statistically significant. More tests for the distant tail crossings are needed to verify the applicability of this model to the tail region of the MP. The subsolar distance and the flaring of the BS are controlled by the four variables D_p, B_z, β and M_{ms}.

We have also noticed that the subsolar distance r_0 of BS is weakly dependent on B_z, whereas for the MP the B_z dependence is stronger. The model has been tested successfully both for normal and for extreme solar wind conditions when the solar wind number density and Mach number can drop to much smaller than normal values [*Lin et al.*, 2000].

In summary, we have derived new models for the size and shape of the Earth's MP and BS, based on a criterion for the selection of crossing events and their corresponding up-stream solar wind parameters. Our work emphasizes the importance of accurate interplanetary parameters in making predictions. Although these models are derived using crossings under normal solar wind condition, they have also been applied to extreme solar wind conditions with reasonable success. During the coming solar maximum, more events under extreme solar wind conditions can be used to test the models

ACKNOWLEDGEMENTS

This work was supported by grant NSC 89-2111-M-008-070 to the National Central University. The authors thank NASA/NSSDC and NOAA/NGDC for data used in this study. JKC thanks A. V. Dmitriev for useful comments.

REFERENCES

Beard, D. B., The interaction of the terrestrial magnetic field with the solar corpuscular radiation, J. Geophys. Res., 65, 3559, 1960.

Beard, D. B., Interaction of the terrestrial magnetic field with the solar corpuscular radiation, 2, Second-order approximation, J. Geophys. Res., 67, 477, 1962.

Behanon, K. W., Mapping of the Earth's bow shock and magnetic tail by Explorer 33, J. Geophys. Res., 73, 907, 1968.

Bennett L., M. G. Kivelson, K. K. Khurana, L. A. Frank, and W. R. Paterson, A model of the earth's distant bow shock, J. Geophys. Res., 102, 26927, 1997.

Binsak, J. H., and V. M. Vasyliunas, Simultaneous IMP 2 and OGO 1 observations of bow shock compression, J. Geophys. Res., 73, 429, 1968.

Boardsen, S. A., T. E. Eastman, T. Sotirelis, and J. L. Green, An empirical model of the high-latitude magnetopause, J. Geophys. Res., 105, 23193-23220, 2000.

Cairns, I. H., and C. L. Grabbe, Towards an MHD theory for the standoff distance of Earth's bow shock, Geophys. Res. Lett., 21, 2781, 1994.

Cairns, I. H., and J. G. Lyon, MHD simulations of Earth's bow shock at low Mach numbers: Standoff distances, J. Geophys. Res., 100, 17173, 1995.

Cairns, I. H., D. H. Fairfield, R. R. Anderson, V. E. H. Carlton, K. I. Paularena, and A. J. Lazarus, Unusual locations of: Earth's bow shock on September 24-25, 1987: Mach number effects, J. Geophys. Res., 100, 47, 1995.

Cairns, I. H., D. H. Fairfield, R. R. Anderson, K. I. Paularena, and A. J. Lazarus, Reply to comment on 'Unusual locations of Earth's bow shock on September 24-25, 1987: Mach number effects," J. Geophys. Res., 101, 7679, 1996.

Chao, J. K., Importance of accurate interplanetary B_z and D_p for the prediction of the size and shape of Earth's magnetosphere, AGU Fall Meeting, San Francisco, CA, F592, 1997, A supplement to Eos, Transactions, AGU, vol. 78, No. 78, No. 46, Nov. 18, F592, 1997

Chao, J. K., C.-H. Lin, D. J. Wu, M. Kessel, S. H. Chen, and R. Lepping, A model for the size and shape of the Earth's bow shock, EOS, Transactions, AGU 1999 Fall Meeting, Vol. 80, No. 46, F897, 1999.

Chapman, S., and V. C. A. Ferraro, A new theory of magnetic storm, I, The initial phase, J. Geophys. Res., 36, 77, 1931.

Dmitriev, A. V., J. K. Chao, and Y.-H. Yang, Saturation of the Bz-influence for the geosynchronous magnetopause crossings, Proceedings of the Seventh Conference on Atmospheric Science in Taiwan, September 25-27, 2001

Dryer, M., and G. R. Heckman, On the hypersonic analogue as applied to planetary interaction with the solar plasma, Planet. Space Sci., 15, 515, 1967.

Egidi, A., V. Formisano, F. Palmiotto, P. Saraceno, and C. Moreno, Solar wind and location of shock front and magnetopause at the 1969 solar maximum, J. Geophys. Res., 75, 6999, 1970.

Fairfield, D. H., Average and unusual locations of the Earth's magnetopause and bow shock, J. Geophys. Res., 76, 6700, 1971.

Farris, M. H., and C. T. Russell, Determining the standoff distance of the bow shock: Mach number dependence and use of models, J. Geophys. Res., 99, 17681, 1994.

Farris, M. H., S. M. Petrinec, and C. T. Russell, The thickness of the magnetosheath: Constraints on the polytropic index, Geophys. Res. Lett., 18, 1821, 1991.

Formisano, V., Orientation and shape of the Earth's bow shock in three dimensions, Planet. Space Sci., 27, 1151, 1979.

Formisano, V., P. C. Hedgecock, G. Moreno, J. Sear,and D. Bollea, Observations of Earth's bow shock for low Mach numbers, Planet. Space Sci., 19, 1519, 1971.

Formisano, V., P. C. Hedgecock, G. Moreno, F. Palmiotto, and J. K. Chao, Solar wind interactions with the Earth's magnetic

field, 2, Magnetohydrodynamic bow shock, J. Geophys. Res., 78, 3731, 1973.
Gosling, J. T., J. R. Asbridge, S. J. Bame, and I. B. Strong, Vela 2 measurements of the magnetopause and bow shock positions, J. Gecpys. Res., 72, 101, 1967.
Greenstadt, E. W., D. P. Traver, F. V. Coroniti, E. J. Smith, and J. A. Slavin, Observations of the flank of Earth's bow shock to -110 R_E by ISEE3/ICE3, Geophys. Res. Lett., 17, 753, 1990.
Holzer, R. E., and J. A. Slavin, Magnetic flux transfer associated with expansions and contractions of the dayside magnetosphere, J. Geophys. Res., 83, 3831, 1978.
Kawano, H., S. M. Petrinec, C. T. Russell, and T. Higuchi, Magnetopause shape determinations from measured position and estimated flaring angle, J. Geophys. Res., 104, 247, 1999.
Kuznetsov, S. N., and A.V. Suvorova, An empirical model of the magnetopause for broad ranges of solar wind pressure and B_z IMF, in Polar Cap Boundary Phenomena, edited by J. Moen et al., p. 51, Kluwer Acad., Norwell, Mass., 1998.
Lepidi S., Villante U., Lazarus A. J., Szabo A., and Paularena K., Observations of bow shock motion during times of variable solar conditions, J. Geophys. Res., 101, 11107, 1996.
Lin, C.-H., J. K Chao, Y.-H. Yang, and X. Y. Wang, Predictions of the Earth's Bow Shock Crossings on May 11-12, 1999, COSPAR Colloquium, Space Weather Study Using Multi-point Techniques, September 27-29, 2000.
Midgley, J. E. and L. Davis Jr., Computation of the bounding surface of a dipole field in a plasma by a moment technique, J. Geophys. Res., 67, 499, 1962.
Nemecek, Z. and J. Safrankova,The Earth's shock and magnetopause position as a result of solar wind - magnetosphere interaction, J. of Atmospheric and Terr. Phys., 53, 1049, 1991.
Peredo, M., J. A. Slavin, E. Mazur, and S. A. Curtis, Three-dimensional position and shape of the bow shock and their variation with Alfvenic, sonic and magnetosonic Mach numbers and interplanetary magnetic field orientation, J. Geophys. Res., 100, 7907, 1995.
Petrinec, S. M., and C. T. Russell, An empirical model of the size and shape of the near-Earth magnetotail, Geophys. Res. Lett., 20, 2695, 1993.
Petrinec, S. M., and C. T. Russell, Near-Earth magnetotail shape and size as determined from the magnetopause flaring angle, J. Geophys. Res., 101, 137, 1996.
Petrinec, S. M., P. Song, and C. T. Russell, Solar cycle variations in the size and shape of the magnetopause, J. Geophys. Res., 96, 7893, 1991.
Roelof, E. C., and D. G. Sibeck, Magnetopause shape as a bivariate function of interplanetary magnetic field B_z and solar wind dynamic pressure, J. Geophys. Res., 98, 21,421, 1993.
Russell, C. T., Planetary Bow Shocks, in Collisionless Shocks in the Heliosphere: A Tutorial Review, Geo. Mono. Ser.,35, edited by B. T Tsurutani and R. G. Stone, pp. 109-130, AGU, Washington, D.C., 1985.
Russell, C. T., and S. M. Petrinec, Comment on "Unusual locations of Earth's bow shock on September 24-25, 1987: Mach number effects", J. Geophys. Res., 101, 7677, 1996.
Russell, C. T., J. V. Olson, R. E. Holzer, and E. J. Smith, OGO 3 search coil magnetometer data correlated with the reported crossing of the magnetopause at 6.6 RE by ATS I, J. Geophys. Res., 73, 5769, 1968.
Shue J. H., J. K. Chao, H. C. Fu, C. T. Russell, P. Song, K. K. Khurana, and Singer H. J., A new functional form to study the solar wind control of the magnetopause size and shape, J. Geophys. Res., 102, 9497, 1997.
Shue J.-H., P. Song, C. T. Russell, J. T. Steinberg, J. K. Chao, G. Zastenker, O. L. Vaisberg, S. Kokubun, H. J. Singer, T. R. Detman, and H. Kawano, Magnetopause location under extreme solar wind conditions, J. Geophys. Res., 103, 17,691, 1998.
Sibeck, D. G., R. E. Lopez, and E. C. Roelof, Solar wind control of the magnetopause shape, location, and motion, J. Geophys. Res., 96, 5489, 1991.
Slavin J., A. Szabo, M. Peredo, R. P. Lepping, R. J. Fitzenreiter, K. W. Ogilvie, C. J. Owen, and J. T. Steinberg, Near-simultaneous bow shock crossings by WIND and IMP 8 on December 1, 1994, Geophys. Res. Lett., 23, 1207, 1996.
Slavin, J. A., and R. E. Holzer, Solar wind flow about the terrestrial planets, 1, Modeling bow shock position and shape, J. Geophys. Res., 86, 11,401, 1981.
Slavin, J. A., R. E. Holzer, J. R. Spreiter, and S. S. Stahara, Planetary Mach cones: Theory and observation, J. Geophys. Res., 89, 2,708,1984.
Song, P., C. T. Russell, T. I. Gombosi, I. R. Spreiter, S. S. Stahara, and X. X. Zhang, On the processes in the terrestril magnetosheath : 1. Scheme development, J. Geophys. Res., 104, 22345-22355, 1999.
Spreiter, J. R., A. L. Summers, and A. Y. Alksne, Hydromagnetic Flow Around the magnetosphere, Planet. Space Sci., 14, 223, 1966.
Spreiter, J. R., and A. W. Rizzi, Aligned magnetohydrodynamic solution for solar wind flow past the earth's magnetosphere, Acta Astronaut, 1, 15, 1974.
Villante, U., Evidence for a bow shock at ~ 400 R_E: Pioneer 7, J. Geophys. Res., 81, 1441, 1976.
Yang, Y.-H., J. K. Chao, C.-H. Lin, J.-H. Shue, X.-Y. Wang, P. Song, C. T. Russell, R. P. Lepping and A. J. Lazarus, Comparison of three magnetopause prediction models under extreme solar-wind conditions, J. Geophys. Res., (in press), 2001.

Magnetospheric Observations and Modeling Session

INTERPLANETARY SHOCK EFFECTS ON THE NIGHTSIDE AURORAL ZONE, MAGNETOSPHERE AND IONOSPHERE

X.-Y. Zhou and B.T. Tsurutani

Jet Propulsion Laboratory, California Institute of Technology, 4800 Oak Grove Drive, Pasadena, CA91109, USA

ABSTRACT

Responses of the nightside magnetosphere and auroral zone to interplanetary shocks are studied using WIND solar wind data and POLAR UV imaging data. It is found that the nightside magnetospheric/ionospheric response depends on the interplanetary magnetic field upstream of the shock. When the IMF B_z is strongly southward upstream (up to 1.5 hr) of the shock, a substorm expansion phase is triggered by the shock. If the IMF B_z is ~0, the interplanetary shock triggers a pseudobreakup. If the B_z is northward, the shock does not trigger nightside auroral activity. Shock compression effects on the magnetotail are discussed, as well as potential substorm triggering mechanisms. A schematic model for tail plasma "loading" and "unloading" is presented. One conclusion of this paper is that it may now be possible to predict what type of auroral activity occurs after interplanetary shock impingement on the magnetosphere/magnetotail. A "dripping, tilting bucket" model is presented to explain the observations. The model incorporates the possibility that the magnetotail has a nonlinear response to unusually strong interplanetary shocks.

INTRODUCTION

Interplanetary shocks have been known to trigger magnetospheric substorms since the 1970's. These earlier results reported that substorm triggerings are associated with shock intensities (Schieldge and Siscoe, 1970; Kawasaki et al., 1971), and interplanetary magnetic field (IMF) preconditions (Burch, 1972; Kokubun et al., 1977). Kokubun et al. (1977) noted that ~43% of interplanetary shocks triggered substorm expansion phases.

In this paper, WIND interplanetary and POLAR UV imaging data are used to characterize interplanetary shocks and nightside auroras, respectively. A more complete set of observational results can be found in Zhou and Tsurutani (2001). The consequences of shock compression of the magnetotail are discussed in light of previously published substorm triggering models. A schematic model is presented to explain the flow of energy from the interplanetary space to the magnetosphere and to the ionosphere for shock-triggered substorm expansion and pseudobreakup events.

OBSERVATIONS

Interplanetary shocks occurring in 1997 and 1998 are used in this study (see also Zhou and Tsturutani, 2001). The resultant geomagnetic events are classified into three types based on the nightside auroral response. They are: 1) intense nightside auroral brightening/substorm expansion phase onsets; 2) pseudobreakups, and 3) auroral quiescent events. Figure 1 shows three interplanetary shocks observed by WIND in the upstream solar wind in 1997 and 1998. Shown in the Figure are shocks that trigger: a) an intense nightside auroral brightening (substorm?), b) a pseudobreakup event and c) a quiescent event (no enhanced aurora). In each panel from top to bottom are the IMF magnitude, the IMF B_z component, the solar wind plasma temperature, proton density, and velocity, the solar wind ram pressure (defined as $P_{ram}=\rho V^2$, where ρ is the mass density of protons and V is solar wind speed), and geomagnetic AE indices. Vertical dashed lines indicate the times of shock occurrence.

For the Sep 24, 1998 shock event (panel a), the IMF B_z average over the 1.5 hr preceding the shock (henceforth called the precondition) is -1.8 nT. At the shock, B_z turned northward from ~-1 to 3 nT. P_{ram} increased from ~4 to 16 nPa. The shock triggered a substorm further expansion as shown by the bottom panel, AE increased from ~700 to 1500 nT. For the Jan 10, 1997 shock event, the IMF B_z precondition was +0.3 nT. The IMF B_z turned northward from ~+1 to +5 nT and P_{ram} increased from ~2 to 5 nPa across the shock. AE increased from ~20 to 50 nT. For the June 25, 1998 shock event, the IMF B_z precondition was +10 nT. At the shock, the IMF B_z became enhanced, increasing from ~9 to 13 nT. The P_{ram} increased from ~4 to 7 nPa. The geomagnetic activity was very low, as shown by AE < 80 nT before 1714 UT.

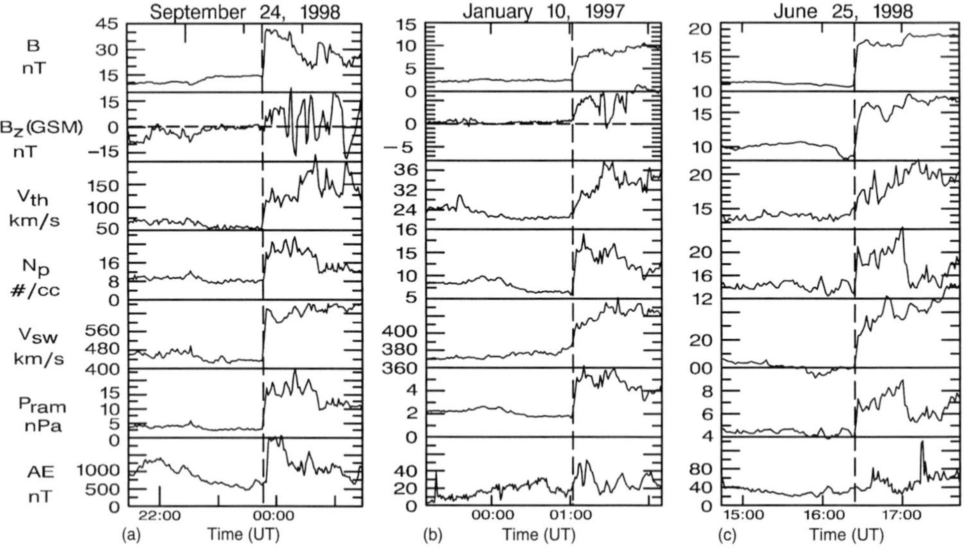

Figure 1. Interplanetary shock events detected by WIND. The WIND magnetic field and plasma data have been "time shifted" to the magnetopause. For the September 24, 1998 shock event (panel a), WIND was in the upstream solar wind at about (184, 13, -9 Re) in GSM coordinates. Panel (b) shows the January 10, 1997 event. WIND was located at (85, -55, -22 Re). Panel (c) shows the June 25, 1998 shock event. WIND was at (69, 66, -1 Re).

Figure 2 shows nightside auroral regions of northern hemisphere in the geomagnetic coordinates. Panels 1 to 3 are auroral images for the substorm "further expansion" event, pseudobreakup event and the quiescent event, respectively. Images in panels (a) are those just after shock arrival to the magnetosphere. For the Sep 24, 1998 event (Panel 1), before the shock arrival there was substorm activity at ~2100 UT (not shown), which presumably corresponded to the "precursor" southward IMF B_z upstream of the shock. After the shock arrival, there was a substorm auroral further expansion at ~2347 UT (panel b). The aurora near ~21 LT intensified and expanded poleward to ~73° Mlat at ~2354 UT (panel e). The corresponding peak AE index was ~1500 nT, as shown by Figure 1(a).

An auroral pseudobreakup event is shown in Figure 2, Panel 2. Approximately 12 min after the shock arrival at the magnetosphere (panel c), an auroral spot occurred at midnight. This spot decayed after ~3 min and intensified again at ~0134 UT as shown by Panels 2(d) and (e). Then it decayed (not shown) (see Arballo et al., 1998 for a more detailed discussion of this "quasiperiodicity"). This auroral spot is considered as a pseudobreakup event based on our studies in Zhou et al. (2000) and Zhou and Tsurutani (2001).

Panel 3 shows a quiet nightside auroral oval. There were no near-midnight auroral brightenings within more than half an hour after the shock arrival. Correspondingly, the AE index was very low (shown in Figure 1(c)). We call this a "quiescent" event.

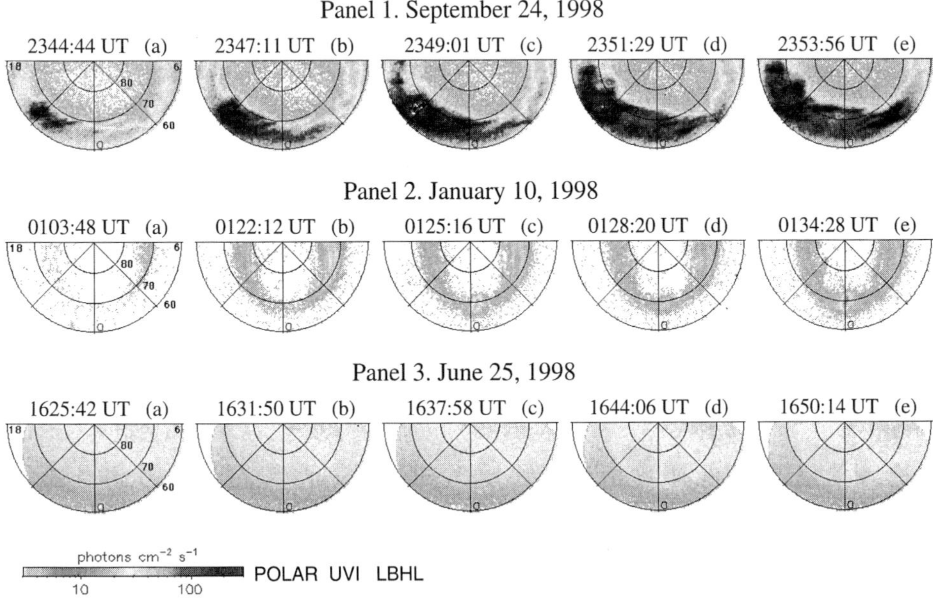

Figure 2. Nightside auroral activity during the September 24, 1998 (Panel 1), January 10, 1997 (Panel 2), and the June 25, 1998 (Panel 3) shock events. For each image, dawn is on the right and midnight at the bottom. The time sequence goes from (a) to (e) for each event. The images are taken from the LBHL filter (~ 170 nm) with a 36.8 sec exposure time.

In this study and the study of Zhou and Tsurutani (2001), the IMF "precondition" is presented as an average of IMF B_z (B_s, B_n) preceding interplanetary shocks. To identify the best temporal lengths of IMF precursors, 0.5-6.0 hr averages of IMF B_z were calculated and studied (Zhou and Tsurutani, 2001). It was found that a ~1.5 hr precursor interval has the highest correlation coefficient (0.78) between the IMF B_s and AE indices. Thus, this time interval has been used as a representative timescale in this analysis.

For all 18 events studied, results similar to the three events of Figure 1 and 2 were found. Our general findings are: 1) with southward IMF B_z preconditions, substorm expansion phases were triggered by interplanetary shocks; 2) with IMF B_z ~0 preconditions, pseudobreakups occurred; 3) with extreme northward IMF preconditions, there was no midnight auroral activity triggered by interplanetary shocks.

INTERPLANETARY SHOCK EFFECTS ON THE MAGNETOTAIL

Compression Effects of Interplanetary Shocks on the Magnetotail

Below, we discuss the effects of solar wind compression of the near-Earth magnetotail. Pressure balance between the solar wind and the tail geomagnetic field is assumed. The tail lobe geometry is primarily determined by the component of solar wind ram pressure normal to the tail magnetosphere boundary and secondarily, by solar wind static pressure. The equilibrium expression is (Fairfield et al., 1989):

$$\frac{B_L^2}{8\pi} = (n_p + n_{He})V_{SW}^2 \sin^2\alpha + nk(T_e + T_p) + \frac{B_{SW}^2}{8\pi} \tag{1}$$

Here B_L is the tail lobe magnetic field strength (we assume that the lobe plasma pressure is negligible), α is the tail flaring angle (the angle between the solar wind flow direction and the tangent to the magnetopause surface), T_e and T_p are the interplanetary electron and proton temperatures, and B_{SW} is the interplanetary field strength. n is the plasma density and k is the Boltzmann constant. The first term on the right-hand side is the component of the solar wind ram pressure perpendicular to the tail magnetopause. The second and third terms are the solar wind plasma thermal pressure and magnetic pressures. In this paper, we have called the sum of the latter two terms the "solar wind static pressure".

Figure 3 shows the magnetopause positions in the X-Y GSM plane before and after the interplanetary shock compression during the Sep 24, 1998 event. The calculation is based on the Petrinec and Russell (1996) model (which only considers the solar wind ram pressure). At X = -15 R_e (down tail), the magnetopause radius is reduced from 19 to 13.7 R_e. By adding the static pressure, we calculate that this radius is reduced further to 13.3 R_e. Assuming that magnetic flux in the tail lobes is conserved (no extra flux is added to or reconnected in the tail), the radius of the tail lobe is reduced from 19 to 13.3 R_e (at X = -15 R_e). The magnetic field in the lobes will thus increase by a factor of ~2.1.

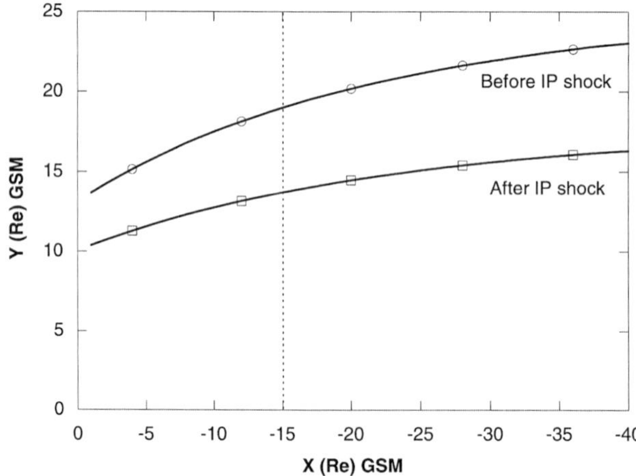

Figure 3. The tail magnetopause position (X ≤ 0) during the Sep 24, 1998 event. At X = - 15 Re in the tail, the radius of the magnetopause is decreased from 19 Re to 13.7 Re by the solar wind ram pressure.

The magnetopause configuration and tail cross-section are sketched in Figure 4 (a) prior to and (b) after shock compression. The stronger the interplanetary shocks are, the stronger the magnetotail compression. As the lobe magnetic field strength is increased, the cross-tail current density will increase accordingly (as shown in Figure 4b). This follows due to the equivalence between the lobe field strength and the currents confining the fields.

Current Sheet Disruption

As proposed by Lui et al. (1990), the current sheet disruption can be triggered by the kinetic cross-field streaming instability (KCSI). This instability can set in when the drift speed between electrons and ions are sufficiently large such that electron Landau damping is overcome by the ion contribution to growth. The cross-tail current will be enhanced by the magnetotail compression. However, calculations (and measurements) are needed to determine whether the onset of the KCSI instability takes place during such compressions or not.

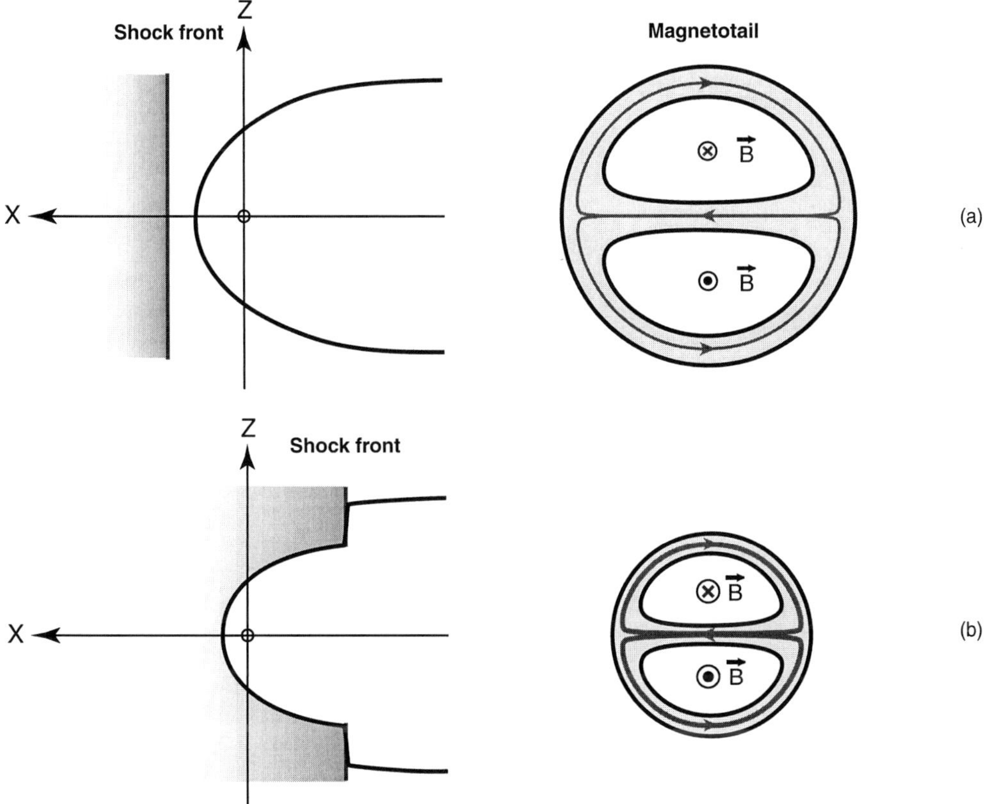

Figure 4. The magnetopause and magnetotail configurations (a) prior to and (b) after interplanetary shock compression. In the left panels, meridional planes are shown in GSM coordinates. The right panels are tail cross-sections viewed from the tail. The tail currents are shown in the GSM Y-Z plane.

X-Line Formation

When the IMF turns southward, dayside magnetic reconnection is activated. Dayside magnetic flux is transported over the polar caps to the outer portions of the tail lobes. The accumulated magnetic flux in the tail lobes will change the magnetic configuration including thinning of the plasma sheet and earthward motion of current sheet (Coroniti and Kennel, 1972; McPherron, 1991). When interplanetary shocks compress the magnetotail, the lobe magnetic field will increase further, especially the B_x component. The magnetotail is effectively stretched. Coroniti (1985) and Baker and McPherron (1990) postulated that when the vertical component of magnetic field across the plasma sheet becomes sufficiently small, ions in the cross-tail current no longer behave adiabatically, and magnetic reconnection begins in the central plasma sheet. Then the rate of reconnection increases explosively. It has been indicated (de la Baeujardiere et al., 1991 and Blanchard et al, 1997) that substorm expansion phase is associated with increased magnetic reconnection the magnetotail. This scenario can be studied as well using shock events as time markers.

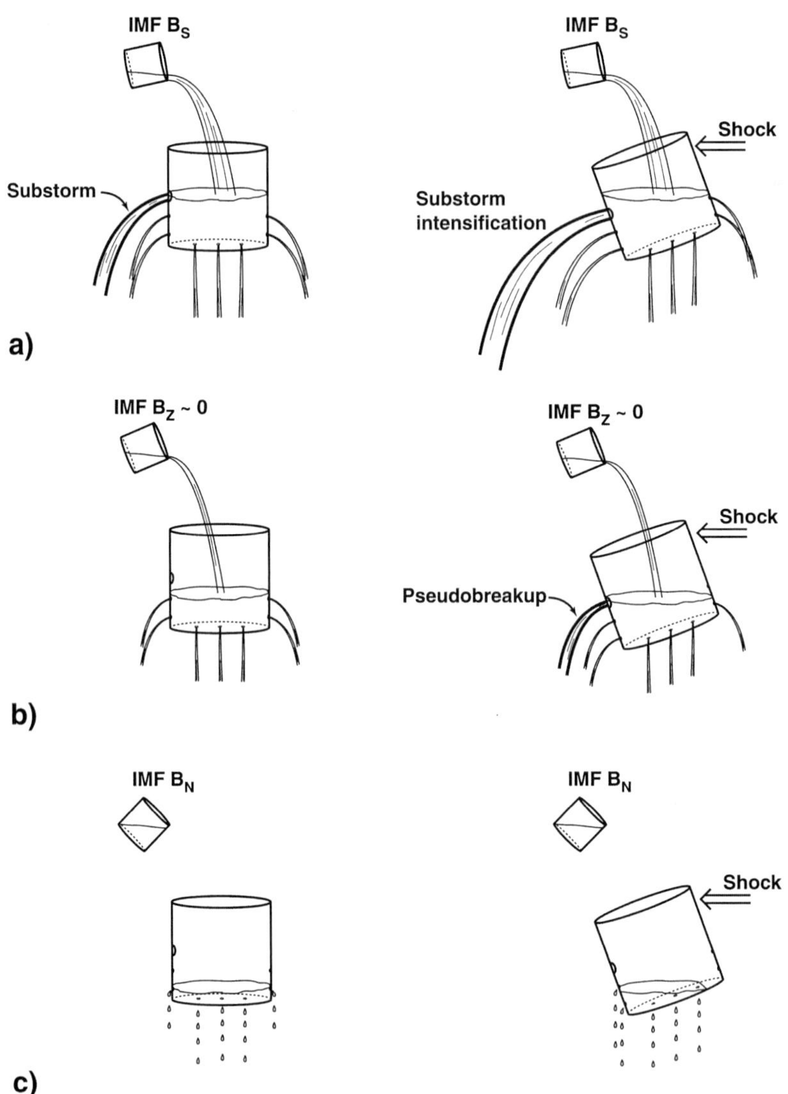

Figure 5. A schematic of the Dripping, Tilting Bucket model.

A DRIPPING, TILTING BUCKET MODEL

To illustrate the findings in this paper and our conceptual ideas, a schematic is provided in Figure 5. The left hand side of the figure shows the IMF B_z preconditions. The liquid levels represent the amount of the solar wind energy input into the plasma sheet by dayside magnetic reconnection (the small pail pouring water into the bucket) before the interplanetary shock arrives at the dayside magnetopause. The greatest rate of energy input occurs when the IMF has a large southward component within ~1.5 hr (top left panel), less so when $B_Z \approx 0$ (middle left panel), and the least, when the IMF has a completely northward component (bottom left panel).

The energy flows out of the bucket in two ways, through the spigot and through holes in the bucket. The energy flow through the spigot represents energy into the magnetospheric/ionospheric system: storms, substorms and pseudobreakups. The "leakage" through holes represents energy dissipation down tail or into the magnetosheath.

This "dripping" or leakage is implied from the limited ~ 1.5 hr "priming" found in the Zhou and Tsurutani (2001) study and other previous works (Arnoldy (1971), Tsurutani and Meng (1972) and Meng et al. (1973) and many others). If the energy was stored in the plasma sheet for much longer time intervals, the relationship found here and the previously cited works showing high IMF B_s-AE correlations would not be present. What exactly are these loss processes? At this time we don't exactly know. But several possibilities are internal dissipation and down-tail energy flow. Ho and Tsurutani (1997) have argued that deep tail magnetic reconnection is not related to substorm activity, so this "magnetotail sloughing" may be one possible dissipation mechanism.

On the right hand side of Figure 5 are the interplanetary shock effects. The shock "tilts" the bucket, i.e., the shock triggers current sheet disruption or reconnection in the tail as discussed above. The stronger the shock, the greater the tilting, i.e., the larger amount of energy will be released from tail to the ionosphere. In the top right panel, the shock (bucket tilting) leads to substorm "intensification". At this time, more energy is pouring out from the spigot (there is more energy going into the ionosphere than the solar wind is transferring to the plasma sheet).

For cases where the plasma sheet energy storage level is relatively low (middle right panel), i.e., for the precondition to IMF $B_Z \sim 0$, a moderate shock leads only to a pseudobreakup (small substorm). A stronger shock corresponds to greater bucket tilting, and a substorm.

For very low energy storage (bottom panels), a moderate shock leads to no energy output. However, for a very strong shock, greater bucket tilting will occur. We speculate that a pseudobreakup or even a substorm are possible outcomes.

In the above model, the IMF B_z (dayside magnetic reconnection) precondition plays the role in the filling of the bucket. The shock compression and cross-tail current sheet microinstabilities are represented by the tilting of the bucket. The "dripping, tilting bucket" model has been constructed to explain the shock/magnetosphere/ionosphere relationships found in this paper.

FINAL COMMENTS

In this paper and in Zhou and Tsurutani (2001), we discuss plasma loading into the tail by "precursor" IMF B_s fields. This plasma is "loaded" into the tail and then "unloaded" into the magnetosphere triggered by the interplanetary shock compression effects. However, of the 18 shock events studied, there was on "anomalous" event, that of Sep 24, 1998. For this event, Zhou and Tsurutani (2001) found that there was anomalously high energy (AE) output for the IMF B_s "precursor" input. We have shown in this paper that the ram pressure increase across this shock was anomalously high (a factor of 4 increase), that highest of the 18 events studied. Schieldge and Siscoe (1970), Kawasaki et al. (1971) and Kokubun et al. (1977) have also previously noted a geomagnetic activity dependence on shock intensity. Thus it is possible that for very strong shocks, the magnetotail responds in a nonlinear fashion. As one possible scenario, additional energy may be injected into the magnetosphere due to enhanced tail magnetic reconnection (Figure 5a indicated such a potential scenario). In the near future, we will examine strong shock (Mach 3 to 4) versus weak shock effects on the magnetosphere.

ACKNOWLEDGMENTS

Portions of this paper represent work done at the Jet Propulsion Laboratory, California Institute of Technology, Pasadena, under contract with the National Aeronautics and Space Administration. T. Araki provided the ground based AL and AE indices, magnetometer data and K. Ogilvie and R.P. Lepping the WIND SWE and MFI data. J.K. Arballo helped in the data processing and software support. X.-Y. Zhou would like to thank the National Research Council for the award of a Resident Associateship at the Jet Propulsion Laboratory.

REFERENCES

Arnoldy, R.L., Signature in the interplanetary medium for substorms, *J. Geophys. Res.,* 76, 5189 (1971).

Baker, D.N., and R.L. McPherron, Extreme energy particle decreases near geostationary orbit: a manifestation of current diversion with the inner plasma sheet , *J. Geophys. Res.*, 95, 6591 (1990).

Blanchard, G.T., L.R. Lyons, and O. de la Beaujardiere, Magnetotail reconnection rate during magnetospheric substorms, *J. Geophys. Res.,* 102, 24303 (1997).

Borovsky, J.E., M.F. Thomsen, R.C. Elphic, The driving of the plasma sheet by the solar wind, *J. Geophys. Res.,* 103, 17617 (1998).

Burch, J.L., Preconditions for the triggering of polar magnetic substorms by storm sudden commencements, *J. Geophys. Res.,* 77, 5629 (1972).

Chang, C.L., H.K. Wong, and C.S. Wu, Electromagnetic instabilities attributed to a cross-field ion drift, *Phys. Rev. Lett.,* 65, 1104 (1990).

Coroniti, F.V., and C.F. Kennel, Changes in the magnetospheric configuration during the substorm growth phase, *J. Geophys. Res.,* 77, 3361 (1972).

Coroniti, F.V., Explosive tail reconnection: the growth and expansion phases of magnetospheric substorms, *J. Geophys. Res.,* 90, 7427 (1985).

de la Beaujardiere, O., L.R. Lyons, and E. Friis-Christensen, Sondrestrom radar measurements of the reconnection electric field, *J. Geophys. Res.,* 96, 13907 (1991).

Fairfield, D.H., D.N. Baker, J.D. Craven, R.C. Elphic, J.F. Fennell, L.A. Frank, I.G. Richardson, H.J. Singer, J.A. Slavin, B.T. Tsurutani, and R.D. Zwickl, Substorms, plasmoids, flux ropes, and magnetotail flux loss on March 25 1983: CDAW 8, *J. Geophys. Res.,* 94, 15135 (1989).

Ho, C. M., and B.T. Tsurutani, Distant tail behavior during high speed solar wind streams and magnetic storms, *J. Geophys. Res., 102*, 14165 (1997).

Kawasaki, K., S.-I. Akasofu, F. Yasuhara, and C.-I. Meng, Storm sudden commencements and polar magnetic substorms, *J. Geophys. Res.,* 76, 6781 (1971).

Kokubun, S., R.L. McPherron, and C.T. Russell, Triggering of substorms by solar wind discontinuities, *J. Geophys. Res.,* 82, 74 (1977).

Lennartsson, W., and E.G. Shelley, Survey of 0.1- to 16-keV/*e* plasma sheet ion composition, *J. Geophys. Res.,* 91, 3061 (1986).

Lui, A.T.Y., A. Mankofsky, C.-I. Chang, K. Papadopoulos, and C.S. Wu, A current disruption mechanism in the neutral sheet: a possible trigger for substorm expansions, *Geophys. Res. Lett.,* 17, 745 (1990).

McPherron, R.L., Physical processes producing magnetospheric substorms and magnetic storms, in *Geomagnetism*, vol., edited by J.A. Jacobs, pp. 593, Academic, San Diego, Calif., (1991).

Meng, C.-I., B. Tsurutani, K. Kawasaki, and S.-I. Akasofu, Cross-correlation analysis of the AE index and the interplanetary magnetic field B_z component, *J. Geophys. Res.,* 78, 617 (1973).

Petrinec, S.M., and C.T. Russell, Near-Earth magnetotail shape and size as determined from the magnetopause flaring angle, *J. Geophys. Res.,* 101, 137 (1996).

Schieldge, J.P., and G.L. Siscoe, A correlation of the occurrence of simultaneous sudden magnetospheric compressions and geomagnetic onsets with selected geophysical indices, *J. Atmos. Terr. Phys.,* 32, 1819 (1970).

Tsurutani, B.T., and C.-I. Meng, Interplanetary magnetic-field variations and substorm activity, *J. Geophys. Res.,* 77, 2964 (1972).

Zhou, X.-Y., and B.T. Tsurutani, Interplanetary shock triggering of geomagnetic activity: Substorms, pseudobreakups and quiescent events, *J. Geophys. Res.,* 106, 18957 (2001).

Zhou, X.-Y., B.T. Tsurutani, W.D. Gonzalez, and A.J. Lazarus, The solar wind depletion event of 26 April 1999: Triggering of a midnight auroral event, *Geophys. Res. Lett.,*27, 4025 (2000).

DEVELOPMENT OF AN INTEGRATED PREDICTIVE MHD SPACE WEATHER MODEL FROM THE SOLAR SURFACE TO THE EARTH'S UPPER ATMOSPHERE

C. R. Clauer[1], T. I. Gombosi[1], K. G. Powell[2], Q. F. Stout[3], G. Toth[1], D. DeZeeuw[1], A. J. Ridley[1], R. A. Wolf[4], R. G. Roble[6], and T. E. Holzer[5]

[1] *Space Physics Research Laboratory, University of Michigan, Ann Arbor, MI.*
[2] *Department of Aerospace Engineering, University of Michigan, Ann Arbor, MI*
[3] *Department of Electrical Engineering and Computer Science, University of Michigan, Ann Arbor, MI*
[4] *Department of Physics and Astronomy, Rice University, Houston, Texas*
[5] *High Altitude Observatory, National Center for Atmospheric Research, Boulder, CO*

ABSTRACT

Taking advantage of the advent of massively parallel computers, sophisticated solution-adaptive techniques, and recent fundamental advances in basic numerical methods we have developed a high performance, adaptive-scale MHD code capable of resolving many of the critical processes in the Sun-Earth system which range over more than 9 orders of magnitude in scale size. The development of such models are of increasing importance as the impact of space weather on vulnerable technological systems increases, and too, as the severity of space weather increases with solar maximum. There is an increasing need to develop physics-based, high performance models of the Sun-Earth system – from the solar surface to the Earth's upper atmosphere – which can operate faster than real time and which can provide reliable predictions of the near Earth space environment based upon solar observations and upstream solar wind measurements. We report the status of the prototype development of a Comprehensive Space Environment Model for this purpose, to be composed of a core based upon the Michigan MHD code coupled to a high performance inner magnetosphere model and ionosphere/upper atmosphere model. The inner magnetosphere model and ionosphere/upper atmosphere models are derived from the Rice Convection Model (RCM) and the Thermosphere Ionosphere Electrodynamic General Circulation Model (TIEGCM) respectively, both reformulated for parallel operation using advanced numerics on adaptive grids with significantly higher spatial resolution.

INTRODUCTION

The Sun-Earth system is an extremely complex natural system involving many different interacting elements. Today, the science of space plasma physics has matured to the level of being able to describe many of these complex interactions and to model them. A major goal now is to unify our understanding into a more comprehensive mathematical framework that can simulate and predict the properties of this system (space weather).

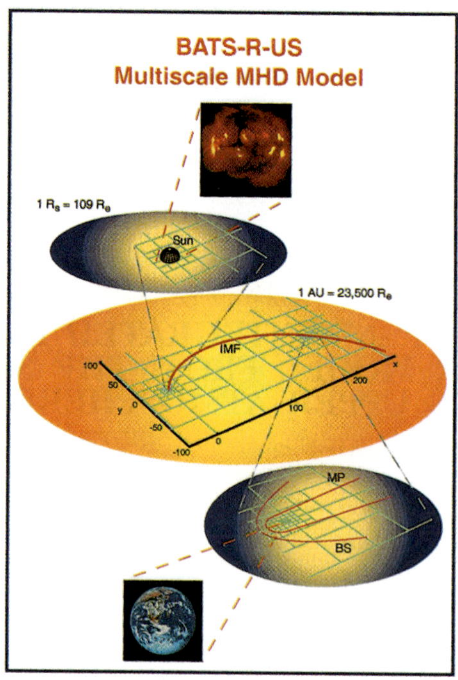

Fig. 1. The Michigan MHD Model. Note that the model spans multiple scales, from a few hundred kilometers in the ionosphere to the radius of the Earth (R_e), to the radius of the Sun (R_s), to the Sun-Earth distance (AU).

This goal is made all the more relevant and timely because of the increased need for a predictive space weather model. "Space Weather" has been used to refer to the conditions on the Sun and in the solar wind, magnetosphere, ionosphere, and thermosphere that can influence the performance and reliability of space-borne and ground based technological systems or can endanger human life or health (*Wright et al.*, 1995). For example, satellites experience the disruptive effects of energetic particles and differential electrical charging during various conditions; astronauts are vulnerable to energetic radiation that may occur at space station altitudes; navigation signals from the GPS satellites are affected by irregularities in the ionosphere that occur under some conditions; massive disruptions in electric power distribution systems can be triggered by geomagnetic storms; and the Skylab space station crashed into the atmosphere because of the increased atmospheric drag due to the atmospheric heating and consequent expansion which resulted from several unanticipated severe geomagnetic storms; to name a few consequences of space weather.

To build a predictive, physics-based, space weather computer model, however, requires that we address and solve several difficult problems. It is necessary to integrate and synthesize the work of many scientists in many disciplines. Further, translating this understanding into models capable of representing the large range of scales for critical phenomena in this system requires very advanced numerical technology. In order to produce models that could eventually be utilized to predict space weather for operational purposes, there is also a need to run the models at speeds faster than real time. This requires that we achieve considerable high performance computing efficiency.

This paper describes the development of a comprehensive space environment model (CSEM) to be used as a prototype space weather code. This model utilizes the Michigan solution adaptive MHD space plasma code as a core to model the region from the solar surface to the Earth's upper atmosphere. This requires that we address physical processes which occur over 9 orders of magnitude in scale size. A schematic representation of the physical domain described by the CSEM is shown in Figure 1.

The upper atmospheric boundary of the CSEM will be provided by a high performance parallel reformulation of the Thermosphere-Ionosphere-Electrodynamic General Circulation Model (TIEGCM) developed at the National Center for Atmospheric Research (NCAR) (*Roble et al.*, 1988; *Roble and Ridley*, 1994). This model determines the atmospheric composition, temperature, and flow pattern over the whole Earth from 95 km to 500 km altitude, a region of the atmosphere which is greatly affected by solar emissions and interactions with the magnetosphere. *Richmond* (1992) extended the original TI-GCM to include self-consistent electrodynamic interactions between the magnetosphere, ionosphere, and thermosphere.

The inner magnetosphere is a region is crucially important from a space-weather point of view, because it is home to most of the world's fleet of working spacecraft, including hundreds of communications spacecraft in geosynchronous orbit and the Global Positioning System (GPS), which occupies a beehive of orbits inside geosynchronous. The physics of the region is complicated, because it contains overlapping particle distributions with a wide range of energies and characteristics and as such, is not properly modeled by ideal MHD. These different coexisting particle populations cannot be treated as a single fluid, because they all move differently and the dynamics are dominated by gradient and curvature drifts. The inner magnetosphere portion of the CSEM will be derived from the Rice Convection Model (RCM), built specifically to treat this unique and complex region. The RCM represents the particles in terms of about thirty separate fluids. Its equations and numerical methods have been specifically designed for accurate treatment of the inner magnetosphere (*Jaggi and Wolf*, 1973; *Harel et al.*, 1981; *Wolf*, 1983; *Erickson et al.*, 1991), including the flow of electric currents along magnetic field lines to and from the conducting ionosphere.

Description of the Sun-Earth System.

The Sun's atmosphere, the solar corona, is so hot ($> 10^6 K$) that in regions where the solar magnetic field is insufficiently strong, the corona undergoes a rapid expansion, filling all of interplanetary space with a superfast magnetized plasma flowing radially outward from the Sun. As this flowing plasma, which is called the solar wind, passes the Earth, it interacts strongly with the geomagnetic field, forming a bow shock upstream and severely compressing the field on the dayside of the Earth, and drawing it out into a long, comet-like tail on the nightside. In ideal MHD, the confined region of geomagnetic field forms a diamagnetic cavity in the solar wind flow and this cavity is called the Earth's magnetosphere.

The solar wind not only confines the terrestrial magnetic field within the magnetospheric cavity, but it also transfers significant mass, momentum, and energy to the magnetosphere, as well as to the ionosphere and upper atmosphere. The primary mechanism for energy coupling is based upon the concept of magnetic merging and the open magnetosphere proposed by *Dungey* (1961). One dramatic consequence of this interaction between the solar wind and the magnetosphere is the generation of the aurora at high latitudes in both the northern and southern hemispheres. Another consequence is the production of a variety of complex electric current systems, ranging from a sheet of current flowing on the magnetopause boundary between the solar wind and magnetosphere, to an enormous ring of current flowing around the Earth in the inner magnetosphere, to currents flowing throughout the ionosphere and connecting along magnetic field lines to magnetospheric currents systems. Yet another result of the solar-wind/magnetosphere interaction is the production of populations of very energetic particles that are stored in the magnetosphere and

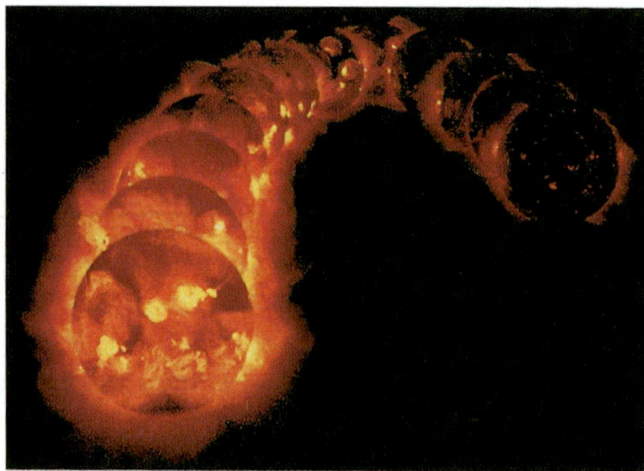

Fig. 2. The 12 X-ray images of the Sun obtained by the Yohkoh satellite between 1991 (left) and 1995 (right) at 90-day increments provide a dramatic view of how the solar corona changes from solar maximum to solar minimum. As we approach solar maximum anticipated in 2001 or 2002, the reverse progression will occur. (Figure courtesy of Lockheed Martin.)

precipitated into the upper atmosphere. Both the electric currents and the energetic particles can have severe consequences for a number of human activities being carried out in various locations, all the way from the ground to space. It is the variation over time of these electric current systems and energetic particle populations in the geospace environment that modulates the impact of what we refer to as space weather on various human activities.

The sources of the space weather disturbances observed in the space around the Earth lies ultimately in the sun and its variability. Solar disturbances propagate through the ambient solar wind and those which produce large extended periods of interplanetary magnetic field which is antiparallel to the geomagnetic field intensify the energy coupled to the magnetosphere thereby producing magnetic storms. Some geomagnetic storms can be associated with the regular 27 day appearance of long lived coronal holes which are prevalent during the declining phase of the solar cycle and at solar minimum. High speed solar wind from these holes interact with the slower solar wind to produce shock and interaction regions which, in turn, produce geoeffective configurations of the interplanetary magnetic field. Other storms are the result of large coronal mass ejections in which large regions of dense material are ejected from the sun to form huge magnetic flux ropes which may interact with the Earth's magnetosphere. A major problem is to determine the evolution of solar produced disturbances as they move from the solar surface and to determine if and when they will impact the terrestrial magnetosphere, and to predict the form of the disturbance at the Earth to determine the degree of mass, momentum and energy transfer to the magnetosphere and the associated space weather which would result.

Time scales important to space weather range from minutes to decades. The longest time scales that are usually considered important to space weather are the 11-year solar activity cycle and the 22-year solar magnetic cycle (Figure 2). Near the minimum of the solar activity cycle, the solar wind is nearly completely dominated by a quasi-steady outflow, although there are occasional ejections of plasma and magnetic field from the near-equatorial sun where the plasma is generally confined by the magnetic field. These so-called coronal mass ejections (CMEs) expand into interplanetary space and become integrated into the overall solar wind flow, but they are characterized by closed magnetic field structures oriented nearly perpendicular to the ecliptic plane, in contrast to the background solar wind magnetic field, which on the average lies very nearly in the ecliptic plane. The magnetic field structures associated with these CMEs, called magnetic clouds, as well as their generally enhanced dynamic pressure

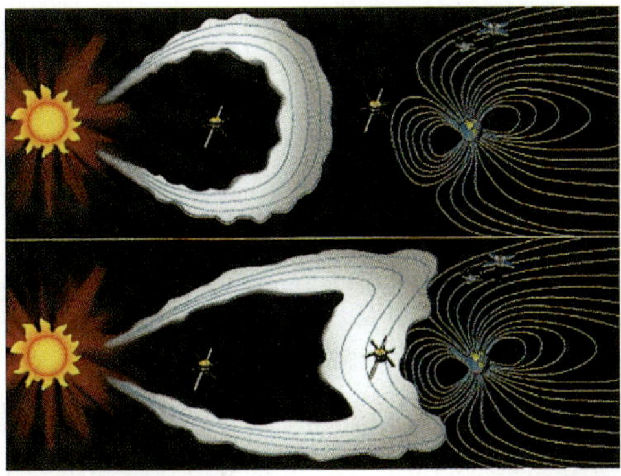

Fig. 3. The interaction of the magnetosphere with an expanding magnetic cloud. (Illustration courtesy of the NASA International Solar Terrestrial Program.)

(ρV^2, where ρ is the mass density and V is the bulk velocity), lead to a significantly stronger interaction with the magnetosphere than that exhibited by quiet solar wind. The major mechanism for energy coupling between the solar wind and the magnetosphere is magnetic merging on the dayside. This is substantially enhanced when the interplanetary magnetic field (IMF) contains a component antiparallel to the Earth's magnetic field, i.e. southward. The CME interaction with the magnetosphere is illustrated in Figure 3. The magnetic cloud generated by the CME approaches the quiet magnetosphere in the top frame. In the bottom frame the cloud initiates a stronger interaction, which generates stronger magnetospheric current systems and larger, more energetic magnetospheric particle populations – a phenomenon which is called a geomagnetic storm. As solar activity increases, the frequency of CMEs is substantially increased, and the "severity of space weather" is concomitantly increased.

Advanced Numerical MHD Modeling

Developed with support from the NASA High Performance Computing and Communications and from NSF, the University of Michigan adaptive-scale MHD code was designed from the ground up to capitalize on modern numerical methods, solution-adaptive techniques, and massively parallel implementation. The Michigan code, whimsically called BATS-R-US for (Block Adaptive-Tree Solar-wind Roe-type Upwind Scheme), has been used successfully to simulate the environment from the solar surface to the top of the Earth's upper atmosphere.

The MHD algorithm was designed to take advantage of advances in upwind methods, approximate Riemann solvers, and limited solution reconstruction. A cell-centered upwind finite-volume formulation is adopted to solve the governing ideal MHD equations in divergence form. The limited solution reconstruction of *van Leer* (1979) is used to ensure second-order accuracy away from discontinuities, while simultaneously providing the stability required for monotonic non-oscillatory solutions. In addition, the user can choose from three approximate Riemann solvers to evaluate the numerical flux function. Finally, an explicit multi-stage method is used to integrate the ordinary differential equations that result from the upwind spatial discretization of the PDEs (*van Leer et al.*, 1989). The resulting finite-volume scheme solves for the hydrodynamic and electromagnetic effects in a tightly coupled manner, provides accurate resolution of discontinuous solutions and complicated wave structures, and works equally well across a range of several orders of magnitude in plasma β. The details of the scheme are discussed by *Powell*

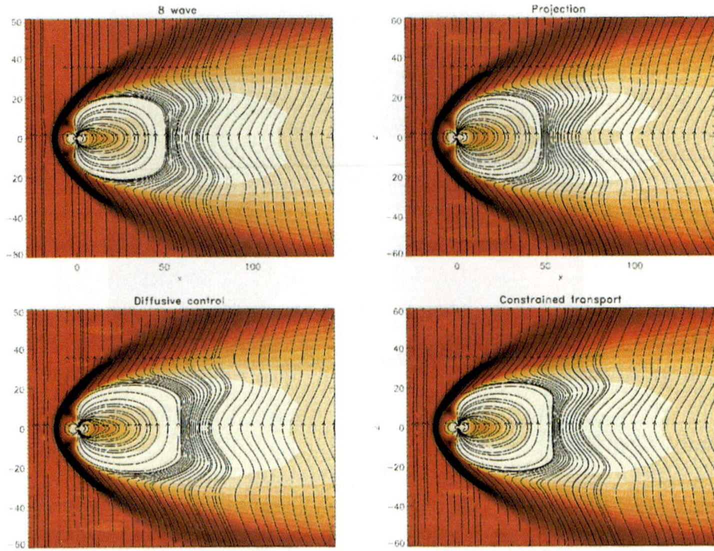

Fig. 4. Noon-midnight meridional cross section of the simulated terrestrial magnetosphere for northward interplanetary magnetic field (IMF) conditions. Four divergence B control methods were used (from top left clockwise): 8 wave method (*Powell*, 1994), the projection scheme (*Brackbill and Barnes*, 1980), constrained transport (*Balsara and Spicer*, 1999) and the diffusive divergence B control technique (*Linde and Malagoli*, 2001). The simulations were carried out on identical grid (370,000 cells, smallest cell size: 0.25 R_E, largest cell size: 8 R_E). The color shading represents the plasma density, solid lines are magnetic field lines.

et al. (1999).

Users of BATS-R-US can also select from 4 different methods to control the $\nabla \cdot \mathbf{B}$ truncation errors. In addition to the 8-wave method (resulting from the symmetrizable form of the MHD equations) which explicitly tracks the $\nabla \cdot \mathbf{B}$ error, three additional approaches have been implemented: (1) the projection scheme (*Brackbill and Barnes*, 1980) which is solved with a parallelized Krylov-type iterative solver (*Toth*, 2000); (2) the diffusive $\nabla \cdot \mathbf{B}$ control technique recently proposed by *Linde and Malagoli* (2001); and (3) a conservative (for all MHD variables) constrained transport (CT) method. The latter scheme is a generalization of the *Balsara and Spicer* (1999) method to AMR grids. The resolution changes and the restriction and prolongation operators are treated with a new algorithm developed by *Toth and Roe* (2001). This AMR-MHD-CT scheme is different from the original CT method proposed by *Evans and Hawley* (1988) which treats the electromagnetic terms, such as $\mathbf{j} \times \mathbf{B}$, in a non-conservative form. Since the $\mathbf{j} \times \mathbf{B}$ force is dominant in many space plasma regions, the *Evans and Hawley* (1988) method may not provide accurate discontinuous solutions (such as shocks as current sheets). A comparison of results obtained with the four divergence B methods is shown in Figure 4.

Solution adaptation is a powerful tool for resolving a problem with disparate length scales. By using a solution-adaptive grid, one can avoid under-resolving high-gradient regions, or, conversely, over-resolving low-gradient regions at the expense of more critical regions, and thereby save several orders of magnitude in computing resources for many problems. Length scales in the Sun-Earth system range from tens of kilometers in the ionosphere to the Earth-Sun distance (1.5×10^{11} m); time scales range from a few seconds near the Sun to the expansion time of the solar wind from the Sun to the Earth ($\sim 10^5$ s). These problems need solution-adaptive schemes — a simple Cartesian mesh would grossly under-resolve much of the problem, while over-resolving relatively uninteresting regions.

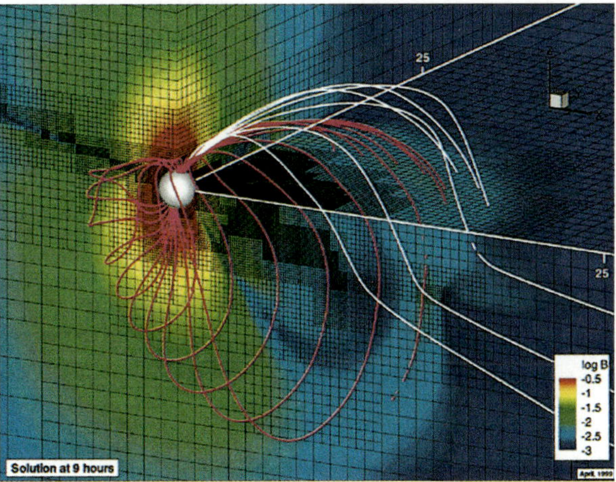

Fig. 5. Simulation of a coronal mass ejection (CME) illustrating the 3-D adapted grid structure to resolve multiple-scale phenomena. Magenta lines represent compressed closed field lines ahead of the CME, while white lines show the distorted structure of open draped magnetic field lines. The color code represents $log(B)$.

The approach to adaptation taken in the BATS-R-US code is one of self-similar adaptive blocks *Stout et al.* (1997). This approach was designed with performance on massively parallel machines in mind. The basic unit of data is a three-dimensional block of grid cells. A block has a size that is set at run-time; a typical block might have $8 \times 8 \times 8$ cells. The data in the block are stored in i, j, k-indexed arrays — a data structure that is conducive to high cache performance. The code is designed to optimize serial performance on single processors of the target machine while the overall adaptive block scheme also reduces communication to provide high parallel performance. An initial grid is composed of a number of blocks, all at the same level of refinement. Then, in regions that appear under-resolved (as defined by suitable adaptation criteria), a block is refined — each of its eight octants becomes a block with the same number of cells as the original block. In regions that appear over-resolved (again, as defined by suitable adaptation criteria), eight blocks can be coalesced into one. As an example we show in Figure 5 a 2D cut through the 3D grid taken from a calculation of a coronal mass ejection *Stout et al.* (1998).

Grids like those shown in Figure 5 go a long way towards resolving the disparate scales in a problem. Each level of refinement in the grid introduces cells that are smaller by a factor two in each dimension from those one level higher in the grid. Typical calculations have 5 - 10 levels of refinement; some calculations have more than 20 levels of refinement. In the case of 20 levels of refinement, the finest cells on the mesh are more than one million times smaller in each dimension than the coarsest cells on a mesh.

Massively parallel machines entice users with a factor of 512, or 1024, or even more in CPU and memory resources than single-processor machines. Capitalizing on the promise of these resources is, however, not always straightforward. In general, researchers have had very poor luck with "automatically parallelizing" their codes, or, more generally, with porting legacy codes to this class of machines. The method of *domain decomposition*, i.e. the partitioning of the problem by dividing the computational domain into sections and farming the separate sections off onto separate processors, is the most practical approach for many physical systems requiring the solution to partial differential equations. However, codes that were designed with single-processor computing in mind may have inherent limits on their scalability via this approach. For example, such codes may achieve speed-ups for 16, or 32, or 64 processors, but with additional processors, the codes not only fail to speed up, but actually slow down. These inherent limitations can arise from a variety of sources:

- underlying basic algorithms that are global in nature, resulting in high communication costs;
- underlying data structures that are expensive to partition or to update in a parallel fashion;
- underlying processes that are inherently serial.

In order to avoid any of these limitations, the BATS-R-US code was designed from the ground up with parallelism in mind. The underlying basic algorithm that was chosen is highly local in nature, resulting in low communication overhead. The adaptive-block data structures on which the code was built allow easy, natural partitioning of the data, and greatly facilitate load-balancing, a crucial element of truly scalable computing. Thus, most computation is done within a data block and all data blocks look the same to a processor, whether they are a large block or a small block. Balance and efficiency are achieved by partitioning the blocks among the available processors. The design was carried out in such a way that even the adaptation of the grid could be carried out in parallel.

The results of this bottom-up design for scalability are shown in Figure 6 (left panel). The figure shows code scaling on two different Cray T3E systems. A fixed problem size per processor of 16 blocks was used. The scale-up for both machines is nearly 100% efficient. Nearly identical results have been obtained on an IBM SP2. The code is written in Fortran 90 with message-passing via MPI, and hence is portable to a wide range of machines, from integrated shared-memory systems to networks of workstations. For each target architecture, simple single-processor measurements are used tune the size of the adaptive blocks. In Figure 6 (right panel) we also show the performance obtained on a number of parallel architectures without any particular tuning.

Elements of a Comprehensive Space Environment Model

The Comprehensive Space Environment Model (CSEM) which we are developing is based upon the integration of three existing models. The core of the model will be the BATS-R-US advanced MHD space plasma simulation code. The adaptive MHD model will start from flow conditions near the Sun and will track the ensuing interplanetary evolution out to the Earth and beyond. This entails simulating the dynamic interactions of transient structures injected into a pre-existing, structured and quasi-steady solar wind flow. The initial conditions will be specified by observation and a solar coronal model. The interaction of the solar wind flow with the Earth's dipole generates the Earth's magnetosphere and this domain will be simulated also by the BATS-R-US code.

The Rice Convection Model (RCM) which provides the best specification of the inner magnetosphere, of particular importance to space weather because of the many satellites which are deployed in this region, has been reformulated to utilize modern advection algorithms together with a high performance parallel implementation. The physics of this region is complex since it contains overlapping particle distributions with a wide range of energies and characteristics. These different coexisting particle populations cannot be treated as a single fluid, because they all move differently. The RCM represents the particles in terms of about thirty separate fluids. Its equations are specifically designed for the accurate treatment of the inner magnetosphere including the flow of electric currents along magnetic field lines to and from the conducting ionosphere.

Figure 7 shows the spherical RCM grid and an idealized Hilmer-Voigt magnetic field model mapped to the equatorial plane on top of the MHD cartesian grid. For the coupling between the two models, the RCM boundary condition is provided by the MHD model at the last closed field line. The pressure distribution computed by the

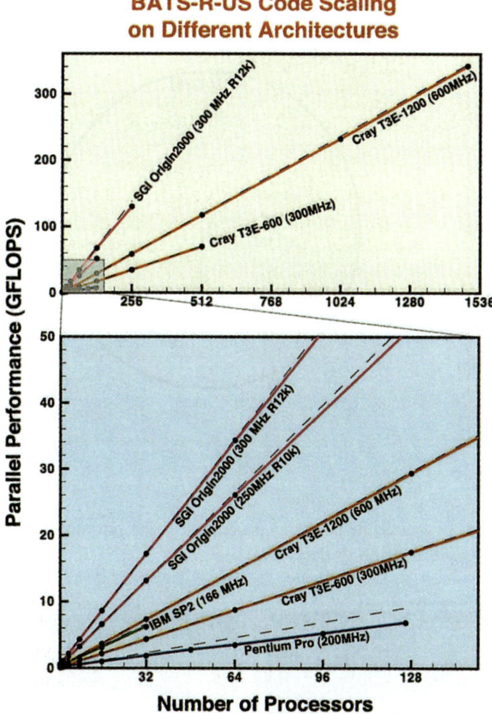

Fig. 6. Parallel performance of BATS-R-US code for a variety of parallel architectures. The dashed line indicates ideal scale-up performance based on single node performance and solid lines indicate actual performance achieved on each of the machines.

RCM is fed back to the MHD along the RCM grid and then interpolated to the MHD grid. If the comparison between the RCM pressure distribution and MHD pressure distribution is different, the MHD pressure is "nudged" toward the RCM value using

$$P_{MHD}^{n+1} = P_{MHD}^{n} + \frac{\Delta t}{\tau}(P_{RCM} - P_{MHD}^{n}).$$

τ is usually on the order of $100\Delta t$ so with each iteration there is a $\approx 1\%$ change made to the MHD pressure on closed field lines within about $15R_e$.

The results of the coupling can be seen in Figures 8 and 9. The simulation was run using 343,000 computational cells with the smallest cells measuring 0.5 R_e on a side. The input IMF conditions contained $B_z = -5$ nT and the simulation was run to steady-state. The MHD-RCM coupling was done once every 60 s. The results from the coupled model produce a more realistic pressure distribution in the inner magnetosphere and the field-aligned current distribution measured at the ionosphere also shows more realistic Region 1 and Region 2 currents.

In addition, the Thermosphere-Ionosphere-Electrodynamic General Circulation Model (TIE-GCM) developed at

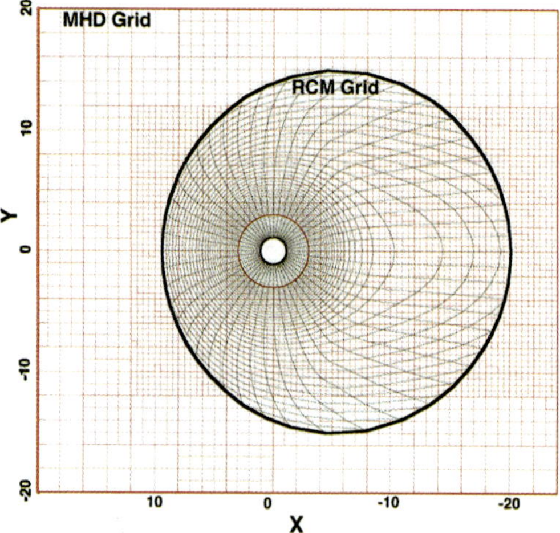

Fig. 7. Projection of a Hilmer-Voigt magnetic field model projected to the equatorial plane along the RCM spherical grid system and mapping onto the MHD Cartesion grid system.

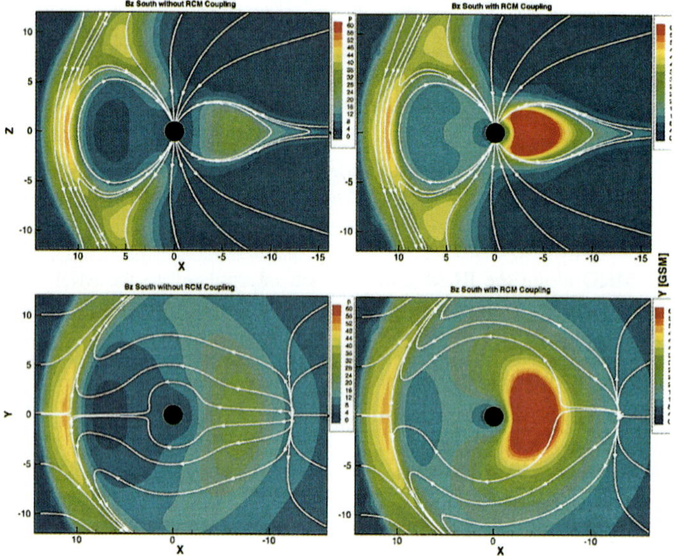

Fig. 8. The left panels show the $Y = 0$ meridian view of magnetic field lines and pressure contours (top) and the $Z = 0$ equatorial plane view of stream lines and pressure contours for the simulation (bottom) using the MHD code alone. The right panels show the same plots for the simulation using the coupled MHD and inner magnetosphere model derived from the RCM. The coupled simulation produces higher pressure in the inner magnetosphere, particularly in the night sector.

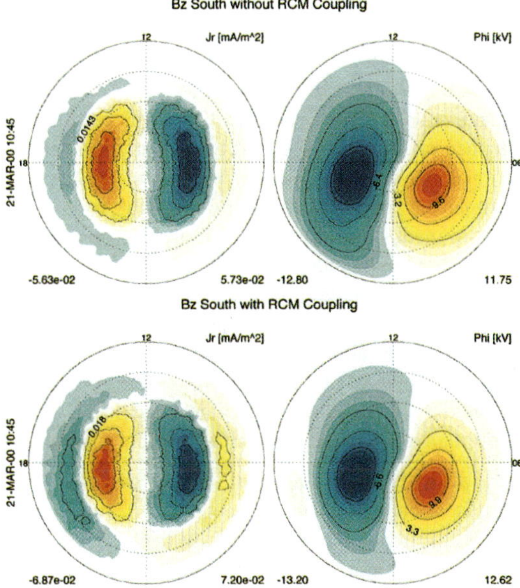

Fig. 9. The top panels show the ionospheric projection of field-aligned current distributions (left) and electric potential patterns (right) for the simulation using the MHD code alone. The bottom panels show the same plots for the simulation using the coupled MHD and inner magnetosphere model derived from the RCM. Inward current is shown as blue and upward current is shown as yellow/red on the left panels. The left panels show the region 1 and 2 field-aligned currents. The Region 2 currents are stronger and more well defined in the coupled simulation.

NCAR is being reformulated as a high performance code with modern advection algorithms and will be integrated into the CSEM to provide the best description of the lower boundary of the Sun-Earth system. The TIE-GCM calculates the dynamics of the upper atmosphere and determines the composition, temperature, and flow pattern over the whole Earth from 95 km to 500 km altitude, a region of the atmosphere which is greatly affected by solar emissions and interactions with the magnetosphere. It also computes the neutral winds, ion drifts, neutral, ion, and electron temperatures, and the neutral and ion densities. The model solves for the ionospheric structure giving global distributions of five ion species.

To do the reformulation of the RCM and TIE-GCM we have had to develop a spherical adaptive block structure and the rendezvous mechanisms to interpolate between the spherical block grid and the BATS-R-US cartesian block grid. The rendezvous methods have been tested and we have completed the reformulation and coupling of the inner magnetosphere model to the MHD. The reformulation of the TIE-GCM code is under way with the hope that it will be completed this year. At present, we have coupled the existing TIE-GCM code with the MHD to test the coupling strategy. We have done several science runs using the coupled code to demonstrate the effects of the neutral winds to feed back an emf to the magnetosphere (flywheel effect).

Conclusions

Continued research into the development of high-performance numerical algorithms combined with the continued development in high-performance hardware will make it possible to achieve a predictive space weather model in the near future. The present Michigan code has been run on a Cray T3E with 512 processors and has achieved performance at about 2 times faster than real time using 2 million cells in the CME simulation for the entire Sun-Earth system. However, for operational and realistic space weather purposes, considerable improvement must be achieved. Much greater resolution, requiring approximately 10 million cells, is required, and the code must achieve 2 times faster than real time on a more modest machine – say 16 or 32 processors. Further, the code requires a number of improvements in the physics. These include a more realistic ionosphere and upper-atmospheric model, rotating Earth with a tilted dipole, as well as the inclusion of a model of the inner magnetosphere where the radiation belts and ring current are significant space weather features during magnetic storms. Thus, the challenge over the next few years is to improve the computational performance by about a factor of 100 while also including more complete and increasingly complex physical processes. Some of the anticipated performance improvement will be accomplished by improvements in computational hardware capabilities, and further improvement will be accomplished by improvements in our numerical algorithm technology and software design.

Validation of space weather codes is also a crucial component of the National Space Weather Program. Metrics have been developed to test the output of codes and to also monitor the improvement in our ability to simulate space weather disturbances. Various ionospheric measurements have been selected as metrics because of the large array of instruments deployed globally which can be used in validation studies. Ground based magnetograms were simulated using the Michigan MHD code and compared with observations on March 19 - 20, 1999 (*Ridley et al.*, 2001a) and the cross track ion plasma drift measured by the low altitude DMSP polar orbiting satellite was simulated for several selected events in another recent "metrics challenge" (*Ridley et al.*, 2001b).

ACKNOWLEDGEMENTS

This work has been supported by the NSF Knowledge and Distributed Intelligence Grant ATM-9980078, NASA HPCC CAN NCCS5-146, and a NSF-NASA-AFOSR Interagency Grant NSF-ATM-9318181.

REFERENCES

Balsara, D. S., and D. S. Spicer, A staggered mesh algorithm using high order Godunov fluxes to ensure solenoidal magnetic fields in magnetohydrodynamic simulations, *J. Comput. Phys.*, *149*, 270, 1999.

Brackbill, J., and D. Barnes, The effect of nonzero $\nabla \cdot \mathbf{B}$ on the numerical solution of the magnetohydrodynamic equations, *J. Comput. Phys.*, *35*, 426, 1980.

Dungey, J. W., Interplanetary magnetic field and the auroral zones, *Phys. Rev. Lett.*, *6*, 47, 1961.

Erickson, G. M., R. W. Spiro, and R. A. Wolf, The physics of the harang discontinuity, *J. Geophys. Res.*, *96*, 1633–1645, 1991.

Evans, C. R., and J. F. Hawley, Simulation of magnetohydrodynamic flows: A constrained transport method, *Astrophysical Journal*, *332*, 659, 1988.

Harel, M., R. A. Wolf, P. H. Reiff, R. W. Spiro, W. J. Burke, F. J. Rich, and M. Smiddy, Quantitative simulation of a magnetospheric substorm 1, Model logic and overview, *J. Geophys. Res.*, *86*, 2217–2241, 1981.

Jaggi, R. K., and R. A. Wolf, Self-consistent calculation of the motion of a sheet of ions in the magnetosphere, *J. Geophys. Res.*, *78*, 2842, 1973.

Linde, T., and A. Malagoli, On local $\nabla \cdot \mathbf{B}$ control in ideal MHD simulations, *J. Comput. Phys.*, 2001, in press.

Powell, K. G., An approximate Riemann solver for magnetohydrodynamics (that works in more than one dimension), Tech. Rep. 94-24, ICASE, Langley, VA, 1994.

Powell, K. G., P. L. Roe, T. J. Linde, T. I. Gombosi, and D. L. DeZeeuw, A solution-adaptive upwind scheme for ideal magnetohydrodynamics, *J. Comput. Phys.*, 1999, in press.

Richmond, A. D., Assimilative mapping of ionospheric electrodynamics, *Adv. Space Res.*, *12*, 59, 1992.

Ridley, A. J., D. L. DeZeeuw, T. I. Gombosi, and K. G. Powell, Using steady state MHD results to predict the global state of the magnetosphere-ionosphere system, *J. Geophys. Res.*, 2001a, in press.

Ridley, A. J., K. C. Hansen, G. Toth, D. L. DeZeeuw, T. I. Gombosi, and K. G. Powell, University of Michigan MHD results of the GGCM metrics challenge, *J. Geophys. Res.*, 2001b, submitted.

Roble, R. G., and E. C. Ridley, A thermospheric-ionospheric-mesospheric-electrodynamics general circulation model (TIME-GCM), *Geophys. Res. Lett.*, *21*, 417–420, 1994.

Roble, R. G., E. C. Ridley, A. D. Richmond, and R. E. Dickinson, A coupled thermosphere/ionosphere general circulation model, *Geophys. Res. Lett.*, *15*, 1325–1328, 1988.

Stout, Q., D. L. DeZeeuw, T. I. Gombosi, C. P. T. Groth, H. G. Marshall, and K. G. Powell, Adaptive parallel computation of a grand-challenge problem: Prediction of the path of a solar coronal mass ejection, in *Proc. Supercomputing'98*, 1998.

Stout, Q. F., D. L. DeZeeuw, T. I. Gombosi, C. P. T. Groth, H. G. Marshall, and K. G. Powell, Adaptive blocks: A high-performance data structure, in *Proc. Supercomputing'97*, 1997.

Toth, G., The $\nabla \cdot \mathbf{B}$ constraint in shock capturing magnetohydrodynamic codes, *J. Comput. Phys.*, *161*, 605, 2000.

Toth, G., and P. L. Roe, Divergence- and curl-preserving prolongation formulas, *J. Comput. Phys.*, 2001.

van Leer, B., Towards the ultimate conservative difference scheme. V. A second-order sequel to Godunov's method, *J. Comput. Phys.*, *32*, 101–136, 1979.

van Leer, B., C. H. Tai, and K. G. Powell, Design of optimally-smoothing multi-stage schemes for the Euler equations, no. AIA-89-1933-CP, Buffalo, New York, 1989.

Wolf, R. A., The quasi-static (slow-flow) region of the magnetosphere, in *Solar Terrestrial Physics*, edited by R. L. Carovillano, and J. M. Forbes, pp. 303–368, D. Reidel Publishing, Hingham, MA, 1983.

Wright, J. M., T. J. Lennon, R. W. Corell, N. A. Ostenso, W. T. Huntress, J. F. Devine, P. Crowley, and J. B. Harrison, *National Space Weather Program: The Strategic Plan, FCM-P30, 1995*, Off. Fed. Coord. Meteorol. Serv. Supp. Res., Washington, D. C., 1995.

SUBSTORMS AND MAGNETIC STORMS FROM THE SATELLITE CHARGING PERSPECTIVE

J. F. Fennell, J. L. Roeder and H. C. Koons

The Aerospace Corporation, Los Angeles, CA, 90009, USA

ABSTRACT

Substorms and magnetic storms generate significant space weather effects in the inner magnetosphere. They change the dose rates experienced by satellites in many orbits and are directly linked to the occurrence of satellite charging. Substorms inject hot plasma into the nightside magnetosphere. The drifting electron component of this hot plasma can charge the surfaces of the satellites leading to electrostatic discharges, the associated satellite anomalies and sometimes failures. These occur in regions that are consistent with the expected motions of the substorm injected particles. The high-energy electron enhancements following many magnetic storms can be sufficient to cause charging of shielded satellite elements. Not all magnetic storms result in flux enhancements sufficient to cause such internal charging. Because the induced voltages from the internal charging are usually not directly measured, the anomalies they cause are more difficult to link to the space environment and magnetic storms. However, the anomaly statistics are sufficient to show linkage in a several cases.

INTRODUCTION

The study of magnetic storms and substorms has been a focus of the scientific community for a long time. Much effort has gone into trying to determine the cause and effect relationships for storms and substorms and the resultant changes in the magnetosphere observed by satellites and ground observatories. The space weather community uses the results of these studies in their attempts to predict the occurrence of the storm and substorm events and to try to gauge the seriousness of their consequences. Much of the work in the space weather community is focused on the development of predictive models. The driving force behind this effort is the conviction that the storms and substorms cause serious problems for space borne technology that we rely on for many services. It is also driven by the assumption that knowledge of where and when such environmental events will occur can be used to reduce or manage the problems they cause. In this paper, we present the results of different studies of phenomena that can have deleterious effects on space assets in the inner magnetosphere. In particular, the focus of this paper is to identify satellite charging as a threat to spacecraft and to show the link between such charging, magnetic storms, and magnetospheric substorms.

SATELLITE CHARGING

Satellite charging is a simple concept, and its analog is easily experienced by anyone who received a shock after walking across a rug on a dry day. A similar spark or electrostatic discharge (ESD) can occur on a satellite when its surfaces or interior elements become charged relative to the space plasma or to neighboring satellite components. The energy from ESD can be coupled into electronics causing upsets and damage.

The problems caused by charging on satellites have been compared to other environmental effects in a recent study (Koons, et al., 2000). Some of the main results of this study are summarized in Table 1, which indicates that satellite charging is responsible for more than half (161 out of 198) of the environment-related anomalies. The study also showed (see Table 2 of Koons et al., 2000) that ESD caused about 50% of the lost or terminated missions associated with environmental effects. Thus, the issue of satellite charging is serious, from the perspective of the threat it poses for satellites with orbits in the inner magnetosphere. It is important to understand where and why satellites charge and how the charging is related to magnetospheric and solar-terrestrial processes.

Table 1. Distribution of Records by Anomaly Diagnosis.

Diagnosis	Number of Records
ESD- Internal Charging	74
ESD- Surface Charging	59
ESD - Uncategorized	28
Single Event Effects	85
Damage	16
Micrometeoroid/Debris Impact	10
Miscellaneous	26

(After Koons et al., 2000)

Satellite Surface Charging

The low-energy component of the space plasma incident on a satellite does not penetrate into the satellite materials but resides near the surface. The particles and solar UV generate secondary electrons from the surfaces. The satellite surface aquires a charge until the net current to the surface is zero. This is presented schematically in Figure 1 where the charges flowing to and from a body immersed in the space plasma are shown. The "satellite," in this case, is an insulator. Charge cannot move from the sunlit side to the dark side or vice versa. The surfaces charge until the net current is zero at each element of surface area. This usually means the sunlit areas are slightly positive and the shadowed areas are negative relative to the plasma.

If the surface is a conductor, the net current to the whole surface still becomes zero but the potential would be uniform and either positive or negative relative to the plasma. Plasma electrons have high speed compared to ions of similar energy and are usually the dominant source of initial plasma current to a satellite. The photo and secondary electron currents are usually higher than the plasma electron current, during average conditions. If the satellite is in a "hot" plasma (average electron energy ≥ few hundred eV), its shadowed regions will generally charge negative to significant potentials, sometimes several kilovolts. At times, even the sunlit regions of the body can charge to significant levels.

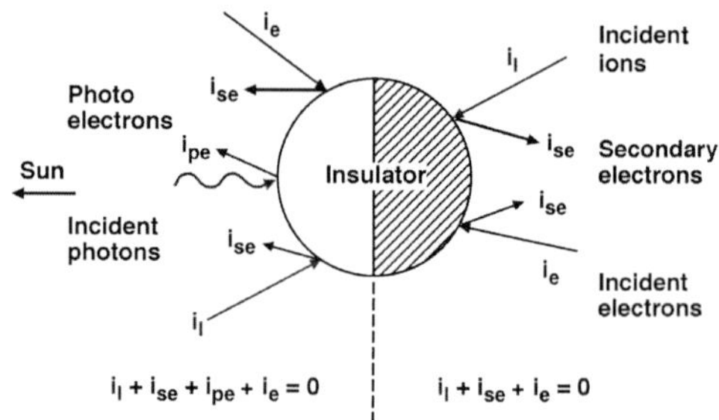

Fig. 1. Cartoon showing the currents that flow to/from an object in the space plasma.

Because the secondary and photoelectron currents are different for every material, satellites constructed from many different conductive and nonconductive materials will have a range of surface potentials. The differences in potential between adjacent materials, such as thermal blankets and metallic structure, can lead to local electrical stress. This can result in arcs. A surface material can discharge into space (a so-called "blow-off" discharge) or to structure-ground. The ESD currents can electromagnetically couple into electronic circuits and subsystems, causing mischief or damage as noted above.

Surface Charging Environment.

The plasma electrons in charging regions usually have characteristic energies from a few hundred eV to a several keV. Above 25 to 30 keV the electrons start to penetrate very thin materials such as paints and start to generate internal charging of thick materials or the underlying structure. In regions where the plasma is very dense, it is usually "cold" and doesn't cause significant charging. The ionosphere and the plasmasphere are such regions. If the plasma is very dilute (density < 0.1 cm^{-3}) photoemission dominates and a satellite with a conductive surface will have a positive potential. This occurs, for example, in the magnetotail lobes.

Thus, significant surface charging occurs where the plasma is "hot", such as in the near-Earth plasma sheet and its extension to lower altitudes in the auroral regions. Plasma-sheet electrons have average energies of a few hundred eV to several keV and densities of ~0.1 to ~1 cm^{-3}. During substorms, hot plasma is injected into the nightside high-altitude equatorial regions as indicated in Figure 2. Gradient-curvature and convective drifts of particles in the magnetosphere's magnetic and electric fields cause the "hot" electrons to drift towards dawn. These freshly injected electrons cause dramatic changes in satellite charging levels. The direction of electron drift leads one to predict that the greatest negative charging levels will be observed beyond the plasmasphere in the pre-midnight through dawn region of the magnetosphere.

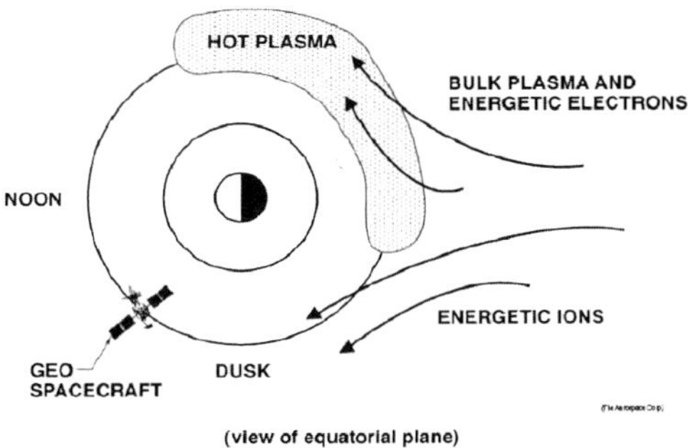

Fig. 2. Cartoon of substorm plasma injection.

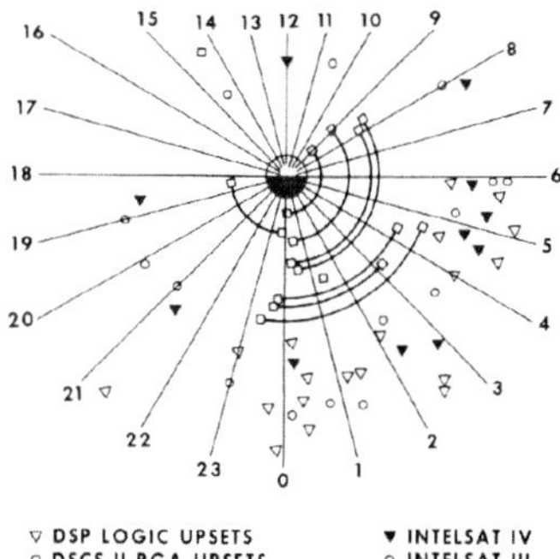

Fig. 3. Local time dependence of anomalies observed on geosynchronous satellites (after McPherson and Schober, 1976).

The auroral region, at altitudes from a few hundred km upward, has a mixed plasma, combining low-density, high-temperature electrons from the equator with cool, high-density ionospheric electrons. During substorms, electric potential drops occur along auroral magnetic-field lines. Density "cavities" can appear and average electron energies rise at ionospheric altitudes because hot electrons are accelerated downward to provide the auroral current. The combination of lower background density and increased electron energies causes satellites to charge in the low altitude auroral regions (Anderson and Koons, 1996), especially in the shadowed and wake regions of satellites. When the background density is sufficiently low, the whole satellite can charge if it is eclipsed, otherwise the shadowed portions will charge.

Equatorial Satellite Surface Charging

Early anomaly studies showed spatial distributions reminiscent of the electron drift patterns sketched in Figure 2. An example of these anomaly patterns is shown in Figure 3. The anomalies were on geosynchronous satellites with each plotted at the local time it occurred. (Radial position is arbitrary.) Most of the anomalies occurred in the 2300 to 0600 local time region. Plots like Figures 2 and 3 showed that anomalies were probably caused by satellite charging. Such equatorial surface charging has been linked to substorm plasma injections specifically and magnetic activity in general.

An example of substorm associated charging of the SCATHA satellite (Fennell, 1982) is shown in Figures 4 and 5. Figure 4 shows spectrograms of the SCATHA plasma data. A substorm plasma injection occurred near 0040 UT. The satellite charging is identified by the bright, low energy bound on the ion spectrogram, caused by the ions being accelerated into the plasma instrument by the satellite potential. These ion "acceleration" signatures show that the satellite was charged negative relative to the plasma. The electron fluxes (Figure 4 top) are reduced and the spectrum is shifted by the effective "retarding" potential of the satellite. These data were used to generate the temporal profile of satellite potential in the top panel of Figure 5. The bottom panel in Figure 5 shows the potential of a thin Kapton sample on SCATHA. The Kapton started to charge at substorm onset, and its potential, relative to the satellite frame, initially increased while the frame potential stayed low. As the satellite entered the Earth's shadow near 0046 UT, the frame charged to high levels as shown in both Figures 4 and 5. At shadow entry, the differential potential between the Kapton and the satellite frame decreased rapidly. The sequence was reversed as SCATHA exited the eclipse. Electrostatic discharges were detected during the rapid changes in potentials associated with the eclipse entry and exit.

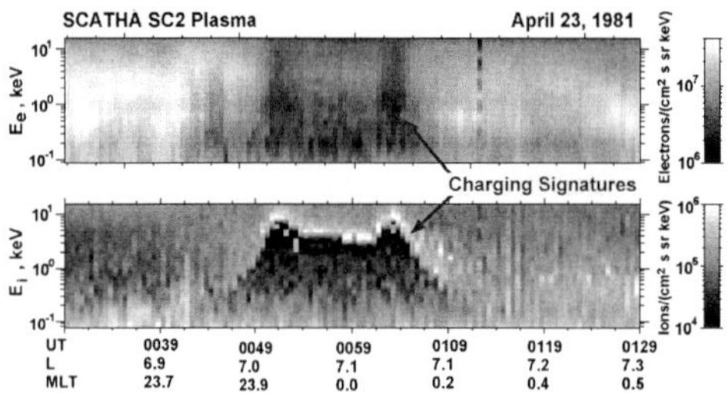

Fig. 4. SCATHA plasma spectrogram showing evidence of satellite charging in both electrons (top panel) and ions (bottom panel).

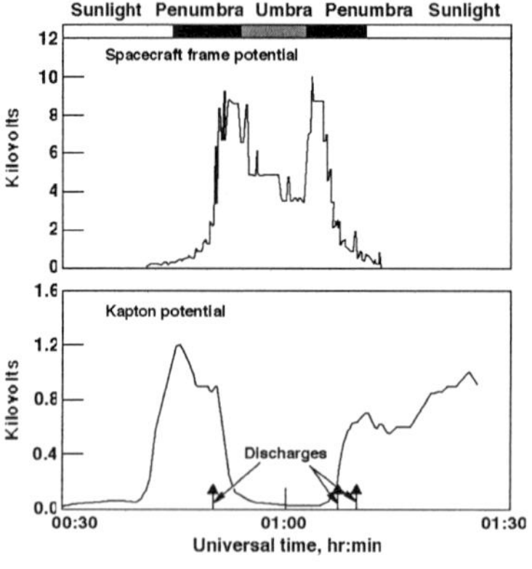

Fig. 5. Example of substorm related charging near midnight. The spacecraft frame potential is shown in the top panel and the potential of a thin Kapton sample is shown in the bottom panel. The potentials were negative.

This example contains many of the common features of surface charging observed by SCATHA. These are: (1) Each dielectric material and the satellite frame responded differently; (2) Discharges tended to occur when the potentials were changing rapidly; and (3) The potentials were never stable during an event, whether it occurred during eclipse or not. Data similar to those in Figure 5 were used to produce statistical maps of surface charging for the SCATHA orbit. An example is

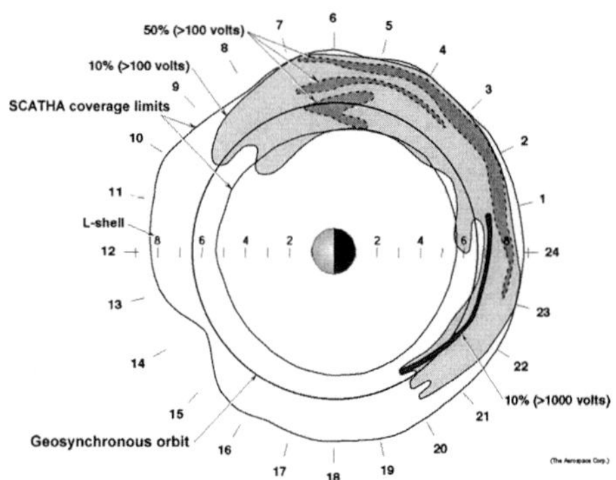

Fig. 6. Location of surface charging as determined by SCATHA.

shown in Figure 6. This map shows that satellite charging in the near-geosynchronous orbit region follows the same pattern as that expected for the drifts of the 10's of keV electrons.

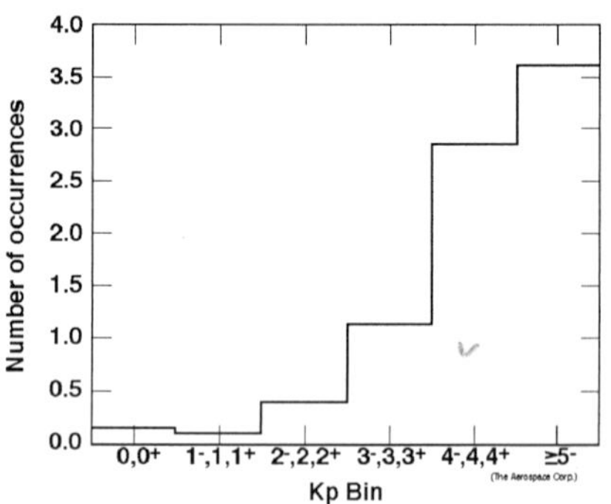

Fig. 7. Anomaly occurrence versus Kp. (after Spence et al., 1993)

Charging-related anomalies on high altitude satellites increased with heightened magnetic activity. The increased occurrence of anomalies with Kp is shown in Figure 7. The steep rise in anomaly occurrence with increasing Kp means that the satellite anomalies are associated with the high levels of magnetic activity usually experienced during substorms and magnetic storms. The local time pattern and Kp dependence of anomalies and charging, plus the direct observation of satellite charging in response to substorms all link the charging and anomalies to the substorm process. However, we emphasize that while surface charging can be tied to substorms, not all substorms will lead to satellite charging and not all charging will lead to anomalies.

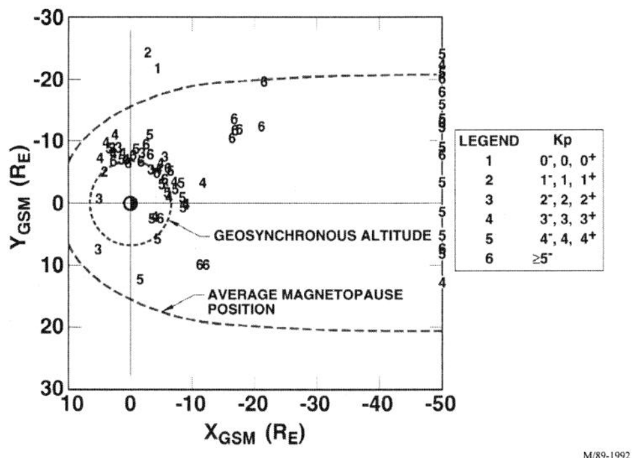

Fig. 8. HEO anomalies mapped to magnetic equator. Symbols indicate Kp value. (after Spence et al., 1993)

High-Altitude Off-Equator Satellite Surface Charging

Observers recognized that auroral displays were associated with disturbances in the high latitude geomagnetic field. Over time, it became obvious that the auroral emissions and the currents were the result of enhanced precipitation of energetic electrons associated with the substorm plasma injections observed at geosynchronous orbit. Finally, the auroral substorm, magnetic substorm, and plasma injections were recognized as different aspects of the same processes called a magnetospheric substorm.

Given these linkages, one might expect that the hot electrons that reached the atmosphere at high latitudes could charge satellites flying through those regions. Under appropriate conditions, "hot" auroral electrons do charge low polar-orbiting satellites. This has been established by DMSP satellite charging observations (Gussenhoven, et al., 1985, Anderson, 2001) and the occurrence of an anomaly associated with such charging (Anderson and Koons, 1996). These studies identified the conditions for satellite surface charging at low altitudes (~850 km) in the auroral zones. i.e., if the plasma density is low, the average energy of precipitating electrons is high and the satellite is eclipsed by the Earth or self shadowed, then it can charge. The substorm processes cause the development of potential drops along the magnetic field lines, enhance the precipitating electron fluxes. These potential drops are regions of downward electron acceleration, ion upflow and lowered background plasma density. For some substorms, the changes in the average electron energy and the plasma density can be extreme. Figures 3 and 4 of Anderson and Koons [1996] showed such conditions where the DMSP F13 satellite charged to several hundred volts and a subsystem on the satellite experienced anomalous operation.

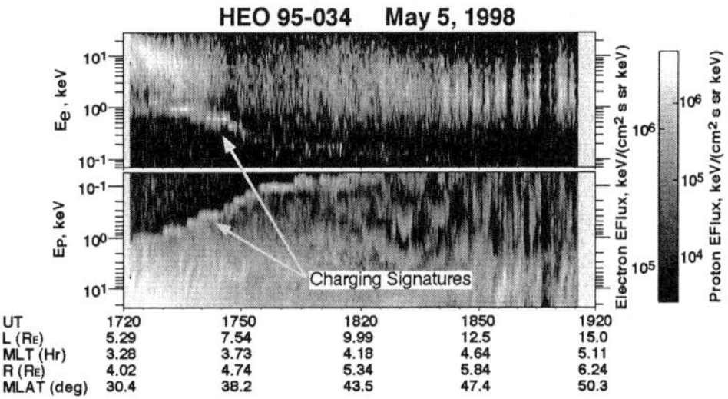

Fig. 9. Plasma electrons (top) and ions (bottom) from the Aerospace HEO 95-034. The data show charging signatures during 1720-1820 UT. The satellite was on field lines that map to regions extending below and above geosynchronous altitudes.

The plasma that maps to the auroral regions exists all along the high-latitude field lines. Any satellite that intercepts these field lines is also connected to the equatorial charging regions and can experience surface charging. This was substantiated by the occurrence of anomalies on HEO/Molniya orbit satellites. Their orbits have high apogees and latitudes (~40,000 km and 63° respectively). They cross magnetic field lines that map to the equator from well inside geosynchronous orbit to significant distances from the Earth, corresponding to the equatorial crossing points that span auroral field lines. Projecting HEO satellite positions along magnetic field lines to the equator for each observed anomaly results in a local time and distance pattern like that shown in Figure 8. The resultant spatial distribution of the HEO anomalies mirrors the pattern expected for substorm injected electrons and satellite charging near the magnetic equator. This pattern is evidence that these satellites were experiencing surface charging related anomalies.

Figure 9 shows an example of charging that was observed on a HEO satellite as it flew poleward at high altitude. When the plasma instrument was turned on at 1720 UT the satellite was already charged. The charging level decreased as the satellite went to higher latitude and larger L. Charging signatures are visible in both the electron and ion data. The satellite structure potential is negative relative to the plasma, as the ion "acceleration" feature indicates. Similarly, the bright low-energy bound in the electron spectrogram is caused by photo emission and/or secondary electron emission by a surface that is charged negatively relative to the satellite structure. The electrons emitted from the charged surface leave with an energy equal to the surface's potential. They are detected by the plasma analyzer at an energy that is equal to the difference between the potential of the emitting surface and the structure ground. The magnitude of the potential of the electron-emitting surface relative to plasma would be the sum of the two potential estimates. For this case it was as much as 1.7 kV. Thus, there were different parts of the satellite at different potentials.

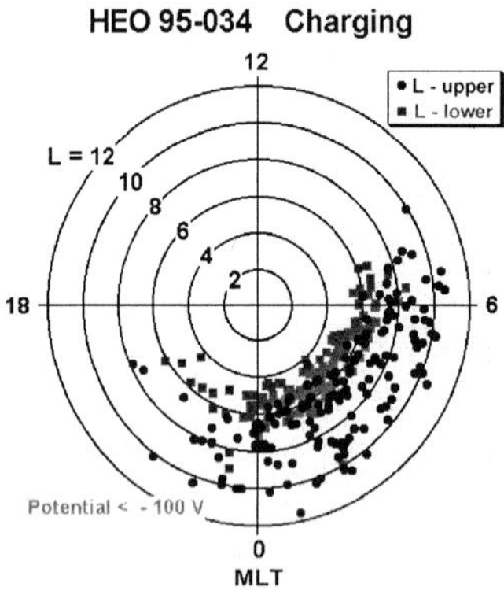

Fig. 10. Occurrence of >100 V satellite frame potentials in HEO/Molniya orbit. The symbols mark the upper () and lower() bounds in L for each charging interval.

Examination of many orbits of HEO plasma data provides a map of the regions where off equatorial satellite charging occurs. Figure 10 is a map for intervals when the HEO frame was charged to more than 100 volts. The squares [ ] and dots [ ] correspond to the lower and upper bounds in L, respectively, of the charging region during a traversal. The spatial pattern of the charging is consistent with that in Figure 6 except it extends to higher L. The lower L bound of charging starts just inside geosynchronous orbit and extends somewhat lower than the region covered by SCATHA. The upper L bound of charging extends into the auroral

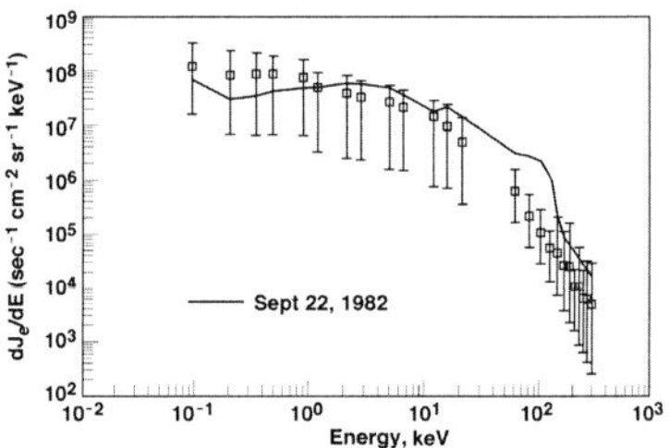

Fig. 11. Comparison of a "Worst Case" plasma electron spectrum and an average electron spectrum. Note the similarities.

regions. This is consistent with our present understanding of the spatial regions accessible to the few to 10's of keV electrons that are injected into the nightside inner magnetosphere during magnetospheric substorms.

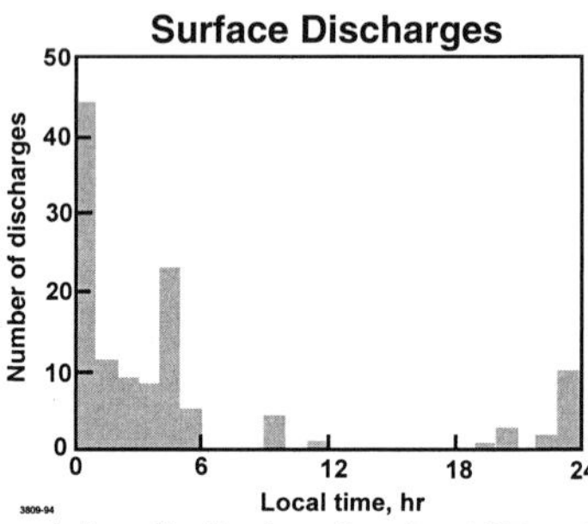

Fig. 12. Local time distribution of surface ESD on SCATHA

Complexities of Surface Charging

Figures 4 and 5 show both the apparent simplicity and inherent complexity of the charging process. The correlation of the satellite frame charging with the increased mean energy of substorm-injected electrons is, at first look, quite simple. However, Figure 5 shows that tracking the satellite frame potential is not the whole answer. The potentials of dielectric materials do not track the satellite frame potential but respond in their own way. The differential potentials that develop are complex. In fact, they are more complex than even these figures indicate (see Mizera, et al., 1980; Leung, et al., 1982). The hazards caused by spacecraft charging result from complex interactions between the space environment and materials and the resultant ESD and electronics.

The shape of the electron distribution function is important to surface charging. At low energies, the secondary-electron yield from surfaces is high. If the low-energy flux is large, it may prevent spacecraft from charging compared to environments with identical high-energy fluxes but small low-energy fluxes. This makes it difficult to predict whether satellites with mixed surface materials will charge, whether ESD will occur and whether a satellite's electronics will respond. Figure 11 provides an example of what the space weather community is up against in trying to predict satellite charging. The solid line shows the electron spectrum that was observed during a severe sunlight-charging event on SCATHA (Koons, et al., 1988).

An "average" electron spectrum, taken at the same spatial coordinates on 15 different non-charging days, is shown for comparison. The days were chosen to be representative of normal conditions. The vertical bars indicate the range of flux variability during the 15 days. One sees that the extreme charging environment differs little from the maximum in normal daily variations. It is only slightly higher in the 10-100 keV range. Yet, the response of SCATHA to this difference was quite extraordinary.

We show one final observation that ties the concept of substorm electron injections with surface charging and satellite anomalies. Figure 12 shows the local time distribution of ESD during surface charging events on SCATHA. The local time pattern is much like that in Figures 2, 3, 6, 8, and 10. These link the observations of surface charging with predictions of how injected electrons drift, with observations of ESD noise and with satellite anomalies.

Internal Charging

What is internal charging? It is simply the deposition of charge on the internal elements of a satellite by electrons with sufficient energy to penetrate through the satellite skin. The electrons can deposit their charge in thick dielectrics near the surface of the satellite, in the interior, or on isolated conducting structures inside the satellite. In any case, if the leakage path to ground is sufficiently resistive the charge can build up over time until ESD occurs. If the charge is on a conductive element, all the charge is removed from the element by the discharge and deposited in surrounding, usually grounded, elements. If the charge is in a dielectric it may "break down" from high voltage stress and some (not all) of the charge will flow away. The energy in the discharge can be coupled into electronics as a fast signal or can over-voltage devices and damage them. Internal charging can lead to satellite anomalies by this mechanism. Most of the time the satellites can recover from the anomaly. In rare cases, the anomaly can impair vehicle operations or even be fatal.

As Table 1 shows, internal charging causes a significant fraction of charging related anomalies. As surface charging was being established as a serious threat to satellites, it was realized that electrons with energies sufficient to penetrate significant thickness of satellite materials existed in the inner magnetosphere and could cause internal charging (or bulk charging as it is sometimes called) of satellites.

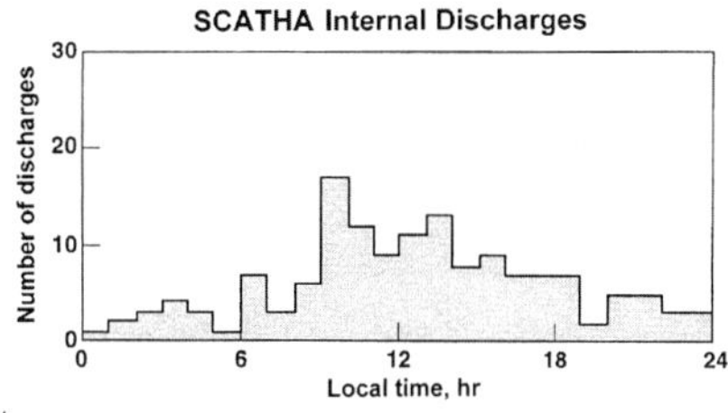

Fig. 13. Local time distribution of internal ESD on SCATHA. (after Koons and Gorney, 1991)

<u>Internal Charging Observations</u>

SCATHA data hinted that internal charging might be a cause of satellite problems (Koons, et al. 1983). Figure 13 shows the local time distribution of internal ESD pulses that were detected by SCATHA. Note that the distribution of internal discharges had a broad peak near local noon. This may result from the fact that a near geosynchronous satellite is on lower L values near noon than near midnight, because of the asymmetric magnetospheric magnetic field. The penetrating electron fluxes tend to peak inside geosynchronous orbit and thus would more rapidly charge the interior of the satellite when it was at lower L near noon.

Once one accepts the existence of internal charging and recognizes that it can lead to ESD, a reexamination of Figures 3 and 8 leads one to suspect that some of the anomalies plotted there may have been caused by internal charging. For example, in Figure 3 there are a few

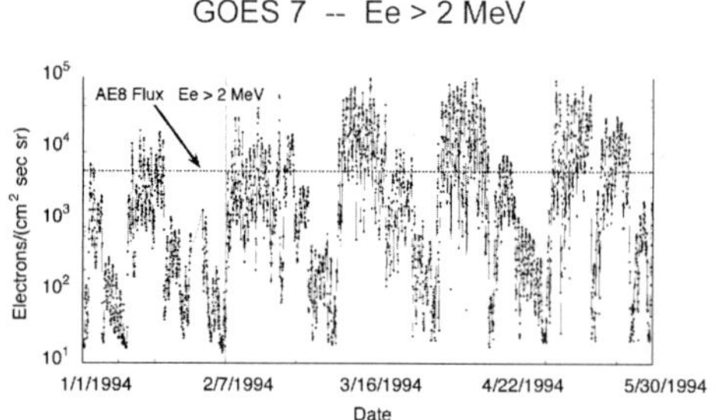

Fig. 14. Variation in the energetic electron fluxes at geosynchronous orbit during a period of successive high speed solar wind streams in 1994.

anomalies in the noon sector. Similarly, a few anomalies occurred on the HEO/Molniya satellites (Figure 8) in the noon sector. So far, there have been no observations of high levels of surface charging in the noon sector. This is consistent with the fact that the substorm injected electrons have reduced fluxes by the time they drift to noon. It is most likely that there were internal charging anomalies on satellites from the beginning but that they were not recognized as such initially.

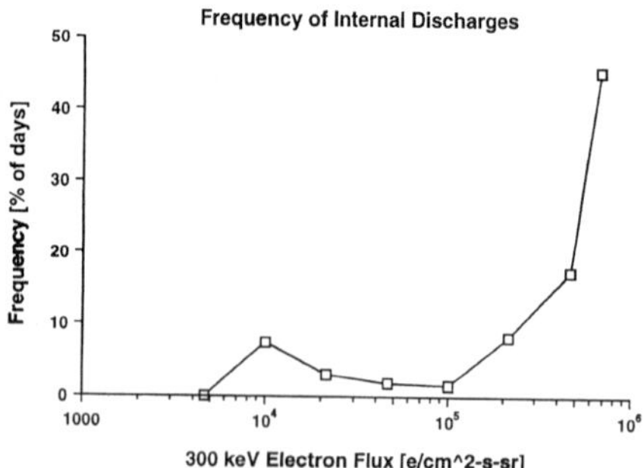

Fig. 15. Comparison of SCATHA anomalies with energetic electron fluxes.

Causes of Internal Charging-Magnetic Storms

A common source of energetic electron enhancements in the inner magnetosphere is magnetic storms. Magnetic storms are often generated by CME's (coronal mass ejections), which appear as "magnetic clouds" with high bulk speed and a structured magnetic field. The interplanetary magnetic field of the cloud rotates in direction relative to the Earth-Sun line, often presenting a strong southward magnetic component. These changes in the field direction can take hours, which means the interplanetary magnetic field impinging on the Earth's magnetic field can be southward for hours simultaneous with a high speed solar wind and sometimes elevated solar wind density. This combination causes an efficient coupling of the solar wind energy into the magnetosphere. The energy is coupled into the plasma in the magnetosphere and enhances the transport and energization processes there. This can result in a large enhancement in the particle fluxes drifting into and through the magnetosphere from the nightside. They carry an enhanced current that opposes the Earth's magnetic field leading to a reduction of the field at the Earth's surface. This is called a ring current and is characterized by a magnetic disturbance index D_{ST}. The geoeffectivness of a CME or magnetic cloud associated magnetic storm can, in some sense, be quantified by the magnitude of D_{ST}. As D_{ST} rapidly drops (during the main phase of the storm), the energetic electron fluxes are often reduced significantly in the inner magnetosphere (Blake, et. al. 1997). As the D_{ST} recovers, if the interplanetary conditions are just right, the energetic electron fluxes will increase by orders of magnitude over their pre-storm values.

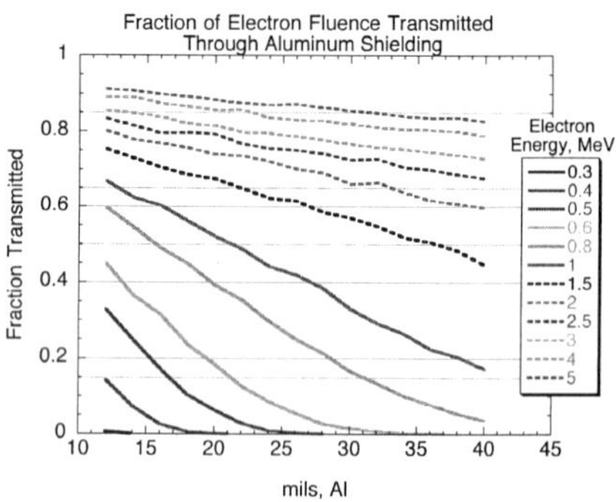

Fig. 16. Fractional reduction in electron flux by Aluminum shielding.

Inner magnetosphere energetic electron fluxes ($E_e \geq 300$ keV) are highly variable and their enhancements have been associated with high-speed solar wind streams (Paulikas and Blake, 1979) as well as to magnetic storms, as noted above. Recently, it has been shown that such enhancements in the energetic electron fluxes requires not only an enhanced solar wind velocity but also a southward component of the interplanetary magnetic field (Blake, et al., 1997). There is evidence that the energetic electron flux levels may also be related to the fluctuations in the solar wind velocity (Li and Temerin,

2000 and references therein). It is argued that these variations drive the diffusive transport of electrons into the inner magnetosphere causing large the variations (orders of magnitude) in the electron fluxes near geosynchronous orbit. Figure 14 shows an example of such electron flux variability. During early 1994, there was a nearly periodic arrival of high-speed solar wind streams at Earth. The energetic electron fluxes varied by orders of magnitude. In particular, they exceeded the predicted average levels by more than an order of magnitude for days at a time. Some satellites experienced anomalies during this period that were ascribed to internal charging.

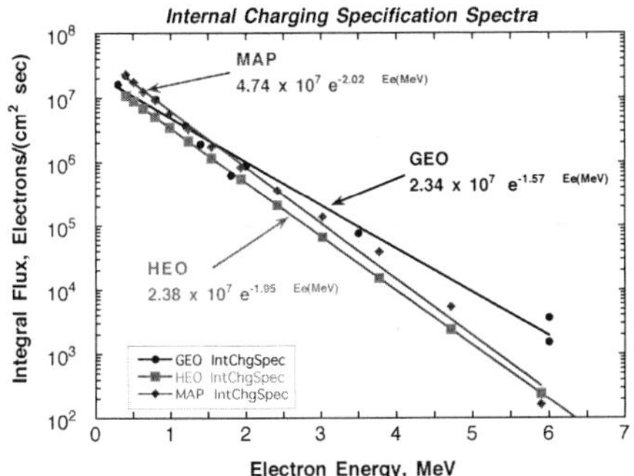

Fig. 17. Examples of worst case 10-hour-average electron spectra for three different orbits.

Relation Between Internal-Charging, ESD and Penetrating Electrons

Both SCATHA (Fennell, 1982) and CRRES (Johnson and Kierein, 1992) carried science and engineering instrumentation that could measure ESD from charging as well as the electron fluxes that could cause it. Figure 15 shows one example of the kind of data obtained. SCATHA observed an increased frequency of internal discharges with increased energetic electron flux. Both SCATHA and CRRES showed that if average fluxes of >300 keV electrons were greater than 10^5 electrons/(cm^2 s sr) the rate of internal discharges increased dramatically. Frederickson (1992) indicated that a ten-hour-average penetrating-electron flux greater than 10^5/(cm^2 sec) was a possible reference level for the onset of discharges from internal charging. This level has been adopted as the maximum average flux that should be allowed to penetrate into the interior of a satellite (Fennell, et al, 2000, and references therein).

Whether discharges from internal charging occur or not depends on the amount of shielding available to protect sensitive satellite circuitry. Figure 16 shows how shielding reduces the level of electron fluxes that can reach a satellite's interior. The peak levels of electron fluxes depend on interplanetary conditions and magnetospheric processes. The maximum long-duration electron fluxes experienced by a satellite depends on its orbit. A satellite that spends a long time in the heart of the radiation belts, as the GPS satellites do, will experience very high fluxes and require very thick shielding to protect them from internal charging.

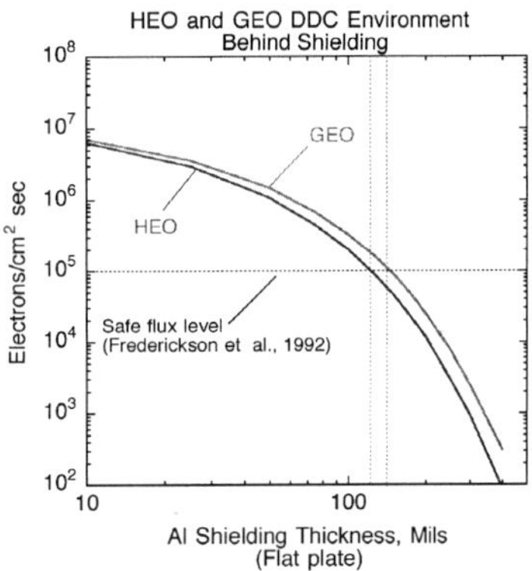

Fig. 18. Shielding required to protect HEO and GEO satellites from the worst case average electron spectra.

Internal Charging Specifications

The major unknown for internal charging is what the worst electron fluxes could be. For example, what is the result of a "100-year" magnetic storm? Since we have only been making space plasma measurements for slightly more than 30 years, it is possible we haven't experienced the "worst case" storm, yet. We have had continuous monitoring of the energetic particle fluxes in the inner magnetosphere only in the last decade and then only for $L \geq 4$ near the magnetic equator. To date, the energetic particle measurements needed to specify the extreme conditions have not been routinely taken throughout the inner magnetosphere where internal charging is a problem.

One can intercompare measurements taken by different satellites to try to infer what the worst case fluxes could be. Fennell, et al. (2000) has done this using CRRES, HEO, GPS and geosynchronous energetic electron data.

Storm time data from these spacecraft were examined and it was found that the great magnetic storm of March 1991 was a good representation of a worst case storm (to date) for a range of L from geosynchronous earthward. Fennell, et al. (2000) used those data to generate worst case average spectra for a few special satellite orbits. A 10-hour averaging interval was selected, after Frederickson, et al. (1992). The resultant 10-hour average spectra for geosynchronous (GEO), HEO/Molniya (HEO), and a lunar transfer phasing trajectory (MAP) orbits are shown in Figure 17. The kind of shielding required to protect satellites in HEO and geosynchronous from suffering internal charging problems can derived from these spectra. The results of such a calculation are shown in Figure 18 for HEO and GEO. The line at 10^5 electrons /(cm^2 sec) is a reference level at which internal discharges may start to occur. Whether they do or not, and whether they cause sever problems for satellites in those orbits or not, depends on more than the electron fluxes. The technology and engineering design practices used during spacecraft development also play a role but that is outside the scope of this paper.

Figure 18 represents the kind of shielding levels that are required to withstand expected "worst case" fluxes that can be generated by a magnetic storm. However, these are the worst fluxes that have been observed over the last 12 years or so and they could be exceeded by the next large storm. Data are still being taken and past measurements are being examined using "extreme event" statistics to try to determine if the levels in Figure 17 are consistent with “100-year” storm levels (Koons, 2001).

DISCUSSION

The linkage of substorms with surface charging and magnetic storms with internal charging is clear, as noted above. The difficulty involves (1) predicting when storms and substorms will occur, (2) predicting the particle environment that will result and (3) predicting whether the environment will cause a problem for a given satellite.

Predicting substorms isn't possible at present. While there is a description of what occurs once a substorm starts, the onsets have not been successfully predicted. We have clues that a substorm may occur, such as the increase in magnetic flux in the magnetotail lobes and the stretching of the tail exhibited by the change from a dipolar to non-dipolar field in the near geosynchronous regions. There are corresponding changes in the particle distributions. However, the magnetosphere can return to its unstressed state without a substorm. When substorms do occur, we cannot predict the onset time with any accuracy. We also cannot predict the changes in the particle distributions that cause the surface charging. For example, we cannot predict the electron spectral shapes or the exact spatial regions that will experience the hot electron fluxes that cause charging. Finally, and most important, we cannot predict whether a specific satellite will suffer problems from a given substorm environment. There are too many imponderables. Two "identical" satellites often respond to the same environment in different ways. This is because they are always different from the perspective of environmental interaction.

The ability to predict magnetic storms is much greater than for predicting substorms. However, we are far from making reliable predictions. The recent SOHO observations and modeling have taken us a long way towards predicting whether a CME will strike the Earth's magnetosphere. Future advances will raise our success rate for predicting the arrival of CME effects at Earth. However, we still do not know whether a particular CME will be geoeffective or to what degree. Work on predicting the field geometry of the "cloud" as it arrives at Earth is progressing. It will be years before we can predict whether a CME will have a small, moderate or large effect at Earth. At present, all earthward directed halo-CME's are expected to have large affects, if we believe the news releases. That is obviously not true. Many have no effect at all. Since magnetic storms usually have many associated substorms, they are, in some sense, also a source of surface charging events. Even if we ignore the storm-related substorms and focus only on the possible storm enhancements of the energetic electrons that cause internal charging, we are still left with difficulties.

At present, we could just track D_{ST} and attempt to use it as proxy for enhancements of the energetic electrons. The L position of the post-storm peak in the energetic electron fluxes appears to track D_{ST} fairly well with a delay of a day or more from the time D_{ST} has reached its minimum value (Tverskaya, et al., 2000). What we cannot predict are the spectral shapes and flux levels as a function of L. For geosynchronous satellites, one could use near-real-time measurements to track the spectra and make running estimates of the flux behind different shielding thickness. Individual satellite operators could then track the levels that they feel are important to their systems, based on experience. However, such near-real-time data is not yet readily available.

It is clear that steady progress is being made in understanding the relationship between magnetospheric processes and charging related effects on satellites. There is also progress toward predicting the occurrence of storms and being able to predict and now-cast whether the storm related electron flux changes are approaching critical levels. We are a long way from providing useful predictions to the satellite operators at the high level of confidence they require. This is especially true for surface charging where we cannot predict substorm onsets and resultant environmental changes with any degree of reliability. The substorm-related surface charging problems is a big challenge for the whole space-weather community and is likely to remain so for the near future.

ACKNOWLEDGEMENTS:

We would like to thank P. C. Anderson for useful discussions about low altitude surface charging. This work was supported, in part, by the Aerospace Corporation investment program under U. S. Air Force contract F04701-93-C-0094 and in part by NASA LWS grant NAG5-10972 and NASA NRA grant NAG5-9037

REFERENCES

Anderson, P. C., and H. C. Koons, Spacecraft charging anomaly on a low-altitude satellite in an aurora, J. Spacecraft Rockets, 33, 734, 1996.

Anderson, P. C., A survey of spacecraft charging events on the DMSP spacecraft in LEO, in press, ESA SP-476, 2001.

Blake, J.B., D. N. Baker, N. Turner, K. W. Ogilvie, and R. P. Lepping, Correlation of changes in the outer-zone relativistic electron population with upstream solar wind and magnetic field measurements, Geophys. Res. Lett., 24, 927, 1997.

Fennell, J. F., "Description of P78-2 (SCATHA) Satellite and Experiments," in The IMS Source Book, C. T. Russell and D. J. Southwood, Eds., Am. Geophys. Union, Washington, D. C., 1982.

Fennell, J. F., H. C. Koons, M. Chen and J. B. Blake, Internal charging: A preliminary environmental specification for satellites, IEEE Trans. on Plasma Science, 28, 2029, 2000.

Frederickson, A. R., et al., Characteristics of spontaneous electrical discharging of various insulators in space radiation, IEEE Trans. on Nuclear Science, 39, 1773, December 1992.

Gussenhoven, M. S., D. A. Hardy, F. Rich, W. J. Burke, and H. -C. Yeh, High level charging in the low-altitude polar auroral environment, J. Geophys. Res., 90, 11000, 1985.

Johnson, M. H., and J. Kierein, Combined release and radiation effects satellite (CRRES): Spacecraft and mission, J. Spacecraft Rockets, 29, 556, 1992.

Koons, H. C., Summary of environmentally induced electrical discharges on the P78-2 (SCATHA) satellite, J. Spacecraft Rockets, 20, 425, Sept. 1983.

Koons, H. C., P. F. Mizera, J. L. Roeder, and J. F. Fennell, "Severe Spacecraft-Charging Event on SCATHA in September 1982, J. Spacecraft and Rockets, 25, 239, 1988.

Koons, H. C. and D. J. Gornery, The relationship between electrostatic discharges on spacecraft P78-2 and the electron environment, J. Spacecraft Rockets, 28, 683, 1991.

Koons, H. C., J. E. Mazur, R. S. Selesnick, J. B. Blake, J. F. Fennell, J. L. Roeder, and P. C. Anderson, The impact of the space environment on space systems, in Proceedings of the 6th spacecraft Charging Technology Conference, Air Force Research Laboratory, in press, 2001.

Koons, H. C., Statistical analysis of extreme values in space science, J. Geophys. Res. 106, 10,915, 2001.

Leung, M. S., P. F. Mizera, and R. M. Broussard, Space effects on physical properties of materials, Aerospace Corp., ATR-82(8378)-1, Nov. 1982.

Li, X, and M. Temerin, Review of the electron radiation belt, Sp. Sci. Rev, in press, 2000.

McPherson, D. A., and W. R. Schober, Spacecraft charging at high altitudes: the SCATHA satellite program, in Spacecraft Charging by Magnetospheric Plasmas, A. Rosen Ed., Progress in Astronautics and Aeronautics, 47, p15, 1976.

Mizera, P. F., H. C. Koons, E. R. Schnauss, D. R. Croley, H. K. Kan, M. S. Leung, N. J. Stevens, F. Berkopec, J. Staskus, W. L. Lehn, and J. E. Nanewicz, First results of material charging in the space environment, App. Phys. Lett., 37, 276, 1980.

Paulikas, G. A., and J. B. Blake, Effects of the solar wind on magnetospheric dynamics: energetic electrons at the synchronous orbit, in Quantitative Modeling of Magnetospheric Processes, W. P. Olson, Ed., 180, Am. Geophys. Union, Washington D. C., 1979.

Spence, H. E., J. B. Blake, J. F. Fennell, "Surface Charging Analysis of High-Inclination, High-Altitude Spacecraft: Identification and Physics of the Plasma Source Region," IEEE Trans. Nucl. Sci., 40,1521, 1993.

Tverskaya, L. V., N. N. Pavlov, J. B. Blake, R. S. Selesnick, and J. F. Fennell, Predicting the L-position of the storm-injected relativistic electron belt, Adv. Sp. Res., in press, 2000.

PROPAGATION OF SUDDEN IMPULSES IN THE MAGNETOSPHERE: LINEAR AND NONLINEAR WAVES

Dong-Hun Lee[1] and Mary K. Hudson

Dept. of Physics and Astronomy, Dartmouth College, Hanover, NH 03755, USA

ABSTRACT

Dynamics of MHD wave propagation is studied in the magnetosphere when strong initial impulses are assumed at the magnetopause. In a linear wave approximation, we will present simulations results, which show the strong effects of refraction in the magnetosphere. In a nonlinear approximation, we adopt the approach of simple waves which is often used in ordinary fluids. We obtain an exact MHD solution for the velocity profile in space and time when arbitrary initial perturbations are assumed in a simplified model.

INTRODUCTION

Interplanetary shocks and solar wind discontinuities can give rise to relatively large-scale perturbations in the magnetosphere. This impact at the dayside magnetopause is expected to launch sudden impulses (SI) into the magnetosphere. It has been more than three decades since hydromagnetic waves were adopted to understand the nature of these electromagnetic pulses in the magnetosphere (Francis *et al.*, 1959; Wilson and Sugiura, 1961; Nishida, 1964; Stegelmann and von Kenschitzki, 1964; Tamao, 1964; Burlaga and Ogilvie, 1969). In the ground-based observations, the onset time distributions and their observational characteristics were investigated and well summarized by Araki (1977, 1994). Ionospheric responses associated with the SI were also studied by making assumptions about the earth-ionosphere waveguide system (Kikuchi and Araki, 1979a,b). The spatial and temporal structures have also been investigated by simultaneous observations on satellite measurements as well as ground-based magnetometers (e.g., Knott et al., 1982, 1985; Nopper et al., 1982; Wilken et al., 1982; Nagano and Araki, 1984; Wedeken et al., 1986; Petrinec et al., 1996; Araki et al., 1997). Current observations, however, have not revealed the details of the SI propagation on its way from outer space to the ionosphere. Since a limited number of satellites are not capable of tracing the propagation of disturbances, and ground-based measurements are allowed to see only secondary signals modified by the ionosphere, it becomes important to study the propagation in terms of numerical and analytical tools. Recently, hydromagnetic wave properties of the SI have been studied both numerically (Lee and Kim, 2000; Lee and Hudson, 2000) and analytically (Lee et al., 2000). In this paper, we will discuss the time-dependent properties of the SI propagation in the magnetosphere in terms of linear and nonlinear MHD waves.

LINEAR AND NONLINEAR APPROXIMATIONS

The two different approaches have their own merits. Linear waves allow us to easily develop a numerical model which can include the effects of curved geometry and inhomogeneity. However,

1 On leave from *Kyung Hee University, Kyunggi, 449-701 Korea*

these waves are no longer valid when the convective motion becomes important. Nonlinear waves can carry full information about the perturbation, but often prohibit us from obtaining self-consistent solutions. In order to examine when/where these two approximations are differentiated, it is implicit to begin with the source properties of perturbations (Lee et al., 2000). From the equation of motion in MHD, $\partial v/\partial t + \vec{v}\cdot\vec{\nabla} v \approx v/\tau_1 + v^2/x \sim v/\tau_1 + v/\tau_2$ is obtained, where the second term introduces the nonlinear effect. When the amplitude v is not sufficiently small or when the period τ_1 of the local variations is comparable to the convection timescale $\tau_2 \sim x/v$, nonlinear effects are expected to appear. In an initial-valued problem with a finite size system, the appearance of such nonlinear effects would largely depend on the source properties. If the source perturbation is strongly impulsive but still small enough, satisfying $\tau_1 \ll \tau_2$ and a relatively small v, it would provide a broadband input and excite a lot of disturbances which can be approximated as linear waves. If the source provides a DC perturbation over a relatively long timescale, which results in slow variations at a local point ($\tau_1 \sim \tau_2$), the propagation of such disturbances should include the nonlinear effect. In the following sections we will present 1) the numerical results in a linearized dipole model, which includes the effects of curved geometry and inhomogeneity in the magnetosphere, and 2) the analytical results in a fully nonlinear box model, which rather focuses on the steepening process. As discussed above, the linear model is expected to approximate the leading edge of the SI propagation at a very early phase, where we discuss how refraction and reflection occur in a highly structured magnetosphere. In the nonlinear approach, we introduce a simple wave solution, which is full nonlinear in a box-like geometry. Based on this time-dependent analytical solution, we will discuss the possibility of shock formation inside the magnetosphere when various timescales are assumed for the SI.

LINEAR MODEL

This numerical experiment is based on the recent three-dimensional dipole model (Lee and Hudson, 2000) which adopted different boundary conditions at different local times. The inner boundary is at (L=2) and the outer boundary is at (L=10.5), and perfect reflecting boundary conditions are used at the ionospheres, which are assumed to be at an altitude of 0.5 R_E. We apply different boundary conditions at the dayside and nightside outer regions, respectively. At the dayside magnetopause the Alfven speed drops from $V_M = 600$ km/s in the magnetosphere to $V_S = 60$ km/s in the solar wind. Thus the reflection coefficient at the dayside boundary is given by $r = (V_M - V_S)/(V_M + V_S)$ when waves are propagating out of the magnetosphere. With this boundary condition, wave energy can escape from the magnetosphere across the dayside magnetopause. In the nightside region, we assume a free boundary condition, that there is no reflection from the nightside magnetosphere (Lee and Hudson, 2000). The two linearized wave equations in a cold plasma (Lee and Lysak, 1989) are solved in the dipole coordinates $(\hat{\mu}, \hat{\nu}, \hat{\phi})$ similar to those used by Chen and Hasegawa (1974). Here $\hat{\nu}$ is normal to the field line pointing outward, $\hat{\phi}$ is the usual azimuthal direction ($\hat{\mu} = \hat{\nu} \times \hat{\phi}$). We assume a propagating impulse at the dayside magnetopause. This impulse is given by E_ϕ which corresponds to a radial compression. Disturbances are assumed to be symmetric in both latitude and local time and the size of an impulse is assumed to be about $5\,R_E$ x $10\,R_E$ in each direction at the dayside magnetopause. We have investigated time histories at various locations. A set of grid points has been selected across the magnetic shells near the equator and the polar ionosphere slightly above the ionosphere as shown Figures 1 and 2.

In Figure 1 of the dayside equator, the impulse is propagating across the equatorial magnetic shells, which are evident in the compressional components (B_μ, B_ν, E_ϕ). The transverse components (B_ϕ, E_ν) stay quiet since they have nodes at the equator due to the symmetry of the initial impulse. The wave propagation becomes slower inside the plasmasphere (L<6) owing to the relatively small Alfven speed. It takes $t \approx 70$ s until the signal reaches the inner magnetic shell near $L = 2.4$. In the dayside polar region, Figure 2 shows that the earliest signals are found around $t \approx 40$ in all components, which indicates that the travel time from the subsolar point to the polar region can be shorter than that to the deep plasmasphere near the equator. Thus effects of refraction are very large in the magnetosphere in the sense that the geometric path between the source and the inner plasmasphere is shorter than that between the source and the polar ionosphere, which is consistent with the previous numerical studies (Lee and Kim, 2000). It is interesting to note in Figure 2 that the compressional components almost simultaneously arrive in the polar region except for part of the plasmaspheric region, while the transverse components become differentiated in latitude. In the transverse components in Figure 2, the earliest signals are found near $L=6$ where the Alfven speed become largest in the magnetosphere. This feature can be explained by the wave coupling between the compressional and transverse waves. When the initial impulse propagate radially inward from the magnetopause, transverse waves are continuously excited and start propagating along each field line.

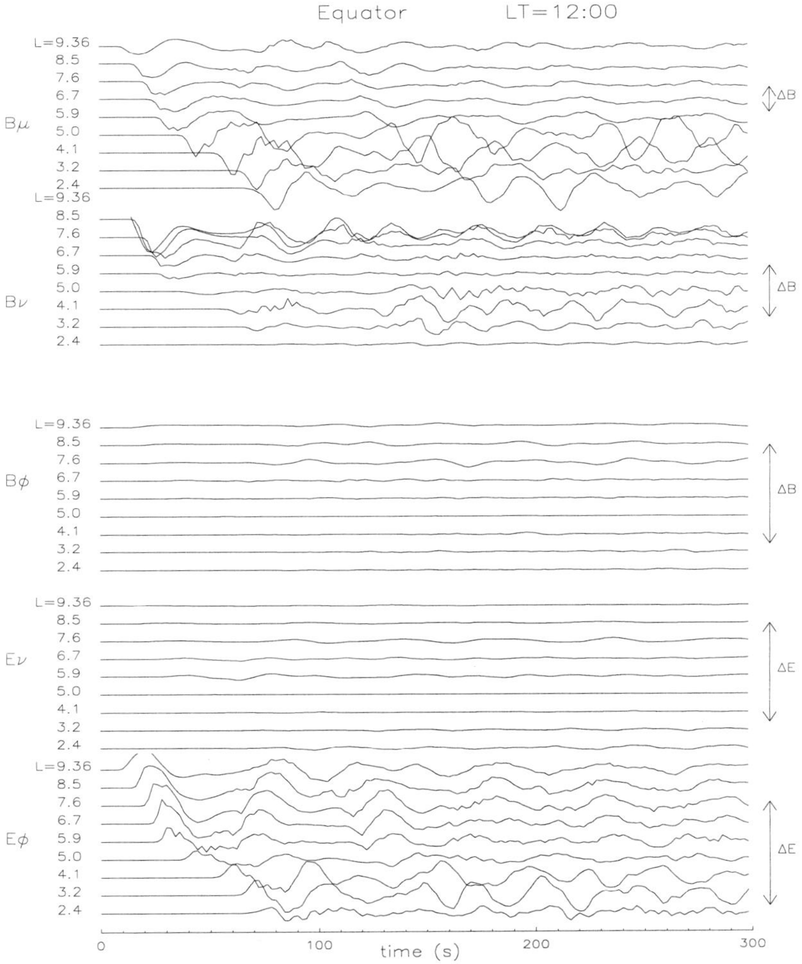

Figure 1. Time histories of the magnetic and electric fields at the dayside equatorial region when the local time is LT=12:00. Arbitrary units are represented by ΔB and ΔE, respectively.

Our results suggest that the path of the least travel time to the polar ionospheric region is along the magnetic field line located near the outer edge of the plasmapause where the Alfven speed is largest in the outer magnetosphere. For higher latitudes, the wavefronts become slightly delayed. Compressional components in Figure 2 are sensitive not to the direction of the magnetic field, but to the magnitude, which is associated with the Alfven speed. Thus the travel paths become independent of the field lines and the arriving signals are almost in phase over different latitudes in the polar region except for the deep plasmasphere. On the other hand, the transverse waves in the polar region have phase shifts at different latitudes since they are guided along different field lines. It is interesting to note that the leading edges of B_ϕ and E_ν in Figure 2 show polarization reversals near $L=6.7$ in addition to the continuous phase shifts in arrival times.

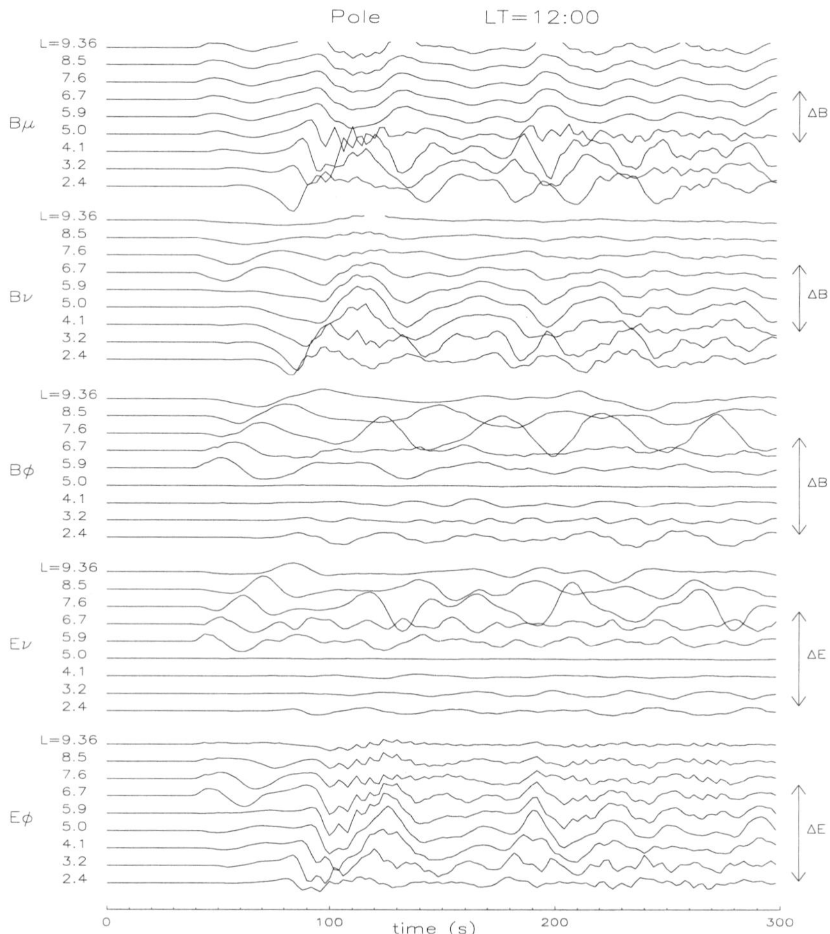

Figure 2. Time histories of the magnetic and electric fields at the dayside polar region (LT=12:00).

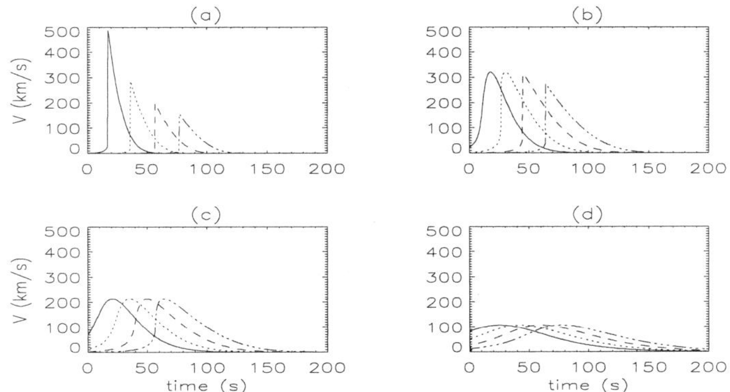

Figure 3. Time histories of the velocity for different τ_o: (a) τ_o=10, (b) τ_o=20, (c) τ_o=30 and (d) τ_o=60 s. The solid, dotted, dashed, and dash-dotted lines represent the time histories at the different locations x=3,5,7, and 9 R_E, respectively.

NONLINEAR MODEL

In the MHD equations in a one-dimensional inhomogeneous plasma. All variables are assumed to be functions of x and t only. The magnetic field is also assumed to be perpendicular to the x direction. If the perturbation source is given by $X(t)$ at a certain point, $v=\dot{X}(t)$ represents the velocity at the same point. The simple wave solutions (Lee et al., 2000) become

$$x = X(\tau) + (t-\tau)\{\dot{X}(\tau) \pm C_T[\dot{X}(\tau)]\}, \quad v = \dot{X}(\tau) \tag{1}$$

$$\frac{C_t}{V_{Ao}} = \frac{v-v_o}{2V_{Ao}} - \frac{C_{So}^2}{2^{1/3}V_{Ao}^2}\left(\frac{v-v_o}{V_{Ao}}\right)^{1/3} \tag{2}$$

$$v_o = \mp 2V_{Ao}\left(1+\frac{C_{So}^2}{V_{Ao}^2}\right)^{3/2} \tag{3}$$

where C_{So} and V_{Ao} represent the unperturbed sound and Alfven speeds, respectively. The plus sign corresponds to wave propagation in the direction of the positive x axis, and the minus sign represents wave propagation in the opposite direction. Equation (1) provides an expression for v in parametric form as a function of x and t since τ can be eliminated. If the motion of the source is assumed to be $X(t)=X_o\tanh(t/\tau_o)$, $v=X_o/(\tau_o\cosh^2 t/\tau_o)$, X_o represents the displacement of the boundary and τ_o represents the timescale of that motion. We assume that V_{Ao}=600 and C_{So} =60 km/s, respectively. The magnetopause is assumed to be at x=0. A displacement of the magnetopause of $X_o=1\,R_E$ is assumed to occur for four different timescales: τ_o=10, 20, 30, and 60 s. These qualitatively correspond to perturbations driven by the solar wind that have different rising times. Figure 3 shows the time histories of velocity at various locations such as x = 3, 5, 7, and 9 R_E. It is evident that the wave steepens as it travels owing to the nonlinear effect. As the rise time of the source movement becomes shorter, the steepening occurs in relatively shorter periods. Figures 3a and 3b shows that the shocks may be easily formed if the magnetopause moves toward the Earth with τ_o = 10 - 20 s. When τ_o becomes larger than 30 s for the same displacement, Figures 3c and 3d show that no sharp discontinuities are likely to form in the dayside magnetosphere, which is assumed to have a size of 10 R_E in radial distance. Therefore our results suggest that the formation of shocks strongly depends on the timescale of the boundary movement. When the movement occurs in relatively short periods, the leading edge becomes steep easily. If the compression at the magnetopause is assumed to last less than 2 min (τ_o=30) with the total displacement of about 2 R_E, the wavefront is expected to form sharp discontinuities in the magnetosphere.

ACKNOWLEDGEMENTS

This work was supported by the National Science Foundation grant (ATM 9911975)..

REFERENCES

Araki, T., Global structure of geomagnetic sudden commencements, *Planet. Space Sci.*, 25, 373 (1977).

Araki, T., A physical model of the geomagnetic sudden commencement, in *Solar Wind Sources of Magnetospheric Ultra-Low-Frequency Waves*, Geophys. Monogr. Ser., vol. 81, ed. M. J. Engebretson, K. Takahashi, and M. Scholer, pp. 183-200, AGU, Washington, D.C. (1994).

Araki, T., et al., Anomalous sudden commencement on March 24, 1991, *J. Geophys. Res.,* 102, 14075 (1997).

Burlaga, L. F., and K. W. Ogilvie, Causes of sudden commencements and impulses, *J. Geophys. Res.*, 74, 2815 (1969).

Chappell, C. R., The terrestrial plasma source: a new perspective in solar-terrestrial processes from Dynamics Explorer, *Rev. Geophys.*, 26, 229 (1988).

Chen, L., and A. Hasegawa, A theory of long-period magnetic pulsations, 1, Steady state excitation of field line resonance, *J. Geophys. Res.*, 79, 1024 (1974).

Francis, W. E., M. J. Green, and A. J. Dessler, Hydromagnetic propagation of sudden commencements of magnetic storms, *J. Geophys. Res.*, 64, 1643 (1959).

Kikuchi, T., and T. Araki, Transient response of uniform ionosphere and preliminary reverse impulse of geomagnetic storm sudden commencement, *J. Atmos. Terr. Phys.*, 41, 917 (1979a).

Kikuchi, T., and T. Araki, Horizontal transmission of the polar electric field to the equator, *J. Atmos. Terr. Phys.*, 41, 927 (1979b).

Knott, K., D. Fairfield, A. Korth, and D. T. Young, Observations near the magnetopause at the onset of the July 29, 1977, sudden storm commencements, *J. Geophys. Res., 87*, 5888 (1982).

Knott, K., A. Pedersen, and U. Wedeken, GEOS 2 electric field observations during a sudden commencement and subsequent substorms, *J. Geophys. Res.*, 90, 1283 (1985).

Lee, D. H., and R. L. Lysak, Magnetospheric ULF wave coupling in the dipole model: The impulsive excitation, *J. Geophys. Res.*, 94, 17097 (1989).

Lee, D.-H., and K. Kim, Propagation of sudden impulses in the magnetosphere: Linear waves, *Adv. Space Res.*, 25(7/8), 1531 (2000).

Lee, D.-H, and M. K. Hudson, Numerical studies on the propagation of sudden impulses in the dipole magnetosphere, *J. Geophys. Res.,* in press (2000)

Lee, D. H., K. Kim, and R. L. Lysak, Nonlinear MHD wave propagation in the magnetosphere, *J. Geophys. Res.*, in press (2000).

Nagano, H., and T. Araki, Polarization of geomagnetic sudden commencements observed by geostationary satellites, *J. Geophys. Res.,* 89, 11,018 (1984).

Nishida, A., Ionospheric screening effects and sudden commencements, *J. Geophys. Res.*, 69, 1761 (1964).

Nopper, R. W., Jr., W. J. Hughes, C. G. Maclennan, and R. L. McPherron, Impulse-excited pulsations during the July 29, 1977, event, *J. Geophys. Res.*, 87, 5911 (1982).

Petrinec, S. M., K. Yumoto, H. Luhr, D. Orr, D. Milling, K. Hayashi, S. Kokubun, and T. Araki, The CME event of February 21, 1994; Response of the magnetic field at the Earth's surface, *J. Geomagn. Geoelectr.*, 48, 1341 (1996).

Stegelmann, E. J., and C. H. von Kenschitzki, On the interpretation of the sudden commencement of geomagnetic storms, *J. Geophys. Res.*, 69, 139 (1964).

Tamao, T., A hydromagnetic interpretation of geomagnetic SSC, *Rep. Ionos. Space Res. Jpn.*, 18, 16 (1964).

Wedeken, U., H. Voelker, K. Knott, and M. Lester, SSC-excited pulsations recorded near noon on GEOS 2 and on the ground (CDAW 6), *J. Geophys. Res.*, 91, 3089 (1986).

Wilken, B., C. K. Goertz, D. N. Baker, P. R. Higbie, and T. A. Fritz, The SSC on July 29, 1977 and its propagation within the magnetosphere, *J. Geophys. Res.*, 87, 5901 (1982).

Wilson, C. R., and M. Sugiura, Hydromagnetic interpretation of sudden commencements of magnetic storms, *J. Geophys. Res.*, 66, 4097 (1961).

MULTI-SPACECRAFT STUDIES IN AID OF SPACE WEATHER SPECIFICATION AND UNDERSTANDING

V. Angelopoulos[1], M. Temerin[1], I. Roth[1], F. S. Mozer[1], D. Weimer[2], and M. R. Hairston[3]

[1]*Space Sciences Laboratory, University of California, Berkeley, CA, USA*
[2]*Mission Research Corporation, Nashua, NH, USA*
[3]*W. B. Hanson Center for Space Sciences, University of Texas, Dallas, TX, USA*

ABSTRACT

We present particle modeling in prescribed magnetic and electric fields during the February 17-18, 1998 magnetic storm in an effort to explain observed ion signatures on three spacecraft (POLAR, EQUATOR-S and FAST) which sampled the inner magnetosphere over a wide range of L-shells. The objective is to improve global electric field models, which are crucial to the evolution of the storm-time particle distributions. We back-trace ion distributions from the satellite locations and keep track of charge-exchange losses. The electric field models used are the Volland-Stern, Weimer 96, Weimer 2000 and a version of the Weimer 96 model modified to best fit instantaneous potential measurements made by the electric field instrument on POLAR and the ion drift meter instrument on DMSP satellites. We find that the Weimer 2000 model provides the best agreement with the data. However differences with observations do exist and cannot be accounted for simply by modification of existing models. Explaining those differences requires addition of either nightside injections or global storm-time inductive electric fields.

MOTIVATION

The global electric field distribution in the magnetosphere is a far more elusive quantity than the global magnetic topology. Normally low altitude satellites have been used to obtain the spacecraft potential and model it. Assuming the field lines are equipotentials (an approximation valid only when the allowable uncertainty can be larger than the typical field aligned potential drop of a few kV) the global electric field can then be obtained. Such efforts since the early days of multiple balloon (Mozer et al., 1974) and low altitude satellites (e.g., S3-3) have led to an increasingly improved understanding of the global ionospheric electric field patterns (Heelis et al., 1982). Semi-empirical models like the Volland-Stern model, hereby referred to as the "VS" model, (Volland, 1973; Stern, 1975) started to develop; they were quickly compared and fitted against data under arbitrary activity conditions (Maynard and Chen, 1975). More recently, the availability of the DE and the DMSP datasets has led to more sophisticated models of the average electric potential pattern under a variety of solar wind conditions (Heppner and Maynard, 1987; Rich and Hairston, 1994). A spherical harmonic representation of the fits to the electric potential averages of DE by Weimer (1996) has led to a generalized data-based model, herein termed the "W96" model, which can compute the ionospheric potential pattern under any interplanetary magnetic field orientation. This model was recently enhanced to the "W2k" model, to include the effects of nightside convection increases under substorm conditions using the AL index as an activity proxy (Weimer, 2001). All models above have been derived from large statistical databases and thus cannot proclaim to represent accurately storm time electric fields. In fact, recent evidence from CRESS (Wygant et al., 1998) suggests that such fields penetrate to much lower latitudes than during non-storm periods. The low latitude electric fields probably play a key role in populating the innermost L-shells, but neither their origin nor their effects have been adequately understood because of the lack of electric field measurements in the equatorial inner magnetosphere.

Efforts to understand particle evolution during substorms and storms have been remarkably successful in reproducing gross features of particle distributions in the inner magnetosphere, using the aforementioned simple electric field models and the assumption that the field lines are equipotentials. Specifically, Ejiri et al. (1980) were able to model nose dispersions, one of the most dramatic features of substorm and storm time fluxes at the inner magnetosphere at dusk (Smith and Hoffman, 1974), using a step function for the electric field increase in a dipole field. "Nose" refers to the shape of the dispersion of a particle enhancement seen in a energy versus L-shell spectrogram, predominantly at dusk. Only particles of energy ~10-20 keV make it to low L-shells because under a steady electric field those energies correspond to open trajectories and are constantly replenished with particles from the tail. Lower energies correspond to EXB-dominated closed trajectories and higher energies correspond to gradient/curvature drift-dominated closed trajectories, both inaccessible by tail ions. The long residence in closed trajectories allows losses (such as charge-exchange) to deplete ions of those energies.

The effect of an ion lifetime limited in comparison to the residence time on a drift path was invoked to explain another feature of ion spectra, such as a decrease of flux at a discrete energy around a few keV seen in energy-time spectrograms of satellites in the inner magnetosphere (McIlwain, 1972). The ions which are lost at discrete energies are a subset of the ions with open trajectories (accessible from the tail), but simply take a long time to arrive at the satellite location. The energy of lost particles signifies the transition from EXB-dominated to gradient/curvature drift-dominated orbits. This principle was used by Kistler et al. (1989), Fok et al. (1996) and Jordanova et al. (1999) to explain observed storm time ion spectra on single satellite passes. Charge exchange is the dominant loss process. Kistler et al. (1999) compared the particle fluxes obtained from backtracing bounce averaged particle orbits in a dipole magnetic field and time-dependent models of the electric field. They used the VS and the W96 global electric field models whose time-dependence was obtained from the time dependence of the Kp index (for the VS model) or the solar wind parameters (for the W96 model). The resultant fluxes were compared against storm-recovery phase observations by the EQUATOR-S satellite, with the objective to ascertain the electric field model that results in best agreement with the data. The authors found that the W96 model does a better job in reproducing the spectral features of the ions, but that neither model can accurately predict the energies of the observed minima.

In this work we extend the analysis of Kistler et al. (1999) on the February 17-18 1998 storm, by incorporating the POLAR and FAST satellite observations to our analysis and by using additionally a modified W96 electric field model and the W2k model. Details on the storm evolution can be found in that paper. Our objectives are: 1) to see how the existing picture for a storm studied by one inner magnetospheric spacecraft (EQUATOR-S) changes from the inclusion of data from two other spacecraft, and 2) to test the capability of two new electric field models to reproduce the observed particle features.

TRACING PROCEDURE

We used a guiding center model of the particle motion in a prescribed electric and a dipole magnetic field. The VS and W96 models reproduce the same results as those of Kistler et al. (1999) but for small differences which may be attributed to the exact implementation of the time dependence of the electric field model (interpolation/averaging procedure of Kp and of solar wind input). We start at a given time and a given satellite's dipole latitude and longitude. We backtrace ions which at are locally mirroring at the satellite location (though our guiding center code can study any other pitch angle). Backtracing lasts for 24 hours, or until the ions reach an L-shell of 10. We assume ions are protons. At each time we keep track of the ratio of the instantaneous particle phase space density to the one at the satellite, by updating the particle losses due to charge exchange. This is done, like in Kistler et al. (1999), by using the neutral geocoronal model of Rairden et al. (1986) and the charge-exchange cross section for protons tabulated in Smith and Bewtra (1978). If backtracing stops at L=10 (open orbit) the source phase space density is obtained from a Maxwellian of T=20 keV and N=0.1 cm^{-3}. If backtracing stops at 24hours (closed orbit) then the source distribution (at L<10) was also derived from a Maxwellian with the same properties as at L>10, but assuming adiabatic transport to the backtracing stopping point. Other distributions consistent with FAST storm-time statistical averages, or with CCE typical observations (Kistler et al., 1989) were attempted with similar results. The specifics of our choice of an initial distribution on closed trajectories affects only the depth of the spectral minima, not their position. Our choice of the initial distribution on open trajectories does affect the final particle spectra. The specific choice of temperature and density reflects a visual best match to the data. Each pitch angle, energy and time at the satellite is thus assigned a source phase space density, and a charge-exchange-loss factor computed from the backward tracing. This permits a (forward) assignment of a phase space density at the satellite using those values.

ELECTRIC FIELD MODELS

Analytical expressions from dipole mapping were used to obtain the electric field at any given location from the electric field at the ionosphere as specified by the VS, W96 and W2k models. We implemented the VS model by obtaining the 3-hour Kp index, re-sampling on 1 hour intervals and then computing the electric field at the position of each particle after linearly interpolating to obtain the instantaneous Kp value. The solar wind values to input in the W96 and W2k models were obtained from 10 minute averages in the WIND database projected forward in time to correspond to their values at Earth. The data (IMF By, Bz, Vsw and Nsw) were linearly interpolated at the times of interest before they entered the W96 or W2k model. Similarly, the AL index (provisional) was obtained at 1min resolution from the World Data Center and interpolated at the times of interest before usage by the W2k model.

The modified W96 model (herein referred-to as "W96mod" model) represents an attempt to improve the instantaneous electric field model using available measurements of the polar cap potential by the DMSP and POLAR satellites. The idea is that even though the electric field models may not be accurate representations of the global electric field distribution during storms, they can get closer to reality once their input parameters are "tweaked" such that the results match observed traces of the polar cap potential. To that end we used one POLAR pass over the southern oval (at 0610-0850 UT on Feb. 18) and integrated the electric field measured by the EFI instrument [Harvey et al., 1995] along the spacecraft velocity vector to produce the potential along the orbit, assuming the potential is zero at low latitudes (<55°). After subtracting the corotation potential, a small potential difference was found between the two 55° latitude points at the two legs of the oval pass (inbound and outbound). The difference was assumed to be due to a small temporal change of the polar cap potential over the ~2.5 hr period of the pass. As such it was subtracted from the trace of the potential, by linearly interpolating in time from zero to the full value. In addition to the POLAR data we used data from four DMSP satellites (F11, F12, F13 and F14). The ion drift meter on those satellites provides the ion velocity components perpendicular to the spacecraft motion, which are then used to obtain the potential along the spacecraft track, assuming that the potential is zero at low latitudes. Here too the corotation potential was subtracted and the potential offset between the opposite low latitude points on each pass were assumed to be due to temporal variations and were subtracted after linear interpolation in time.

Figure 1 (bottom) shows the center times of each oval pass by the POLAR and the four DMSP satellites. The top figure shows the number of oval crosses during a sliding two hour period centered on half hour centers (anywhere from 2-6 satellite passes available each time). For these multiple crosses we obtained the value of the W96 potential starting from the average solar wind conditions during the given 2 hour interval and then varied the input parameters (By, Bz and Vsw) so as to minimize the difference with the observed potential. All passes were mostly in a dawn-dusk fashion at various distances from the terminator (depending on dipole tilt), providing a relatively good measure of the full potential drop. The resultant parameters for a best fit to the data varied significantly from the initial solar wind parameters, and mostly in a manner that tended to decrease the total W96 potential drop. This is in agreement with the conclusion of Weimer (2001) that the W96 model overestimates the transpolar cap potential drop during storms. This procedure provides a new set of W96 model input parameters at a half hour resolution, which we linearly interpolate in time to obtain instantaneous values of the electric field at the time and position of interest for particle orbit integrations. The above constitutes our "W96mod" model.

PARTICLE DATA

In addition to EQUATOR-S, two more satellites equipped with thermal plasma instrumentation crossed a large range of L-shells during the recovery phase of the February 17-18, 1998 storm. These were POLAR and FAST. Information on ion (assumed proton) spectra is obtained on POLAR from the HYDRA instrument (Scudder et al., 1995) and on FAST from the ion electrostatic analyzer instrument (Carlson et al., 2001). The energy-time spectrograms of the energy flux seen on the three spacecraft, along with their positions are shown at the bottom three panels on Figure 2.

As explained in Kistler et al. (1999) EQUATOR-S was at the prenoon sector and observed at least two energy minima (one at ~10 keV and another at a few keV) inside of L~7. It is evident in the spectrogram that the average energy of both main minima decreases as L-shell decreases.

POLAR, in the postnoon sector, observed one main minimum at ~10 keV. The minimum increased in energy as the satellite moved to smaller L-shells and attained its maximum energy at the innermost L-shell (8:20 UT). The energy of the minimum decreased thereafter as POLAR moved to the northern hemisphere at higher L-shells. A second minimum is observed at ~9:30 UT at energies of ~15 keV and its energy decreases with time (i.e., L-shell). This minimum might have been present at higher energies at earlier times but would be beyond the energy range of the instrument. Other minima at lower energies are also present but are not as clear at 90° pitch angles. They are much clearly defined at the spectrograms of field-aligned (or field-opposed) particles (not shown). Such lower energy-dispersed ion features have

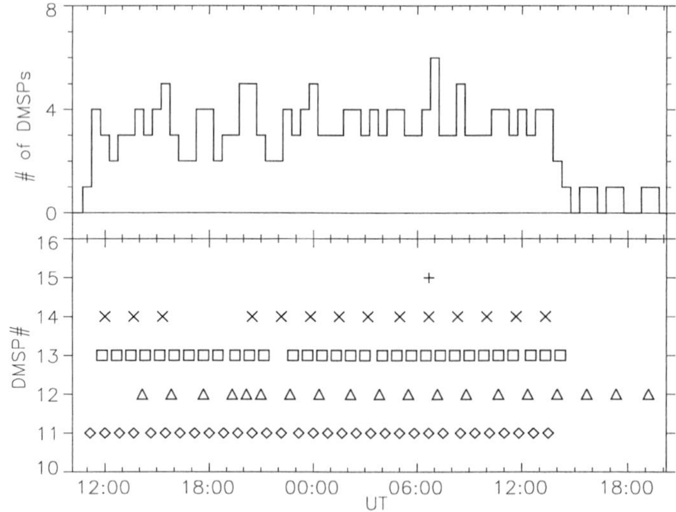

Fig. 1. Bottom: Satellite oval crossings available arranged by satellite. The number denotes DMSP number, except for #15, which denotes POLAR. Number of available passes within 2-hour intervals centered on 1/2 hour centers.

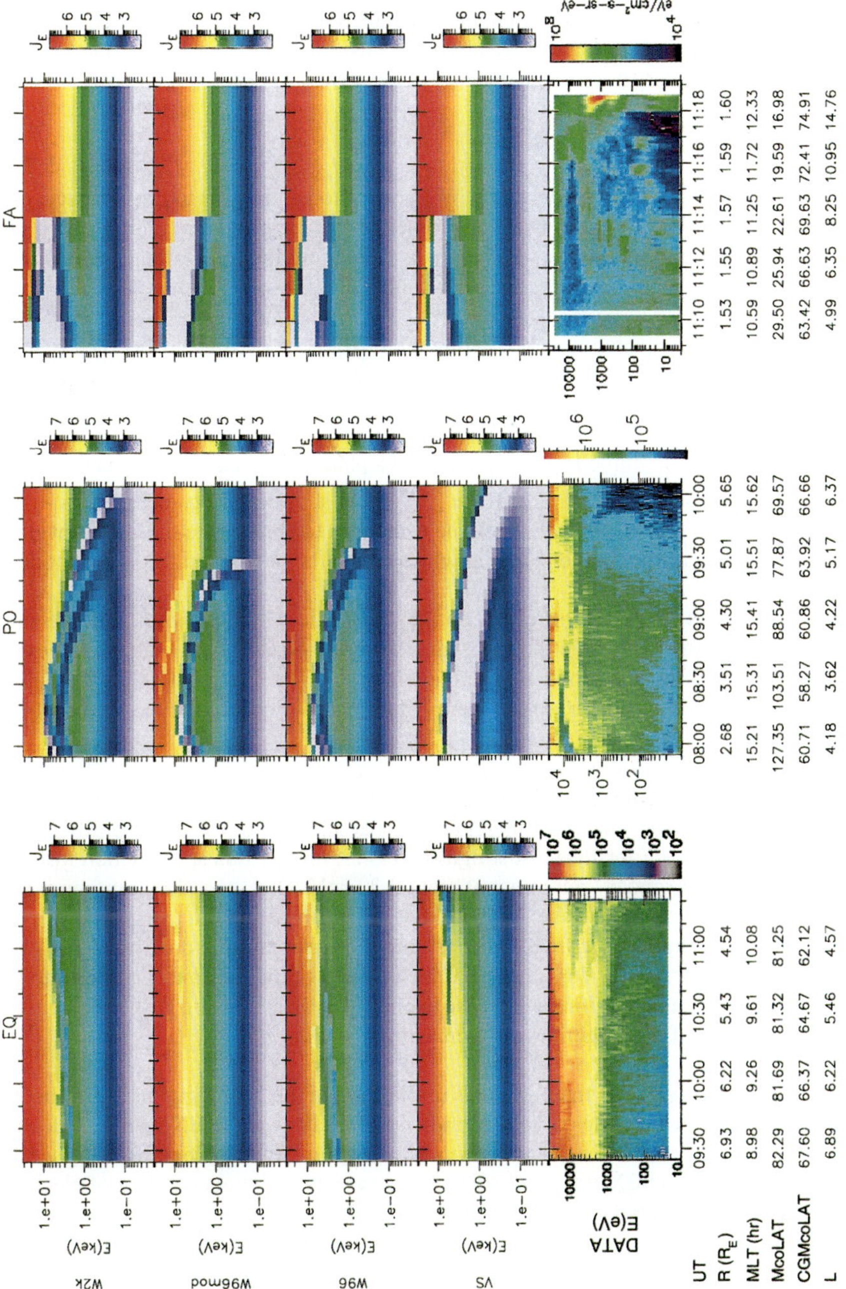

Fig. 2. Data (bottom panels) and model-derived (other panels) energy flux spectrograms at ~90° pitch angle from three satellites (EQUATOR-S, left panels; POLAR: middle; FAST: right). The electric field model (same on each row) is indicated on the left.

been seen also by Freja and other satellites (e.g., Yamauchi et al., 1996) and can be interpreted as night-side injections (Ebihara et al., 2001).

FAST, in the prenoon sector was approaching the polar cusp when it also observed a major flux minimum at ~ 10 keV, between 1110 and 1116 UT. Lower energy enhancements (<1keV) are also interpreted as injections. At 11:16 UT FAST reached the regions adjacent to the cusp and at 1118 UT it entered the cusp, as evidenced by the enhancement of magnetosheath-like (~1keV) plasma population.

RESULTS

We use the aforementioned electric field models (VS, W96, W96mod and W2k) to trace particle orbits in a dipole field and compute, assuming charge-exchange losses, the particle spectrograms at three satellite locations (EQUATOR-S, POLAR and FAST). The results are shown in Figure 2, along with (and in the same format as) the data to which we are comparing our results.

All electric field models do an adequate job in predicting the fluxes seen on FAST. Specifically the 10 keV minimum is reproduced by all electric field models. This is due to ions which spend much time in the inner magnetosphere because their EXB eastward drift is roughly balanced by their westward gradient/ curvature drift. The sharp change in the character of the model energy fluxes is due to the motion of FAST poleward of L=10, which results in immediate assignment of a phase space density consistent with a 20 keV, 0.1cm^{-3} Maxwellian to those points. In reality, the FAST data show a hot magnetospheric-like population prior to entry to the cusp. This layer of hot plasma is thought to come from the tail (Newell and Meng, 1992) due to distortions of the dipole mapping which cannot be accounted for in our model.

On POLAR, the modeled fluxes depend significantly on the electric field model chosen. The VS field produces a broad and deep minimum. The W96 model produces more moderate (and realistic losses) while both the W96mod and the W2k models produce a double minimum. The energy of both minima has the correct trend in terms of L-shell dependence (peak energy at lowest L-shell). However, the minima are at different energies relative to the observed values.

On EQUATOR-S, all models result in a principal minimum at around 10keV in agreement with the principal minimum observed in the data. The W96 model has a broad minimum, which generally is a poorer reproduction of the observed fluxes (not clearly distinguished in the energy flux spectrogram format). The W2k models result in a robust secondary minimum at around 20 keV. This secondary minimum is not clear in the energy flux spectrogram due to the color coding, but it can be seen in the spectra of percent particle loss due to charge-exchange in Figure 3. The position of the two minima does not agree well with the observations. In addition, the L-shell dependence of the position of the minima is contrary to what is seen in the data: While in the data both minima decrease in energy with decreasing L-shell, all our models show an increase in the energy of the minima with decreasing L-shell.

The L-shell dependence of the energy of the minima is understood in the following fashion: The energy minimum represents the energy where the curvature/gradient drift balances the EXB drift. A decrease in L-shell (which in a dipole field results in a decrease in curvature/gradient drift) requires a higher particle energy in order to balance the same EXB drift. This is a behavior relatively independent of electric field model used. The data however shows a decrease in energy with decreasing L-shell. Assuming that the gradient/curvature drifts are modelled accurately enough, this observation suggests that the real EXB drift decreases with decreasing L-shell faster than any model would predict. This is in agreement with the observation of Wygant et al. (1998) of reverse sign, intense fields in the inner magnetosphere during storms. Since the storm data-based W96mod model also shows the same behavior as the other models in terms of L-shell dependence of the location of the minimum, we conclude that low altitude observations

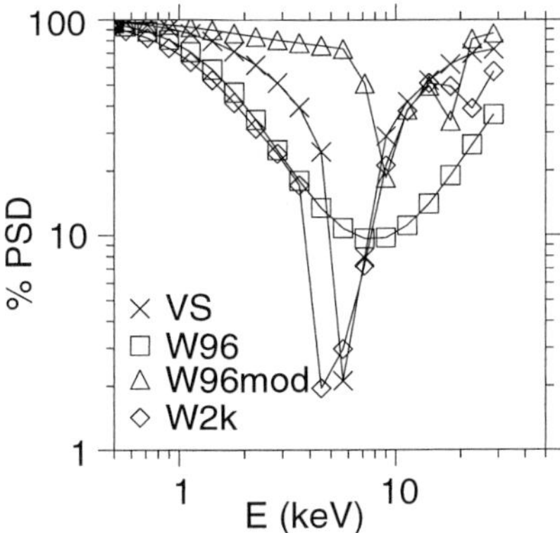

Fig. 3. Spectra of % of ions left at EQUATOR-S (at L=6) after losses by charge exchange. Secondary minima at ~20 keV are evident on spectra derived using W96mod and W2k models.

of electric fields are not sufficient to monitor the high altitude electric field in the magnetosphere. Since low L-shells are being considered, the most likely explanation for the discrepancy is the presence of intense inductive electric fields which are expected to be present during storms.

An alternative explanation of the energy spectrograms may be that indeed the primary ion population exhibits a primary minimum increasing in energy with decreasing L-shell, but superimposed on the spectrogram is the signature of two injections with energy decreasing as a function of L-shell. The two injections start at energies of 5 and 10 keV at 9:30 UT and end at energies 1 and 3 keV respectively. Modelling of this alternative interpretation of the data would require time dependent and localized (inductive) electric field enhancement(s) at the nightside. Similar work (e.g., Li et al., 2000) has shown promise that it can explain single satellite spectrograms. If applied in this case it would have to be consistent with observations on other satellites as well. The above two scenaria are beyond the scope of this paper.

ACKNOWLEDGMENTS

We wish to thank C. W. Carlson for making the FAST data available and J. Scudder for making the POLAR/HYDRA data available for this study. We are grateful to L. Kistler for useful discussions. This work was supported under NASA contract NAS5-30367.

REFERENCES

Carlson, C. W., et al., The electron and ion plasma experiment for FAST, Space Sci. Rev., in press (2001).
Ebihara, Y., et al., Wedge-like dispersion of sub-keV ions in the dayside magnetosphere: Particle simulation and Viking observations, J. Geophys. Res., in press (2001).
Ejiri, M., et al., Energetic particle penetrations into the inner magnetosphere, J. Geophys. Res., 85, 653 (1980).
Fok, M.-C., et al., Ring current development during storm main phase, J. Geophys. Res., 101, 15311, (1996).
Fok, M.-C., and T. E. Moore, Ring current modeling in a realistic magnetic field configuration, Geophys. Res. Lett., 24, 1775 (1997).
Harvey, P., et al, The electric field instrument on the POLAR satellite, Space Sci. Rev., 71, 583, 1995.
Heelis, R. A., et al., A model of the high-latitude ionospheric convection pattern, J. Geophys. Res., 87, 6339 (1982).
Heppner, J. P., and N. C. Maynard, Empirical high-latitude electric field models, J. Geophys. Res., 92, 4467 (1987).
Jordanova, V. K., et al., Simulations of off-equatorial ring current ion spectra measured by Polar for a moderate storm at solar minimum, J. Geophys. Res., 104, 429, (1999).
Kistler, L. M., et al., Energy spectra of the major ion species in the ring current during geomagnetic storms, J. Geophys. Res., 94, 3579 (1989).
Kistler, L. M., et al., Testing electric field models using ring current ion energy spectra from the Equator-S ion composition (ESIC) instrument, Ann. Geophysicae, 17, 1611 (1999).
Li, X., et al., Multiple discrete-energy ion features in the inner magnetosphere: Observations and simulations, Geophys. Res. Lett., 27, 1447 (2000).
Maynard, N. C. and A. J. Chen, Isolated cold plasma regions: observations and their relation to possible production mechanisms, J. Geophys. Res., 80, 1009 (1975).
McIlwain, C. E., Plasma convection in the vicinity of the geosynchronous orbit, in Proceedings of a Symposium on Earth's magnetospheric processes, edited by McCormak, B. M. Dordrecht, Netherlands, D. Reidel Publishing Co, p. 268 (1972).
Mozer, F.S., et al., High-latitude electric fields and the three-dimensional interaction between the interplanetary and terrestrial magnetic fields, J. Geophys. Res., 79, 56 (1974).
Newell, P. T. and C. I. Meng, Mapping the dayside ionosphere to the magnetosphere according to particle precipitation characteristics, Geophys. Res. Lett., 19, 609 (1992).
Rairden, R. L., et al., Geocoronal imaging with Dynamics Explorer, J. Geophys. Res., 91, 13613 (1986).
Rich, F. J., and M. Hairston, Large-scale convection patterns observed by DMSP, J. Geophys. Res., 99, 3827 (1994).
Scudder, J., et al., HYDRA - a 3-dimensional electron and ion hot plasma instrument for the POLAR spacecraft of the GGS mission, Space Sci. Rev., 71, 459 (1995).
Smith, P. H., and R. A. Hoffman, Direct observations in the dusk hours of the characteristics of the storm time ring current particles during the beginning of magnetic storms, J. Geophys. Res., 79, 966 (1974).
Smith, P. H., and N. K. Bewtra, Charge exchange lifetimes for ring current ions, Space Sci. Rev., 22, 301 (1978).
Stern, D. P., The motion of a proton in the equatorial magnetosphere, J. Geophys. Res., 80, 595 (1975).
Volland, H., A semi empirical model of large-scale magnetospheric electric fields, J. Geophys. Res., 78, 171 (1973).
Weimer, D. R. A flexible IMF dependent model of high-latitude electric potentials having "space weather" applications, Geophys. Res. Lett., 23, 2549 (1996).

Weimer, D. R., An improved model of ionospheric electric potentials including substorm perturbations and application to the Geospace Environment Modeling November 24, 1996, event, J. Geophys. Res., 106, 407 (2001).

Wygant, J., et al., Experimental evidence on the role of the large spatial electric field in creating the ring current, J. Geophys. Res., 103, 29527 (1998).

Wygant, J., et al., Polar spacecraft based comparisons of intense electric fields and Poynting flux near and within the plasma sheet-tail lobe boundary to UVI images: an energy source for the aurora, J. Geophys. Res., 105, 18675 (2000).

Yamauchi, M., et al., Meso-scale structures of radiation belt/ring current detected by low-energy ions, Adv. Space Sci., 17, 171 (1996).

Low-Altitude-Satellite Observations and Modeling Session

THE ELECTRON DENSITY DISTRIBUTION IN THE POLAR CAP: ITS VARIABILITY WITH SEASONS, AND ITS RESPONSE TO MAGNETIC ACTIVITY

Harri Laakso and Réjean Grard

ESA Space Science Department, Postbus 299, 2200 AG Noordwijk, The Netherlands

ABSTRACT

Using forty-five months of spacecraft potential measurements gathered with the double probe electric field antenna on the Polar satellite, we study the electron density in the polar cap and investigate its dependence on distance ($2\ R_E < r < 9\ R_E$), season, and geomagnetic activity. The polar cap density increases with magnetic activity at all distances; it is larger during local summer than during local winter, by up to two orders of magnitude at distances of around 2 R_E. The density profile follows a power law of the form r^{-a}. In winter, $a = 3.3$–3.7, which tends to suggest that the escaping ions are accelerated ($a > 3$). However, as the ion composition (particularly, the H^+/O^+ ratio) is known to vary with altitude, the occurrence of ion acceleration cannot be deduced from our observations. In summer, the exponent reaches values of about 6.0, which implies that the density of the polar topside ionosphere is high and the heavy ions are gravitationally bound at low altitudes.

INTRODUCTION

Understanding the structure and dynamics of the Earth's magnetosphere requires information about the spatial distribution and temporal evolution of the electron density, in different geomagnetic conditions. The electron population in the Earth's space environment is controlled by numerous fundamental processes such as energy and momentum transfer from the solar wind, as well as flows of charged particles from internal sources, such as the ionosphere [*Moore and Delcourt*, 1995].

The importance of ionospheric outflows in the magnetosphere has been widely discussed [*Chappell et al.*, 2000]. In the polar regions, a countinuous flux of ions escaping from the ionosphere is detected, and a variety of mechanisms for this polar wind have been invoked to explain the observations [*Ganguli*, 1996; *Schunk*, 2000]. On the nightside, large energy inputs into the ionosphere along auroral field lines can trigger substantial transient ion flows into the magnetotail [*Schunk*, 2000, and references therein]. Recently, it was discovered that resonant ULF waves can also generate ion outflow [*Laakso et al.*, 1998]. On the dayside, the ion cleft fountain supplies ions to the polar cap where they convect tailward [*Lockwood et al.*, 1985].

There are numerous experimental studies and theoretical models of the electron density in the low-altitude polar cap [see *Ganguli*, 1996, and references therein], but still there is a lack of high-altitude measurements. Also, the high-altitude polar cap is a rarefied environment where the spacecraft potential becomes tens of volts positive [*Laakso*, 2000], preventing the detection of the low-energy ions [*Chappell et al.*, 2000]. *Persoon et al.* [1983] used DE 1 data for modeling the density distribution in the polar cap at geocentric distances between 2 and 4.6 R_E. However, in order to increase the statistical significance of their results, they combined observations performed during different seasons. They concluded that the polar cap density varies with distance as $r^{-3.8}$, suggesting a weak acceleration of outflowing ions. However, such a suggestion implicitly assumes a single ion population, whereas the polar cap ionosphere and magnetosphere are known to contain several ion species (H^+, He^+, and O^+ are the major species) and their ratios can significantly vary with altitude [*Su et al.*, 1998].

In this paper, we investigate how the electron density in the polar cap varies with altitude and how this profile is affected by geomagnetic activity and season. The electron density is derived from the spacecraft potential measurements of the EFI experiment on the Polar satellite, and in the present study we utilize 45 months of such measurements in 1996–1999.

MEASURING TECHNIQUE

We use measurements gathered by the electric field instrument (EFI) on the Polar satellite. This satellite was launched on 24 February 1996 into a 90° inclination orbit with a 9 R_E apogee (initially over the northern hemisphere), a 1.8 R_E perigee (initially over the southern hemisphere), and an orbital period of about 18 hours (see Figure 1). The orbital plane rotates about the Earth with respect to the Sun in 12 months so that all local times are covered during a 6-month period.

The Polar EFI experiment [*Harvey et al.*, 1995] consists of three pairs of double probe antennas oriented perpendicular to each other. The difference $\Delta V_{ps} = V_p - V_s$ between the potential of each probe (V_p, where $p = 1$ to 6) and that of the satellite, V_s, is measured. It can be shown that, in a rarefied plasma, the electron density N_e and ΔV_{ps} are related in a way which depends on the temperature of the photoelectrons emitted from the surface of the satellite [*Pedersen*, 1995]. Although the magnitude of the saturation photoelectron current does not play a significant role in this relationship, it is essential that this quantity exceeds the sum of the ambient electron current and bias current which polarizes the sensor with respect to the spacecraft [*Laakso and Pedersen*, 1998]. The ambient electron temperature T_e also affects the value of ΔV_{ps}, but only in a minor way; when T_e varies between 1 and 1000 eV, for example, the derived density differs by only a factor of 2-3 [*Laakso and Pedersen*, 1998].

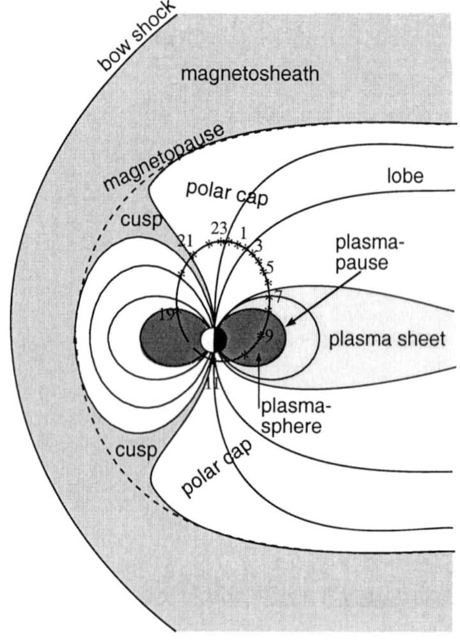

Fig. 1. Projection of the orbit of Polar in the noon-midnight meridian, on 29–30 April 1996, superimposed on a schematic drawing of the magnetosphere.

This technique for determining the ambient electron density fails when the potential of a biased probe becomes negative with respect to the ambient plasma, which happens when the density is more than about 500 cm^{-3}. In a very tenuous environment, the spacecraft potential varies with density as long as there are enough electrons to charge the satellite fast enough compared with the temporal scales of the density variations. The spacecraft potential measurements with EFI are saturated at 67 volts which corresponds to very low densities of below 0.01 cm^{-3}, which can occur occasionally [*Laakso*, 2000]. However, even in such cases, the charging times are quite short, and the density variations can therefore well be monitored by the EFI experiment over the range of 0.01-500 cm^{-3}.

Figure 1 is a schematic representation of the magnetosphere in the noon-midnight meridian. The ellipse shows the orbit of Polar on 29–30 April 1996, and asterisks mark the satellite's position at hourly intervals. In general, the satellite crosses the southern polar cap at distances of about 1.8–2 R_E and the northern polar cap at distances between 5 and 9 R_E.

Figure 2 displays the variation of the electron density, N_e, along the orbit shown in Figure 1. Before 19:00 UT, the satellite is in the plasmasphere (PS), where N_e is larger than 100 cm^{-3}, up to the outbound crossing of the plasmapause (PP), where the density decreases by a few orders of magnitude. An enhancement is observed between 20:30 and 21:00 UT, when the satellite traverses the cusp region. Poleward of the cusp, the satellite enters the high-altitude polar cap, where the density is approximately 0.1 cm^{-3}; however, much lower values can frequently be observed. Towards the end of the interval (10:30–11:30 UT), the satellite passes through the southern polar cap and auroral region (SH); the short duration of this event is due to the high velocity of the satellite and the small dimension of the polar cap at small geocentric distances. There, the observed density is usually significantly larger than over the northern hemisphere because of the eccentricity of the orbit.

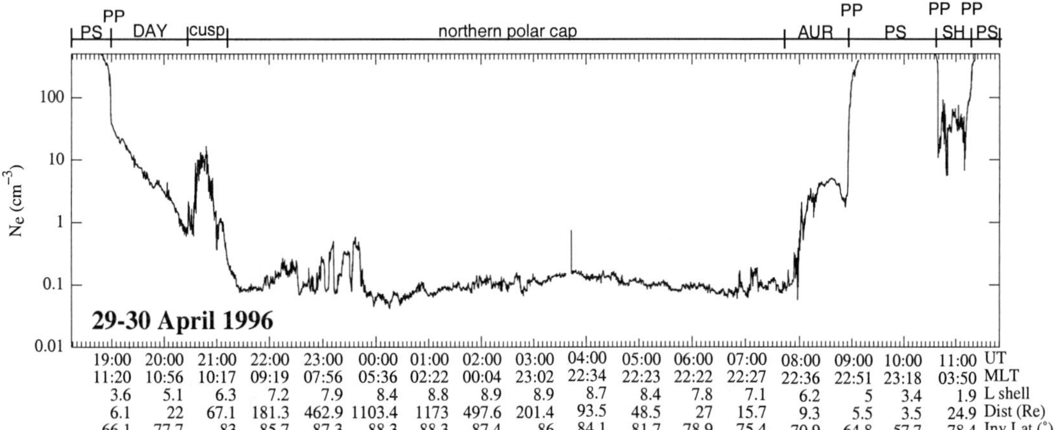

Fig. 2. Electron density profile derived from the Polar spacecraft potential during the interval 29 April 1996, 18:45 UT – 30 April 1996, 11:50 UT (PS - plasmasphere, PP – plasmapause, DAY – dayside magnetosphere, AUR – northern auroral zone, SH – southern polar cap and auroral oval).

STATISTICAL RESULTS

We now study the average density in the polar cap, using the spacecraft potential measurements collected between 1 April 1996 and 31 December 1999. The data are binned with respect to altitude, geomagnetic activity (3-hour Kp index), and season (summer vs. winter).

Effects of geomagnetic activity

We first consider the variation of the average polar cap density with altitude and geomagnetic activity; seasonal effects are ignored. The average electron density distribution in the noon-midnight magnetic local time (MLT) meridian is displayed in Figure 3 (top panels); the azimuthal extent of the noon and midnight sectors is ±1 hour and the sunward direction is to the left. The left-hand panel corresponds to quiet periods (Kp = 0–1) and the right-hand one to disturbed periods ($Kp \geq 3^-$). The bottom two panels are for the electron density distribution in the 06–18 MLT meridian so that the duskside is to the left. The dipole field lines for 65°, 70°, 75°, and 80° invariant latitudes are shown in each plot. The color scale is logarithmic and covers the range from 0.1 to 100 cm^{-3}.

The red (dense) area in the plots represents the plasmasphere (for more details see *Laakso and Jarva* [2001]). The high-altitude polar cap region appears as dark blue, suggesting an average density below 0.1 cm^{-3}. Evaluating the dependence of the electron density distribution on Kp and geocentric altitude over the polar caps requires a careful examination, due to the configuration of the Polar orbit.

It is seen, from the southern polar cap observations, that the density at high latitudes decreases very significantly over a very short altitude range and reaches a level several orders of magnitudes smaller at high altitudes, as observed over the northern polar cap. Assuming radial outflow of a single ion species at constant velocity, the density declines with distance as r^{-3} [*Persoon et al.*, 1983], and thus, the density at 5 R_E distance is expected to be only 1/15 of that at 2 R_E distance. Comparing the low-altitude polar cap densities in the left and right panels reveals furthermore that the geomagnetic activity tends to increase the density, as it will be shown in more detail thereafter.

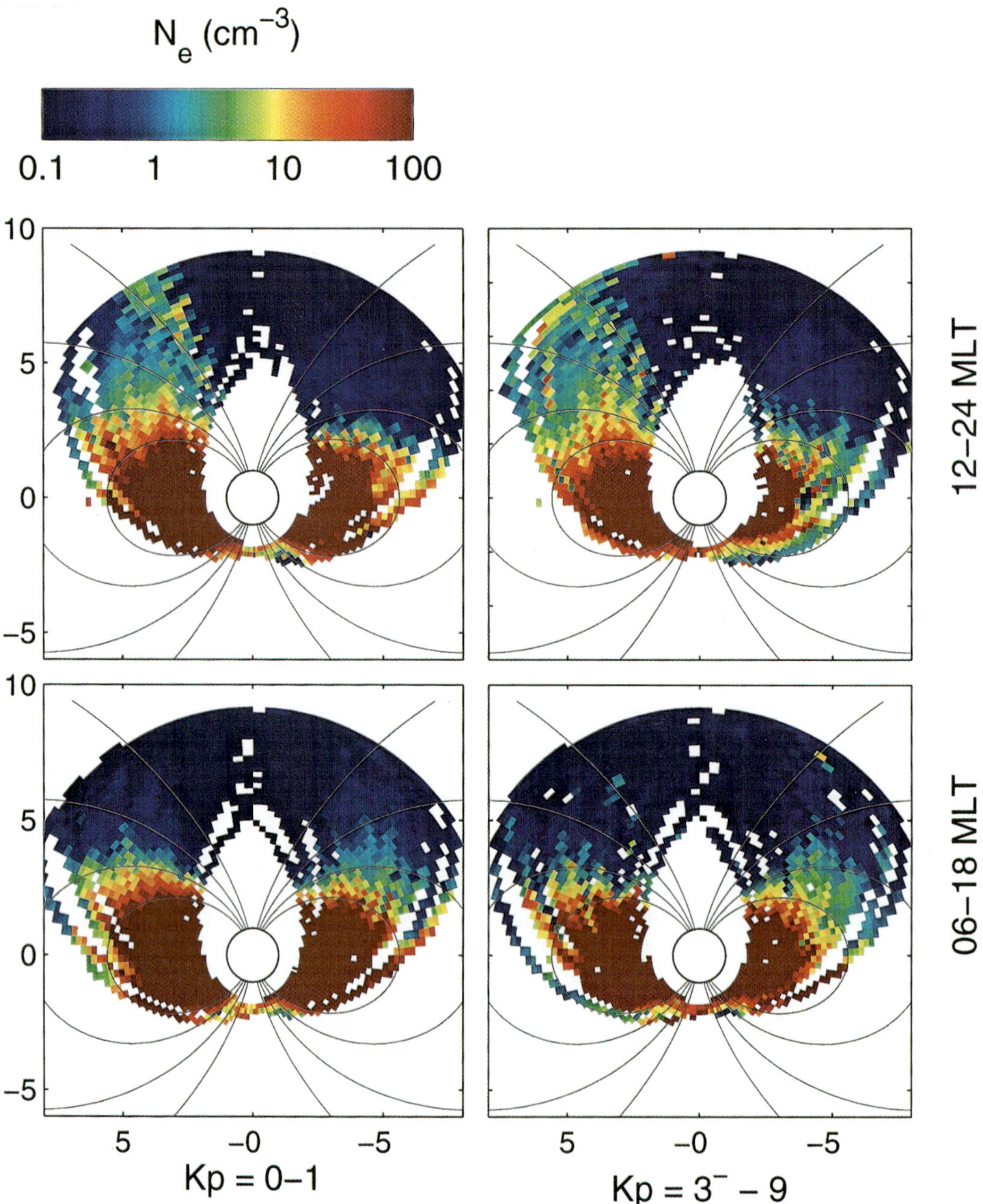

Fig. 3. Average electron density in the 2-hour MLT sectors centred on the 12–24 MLT meridian (top) and the 06–18 MLT meridian (bottom) during quiet (left) and disturbed (right) magnetic conditions. Distances are measured in R_E, and the dipolar field lines are given for invariant latitudes of 65°, 70°, 75°, and 80°.

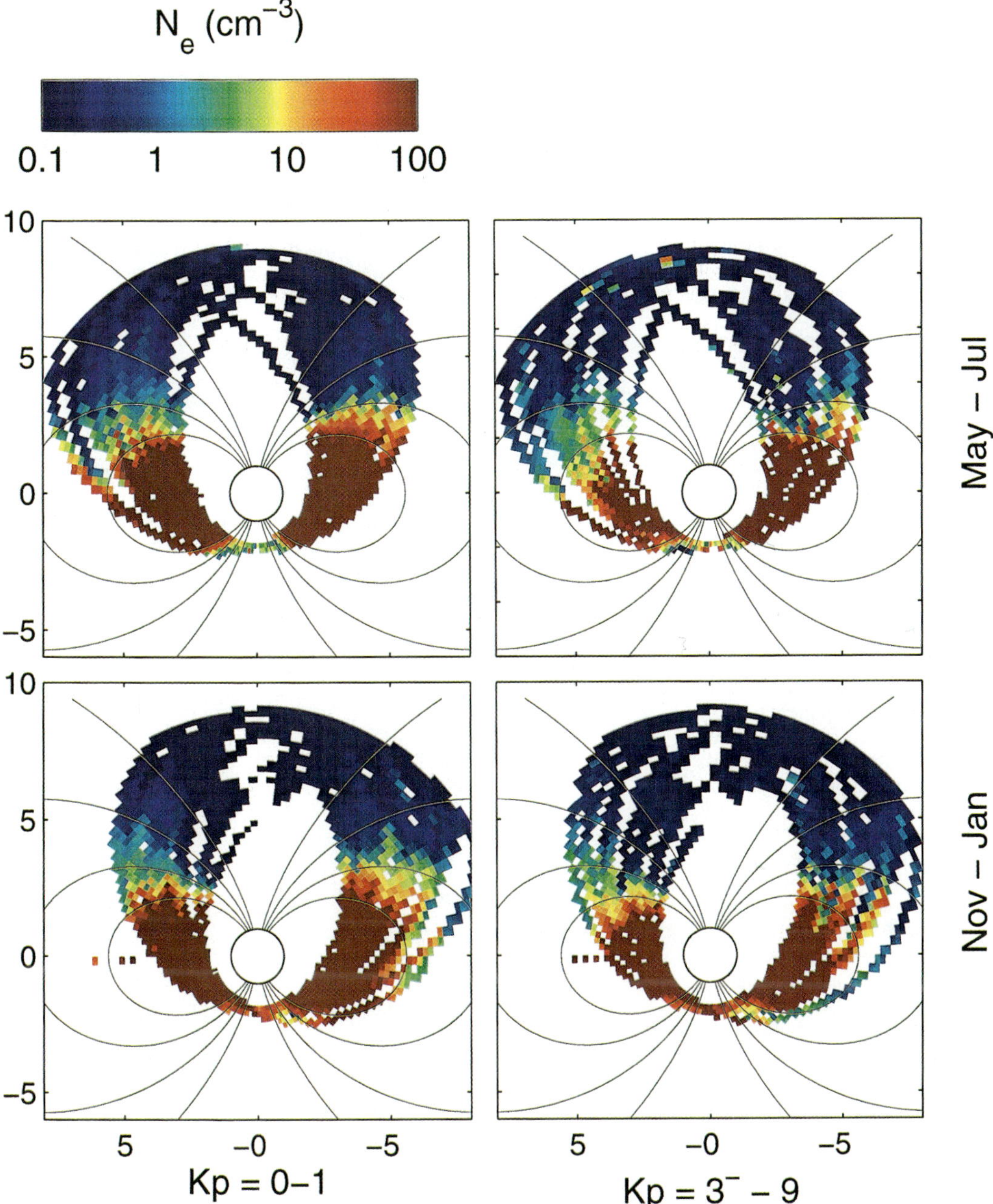

Fig. 4. Same caption as in Figure 3, but for data collected in the 06–18 MLT meridian, during the months of May, June, July (top panels) and November, December, January (bottom).

Seasonal effects

Figure 4 is derived from the same data base, but now the winter and summer results are plotted separately. In other words, the left hand side of Figure 4 is made from the observations gathered during quiet conditions (Kp = 0–1) during 50-day periods centred around the summer (top) and winter (bottom) soltices. Again the color scale does not reveal any major difference between summer and winter in the high-altitude polar cap. On the other hand, the southern low-altitude polar cap densities are markedly different during the two seasons. When the Sun illuminates the southern hemisphere (left), densities of up to ~100 cm^{-3} are detected, whereas in winter the average density does not appear to exceed ~1 cm^{-3}.

The right panels of Figure 4 are relevant to disturbed magnetic conditions. The main difference between left and right panels is seen again in the southern polar cap, where the average density appears to increase significantly during disturbed periods.

VARIABILITY OF THE POLAR CAP DENSITY

The distributions of average electron densities reveals altitude, Kp, and seasonal dependencies in the polar cap. We shall now study each of these features in more detail.

Seasonal variation

The seasonal variation of the polar cap density is better resolved by plotting monthly averages. Only observations for invariant latitude larger than 82° (or L>50) are taken into account. We have in addition separated the data into three Kp ranges: 0 – 1 (quiet), $1^+ – 2^+$ (moderately disturbed), and $4^- – 9$ (strongly disturbed), in order to investigate the effect of geomagnetic activity on the polar cap density. The observations made in the two hemispheres are analysed independently.

The average global densities are presented in Figure 5. The upper and lower panels pertain to the northern (high-altitude) and the southern (low-altitude) polar cap, respectively. The satellite potential yields only densitites which do not exceed 300 cm^{-3}; this value is therefore assumed as a lower limit whenever the density cannot be measured, due to the limitation of the technique. As a result, the real average densities of the low-altitude summer polar cap are necessarily larger than those shown in Figure 5. The low- altitude polar cap density appears to increase by at least one or two orders of magnitude from local winter to local summer, whereas at high-altitude the density increases by a factor of 2–10 only. Comparing the northern and southern hemispheres, one finds that the low to high altitude density ratio is about 10–20 in winter and 200–1000 in summer.

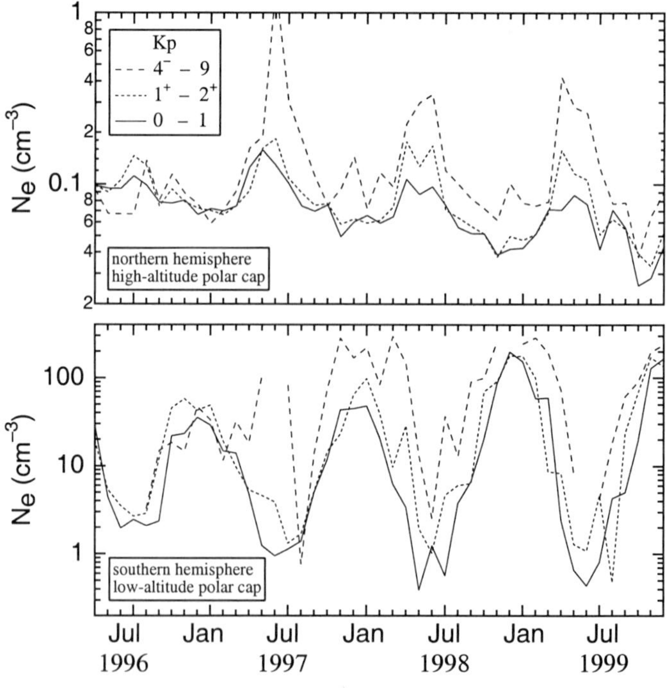

Fig. 5. Monthly averages of the electron density measured in the polar caps for various degrees of magnetic activities.

The geomagnetic activity also affects the density especially at low altitudes (southern polar cap). There, the average density during local summer increases by a factor of 2–10 during disturbed conditions; the effect is more marked during local winter when the density may increase by one or two orders of magnitude. The density enhancements associated with individual events can be much larger than those observed on the average plots. At high altitudes (northern polar cap data), the variation of the average density with magnetic activity is also evident; the density seems to increase by a factor of 2–5 over the whole range of the Kp index, whatever the season.

Effects of Geomagnetic Activity

Figure 6 shows three examples of polar cap density response to a storm sudden commencement (SSC). The Polar satellite was crossing the high-altitude polar cap when an SSC was detected on the ground at 01:43 UT on 13 November 1998 (top panel). The first density enhancement observed 8 minutes later, at 01:51 UT, may be caused by outflowing electrons; it may also result from the compression of the magnetosphere which moves the mantle, characterized by a higher plasma density [*Rosenbauer et al.*, 1975], tailward and closer to the location of the spacecraft. The following enhancement, at 02:54 UT and the larger fluctuations, after 04:00 UT, may be caused by outflowing protons and O^+ ions.

The middle panel of Figure 8 shows similar data for the SSC which occurred at 15:09 UT on 6 July 1999. Again, the density increases immediately after the SSC and larger fluctuations are seen after a delay of 1–2 hours. The bottom panel, on the contrary, displays no density enhancement just after the SSC which was detected on 22 September 1999, at 12:22 UT. Fluctuations similar to those observed in the two previous instances appear with a delay of ~1 hour.

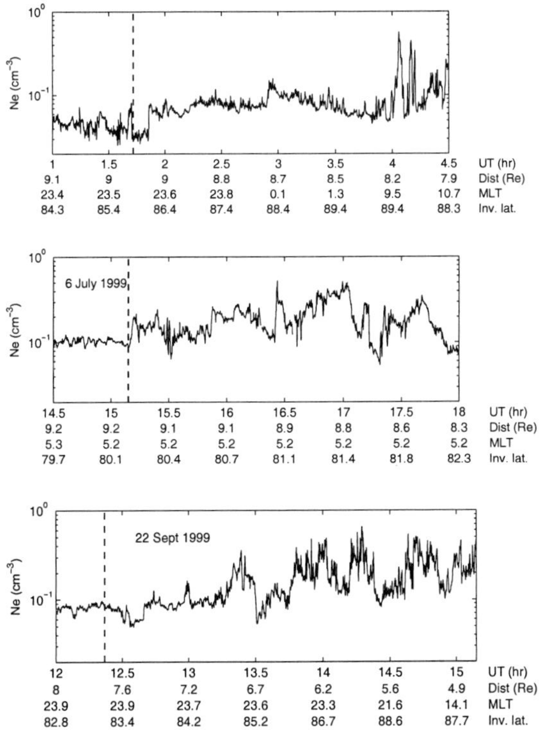

Fig. 6. Examples of electron density fluctuations induced by SSC events (time od occurrence indicated by dashed line).

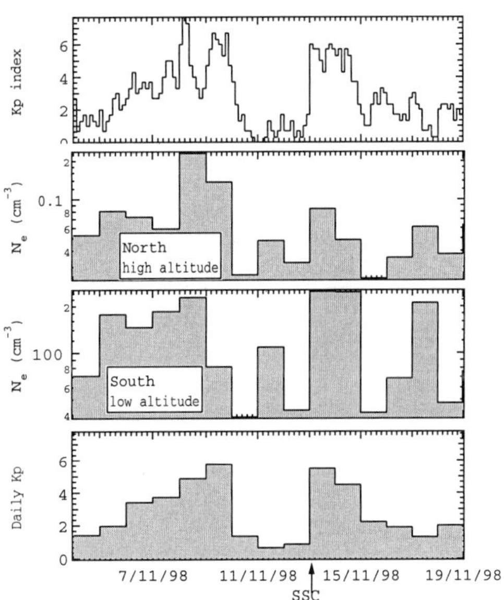

Fig. 7. Daily averages of the electron density in the northern and southern polar cap (second and third panels) during the period of 4–19 November 1998, together with the 3-hour and daily Kp index (top and bottom). The occurrence of an SSC on 13 November 1998 at 01:43 UT is indicated by an arrow.

Next we investigate the daily averaged densities during a storm period of 4-19 November, 1998. The middle panels in Figure 7 display the daily averaged polar cap density over the northern and southern hemispheres, and the 3-hour Kp index and its daily average are shown in the top and bottom panels, respectively. The polar cap densities correlate very accurately, although the two data sets have been collected during different seasons and at different altitudes. The variations of the daily averaged Kp index and average densities are in fair agreement, but not always. Discrepancies are observed on November 12 and 17 when the daily Kp is ~1 and the densities are enhanced in both hemispheres. However, some weak activity can be seen in 3-hour Kp index, which may explain the density enhancements in the polar caps.

In order to study statistically the response of the polar cap to variations of the magnetic activity, we have plotted in Figure 8 the daily averaged density against the daily averaged Kp index. The left and right panels are for the northern (high-altitude) and southern (low-altitude) hemispheres respectively, the upper panels being relevant to local summer and the bottom ones to local winter. We have used here the daily averaged values as the polar cap density seems to increase during magnetic disturbed intervals (e.g. during storms) for an extended period. The scatter of the data points is significant but the fact that the best fit is an increasing function of Kp is unambiguously evidenced in all cases. It appeared in Figure 7 that a weak magnetic activity may, under certain circumstances, raise significantly the polar cap density (see e.g., an enhancement on Nov 11, 1998 in Figure 7), which explains in part the scatter in Figure 8.

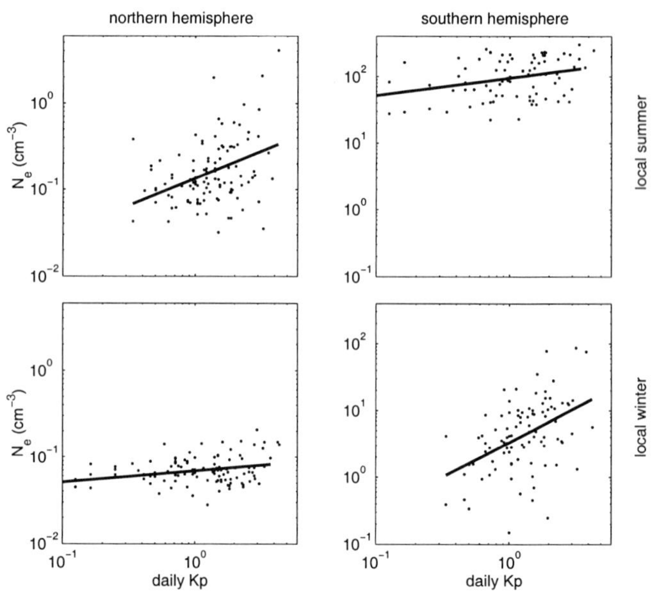

Fig. 8. Daily averaged density against Kp index, during local summer (top panels) and winter (bottom) in the northern (left) and southern (right) polar caps.

<u>Altitude Dependence</u>

The Polar satellite crosses the polar cap in two different altitude ranges, 2 R_E over the southern hemisphere and 4–9 R_E over the northern hemisphere. We shall now combine the observations made in the two hemispheres, under similar magnetic and seasonal conditions, to study the altitude dependence of the polar cap density.

The densities are plotted against distance in Figure 9. In this plot we use 20-minute averages for the high-altitude polar cap and 1-minute averages for the low-altitude polar cap. The short average is selected for the low-altitude passes in order to have enough data points there and this explains partly a large scatter of data points at low altitudes. We have eliminated the observations made above 6.5 R_E where the satellite may often enter the high-latitude boundary layer and where the magnetic field becomes non-dipolar so that it is quite difficult to interpret the altitude dependence of the density. The data are ordered according to the Kp index, as indicated. The exponents of the power law used for the best fits lie between –3.3 and –3.7 in winter, which tend to suggest a weak acceleration of the outflowing ions. However, such a suggestion assumes a single ion population which is not valid for the ionosphere and magnetosphere because the field lines are filled with H^+, He^+, and O^+ ions whose ratios vary strongly with altitude [see e.g., *Su et al.*, 1998].

In summer, on the other hand, the exponents of the best fit lie in the range from –5.8 to –6.0. Such power laws may reflect the fact that a significant portion of the oxygen ions which populate the low-altitude polar cap flow back to the ionosphere and that a small fraction only of the low-altitude ions escape.

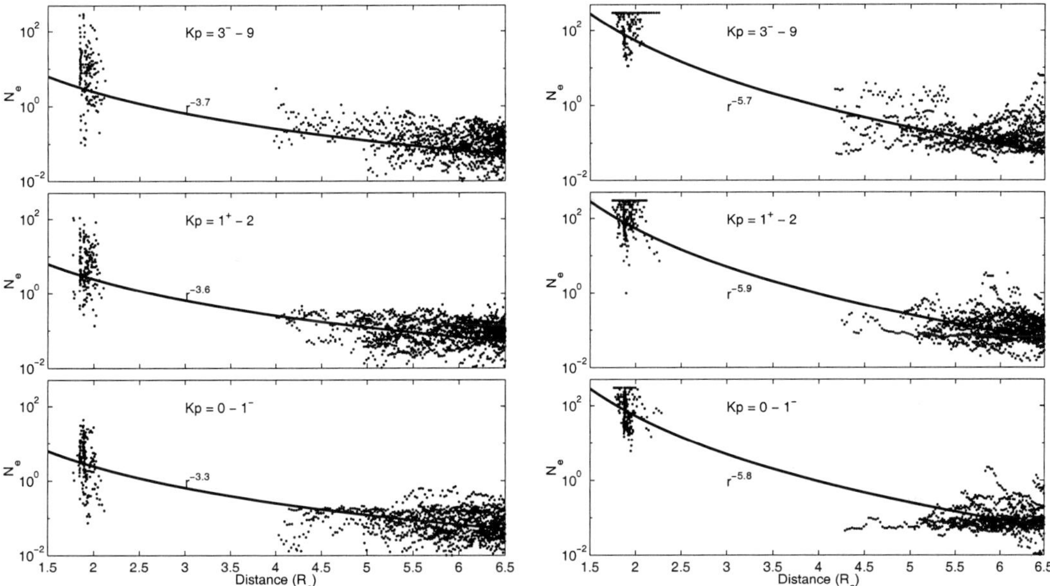

Fig. 9. Plots of average densities against distance for different ranges of Kp index, in winter (left) and in summer (right). The best fits to the data points are represented by power law functions.

SUMMARY

This paper is based on forty-five months of spacecraft potential data gathered with the EFI experiment on the Polar satellite. These observations have been converted into bulk electron densities (for detail, see *Laakso et al.* [2001]). The polar cap is crossed twice per orbit, at altitudes of 1 R_E over the southern hemisphere and 4–8 R_E over the northern hemisphere.

The polar cap density varies systematically with seasons. At low altitudes, the average density is typically 100 cm^{-3} in summer and 1 cm^{-3} in winter. At high altitudes, the density varies respectively between 0.3 and 0.03 cm^{-3}.

The geomagnetic activity also affects the polar cap density. For high Kp, the monthly average increases by one or two orders of magnitude above the quiet-time level; enhancements can naturally be much larger on shorter time scales. The winter densities for high Kp in winter can easily exceed the summer levels in quiet conditions.

The polar cap density decreases with distance as $r^{-3.5}$ in winter and $r^{-6.0}$ in summer. These power laws differ from that expected on the basis of a radial outflow with constant velocity (r^{-3}). At least in summer, the discrepancy is likely due to the fact that a significant part of the low-altitude O^+ ions flows back into the ionosphere and only a fraction of the ions detected near 1 R_E altitude do escape.

REFERENCES

Chappell, C. R., B. L. Giles, T. E. Moore, D. C. Delcourt, P. D. Craven, and M. O. Chandler, The adequacy of the ionospheric source in supplying magnetospheric plasma, *J. Atmos. Sol. Terr. Phys.*, **62**, 421–436, 2000.

Ganguli, S. B., The polar wind, *Rev. Geophys.*, **34**, 311–348, 1996.

Harvey, P. , F. S. Mozer, D. Pankow, J. Wygant, N. C. Maynard, H. Singer, W. Sullivan, P. B. Anderson, R. Pfaff, T. Aggson, A. Pedersen, C.-G. Falthammar, and P. Tanskanen, The electric field instrument on the Polar satellite, *Space Sci. Rev.*, **71**, 583–596, 1995.

Laakso H., Monitoring of the spacecraft potential in space, *J. Atmos. Sol. Terr. Phys.*, submitted, 2000.

Laakso H. and M. Jarva, Position and motion of the plasmapause, *J. Atmos. Terr. Phys.*, *J. Atmos. Sol. Terr. Phys.*, **63**, 1171–1178, 2001.

Laakso, H. and A. Pedersen, Ambient electron density derived from differential potential measurements, in *Measurement Techniques in Space Plasmas: Particles*, edited by R. F. Pfaff, J. E. Borovsky, and D. T. Young, AGU Monograph 102, pp. 49–54, AGU, Washington, DC, 1998.

Laakso H. , D. H. Fairfield, C. T. Russell, J. H. Clemmons, H. J. Singer, B. L. Giles, R. P. Lepping, F. S. Mozer, R. F. Pfaff, K. Tsuruda, and J. R. Wygant, Field-line resonances triggered by a northward IMF , *Geophys. Res. Lett.*, **25**, 2991–2994, 1998.

Laakso, H., R. Pfaff, and P. Janhunen, Polar Observations of Electron Density Distribution in the Earth's Magnetosphere. 1. Statistical results, *Ann. Geophys*, submitted, 2001.

Lockwood, M., M. O. Chandler, J. L. Horwitz, J. H. Waite, Jr., T. E. Moore, and C. R. Chappell, The cleft ion fountain, *J. Geophys. Res.*, **90**, 9736–9748, 1985.

Moore, T. E. and D. C. Delcourt, The geopause, *Rev. Geophys.*, **33**, 175–209, 1995.

Pedersen, A., Solar wind and magnetosphere plasma diagnostics by spacecraft electrostatic potential measurements, *Ann. Geophys.,* **13**, 118–129, 1995.

Persoon, A. M., D. A. Gurnett, and S. D. Shawhan, Polar cap electron densities from DE 1 plasma wave observations, *J. Geophys. Res.,* **88**, 10123–10136, 1983.

Rosenbauer, H., H. Grünwaldt, M. D. Montgomery, G. Paschmann, and N. Sckopke, Heos 2 observations in the distant polar magnetosphere: the plasma mantle, *J. Geophys. Res.*, **80**, 2723–2737, 1975.

Schunk, R. W., Theoretical developments on the causes of ionospheric outflow, *J. Atmos. Sol. Terr. Phys.*, **62**, 399–420, 2000.

Su, Y.-J., J. L. Horwitz, T. E. Moore, B. L. Giles, M. O. Chandler, P. D. Craven, M. Hirahara, C. J. Pollock, Polar wind survey with the Thermal Ion Dynamics Experiment/Plasma Source Instrument suite aboard Polar, *J. Geophys. Res.*, **103** , 29 305 –29 337, 1998.

TWO-LEVEL MESOPAUSE AND ITS VARIATIONS FROM UARS-HRDI TEMPERATURE DATA

S. Thulasiraman and J.B. Nee

Department of Physics, National Central University, Chung Li – 32054, Taiwan.

ABSTRACT

The temperature data from 1992-98 in the height region of 75-105 km from the High Resolution Doppler Imager (HRDI) on board UARS is used to study the mesospheric dynamics. During December solstice (1992), the mesopause is at the high level of ~100 km in the Northern hemisphere with a temperature of 190 K and after 40^0 S, the mesopause changes to low level of 86-88 km with cool temperature (150 K) of 40 K less than the NH high latitude high level mesopause. In the June solstice (1993), the mirror image of this pattern is observed. The two-level mesopause is not observed during equinoxes and a single high level mesopause is observed. The time series analysis of temperature at 99 km shows a decrease of –1.2 K/Yr over the equatorial region during 1992-98.

INTRODUCTION

Mesopause temperatures are important for the occurrence of high latitude phenomena like Noctilucent Clouds (NLC) and Polar Mesospheric Summer Echoes (PMSE). Rocket measurements at high latitudes (Lubken, 1999; Lubken et al., 1999) and Na temperature Lidar measurements at midlatitudes (She et al., 1993; Bills and Gardner, 1993) showed the two-level mesopause structure, the warm high winter and cool low summer mesopause. For the first time, using a ship borne Na Lidar observations from 71^0 S to 54^0 N, von Zahn et al. (1996) observed a two-level mesopause. These measurements are generally location specific and a mean global picture is not clearly understood. The existing models like CIRA or MSISE90 do not reproduce this kind of observed two-level mesopause. Recently, Berger and von Zahn (1999) came up with a model, which reproduces this behaviour. HRDI on board UARS is making regular observations of the daytime temperature structure in the height region of 60 to 105 km (Ortland et al., 1998). In this study, we have used the HRDI level 3 AT version 11 data, which have a vertical resolution of 3 km and considered the data from 75 to 105 km in order to study the mesopause characteristics.

RESULTS AND DISCUSSION

The short period gravity waves, tidal oscillations and planetary waves have their influence on the temperature profiles over any location. These kinds of wave effects can be removed if the data is averaged over a long period. By taking a close look at the HRDI temperature data, we have classified the seasons as, March Equinox (March, April), June Solstice (May-August), September Equinox (September, October) and December Solstice (November-February). The seasonal variation of mesospheric temperatures for December Solstice (1992), is shown in Figure 1. The squares indicate the mesopause height. It can be seen that during December solstice, the mesopause is around 100 km in most of the Northern Hemisphere

(NH) and also up to 40^0 in the Southern Hemisphere (SH). After this latitude, the mesopause level changes to lower level of 86 km and an increasing mesopause height is seen upto 70^0 S where the mesopause height is 88 km with cool temperatures. A temperature difference of nearly 40 K is seen between high latitudes (70^0) in both the hemispheres. A mirror image of this pattern (not shown here) is observed for the June Solstice (1993) with transition occurring at the same latitudes and similar magnitude of temperature difference between the high and low level mesopause. This kind of seasonal variation is consistently observed for all the solstice seasons for which HRDI data (1992-98) is available with high level mesopause temperatures of 180-190 K and low level mesopause temperatures of 140-150 K. The results observed from HRDI data confirm the global two-level mesopause concept first proposed by von Zahn et al. (1996). In their study using a ship borne Lidar, a high (100 +/-3 km) "winter" level extending from 71^0 S to 23^0 N, the low (86 +/- 3 km) "summer" level from 24^0 N until the end of their observations at 54^0 N was observed. The model study by Berger and von Zahn (1999) clearly reproduced this observation data. But when the tidal effects are removed, the model shows the summer low level transition occur at 40^0 latitude. This is consistent with the HRDI observation also, because the seasonally averaged temperatures shown in Figure 1 free from tidal effects and the transition occurs after 40^0 latitude. The dominance of IR cooling processes in combination with the chemical heat release by the major reactions involving O, H and O_3 in the 100 km region is primarily responsible for the high level winter mesopause. The low level summer mesopause is due to the momentum deposition by the breaking gravity waves implicating the role of dynamical processes (Berger and von Zahn, 1999).

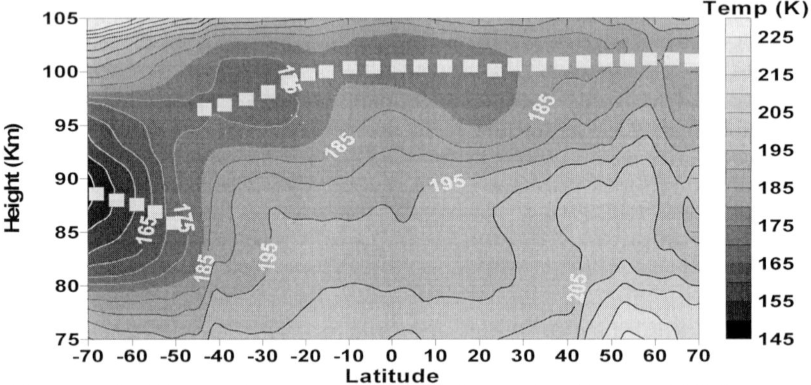

Figure 1. HRDI mean temperatures for the December Solstice (1992) season.

Figure 2 shows the temperature distribution during the March Equinox (1993). It can be seen that the mesopause is around 102 km over equatorial latitudes, 100 km over high latitudes and around 97 km over midlatitudes, with temperatures 170-180 K. Also an inversion of 5 K at 90 km altitude is seen over the equatorial latitudes during March Equinox season. In September Equinox (1993) also, similar kind of height and temperature structure are seen. She et al. (1993) also observed a similar kind of two temperature minima around equinox, implicating an inversion layer near ~ 90 km, suggesting that dynamical and/or chemical heating are possible mechanisms responsible for the inversion. An inversion in the annual mean profile is the result of profiles with double temperature minima and the combined effect of temperature structure transitions from the high winter to the low summer mesopause. Further, the amplitudes of thermal tides are weaker at mid and high latitudes but larger at equatorial latitudes between +/- 40^0, which are strong enough to produce temperature inversion layers (Berger and von Zahn, 1999).

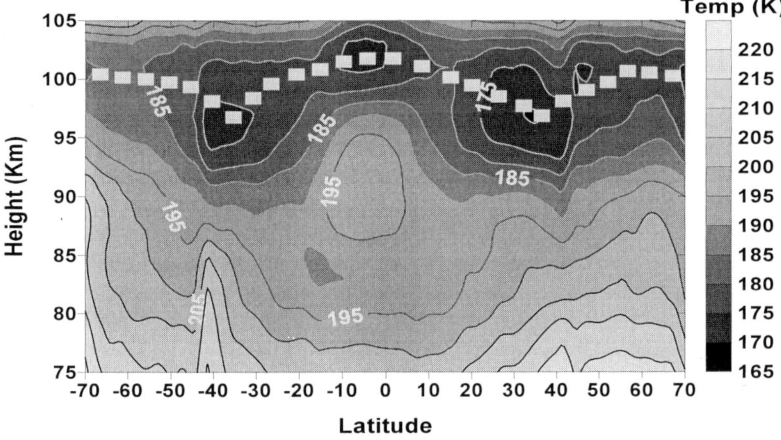

Figure 2. HRDI mean temperatures for the March Equinox (1993) season.

The present study shows that the mesopause height and temperature over the equatorial region has less seasonal variations. The time series of daily averaged temperature from January 1992 to December 1998 at 99 km height for the equatorial (+/- 5^0) region and is shown in Figure 3. A linear fit to the data shows 8.4 K decrease in temperature for the 7 years period (-1.2 K/yr). The possible causes for this decreasing trend are the solar cycle effect, anthropogenic effect and drift in instrument sensitivity. Model studies by Portmann et al. (1995) have predicted a 9-10 K cooling resulting from CO_2 doubling. But the estimated time for atmospheric CO_2 doubling is about a century (Brasseur and Hitchman, 1998). The short-term cooling trend in this study is much larger and the trend must be partly due to instrumental drift and also solar cycle changes. The data used here consists of declining solar flux (1992-96) and increasing solar flux (1997-98). So the declining phase data will have more influence on the trend. This cooling trend is consistent with the cooling trend of –1.4 K/yr at 100 km during 1990-97 (She et al., 1998), -1.0 K/yr cooling at 76 km over the equator (Clancy and Rush, 1989) during 1982-86 and also comparable temperature decrease at the same altitude during 1981-86 (Chanin et al. 1987). Thus the comparable cooling of –1.2 K/yr observed in this study may be due to declining phase of solar cycle, even though the current database is not as long and also due to instrumental drift. HRDI instrumental drift estimation and more data in the increasing phase, after 1997 (proposed TIMED mission will also be valuable) will be useful in better understanding the effect of the solar cycle, if any.

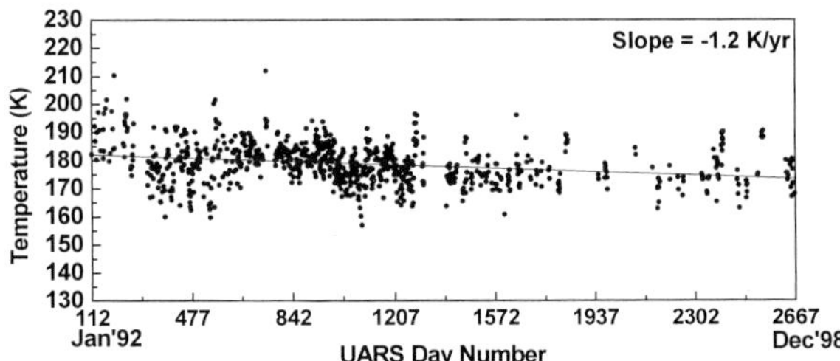

Figure 3. Daily averaged time series of HRDI temperatures at 99km over equator from January 1, 1992 (UARS Day number 112) to December 31, 1998.

SUMMARY

This study confirms the two-level mesopause during the Solstices with high level mesopause around 100 km and low level mesopause around 88 km with temperatures 180-190K and 140-150K respectively. A temperature difference of about 40-45K is seen between the high latitude high and low level mesopause. During equinoxes, only a single level mesopause is seen at around 98 km and an inversion of 5 K around 90 km is seen over the equatorial latitudes. The existing models like CIRA/MSISE90 does not show this two-level mesopause behaviour and HRDI database can be used to improve them.

ACKNOWLEDGEMENTS
We are grateful to the UARS/HRDI science team for providing the data through DAAC/GSFC web archive. We thank the National Space Program Office of the National Science Council, Taiwan, Republic of China, for providing financial assistance through the project NSC89-NSPO(3)-ISUAL-FA0901.

REFERENCES

Berger, U., and U. von Zahn, *J. Geophys. Res.,* 104, 22083-22093, 1999.
Bills, R.E., and C.S. Gardner, *J. Geophys. Res.,* 98, 1011-1021, 1993
Brasseur, G., and M.H. Hitchman, *Science,* 240, 634-637, 1988.
Chanin, M.L., N. Smires and A.Hauchecorne, *J. Geophys. Res.,* 92, 10933-10941, 1987.
Clancy, R.T., and D.W. Rusch, *J. Geophys. Res.,* 94, 3377-3393, 1989.
Lubken, F.-J., *J. Geophys. Res.,* 104, 9135-9149, 1999
Lubken, F.-J., M.J. Jarvis, and G.O.L. Jones, *Geophys. Res. Lett.,* 26, 3581-3584, 1999.
Ortland, D.A., P.B. Hays, and W.R. Skinner, *J. Geophys. Res.,* 103, 1821-1835, 1998
Portmann, R.W., G.E. Thomas, S. Solomon and R.R. Garcia, *Geophys. Res. Lett.,* 22, 1733-1736, 1995
She, C.Y., J.R. Yu and H. Chen, *Geophys. Res. Lett.,* 20, 567-570, 1993.
She, C.Y., S.W. Thiel, and D.A. Krueger, *Geophys. Res. Lett.,* 25, 497-500, 1998.
Von Zahn, U., J. Hoffner, V. Eska, and M.Alpers, *Geophys. Res. Lett.,* 23, 3231-3234, 1996.

Ground-Based Observations and Modeling Session

THE APPLICATION OF HIGH LATITUDE IONOSPHERE RADARS FOR SPACE WEATHER RESEARCH

J. Röttger
Max-Planck-Institut für Aeronomie, 37191 Katlenburg-Lindau, Germany

ABSTRACT

Ionospheric radars are used to investigate the auroral and polar cap regions, where solar wind - magnetosphere - ionosphere coupling is most evident. There are basically four categories of ground-based radars applied for this purpose: The digital ionosondes, coherent backscatter HF radars (F-region), coherent backscatter VHF and UHF radars (E-region), and incoherent scatter VHF and UHF radars. These systems are briefly described and examples of observations are shown, which are regarded to be relevant for space weather research.

INTRODUCTION

Space weather effects on the Earth's upper atmosphere are most pronounced in high latitudes, where the Earth's magnetic field lines couple the ionosphere with those parts of the magnetosphere, which are most sensitively influenced by the solar wind. Groundbased multi-point observations at such high latitudes of the polar caps are not easily achievable. Ionospheric radars, which can sense regions far apart from their locations, cover observations of larger areas in the polar caps. This article describes radar systems of this kind, emphasizing their experimental capabilities. Using radars for studies of the ionosphere and its coupling to the magnetosphere is in itself affected by space weather, since certain radar methods depend strongly on particular states of the ionosphere, which in turn are space weather related. We will not dwell too much on the physics of ionosphere-magnetosphere coupling and its relation to space weather, but just summarize some essentials to be studied with radar observations.

The convective flow of plasma and magnetic flux in the magnetosphere images into the high latitude ionosphere. It has its origin in two processes, namely the dayside solar wind coupling on the magnetopause and the anti-sunward stretching of the magnetic field in the magnetotail. This basically results in the typical two-cell plasma convection pattern, which can be observed in the ionosphere. Depending on the space weather conditions, this convection pattern changes notably and further processes occur. Plasma waves, irregularities and turbulence are initiated and energy and momentum is transferred into the ionosphere and the neutral atmosphere. The plasma convection and the mentioned resulting effects are in essence related to the magnetospheric electric field, which is mapped into the ionosphere. Additionally there are high energy particles, which precipitate from the magnetosphere into the ionosphere. The radars are also effective tools to investigate the effects related to the particle precipitation.

Figure 1 shows sketches of the flow pattern in the northern high latitude ionosphere for different values of the interplanetary magnetic field. Observations of the changes in this plasma convection pattern in turn allow deductions of the coupling mechanisms between the ionosphere, the magnetosphere and consequently the solar wind. Since this plasma convection covers the entire polar cap, multi-point observations are mandatory to observe the large-scale pattern changes.

Embedded in this highly time-variable large-scale process are smaller-scale processes, such as plasma patches, blobs, troughs and small-scale irregularities. To understand the mutual relation between the large- and the small-scale plasma processes, their intrinsic features and their effects on the neutral atmosphere, case studies are required. These can be done at single locations and preferably in multi-instrument programs or campaigns. Typical examples are the studies of the aurora. For an overview of the structure and dynamics of the polar ionosphere, observed by radars and

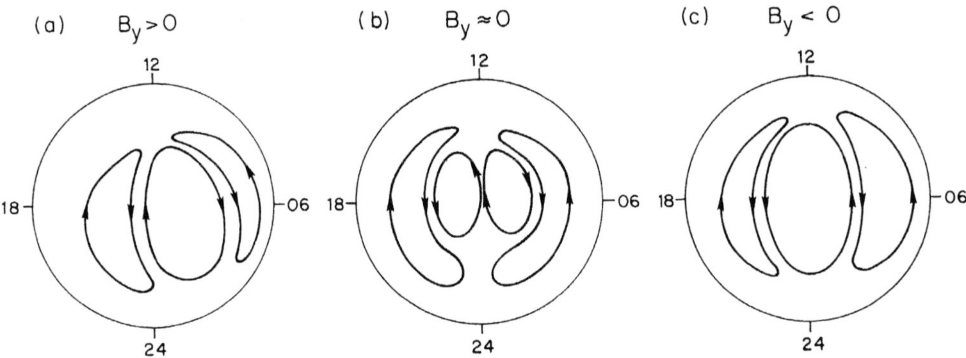

Fig. 1 Examples of plasma flow pattern when the B_z component of the interplanetary magnetic field (IMF) is positive and the B_y component changes from positive to negative (from Cowley et al., 1990).

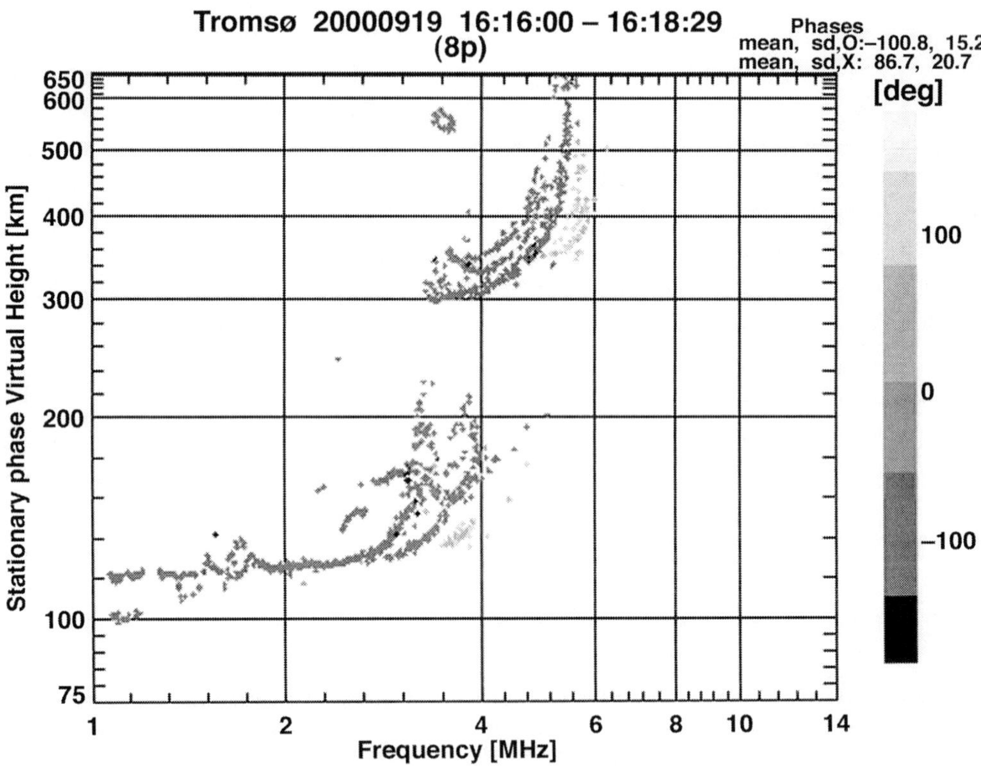

Fig. 2 Ionogram recorded with the Dynasonde in Tromsö/Norway. The original color plot shows that the incidence angle of some of the multiple traces are not from the vertical direction. This is a sign of a patchy ionosphere. Using an inversion procedure the vertical profile of the plasma frequency is obtained (plot provided by courtesy of M.T. Rietveld).

complementary instruments, reference is made to articles in the special issue of the Journal of Geomagnetism and Geoelectricity (Matuura and Kamide, 1995).

There are essentially four types of radars used for ground-based studies of the high-latitude ionosphere and its coupling with the magnetosphere and the solar wind. These are: (1) Ionosondes, operated in the MF and HF bands, (2) Coherent scatter HF radars (F-region), (3) Coherent scatter VHF and UHF radars (E-region), and (4) Incoherent scatter VHF and UHF radars. The main parameters, which are measured and are applicable for space weather assessment, are the plasma density (radar systems 1, 4), the plasma velocity and the electric field (1, 2, 3, 4), the plasma temperature (4) and plasma waves and turbulence (1, 2, 3, 4). The most important contributions of these radar systems are their capability to measure temporal variations of these parameters over particular portions of the ionosphere. Röttger (1999) has summarized the state of the art of such radars.

IONOSONDES

Vertical sounding of the ionosphere with pulsed sweep-frequency transmitters, called ionosondes, has been the traditional technique to obtain the vertical profile of the plasma frequency of the bottom-side ionosphere. In the recent decades advanced techniques have been introduced and these modern systems are usually called digital ionosondes. Fig. 2 shows a digital ionogram measured at Tromsö/Norway in the auroral zone. The multiple traces are signs of an irregular, patchy structure (James and MacDougall, 1997) of the overhead ionosphere during auroral conditions. The irregular structure can be proved by measurements of the incidence angle, which is done with this system. Additionally the vertical and horizontal drift velocity is measured. Cannon et al. (1992) had used digital ionosonde observations to show the dependence of the convection pattern on the interplanetary magnetic field, and Scali and Reinisch (1997) used digital ionosonde data for geomagnetic storm studies. Which relation polar cap patches and blobs have to space weather is still being discussed.

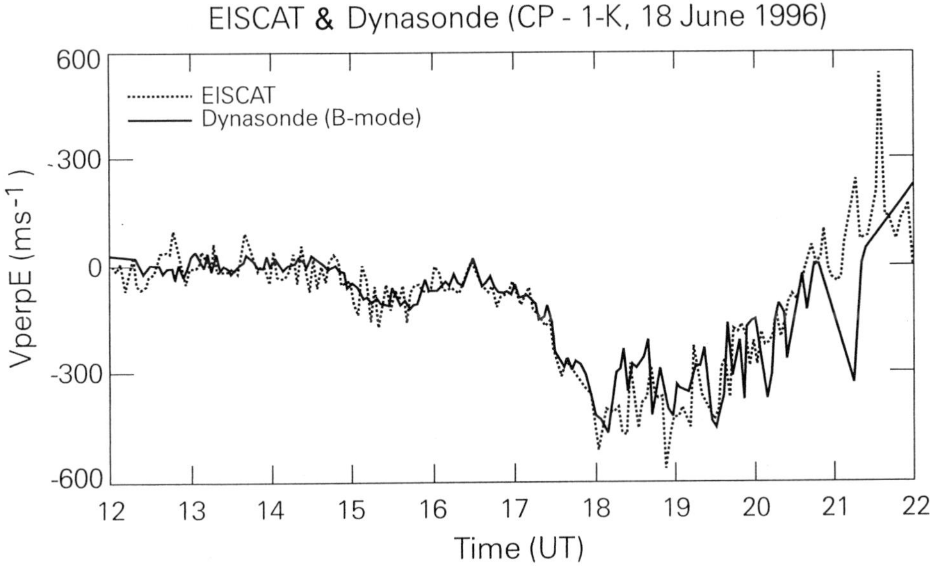

Fig. 3 Plasma drift velocities measured with Dynasonde and the EISCAT incoherent scatter radar showing the pre-midnight enhancement of velocity of the convection pattern (from Sedgemore et al., 1998).

In Fig. 3 the horizontal plasma velocity component in eastward direction is presented, which was measured at EISCAT in northern Scandinavia. This velocity, deduced from the ionosonde observations, compares well with the collocated incoherent scatter measurements of the velocity. It clearly shows the westward velocity increase in the pre-midnight hours, which is consistent with the pattern shown in Fig. 1(b). These ionosonde measurements, thus, are efficient complements to the very elaborate incoherent scatter measurements (see later chapter). The much more simpler observations with ionosondes (Scali et al., 1995), which can be performed on a regular base, allow a continuous record of the diurnal variation of the convection pattern at a fixed latitude as function of magnetic local time.

COHERENT SCATTER HF RADARS

Whereas ionosondes transmit in vertical direction, oblique transmission is used with the so-called HF radars. Reception can be either at a remote location or at the place of the transmitter. Both methods allow the sensing of the ionosphere at large distances. The basic scheme of the possible ray paths is depicted in Fig. 4. The point-to-point method applies sweep frequencies. It is mostly used to study influences of ionospheric disturbances on communication applications (e.g., Angling et al., 1998).

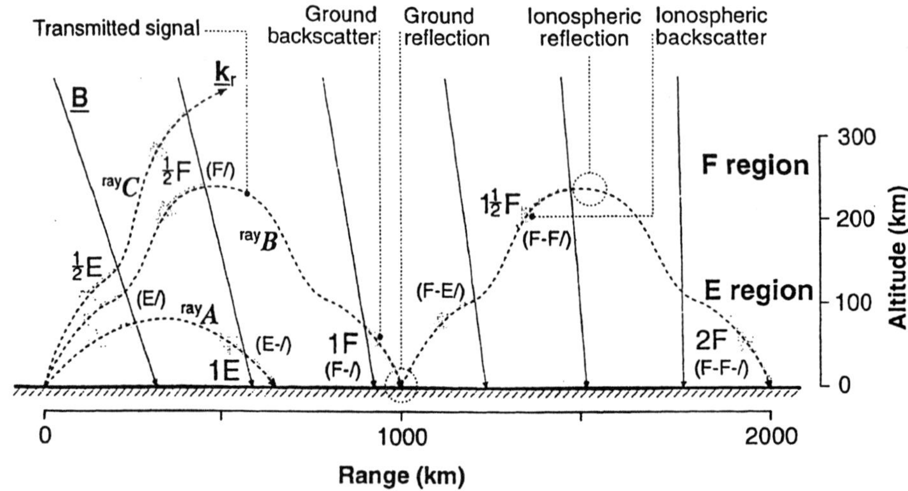

Fig. 4 Schematic ray paths of HF radar signals (dotted curves) in the polar E- and F-region. The Earth's magnetic field is indicated by straight lines (from Milan, 1997).

Essential for space weather observations is backscatter from field-aligned ionospheric irregularities. This coherent backscatter requires perpendicularity of the ray path to the magnetic field direction. Since the inclination of the field lines is very steep in polar latitudes, the perpendicular condition can only be achieved when the ray is sufficiently refracted. For this reason spot frequencies in the HF range 8 MHz to 20 MHz are used, which allow refraction in the ionospheric F-region. By measuring the Doppler-shift of the backscattered signals, the ionospheric drift velocity is determined. The distance to the scattering irregularity is obtained from range gating and the direction is determined by multiple narrow antenna beams. Fig. 5 shows the overlapping beam directions of the two CUTLASS (Collaborative UK Twin Located Auroral Sounding System) radars in Finland and Iceland (Lester et al., 1997). The combination of the measurements of two HF radars yields the horizontal component of the plasma drift. Several of these radar systems are in operation covering a wide area of the northern hemisphere polar cap. A few are operational in the southern polar cap.

Merging the measurements of this SuperDARN (Super Dual Auroral Radar Network) system provides a view of the global plasma convection pattern (Greenwald et al., 1995). In Fig. 6 the velocity field, observed with the American

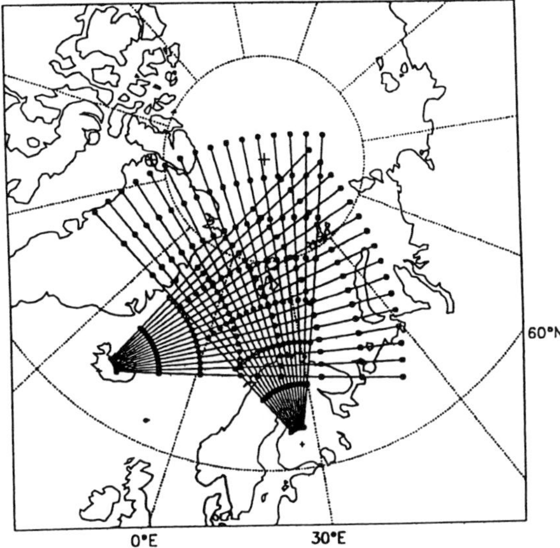

Fig. 5 Beam directions of the CUTLASS HF radars in Finland and Iceland. The combination covers large areas of the auroral zone, the cusp region and parts of the polar cap (from Lester et al., 1997).

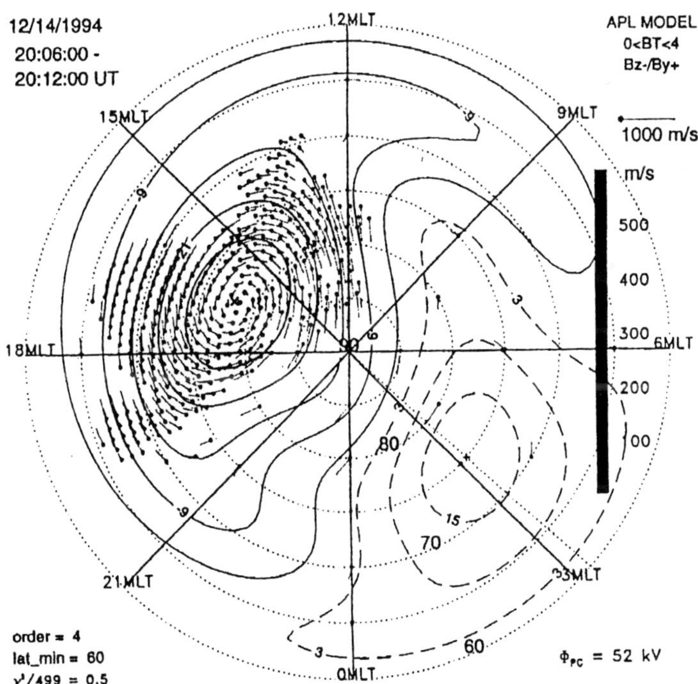

Fig. 6 Velocity vectors measured with the North American SuperDARN system. These are fitted to a statistical convection model, showing the electrostatic potential, indicated by the solid and dotted contours (from Ruohoniemi and Baker, 1998).

and Canadian radars, is fitted to a model. Bristol et al. (1998) obtained similarly convincing results. The shape of the convection pattern and the magnitudes of the velocities are determined by the magnetosphere-ionosphere coupling, which in turn depends on the solar wind. The multi-point observations with SuperDARN are consequently very useful for space weather research.

Direct observations of the magnetosphere with HF radar had also been attempted and signs of ion acoustic waves were reported (Hyssell et al., 1997). Satellite-borne radars also had been used to study the topside ionosphere and its transition into the magnetosphere. The recent space-borne IMAGE mission, using a radar in the frequency range 3 kHz to 3 MHz (Benson et al., 1998), is particularly addressing magnetospheric observations. This will open new directions in space weather research.

COHERENT SCATTER VHF AND UHF RADARS

The HF radars detect backscatter from F-region irregularities. Radars operating in the VHF and UHF bands detect coherent backscatter from field-aligned E-region irregularities in the auroral zone. Ray bending is negligible at these high frequencies. The perpendicular condition can be achieved, since the inclination of the Earth's magnetic field is less steep in the auroral zone than in the polar cap.

At the Japanese Antarctic station Syowa HF and VHF radars can be operated simultaneously (Ogawa, 1996), where the HF radar covers the polar cap and the VHF radar the auroral zone (Fig. 7). In the northern hemisphere, for example, the STARE (Scandinavian Twin Auroral Radar Experiment) VHF radar is operated, and there is the COSCAT (COherent SCATter) system, operated as a hybrid of the EISCAT radars on 929.5 MHz for auroral irregularity studies (Eglitis et al., 1998). Fig. 8 shows some recent results of velocity measurements with STARE, which were interpreted to be signs of field-aligned currents in an aurora event. These currents from the magnetosphere close in the E-region and are a signature of magnetosphere-ionosphere coupling. Most of the studies with VHF and UHF radars are done to understand the plasma waves and turbulence, which are responsible for the backscatter. The intrinsic features of these plasma processes are no direct measures of space weather events, although they are initiated by them. Reviews of the investigations of backscatter from these irregularities in the high latitude ionosphere were written by Haldoupis (1989) and Sahr and Fejer (1996).

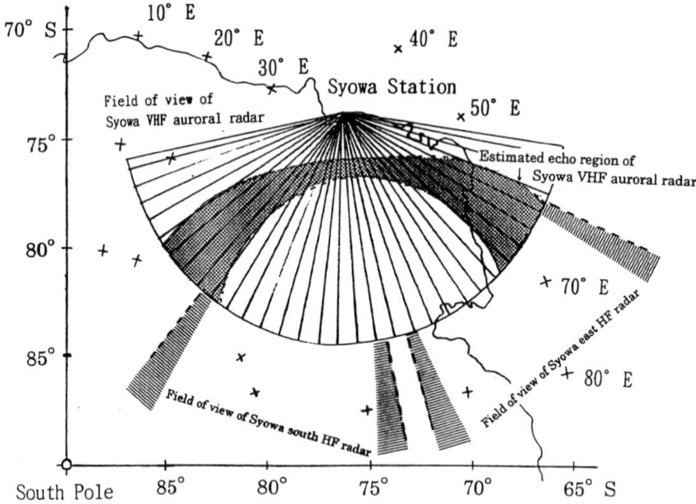

Fig. 7 Field of view of the VHF auroral radar and the extension into the southern polar cap with the Syowa HF radar (from Igarashi et al., 1995).

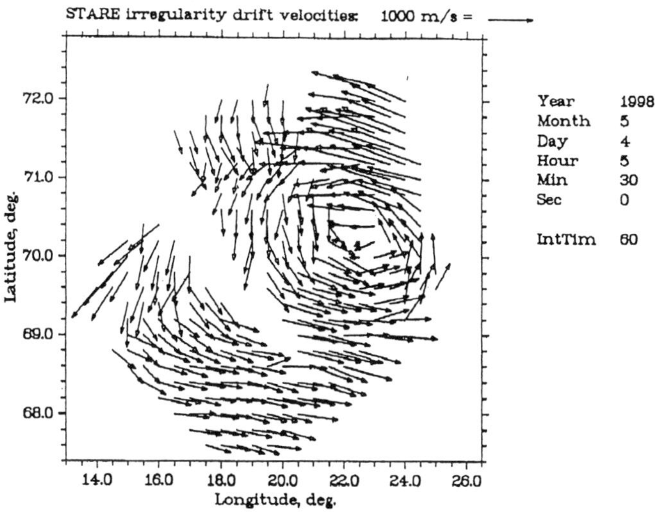

Fig. 8 Ionospheric flow vectors measured with the STARE VHF radar over northern Scandinavia (from Nielsen et al., 1999).

In summary: Coherent scatter radars are proper tools for studying features of plasma physics. The plasma irregularities themselves are used as tracers to measure the bulk plasma drift, which is directly related to the electric field through the measured E x B drift. There are a few major conditions, which have to be obeyed for this method to work. These are: (1) The state of the ionosphere has to be properly disturbed such that plasma irregularities exist which backscatter the radar signals; (2) ray bending by refraction in the F-region has to be adequate (i.e. the background ionosphere is sufficiently ionized) to assure perpendicularity of the HF radar wave vector to the Earth's magnetic field; (3) D-region absorption has to be low when using radars in the HF band. The latter condition (3) is in a certain sense counteracting condition (1), since absorption increases and F-region irregularities are created during disturbed conditions. VHF or UHF radars operated in the incoherent scatter mode are not significantly affected by absorption and irregularities.

INCOHERENT SCATTER VHF AND UHF RADARS

The transfer of energy, momentum, and mass through the magnetospheric system into the polar ionosphere, the thermosphere and the middle atmosphere can appropriately be investigated by the incoherent scatter radar technique. It is, thus, quite useful for space weather research. With this technique it is possible to measure directly the electron density, the electron temperature, the ion temperature and composition as well as the ion velocity. From these basic parameters many related parameters are deduced, such as for example the electric fields, currents, conductivity, particle and Joule heating, heat flux and energy input and the ion-electron production rate, etc. Many of these parameters depend on the space weather. The theory, practice and the science of incoherent scatter is described in articles edited and published by Alcayde (1997). Further articles on radar studies of the high latitude ionosphere, combined with other ground-based and space-borne observations, had been edited and published by Matuura and Kamide (1995) and by Lockwood et al. (1997).

There are three incoherent scatter radar systems, which are applied for space weather studies: The Sondre Stromfjord radar in Greenland (Kelly et al., 1995), the Millstone Hill radar in Massachusetts, USA (Buonsanto and Foster, 1993), and the EISCAT radar systems in northern Scandinavia and on Svalbard (EISCAT, 1996, 1999). In Fig. 9 a sketch of the EISCAT systems is exposed, and in Fig. 10 the two antennas of the EISCAT Svalbard Radar (ESR) are shown. The ESR was particularly designed for studies of solar wind - magnetosphere - ionosphere - thermosphere coupling, since Svalbard is located in the polar cap and the magnetospheric cusp passes over this area.

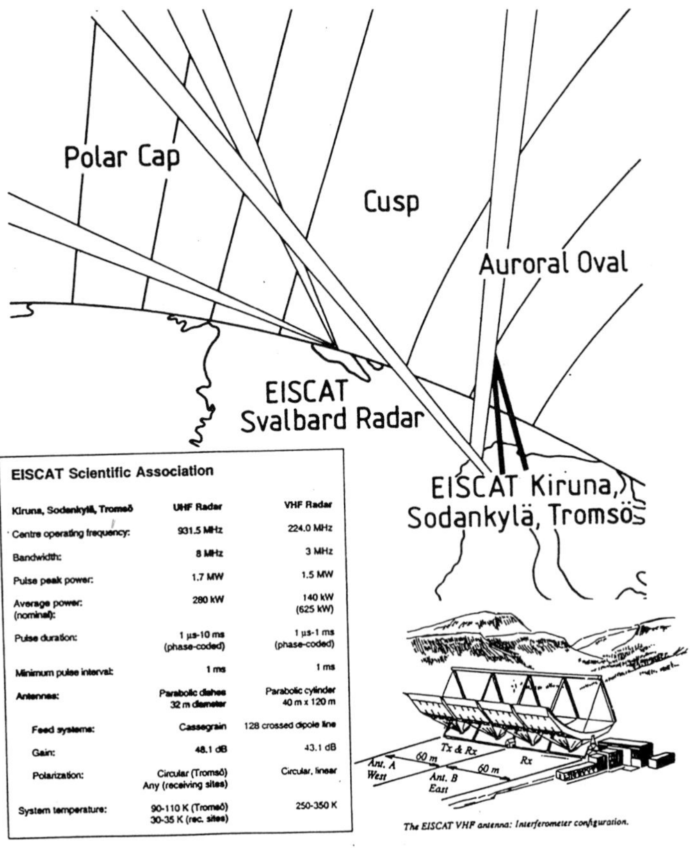

EISCAT Scientific Association

Kiruna, Sodankylä, Tromsö	UHF Radar	VHF Radar
Centre operating frequency:	931.5 MHz	224.0 MHz
Bandwidth:	8 MHz	3 MHz
Pulse peak power:	1.7 MW	1.5 MW
Average power: (nominal):	280 kW	140 kW (625 kW)
Pulse duration:	1 µs-10 ms (phase-coded)	1 µs-1 ms (phase-coded)
Minimum pulse interval:	1 ms	1 ms
Antennas:	Parabolic dishes 32 m diameter	Parabolic cylinder 40 m x 120 m
Feed systems:	Cassegrain	128 crossed dipole line
Gain:	48.1 dB	43.1 dB
Polarization:	Circular (Tromsö) Any (receiving sites)	Circular, linear
System temperature:	90-110 K (Tromsö) 30-35 K (rec. sites)	250-350 K

Fig. 9 The EISCAT radar systems to study the auroral oval, the cusp and the polar cap.

42m fixed dish **32m movable dish**

Fig. 10 The antennas of the EISCAT Svalbard Radar,
which operates on 500 MHz with a peak transmitter power of 1000 kW.

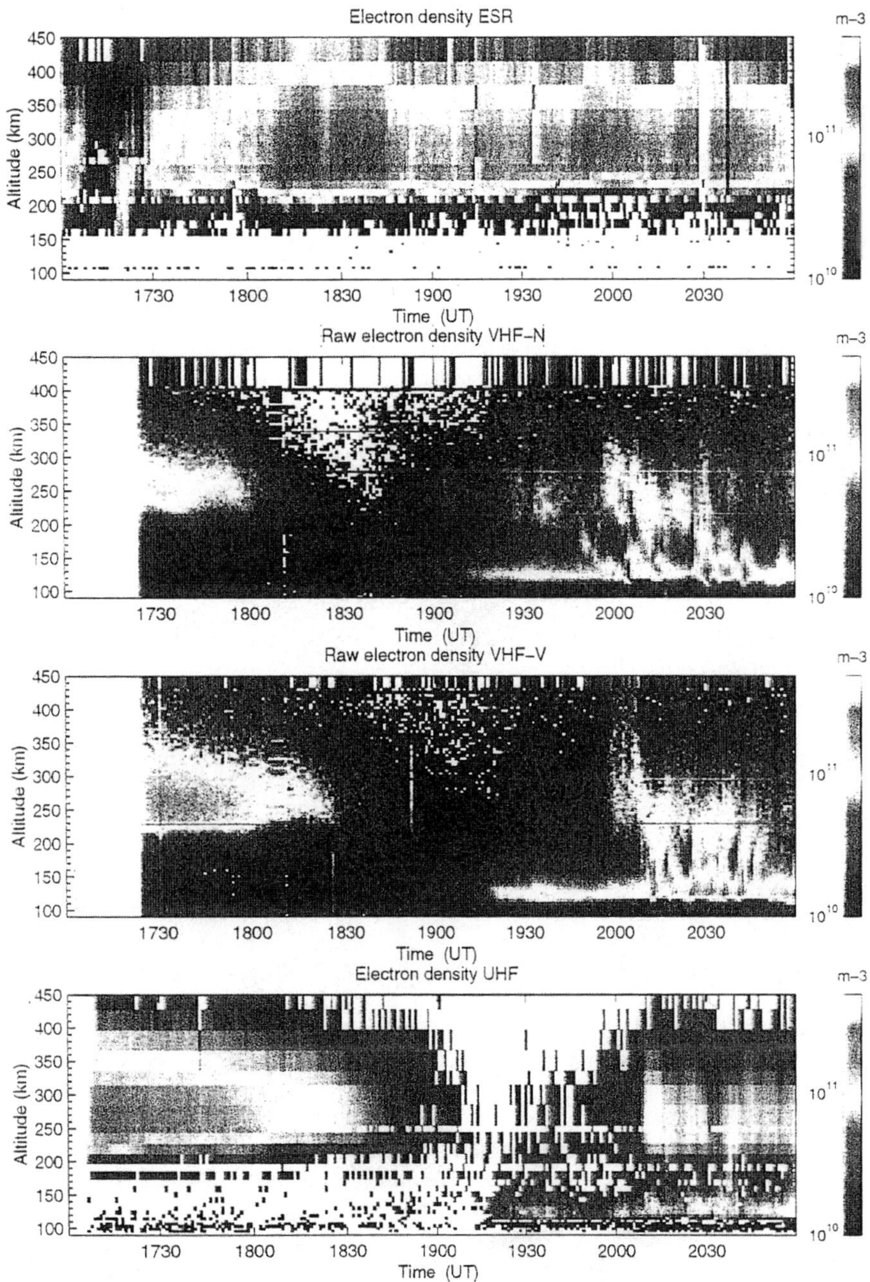

Fig. 11 Electron density profiles observed with the ESR, the VHF and the UHF radars of EISCAT (from Jones et al., 1997; EISCAT, 1999) showing passage of the trough.

It is impossible to present the multitude of investigations with these radar systems, which relate to space weather. The EISCAT Annual Reports (EISCAT, 1996, 1999) provide several representative examples and references to relevant publications. We will just select two samples, which depict the states of calm and disturbed space weather conditions, respectively.

On March 14, 1997, all the EISCAT radars were operated in a common mode. The geomagnetic conditions were extremely calm. During this time the passage of the ionospheric trough was observed. The trough is an ionization depletion occurring in the early evening hours at the low latitude boundary of the polar cap. The electron density profiles in Fig. 11 show the signatures of the trough observed over Svalbard (ESR), north (VHF), vertical (VHF) and south (UHF) of Tromsø. Jones et al. (1997) demonstrated with these observations that the structure of the poleward wall of the trough is related to energy input from localized, transient soft particle precipitation combined with an enhancement of the electron temperature. The apparent equatorward advance of the trough, as detected by these radars, follows as a dynamic response to the movement of this precipitation during calm space weather conditions.

The global structure of the convection pattern is variable at smaller scales, which is a sign of more disturbed space weather. Examples of these disturbances are travelling convection vortices. Fig. 12 shows two such vortices observed on January 6, 1992, with the EISCAT UHF and VHF radar, which were pointing into northward azimuth directions fifteen degrees apart. The velocity measured with the radars is at large and opposite values. The vortex structure, that rotates with opposite sense to those of Fig.12, was also observed by Lühr et al. (1996). Convection vortices are signs of field-aligned currents, which couple the ionosphere and magnetosphere.

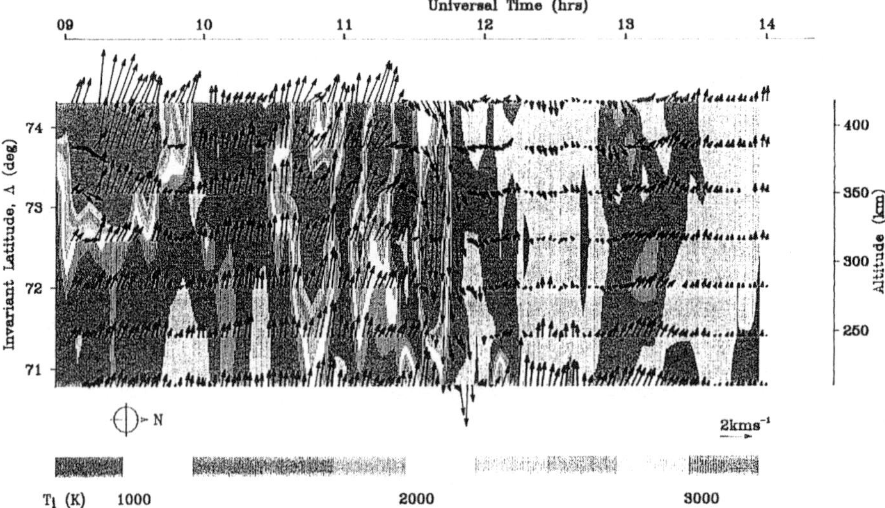

Fig. 12 Line of-site velocities (colour coded) measured on 12 January 1988 with the EISCAT UHF and VHF radar during occurrence of travelling convection vortices. The arrows represent equivalent plasma convection deduced from magnetometer records (from Lühr et al., 1996; EISCAT, 1996).

Passages of the magnetospheric cusp (e.g. Yeh et al., 1990), flux transfer events, magnetopause reconnection and cusp particle precipitation (e.g., Lockwoood et al., 1996), the multitude of effects during magnetospheric substorms and the aurora (e.g. Lanchester et al., 1996) and the energy coupling between the magnetosphere, ionosphere and thermosphere (e.g., Fujii et al., 2000) are further examples of important investigations of space weather effects with incoherent scatter radars. For further profound, though not complete, information on this research the reader is referred to EISCAT Annual Reports (EISCAT, 1996, 1999).

CONCLUSION

The incoherent scatter radars of EISCAT and Sondre Stromfjord (partially also Millstone Hill) are used efficiently for studies of the high latitude ionosphere and how this region is affected by the space weather. Many events have been evaluated to investigate the measured profiles of electron density, electron and ion temperature as well as the ion drift to study the coupling with the magnetosphere and the solar wind. To cover also the interior of the polar cap up to the magnetic pole an even higher latitude radar system would be required, which could suitably be positioned at Resolute Bay in northern Canada, close to the magnetic pole.

The combination of multi-point measurements with the incoherent scatter radars and the coherent scatter radars of the SuperDARN system provide a unique opportunity for the studies of the polar cap convection pattern and its spatial and temporal variations, which lead to the effects of the solar wind on the Earth's magnetosphere and the ionosphere.

Combination with other new instrumentation, such as the planned SPEAR (Robinson, 1998) radar system on Svalbard, and spacecraft, such as Cluster-2 (e.g., Lockwood et al., 1997) and IMAGE/RPI (e.g., Benson et al., 1998), will further enhance the scientific research capabilities to foster understanding of space weather effects on the Earth.

ACKNOWLEDGEMENT

I thank the organizers of this COSPAR Colloquium for their efforts in performing this activity and for inviting me to prepare this paper.

REFERENCES

Alcayde, D., ed., Incoherent Scatter: Theory, Practice and Science, Technical Report EISCAT Scientific Association, 97/53, pp. 1-314 (1997).

Angling, M.J., P.S. Cannon, N.C. Davies, T.J. Willink, V. Jodalen, and B. Lundborg, Measurement of Doppler and Multipath Spread on Oblique High-Latitude HF Paths and their Use in Characterizing Data Modem Performance, Radio Scie., 33, pp. 97-107 (1998).

Benson, R.F., J.L. Green, S.F. Fung, B.W. Reinisch, W. Calvert, D.M. Haines, J.L. Bougeret, R. Manning, D.L. Carpenter, D.L. Gallagher, P.H. Reiff, and W.W.L. Taylor, Magnetospheric Radio Sounding on the Image Mission, Radio Science Bulletin, URSI, 285, pp. 9-20 (1998).

Bristow, W.A., J.M. Ruohoniemi, and R.A. Greenwald, Super Dual Auroral Radar Network Observations of Convection During a Period of Small-Magnitude Northward IMF, J. Geophys. Res., 103, pp. 4051-4061 (1998).

Cannon, P.S., G. Crowley, B.W. Reinisch, and J. Buchau, Digisonde Measurements of Polar Cap Convection for Northward Interplanetary Magnetic Field, J. Geophys. Res., 97, 16877-16885 (1992).

Cowley, S.W.H., A.P. van Eyken, E.C. Thomas, P.J.S. Williams, and D.M. Willis, Studies of the Cusp and Auroral Zone with Incoherent Scatter Radar: The Scientific and Technical Case for a Polar Cap Radar, J. Atmos. Terr. Phys., 52, pp. 645-663 (1990).

Eglitis, P., I.W. McCrea, T.R. Robinson, T. Nygren, K. Schlegel, T. Turunen, and T.B. Jones, New Techniques for Auroral Irregularity Studies with COSCAT, Ann. Geophys., 16, pp. 1241-1250 (1998).

EISCAT, Annual Report 1994-1995, EISCAT Scientific Association, pp. 1-127 (1996).

EISCAT, Annual Report 1996-1997, EISCAT Scientific Association, pp. 1-108 (1999).

Fujii, R., S. Nozawa, S.C. Buchert, and A. Brekke, Energy Coupling between the Magnetosphere, Ionosphere and Thermosphere, Adv. Space Res., 26, pp. 979-984 (2000).

Greenwald, R.A., K.B. Baker, J.R. Dudeney, M. Pinnock, T.B. Jones, E.C. Thomas, J.P. Villain, J.C. Cerisier, C. Senior, C. Hanuise, R.D. Hunsucker, G. Sofko, J. Koehler, E. Nielsen, R. Pellinen, A.D.M. Walker, N. Sato, and H. Yamagishi, DARN/SuperDARN: A Global View of the Dynamics of High-Latitude Convection, Space Scie. Rev., 71, pp. 761-796 (1995).

Haldoupis, C., A Review on Radio Studies of Auroral E-Region Ionospheric Irregularities, Ann. Geophys., 7, pp. 239-258 (1989).

Hyssel, D.L., M.C. Kelley, A.V. Gurevich, A.N. Karashtin, A.M. Babichenko, Y.M. Yampolski, V.S. Beley, and J.F. Providakis, HF Radar Probing of the Lower Magnetosphere, J. Geophys. Res., 102, pp. 4865-4873 (1997).

Igarashi, K., K. Ohtaka, M. Kunitake, T. Tanaka, and T. Ogawa, Development of Scanning-Beam VHF Auroral Radar System, NIPR Symp. Upper Atmos. Phys., 8, pp. 65-69 (1995).

James, H.G., and J.W. MacDougall, Signatures of Polar Cap Patches in Ground Ionosonde Data, Radio Scie., 32, pp. 497-513 (1997).

Jones, D.G., I.K. Walker, and L. Kersley, Structure of the Poleward Wall of the Trough and the Inclination of the Geomagnetic Field above the EISCAT Radar, Ann. Geophys., 15, pp. 740-746 (1997).

Lanchester, B.S., K. Kaila, and I.W. McCrea, Relationships between Large Horizontal Electric Fields and Auroral Arc Elements, J. Geophys. Res., 101, pp. 5075-5084 (1996).

Lester, M., T.B. Jones, T.R. Robinson, E.C. Thomas, T.K. Yeoman, R. Pellinen, A. Huuskonen, H. Opgenoorth, M. Persson, A. Pellinen-Wannberg, and I. Häggström, CUTLASS - A Tool for Co-ordinated Satellite/Ground Based Investigations of the Solar Terrestrial System, ESA SP-1198, pp.191-202 (1997).

Lockwood, M., S.W.H. Cowley, and T.G. Onsager, Ion Acceleration at Both the Interior and Exterior Alfven Waves Associated with the Magnetopause Reconnection Site: Signatures in Cusp Precipitation, J. Geophys. Res., 101, 21501-21515 (1996).

Lockwood, M., M.N. Wild, and H. Opgenoorth, eds., Satellite Ground Based Coordination Sourcebook, ESA SP-1198 (1997).

Lühr, H., M. Lockwood, P.E. Sandholt, T.L. Hansen, and T. Moretto, Multi-instrument ground-based observations of a travelling vortices event, Ann. Geophys., 14, pp.162-181 (1996).

Matuura, N., and Y. Kamide eds., Structure and Dynamics of the Polar Ionosphere, J. Geomagn. Geoelectr., 47, pp. 669-942 (1995).

McCrea, I.W., and M. Lockwood, Incoherent Scatter Radars, ESA SP-1198, pp.239-266 (1997).

Milan, S.E., T.K. Yeoman, M. Lester, E.C. Thomas, and T.B. Jones, Initial Backscatter Occurrence Statistics from the CUTLASS HF Radars, Ann. Geophys., 15, pp. 703-718 (1997).

Nielsen, E., M. Bruns, I. Pardowitz, H. Perplies, and L. Bemmann, STARE: Observations of a Field-Aligned Line Current, Geophys. Res. Lett., pp. 21-24 (1999).

Ogawa, T., Radar Observations of Ionospheric Irregularities at Syowa Station, Antarctica: A Brief Overview, Ann. Geophys., 14, 1454-1461 (1996).

Robinson, T.R., Space Plasma Exploration by Active Radar, Scientific Case, Report Radio and Space Plasma Physics Group, University of Leicester, pp.1-16 (1998).

Röttger, J., Radar Systems in Ionospheric Research, in Modern Radio Science 1999, URSI, M.A. Stuchly ed., pp.213-247 (1999).

Ruohoniemi, J.M., and K.B. Baker, Large-Scale Imaging of High-Latitude Convection with Super Dual Auroral Radar Network HF Radar Observations, J. Geophys. Res., 103, pp. 20797-20811 (1998).

Sahr, J.D., and B.G. Fejer, Auroral Electrojet Plasma Irregularity Theory and Experiment: A Critical Review of Present Understanding and Future Directions, J. Geophys. Res., 101, pp. 26893-26909 (1996).

Scali, J.L., B.W. Reinisch, C.J. Heinselman, and T.W. Bullett, Coordinated Digisonde and Incoherent Scatter Radar F Region Drift Measurements at Sondre Stromfjord, Radio Scie., 30, pp. 1481-1498 (1995).

Scali, J.L., and B.W. Reinisch, Geomagnetic Storm Time Studies Using Digisonde Data, Adv. Space Res., 20,9, pp. 1679-1699 (1997).

Sedgemore, K.J.F., J.W. Wright, P.J.S. Williams, G.O.L. Jones, and M.T. Rietveld, Plasma Drift Estimates from Dynasonde: Comparison with EISCAT Measurements, Ann. Geophys., 16, pp. 1138-1143 (1998).

Yeh, H.C., C. Foster, J.M. Holt, R.H. Redus, and F.J. Rich, Radar and Satellite Observations of the Storm Time Cleft, J. Geophys. Res., 95, pp. 12075-12090 (1990).

MAGNETOSPHERIC SUBSTORMS: AN INNER-MAGNETOSPHERIC MODELING PERSPECTIVE

R. A. Wolf[1], F. R. Toffoletto[1], R. W. Spiro[1], M. Hesse[2], and J. Birn[3]

[1]*Department of Physics and Astronomy, Rice University*
[2]*NASA Goddard Space Flight Center*
[3]*Los Alamos National Laboratory*

ABSTRACT

Kilovolt electrons cause satellite anomalies through surface charging and frequently interfere with the operation of geosynchronous spacecraft. The Magnetospheric Specification Model was designed to specify kilovolt electron fluxes at and near synchronous orbit, and while it has seen limited operational use, its accuracy is not high. One promising approach to improving the accuracy lies in assimilation of real-time-measured geosynchronous fluxes. An initial effort in that direction showed significant, though undramatic, improvements in accuracy. Since major changes in geosynchronous electrons are typically associated with magnetospheric substorms, accurate specification and forecast of geosynchronous electrons is likely to require understanding of the substorm phenomenon, particularly as it impacts the inner magnetosphere. Coupling of the Rice Convection Model (RCM) to a friction-code equilibrium solver now allows solution of a complete set of physical equations for the inner magnetosphere and inner plasma sheet, assuming adiabatic drift and isotropic pressure. In this formulation, the plasma is characterized by the distribution function $f(\lambda,\alpha,\beta)$, where $\lambda=W_K V^{2/3}$ is the isotropic invariant, V is the flux-tube volume, and α and β are Euler potentials. Initial application of this coupled model to the substorm problem has produced the following results:

1. More realistic calculation has verified the standard picture of the substorm growth phase. Enforcing a strong convection electric field across the magnetotail and assuming adiabatic convection (specifically conservation of $f(\lambda,\alpha,\beta)$ along a drift path) leads to storage of magnetic flux in the tail, development of a magnetic-field minimum in the inner plasma sheet, and a thinning and intensification of the current sheet there. Adiabatic convection by itself does not lead to injection of plasma into the inner magnetosphere.
2. From the viewpoint of the inner magnetosphere, the main effect of the substorm expansion phase is the creation of new closed plasma-sheet flux tubes having lower values of $f(\lambda,\alpha,\beta)$ and $pV^{5/3}$ than ordinary plasma-sheet flux tubes. These lower content flux tubes can more easily be injected into the geosynchronous-orbit region and inner magnetosphere.
3. Since the RCM is based on the assumption of adiabatic drift, it does not directly describe the nonadiabatic processes that are essential to the substorm expansion phase, but those processes can be parameterized in the RCM. Different assumptions about the non-adiabatic reduction of f on different flux tubes in the substorm expansion phase lead to different predicted ionospheric electric field patterns. These different patterns, along with differences in the particle distribution functions during injections, may prove useful in clarifying the physics of the expansion phase and choosing between competing substorm scenarios.

INTRODUCTION

The Magnetospheric Specification Model was developed for the U. S. Air Force to specify (nowcast) the fluxes of particles in the 1-100 keV range, with particular emphasis on electrons near geosynchronous orbit (Bales *et al.*, 1993; Lambour, 1994; Wolf *et al.*, 1997). That model used a combination of established physics, empirical models, and real-time data. The established physics consisted of the equations for bounce-averaged gradient/curvature drift, calculated for empirical data-driven model electric and magnetic fields. The MSM, designed in the late 1980's and completed in the early 1990's, is now run routinely by the NOAA Space Environment Center, and its computed fluxes can be accessed on the web (http://sec.noaa.gov/rpc/msm/index.html). The Magnetospheric Specification and Forecast Model (MSFM), an extension of the MSM that was developed in the early 1990's, can be driven by solar-wind data.

The MSM and MSFM have been thoroughly tested by the Air Force. The most extensive testing, with checks against years of geosynchronous data, was by Hilmer and Ginet (2000). Overall, the accuracy of the MSM has been modest, with a mean error in the $\log_{10}$(flux) equal to approximately 0.45.

Kilovolt electrons are introduced into the geosynchronous orbit region in the expansion phase of a substorm, a phenomenon whose causal mechanism was not understood when the MSM was designed. While still incomplete, our understanding of the substorm expansion phase has advanced substantially in the intervening years. The MSM was designed with various data-based "fudges" to avoid dire consequences due to lack of understanding.

There are presently two promising approaches to developing a capability for accurate forecasting and nowcasting of kilovolt electrons at synchronous orbit:

1. Application of data assimilation techniques to allow optimal utilization of real-time measurements of particle fluxes in the region.

2. Development of a first-principles model that accurately represents the substorm phenomenon and its coupling to the inner magnetosphere.

The first effort at data assimilation into the MSM has been completed (Garner *et al.*, 1999). Data from geosynchronous spacecraft were assimilated into the model by the "direct insertion" method. That is, measured fluxes were used to override model-computed values in a small region around the spacecraft, at each model time step. Then the model's computational machinery propagated the correction as the particles drifted in time. The result was a noticeable but undramatic increase in the accuracy of model values, as judged by comparison with measurements made from another spacecraft, data that were not fed into the model. The direct-insertion method is the simplest method of data assimilation. Much more sophisticated data-assimilation techniques are being used in other areas of Earth science, particularly meteorology and oceanography; adapting them to the calculation of magnetospheric quantities will be a major challenge for magnetospheric modeling in the next two decades.

This paper focuses on a second approach, namely the long-term effort to develop a first-principles computational model of the inner and middle magnetosphere that treats magnetic fields self-consistently with potential electric fields and energy dependent drifts. The MSM is clearly a long way from being a complete and defensible model of the region. However, modern computational capabilities are bringing us much closer to the goal.

MODEL

We have coupled the Rice Convection Model with an equilibrium magnetofriction code in order to solve a complete set of differential equations for the inner and middle magnetosphere. As reported in

Toffoletto *et al.* (1996), the coupled code uses the magnetic field from the equilibrium solver to compute plasma drifts in the RCM and the RCM-computed pressure distribution to modify the magnetic field.

The RCM (Harel *et al.*, 1981; Erickson *et al.*, 1991) assumes that the distribution function f is isotropic in velocity space and constant long a magnetic field line and so can be written as a function of Euler potentials α and β, energy, and time. Chaotic ion motion, or some other mechanism, is assumed to keep the distribution function isotropic without changing particle energy. The adiabatic drift equations (Wolf, 1983; Usadi *et al.*, 1996) then imply conservation of the isotropic invariant λ, defined by

$$\lambda = W_K V^{2/3} = \text{constant} \tag{1}$$

Here W_K is the kinetic energy in gyro and bounce motion and V is the volume of a magnetic flux tube containing one unit of magnetic flux:

$$V = \int \frac{ds}{B} \tag{2}$$

The integral extends along the field line from the southern ionosphere to the northern. The RCM advances the particle distribution with a condition equivalent to the form

$$\left(\frac{\partial}{\partial t} + \dot{\alpha}\frac{\partial}{\partial \alpha} + \dot{\beta}\frac{\partial}{\partial \beta}\right) f(\lambda,\alpha,\beta,t) = 0 \tag{3}$$

Atmospheric loss processes can be included on the right side of Eq. 3, but will be neglected for the plasma-sheet ions, under the conditions of interest here. The pressure can be calculated from f through the relation

$$P(\alpha,\beta,t) = \left(\frac{2^{7/2}\pi}{3m^{3/2}}\right) V^{-5/3} \int f(\lambda,\alpha,\beta,t)\lambda^{3/2} d\lambda \tag{4}$$

Note the simple relationship between $PV^{5/3}$ and the distribution function f. The drift velocities, expressed in terms of Euler potentials, are given by

$$\dot{\alpha} = -\frac{\partial H}{\partial \beta}, \quad \dot{\beta} = \frac{\partial H}{\partial \alpha} \tag{5}$$

where the Hamiltonian H is given by

$$H(\lambda,\alpha,\beta,t) = W_K(\lambda,\alpha,\beta,t) + q\Phi(\alpha,\beta,t) \tag{6}$$

and Φ is the electrostatic potential. The Birkeland current down into the ionosphere is given by Vasyliunas' equation

$$J_{\|} = \hat{b}\cdot\vec{\nabla}V \times \vec{\nabla}P \tag{7}$$

The equation of current conservation in the ionosphere is

$$\vec{\nabla}\cdot\left[-\vec{\vec{\Sigma}}\cdot\vec{\nabla}\Phi\right] = J_{\|} \tag{8}$$

where $\vec{\vec{\Sigma}}$ is the ionospheric conductance tensor; for simplicity we have neglected neutral winds.

The friction-code algorithm (Toffoletto *et al.*, 1996, 2000) computes the magnetic field from the equilibrium equations

$$\vec{\nabla} P = \vec{J} \times \vec{B} \tag{9}$$

$$\vec{\nabla} \times \vec{B} = \mu_o \vec{J} \tag{10}$$

$$\vec{\nabla} \cdot \vec{B} = 0 \tag{11}$$

The friction code is called frequently as the coupled code walks along in time, keeping the magnetic field in approximate force-balance with the RCM-computed distribution function. For a more complete description of the coupled RCM and magnetofriction code, see Toffoletto *et al.* (1996).

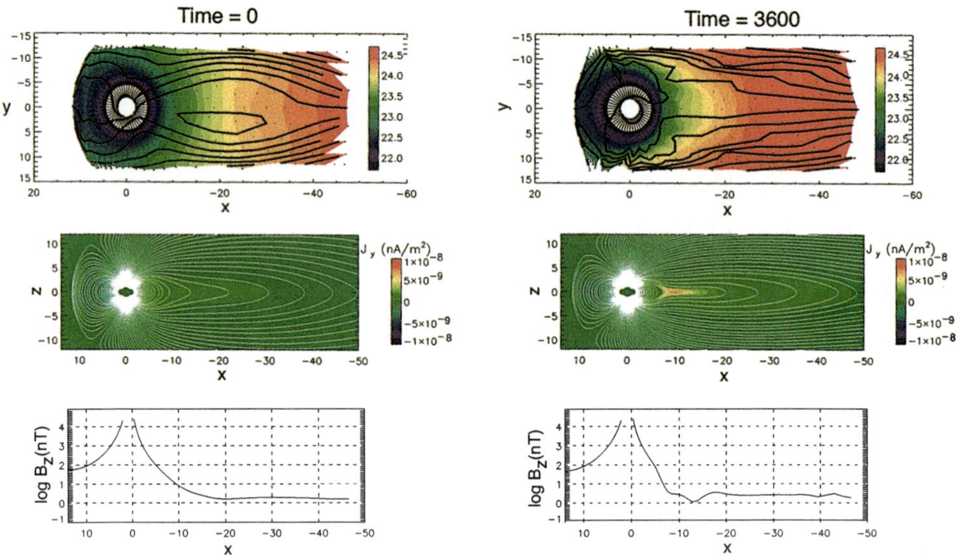

Fig. 1. Initial and final configurations of a coupled RCM/Friction-Code simulation of a substorm growth phase. The right side shows the configuration after an hour of strong convection, with 100 kV potential drop enforced in a dawn-dusk direction across the magnetotail. In the top panel, the colors indicate $\log_{10}(\eta)$, where η is the number of ions per unit magnetic flux and is proportional to both the distribution function and $pV^{5/3}$; the dark curves represent contours of constant electric potential, mapped to the equatorial plane. The middle panels show the noon-midnight meridian plane; color represents current density, while the white curves are magnetic field lines. The bottom panels show $\log_{10}(B_z)$ along the x-axis. In all diagrams, the Sun is to the left.

RESULTS AND IMPLICATIONS

Growth Phase

Figure 1 shows some results from a computer run in which strong convection (100 kV cross-tail potential drop) was imposed. For this run, the plasma sheet was taken to contain ions with λ=4000 eV $(\mathrm{nT/R_E})^{2/3}$

and cold electrons. The magnetic field model for the initial state (left side) comes from a Tsyganenko (1989) model, relaxed to approximate equilibrium. Several features are evident:

1. High-$pV^{5/3}$ flux tubes move into the inner plasma sheet.

2. Inner-plasma-sheet field lines become very stretched and tail-like: when the inner plasma sheet is forced to accept high- $pV^{5/3}$ flux tubes, it does so by decreasing equatorial field strength and thus increasing V, as has been previously shown by Erickson (1992), Hau *et al.*(1989), and Hau (1991).

3. A minimum in the equatorial magnetic field develops in the inner plasma sheet.

4. The current density intensifies there.

Note that maintaining the strong cross-tail potential drop and enforcing the adiabatic condition (Eq. 3) does not, by itself, lead to a substorm expansion phase, with injection of fresh plasma into the inner magnetosphere. In the configuration shown in the right side of Figure 1, inner-plasma-sheet flux tubes contain so much plasma that they cannot be compressed into quasi-dipolar form and injected into the inner magnetosphere. A nonadiabatic process – one that reduces $pV^{5/3}$ or, equivalently, the distribution function – is required before injection can occur.

The need for a nonadiabatic loss process can also be seen by comparing theoretical and observed particle fluxes. Suppose the distribution function in the plasma sheet is given by

$$f_{ps}(\lambda) = \frac{n_{ps} m^{3/2}}{(2\pi k T_{ps})^{3/2}} \exp\left(-\frac{\lambda}{\lambda_{kT}}\right) \tag{12}$$

where λ_{kT} is a constant $(=kT_{ps}V_{ps}^{2/3})$. If we assume that ions drift with no loss by charge exchange or precipitation, then, according to Eq. 3, f is conserved along the drift path of a particle. If the particle drifts to L=6.6 or 4 as part of a ring-current injection process, then f is the same in the ring current as in the plasma sheet, for the same invariant λ. The differential flux J in the ring current $(=2W_K f/m^2)$ can then be written

$$J(W_K) = \left(\frac{3.5\times10^6}{keV\,cm^2\,sr\,s}\right)\left(\frac{W_K}{20\,keV}\right)\left(\frac{n_{ps}}{0.4\,cm^{-3}}\right)\left(\frac{5\times10^7\,{}^\circ K}{T_{ps}}\right)^{3/2} \times$$
$$\times\left(\frac{1}{A^{1/2}}\right)\exp\left[-\left(\frac{W_K}{4.3\,keV}\right)\left(\frac{5\times10^7\,{}^\circ K}{T_{ps}}\right)\left(\frac{V}{V_{ps}}\right)^{2/3}\right] \tag{13}$$

where A is atomic number. Figure 2 shows a plot of flux-tube volume V vs geocentric distance at local midnight for a Tsyganenko (1987) model for Kp = 2, which gives V=1.4 R_E/nT at 15 R_E, 0.06 at 6.6 R_E, and 0.008 at 4 R_E. Thus adiabatic convection of a nominal plasma sheet (n_{ps}=0.4 cm-3, T_{ps}=5x10^7°K, A=1) to geosynchronous orbit would give a differential flux of 2×10^6 kev^{-1} cm^{-2} sr^{-1} s^{-1} for 20 keV protons at geosynchronous orbit, 6×10^6 kev^{-1} cm^{-2} sr^{-1} s^{-1} for 50 keV protons at L=4. (These energies were chosen so that the flux calculated by (13) would be insensitive to errors in the flux tube volumes.) Observed fluxes of 20 keV protons rarely exceed 10^6 kev^{-1} cm^{-2} sr^{-1} s^{-1} at synchronous orbit (Lambour, 1994). Similarly, the effective upper bound on fluxes of 50 keV protons at L=4 in large storms is apparently in the range (1-2)×10^6 kev^{-1} cm^{-1} sr^{-1} s^{-1} (e.g., Lyons and Williams, 1980; Gloeckler and Hamilton, 1987; Hamilton *et al.*, 1988).

Lossless adiabatic convection from the middle plasma sheet at $L \approx 15$ to the inner magnetosphere therefore gives higher ion fluxes than are observed, which suggests that a nonadiabatic process must reduce $f(\lambda)$ somewhere outside geosynchronous orbit. Charge exchange loss does not seem capable of removing the discrepancy, since the lifetimes for 20 keV protons are very long for $L > 6.6$, and days for 50 keV protons at $L=4$.

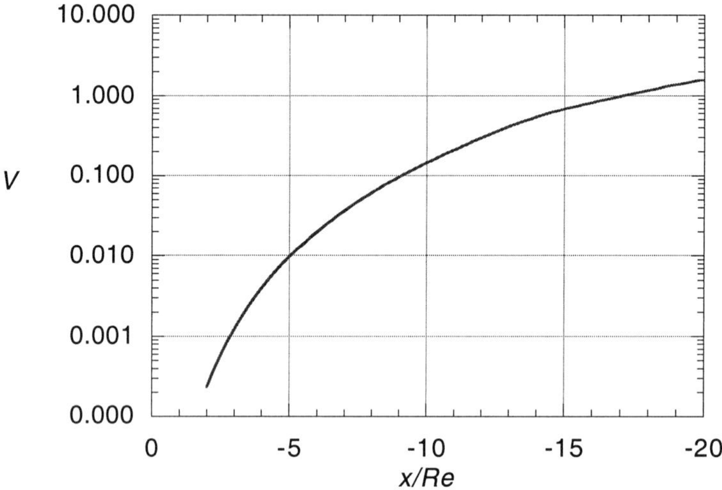

Fig. 2. Flux-tube volume vs. distance from Earth at local midnight, in the Tsyganenko (1987) model for $Kp=2$. These volumes represent integrations from the equatorial plane to the northern ionosphere ($=V/2$).

The Expansion Phase

The last section suggests that some of equations (1)-(11) must be violated if strong convection is to produce injection of fresh particles into the inner magnetosphere, both because strong convection forces the magnetic configuration into the highly-stressed stalemate illustrated on the right side of Figure 1, and because the injection of plasma-sheet particles directly into the ring current would imply higher ring current fluxes than are observed. Since injection into the synchronous-orbit region begins at the onset of the substorm expansion phase, it is natural to associate expansion-phase onset with the violation of the equations.

If the distribution function f is to be dramatically reduced on some flux tubes (i.e., some α, β), then Eq. 3 must be dramatically violated in the expansion phase. That violation is a feature of at least two of the leading substorm models:

1. In the near-Earth-neutral-line model, magnetic reconnection naturally violates Eq. 1 and Eq. 3. For one thing, it causes a particle that had been on a long closed flux tube with large V to suddenly find itself on a shorter tube with much smaller V, but with the same energy. Secondly, after reconnection, the same α and β describe two separate flux tubes — a short closed one and a plasmoid. The total number of particles on the closed tube is, of course, smaller than the number of particles on the original closed tube, which results in a reduction of $\int f \lambda^{1/2} d\lambda$ on the closed tube.

2. In the current-disruption model (Lui, 1990), field lines slip on the plasma, particle drifts are not describable by Eq. 5 and Eq. 6, and λ is not conserved.

To quantitatively model the inner-magnetospheric consequences of the expansion phase, it is necessary to model the violation of the adiabatic condition (Eq. 3). Specifically, we need information on the function $f(\lambda,\alpha,\beta)$ after the non-adiabatic process is finished.

We have carried out an initial effort in that direction for the near-Earth-neutral-line model (Toffoletto *et al.*, 2000), assuming a pattern of reduction of $f(\lambda,\alpha,\beta)$ after the onset of the expansion phase and computing resulting electric and magnetic field patterns. Once f and $pV^{5/3}$ are reduced in a limited sector near local midnight, a new equilibrium is calculated that exhibits a strong dipolarization of field lines in the depleted region. We interpret the near-Earth-X-line model as implying that f is conserved on all field lines that close within 25 R_E of Earth at the end of the growth phase, but that f is reduced on field lines that extended further; the degree of reduction is greater for field lines that initially extended further beyond 25 R_E. One interesting result is the development of jets of plasma flowing away from local midnight in the lower-latitude part of the auroral zone. The jets are caused by the high-content inner-plasma-sheet flux tubes that did not undergo reconnection squirting out of the way, as the newly reconnected dipolarizing flux tubes rush earthwards (Figure 3). (See Toffoletto *et al.* (2000) for more details.) This jetting appears to be an unavoidable consequence of the near-Earth-neutral-line model (including the assumption that non-adiabatic violation of Eq. 3 is confined to field lines that undergo reconnection at an X-line near 25 R_E), unless large field-aligned potential drops decouple the ionosphere from the equatorial plane.

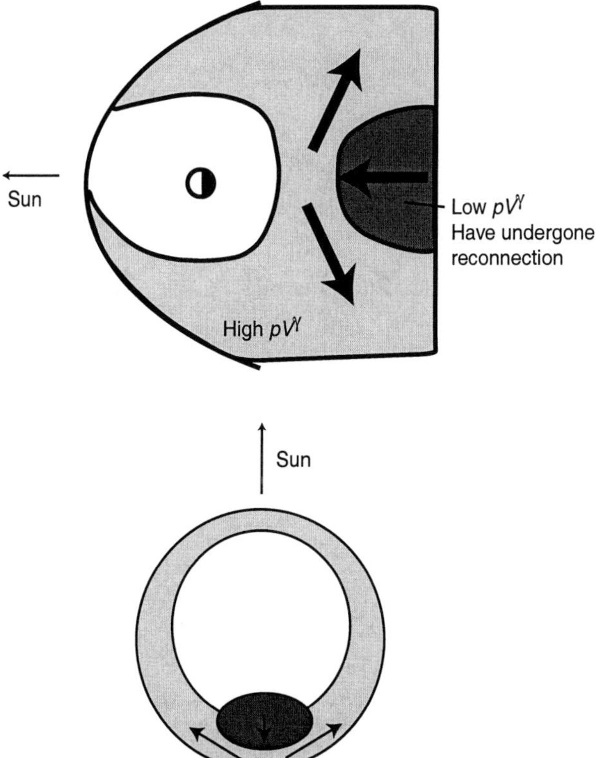

Fig. 3. Illustration of why the near-Earth-neutral-line model implies eastward and westward jetting of plasma away from the center of the substorm. Top, view in magnetic equatorial plane. Bottom, view in ionosphere.

CONCLUDING COMMENTS

The first applications of the coupled Rice Convection Model and friction-code equilibrium solver to the substorm problem have confirmed a basic theoretical understanding of the growth phase. The new computational capability is beginning to allow us to refine models of the expansion phase and to make them more quantitative. The substorm expansion non-adiabatically reduces the particle populations on certain flux tubes, allowing them to be injected into the inner magnetosphere, and quantitative description of this reduction is a major challenge for modeling. One promising approach is to use resistive MHD simulations to estimate the pattern of reduction of f and $pV^{5/3}$, to compute the implied patterns of ionospheric electric field as well as geosynchronous and ring current fluxes, and to compare with observations. An alternative approach would be to try to use measurements of injected particle fluxes to estimate the reduction in the distribution function. The RCM, which is based on the assumption of bounce-averaged adiabatic drift, cannot provide a theory for the nonadiabatic processes that occur in the substorm expansion phase. However, it can now describe the observable inner-magnetospheric consequences of different substorm theories and thus may help answer key questions of substorm physics.

ACKNOWLEDGMENTS

Work at Rice was supported by NSF GEM grant ATM9900983 and by NASA Sun-Earth Connection Theory grant NAG5-9077. The authors are grateful to Trevor Garner whose computer simulations made a significant contribution to the development of the ideas presented here.

REFERENCES

Bales, B., J. Freeman, B. Hausman, R. Hilmer, R. Lambour, A. Nagai, R. Spiro, G.-H. Voigt, R. Wolf, W.F. Denig, D. Hardy, M. Heinemann, N. Maynard, F. Rich, R.D. Belian, and T. Cayton, Status of the development of the Magnetospheric Specification and Forecast Model, in *Solar-Terrestrial Predictions-IV: Proceedings of a Workshop at Ottawa, Canada, May 18-22, 1992*, ed. by J. Hruska, M.A. Shea, D.F. Smart, and G. Heckman, pp. 467-478, NOAA, Environmental Res. Labs, Boulder (1993).

Erickson, G.M., A quasi-static magnetospheric convection model in two dimensions, *J. Geophys. Res.*, *97*, 6505-6522 (1992).

Erickson, G.M., R.W. Spiro, and R.A. Wolf, The physics of the Harang discontinuity, *J. Geophys. Res.*, *96*, 1633-1645 (1991).

Garner, T.W., R.A. Wolf, R.W. Spiro, and M.F. Thomsen, First attempt at assimilating data to constrain a magnetospheric model, *J. Geophys. Res.*, *104*, 25145-25152 (1999).

Gloeckler, G., and D.C. Hamilton, AMPTE ion composition results, *Phys. Scripta*, *T18*, 73-84 (1987).

Hamilton, D.C., G. Gloeckler, F.M. Ipavich, W. Stüdemann, B. Wilken, and G. Kremser, Ring current development during the great geomagnetic storm of February 1986, *J. Geophys. Res.*, *93*, 14343-14355 (1988).

Harel, M., R.A. Wolf, P.H. Reiff, R.W. Spiro, W.J. Burke, F.J. Rich, and M. Smiddy, Quantitative simulation of a magnetospheric substorm 1, model logic and overview, *J. Geophys. Res.*, *86*, 2217-2241 (1981).

Hau, L.-N., Effect of steady-state adiabatic convection on the configuration of the near-Earth plasma sheet, 2, *J. Geophys. Res.*, *96*, 5591-5596 (1991).

Hau, L.-N., R.A. Wolf, G.-H. Voigt, and C.C. Wu, Steady state magnetic field configurations for the Earth's magnetotail, *J. Geophys. Res.*, *94*, 1303-1316 (1989).

Hilmer, R.V., and G.P. Ginet, A Magnetospheric Specification Model Validation Study: Geosynchronous electrons, *J. Atm. Solar-Terrest. Phys*, *62,* 1275-1294 (2000).

Lambour, R.L., Calibration of the Rice Magnetospheric Specification and Forecast Model for the Inner Magnetosphere, Ph. D. thesis, Rice University, Houston, TX (1994).

Lui, A.T.Y., A. Mankofsky, C.-L. Chang, K. Papadopoulos, and C.S. Wu, A current disruption mechanism in the neutral sheet: A possible trigger for substorm expansions, *Geophys. Res. Lett.*, *17*, 745-748 (1990).

Lyons, L.R., and D.J. Williams, A source for the geomagnetic storm main phase ring current, *J. Geophys. Res.*, *85*, 523-530 (1980).

Toffoletto, F.R., R.W. Spiro, R.A. Wolf, M. Hesse, and J. Birn, Self-consistent modeling of inner magnetospheric convection, in *Proc. Third International Conference on Substorms (ICS-3), , ESA SP-389*, edited by E.J. Rolfe, and B. Kaldeich, pp. 223-230, ESA Publications Division, Noordwijk, The Netherlands (1996).

Toffoletto, F.R., R.W. Spiro, R.A. Wolf, M. Birn, and M. Hesse, Computer experiments on substorm growth and expansion, *Proceedings of The 5th International Conference on Substorms, St. Petersburg, Russia, 16-20 May 2000 (ESA SP-443)*, ed A. Wilson, pp. 351-355, European Space Agnecy, Noordwijk, The Netherlands (2000).

Tsyganenko, N.A., Global quantitative models of the geomagnetic field in the cislunar magnetosphere for different disturbance levels, *Planet. Space Sci.*, *35*, 1347-1358 (1987).

Tsyganenko, N.A., A magnetospheric magnetic field model with a warped tail current sheet, *Planet. Space Sci.*, *37*, 5-20 (1989).

Usadi, A., R.A. Wolf, M. Heinemann, and W. Horton, Does chaos alter the ensemble averaged drift equations?, *J. Geophys. Res.*, *101*, 15,491-15,514 (1996).

Wolf, R.A., The quasi-static (slow-flow) region of the magnetosphere, in *Solar Terrestrial Physics*, edited by R.L. Carovillano, and J.M. Forbes, pp. 303-368, D. Reidel, Hingham, MA (1983).

Wolf, R.A., J.W. Freeman, Jr., B.A. Hausman, R.W. Spiro, R.V. Hilmer, and R.L. Lambour, Modeling convection effects in magnetic storms, in *Magnetic Storms*, edited by B.T. Tsurutani, J.K. Arballo, W.D. Gonzalez, and Y. Kamide, pp. 161-172, Am. Geophys. Un., Washington, D. C. (1997).

HIGH-LATITUDE ELECTRODYNAMICS FROM A MULTI-ARRAY NONLINEAR GEOMAGNETIC MODEL

D. Vassiliadis[1], A. J. Klimas[2], B.-H. Ahn[3], R. J. Parks[4], A. Viljanen[5], K. Yumoto[6]

[1] *Universities Space Research Association, NASA/Goddard Space Flight Center, Greenbelt, MD 20771, USA*
[2] *NASA/Goddard Space Flight Center, Code 692, Greenbelt, MD 20771, USA*
[3] *Kyungpook National University, Teachers College, Dept. of Earth Sciences, Taegu, 702-701, Korea*
[4] *Department of Physics, University of Florida, Gainesville, FL 32611, USA*
[5] *Finnish Meteorological Institute, POB 503, Helsinki, FIN 00101, Finland*
[6] *Kyushu University, Dept. Earth & Planet Sci., 33, 6-10-1 Hakozaki Higashiku, Fukuoka, 812-81, Japan*

ABSTRACT

We present a model for the high-latitude geomagnetic disturbances with the goal of studying the time-dependent solar wind-magnetosphere-ionosphere coupling. This nonlinear dynamical model, very different from earlier linear approaches, is based on observations from the WIND and ACE solar wind monitors and the IMAGE and MM210 ground magnetometer arrays. The model performance is evaluated in several activity intervals and shows that the overall amplitude of the disturbance is predicted moderately well. The electrojets and other large-scale spatial structures, however, are not reproduced at a satisfactory level at present, so higher-order terms need to be calculated with an expanded database. A coupling of the model to the KRM electrodynamic model and the Ahn *et al.* [1998] conductance forms the basis of a prediction system for the high-latitude ionospheric circuit.

INTRODUCTION

The geomagnetic state at high latitudes is determined by magnetospheric currents as well as their ionospheric closure [Kelley, 1989] and therefore it is a direct diagnostic of the outer magnetosphere [Nishida, 1978]. Therefore measurement of the surface magnetic field state is important for a number of physical issues including the energy transfer from the solar wind to the magnetosphere and the related question of geoeffectiveness of solar wind structures; the time scales and other dynamical development of the geospace response to solar wind changes; the ionospheric morphology of the global substorm and convection cycles; the propagation and dynamics of traveling convection vortices and other regional and transient structures; etc. Here we model the geomagnetic state in order to follow the magnetospheric dynamics as modulated by the solar wind and estimate the ionospheric energy dissipation following that dynamics. Towards the end of the paper we describe how this geomagnetic model is used as the basis for an electrodynamic modeling of the ionosphere at high latitudes.

Earlier attempts to determine the geomagnetic state have been primarily linear and statistical in nature. Friis-Christensen *et al.* [1985], for example, showed that by classifying the geomagnetic response by the clock angle of the interplanetary magnetic field (IMF), one can recover the patterns of equivalent currents and ultimately relate the activity to the flux transfer cycle and the plasma motion across the geomagnetic field. A similar approach was used in the IZMEM model where the linear response of a large number of mostly high-latitude magnetometers to a standard solar wind key parameter input was computed [Levitin *et al.*, 1981; Papitashvili *et al.*, 1994]. A different line of work has been to first obtain the electric potential distribution from in-situ measurements [e.g., Heppner and Maynard, 1987; Weimer, 1996] from which the large-scale plasma velocity is obtained as tangent to the equipotential lines. Then the magnetic field distribution on the ground is calculated indirectly applying a Biot-Savart law to ionospheric currents and assuming a distant closure profile for the field-aligned currents. Instead here we develop an extension of a nonlinear dynamical approach which has given predictive models for magnetic indices [e.g. Vassiliadis *et al.*, 1995; 1996].

In the sections below we first describe the overall approach and specific model based on the multi-array data; then evaluate the prediction performance of the model in terms of auroral magnetic indices as well as local measurements. Finally we briefly describe the electrodynamic part of the model which allows further modeling of the large-scale distribution of electric fields and currents.

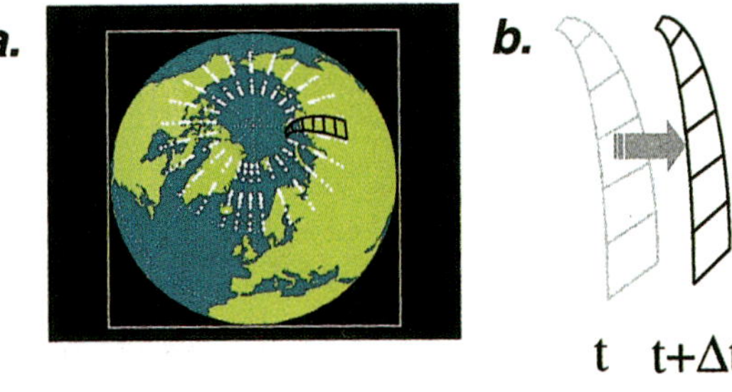

Fig. 1. a. The geographic distribution of IMAGE magnetometers produces a set of "virtual" stations over an annulus at 60-80°. The state variable of the nonlinear model is the vector $\mathbf{B}_x(\lambda,\theta)$ of the North-South component over all latitudes at a local time zone (shown schematically with a narrow grid). b. The time development of the state is given in Eq. (1).

DATA AND METHODOLOGY

In developing the model we have made use of several different magnetometer networks, but here we will discuss results based on only two of them: IMAGE, extending primarily along the Svalbard-Helsinki axis in Fennoscandia, and made available by the Finnish Meteorological Institute; and 5 auroral-zone Siberian and Japanese stations of MM210, an extended network in East Asia, made available by Kyushu University. The latitudinal range of the main array, IMAGE, is approximately 60-80°. The database contains the horizontal field recorded at those stations during January and February of 1995 (although it has been extended significantly at the time this paper was written). These data are binned by local time (as well as by activity later on). In that way we have a coverage of 24x(15+5)=480 "virtual" stations. This provides a more dense grid than obtained from previous modeling or assimilation methods.

Combining data from multiple arrays is useful in several ways: first, during model validation data from one array can be used for training and from the other for testing. Second, combining simultaneous data from several arrays (as well as geomagnetic indices and the recent solar wind key parameters) describes the geomagnetic state in much more detail. IMAGE and MM210 are about 130° apart in longitude so they can constrain the geomagnetic state better. Finally, this results in more accurate modeling in terms of dayside and nightside processes. In combining the two arrays in the way describe below, it is important that the disturbances are scaled correctly: we calculate the quiet-day baseline for each array in a given month from the 5 quietest days.

We develop the field model in a nonlinear analysis approach: the geomagnetic activity is described with a state vector whose dynamics are based on empirical models of the observed activity (in the section on electrodynamics below we supplement this geomagnetic part of the model by several physics-based relations). The state vector contains the horizontal field measured at each station in a fixed local time (Figure 1). Twenty-four of these state vectors, defined at the appropriate equidistant local times, determine the geomagnetic state in an annulus of 60-80° latitude at a resolution of 15° in local time. The time evolution of the state vector at a fixed longitude is determined by its response to either a geomagnetic index or a solar wind input, the latter constructed from the WIND or ACE key parameters. The index input is useful for retrospective studies or, in the future, for nowcasting (when it becomes available in real time), while the solar wind input is being used in forecasting.

The model field $B_x(LT;t)$ for a fixed longitude is then determined by a vector differential equation (the vector containing all IMAGE magnetometers at L as well as the MM210 stations at LT+130°):

$$\frac{dB_x(LT;t)}{dt} = \alpha B_x(LT;t) + \varphi(I(t-T), \frac{dI(t-T)}{dt}, \ldots, \frac{d^{L-1}I(t-T)}{dt^{L-1}}) \qquad (1)$$

A similar equation is used for B_y. The coefficient $1/\alpha$ is a net growth/decay time scale for the ionospheric and field-aligned current systems that determine $\mathbf{B}_x$. The nonlinear function vector ϕ represents the many processes contributing to the coupling between solar wind and high-latitude magnetospheric-ionospheric activity. The input I(t) is either E_{SW}, where

$$E_{SW} \equiv VB_T \sin^4(\theta/2)$$
$$B_T \equiv \sqrt{B_z^2 + B_y^2}$$
$$\theta = \tan^{-1}(B_z/B_y)$$

or the polar cap index, PC, calculated from the Thule magnetometer in Greenland. The delay T in Eq. (1), representing the earliest electrojet response time, is typically of the order of 20 min, but in all of these runs it has been set to 0. In practice, Eq. (1) is discretized in time and becomes a mapping. An analysis of this type of input-output maps has been given for geomagnetic indices [e.g., Vassiliadis *et al.*, 1995; Vassiliadis *et al.*, 1996].

It is important to describe how the coefficients of (1) are determined: for the present value of B_x we find all instances in the database when the field vector was similar. Note that the similarity of two state vectors B_x and B_x' is defined simply as the norm of their difference being less than a predetermined constant. The norm is designed to include the recent history of the input. Those state vectors represent times with a similar geomagnetic activity (as well as solar wind activity, if we are using E_{SW} as a driver) as the present one. We fit the subsequent evolution of those similar historical state vectors using Eq. (1) and solve for the coefficients α and ϕ. Then we use those coefficients with the present state vector (and possibly solar wind) to calculate the derivative of the field, again from (1). Clearly the determination of the coupling coefficients from a very small subset of geomagnetic/solar wind conditions (which changes at each time step) produces a very nonlinear model in general. This is called a local-linear model (because of the linearity of Eq. (1).

January 16, 1995, 11:45 UT

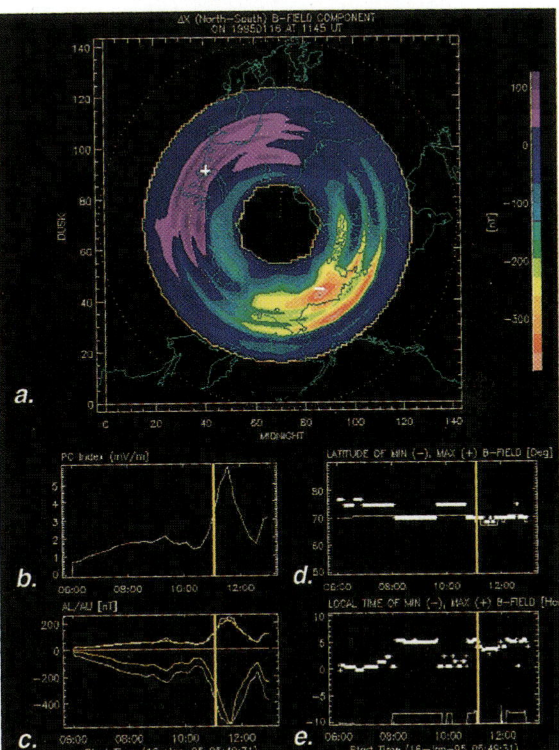

Fig. 2. The graphic output of the geomagnetic model shows the North-South magnetic field component, B_x at high latitudes. Here it is shown during the expansion of the January 16, 1995, substorm. a. A polar view of the B_x component. b. The model input I(t), here the polar cap index, over the course of the substorm. c. We use the minimum and maximum values of B_x to define model "electrojet indices." These are compared with the output of a time series model (red line) [developed by Vassiliadis *et al.*, 1996]. d. The latitude of the minimum and maximum $B_x(\lambda,\theta)$. e. The local time of the same extrema. The same type of output plot is made available on-line with the model driven by real-time ACE data (see text).

The field at any position on the annulus can be obtained by interpolating nearby model outputs. Figure 2 shows the output of the model as it is displayed on-line at http://lep694.gsfc.nasa.gov/ RTSM/rt_predictions.html/ except here it is driven by the historical polar cap index rather than the near-real-time ACE solar wind in the website. The minimum and maximum of the field are defined as model "electrojet indices" analogous to AL and AU (Figure 2c). The positions of the extreme values are used as boundary indices (Figure 2d-e) to quantify the displacement of the polar cap and auroral oval.

MODEL RESULTS AND VALIDATION

The model capability in reproducing geomagnetic events is quantified by comparisons against observations. We first compare magnetic indices calculated from the model output and then examine individual magnetometers. The indices are compared against the Kyoto World Data Center's Quicklook electrojet indices for several levels of geomagnetic disturbance (Note: the Quicklook indices are by definition preliminary and often change as more data become available, so we use them only as a guide). Figure 3 shows the comparison for three recent cases: a northward IMF B_z event (September 15, 1999), a solar wind ejecta event (April 6-7, 2000), and a CME-induced storm (September 22-23, 1999). The 2-hour variations in the first and third event (parts a and c of the figure) are well reproduced. Regrettably any activity with AL<-600 nT is clipped in the nonlinear model, because our database at the time of testing did not contain any events with AL below that level. (Recently higher-activity intervals have been added to the database and a second testing will be conducted.) On the other hand the unusual storm due to ejecta in the solar wind is not predicted well. Only its onset is predicted qualitatively. Overall, however, except for the clipping the agreement between model output and observed preliminary indices is moderately good.

A closer look at the model reveals where the largest uncertainty in model output should be expected and perhaps not surprisingly that region is the central auroral zone (70-73°, between stations BJN and TRO). The average model error is highest in that region. To illustrate that here we examine a moderate substorm that occurred on January 16, 1995 caused by a high-speed stream and lasting ~3.5 hours (Figure 4). Input to the model is provided by key parameters measured by WIND. The model output, the field **B**(LT, θ; t) is calculated from the MM210 stations of the model. That output is evaluated at the locations of the IMAGE magnetometers which are on the dawn side (07:00 LT) at the beginning of the event. The magnetograms are arranged in order of decreasing latitude. The small growth phase and sudden expansion of the poleward magnetometers is reproduced well. In the central part of the auroral zone, however, the activity is predicted with the wrong sign (positive instead of negative) showing that the latitudinal boundary between the eastward and westward electrojets is determined with an error of ~7°. One of us (RJP) has investigated a range of model parameters and found that the prediction error for this substorm does not change significantly. That work has shown the need for a more encompassing database. The recent addition of higher-activity intervals is expected to improve determination of the boundary. Going further south, the last few IMAGE magnetometers are beyond the MM210 range (Figure 4, marked with an 'X"), so the corresponding errors are not evaluated.

NBZ event: September 15, 1999

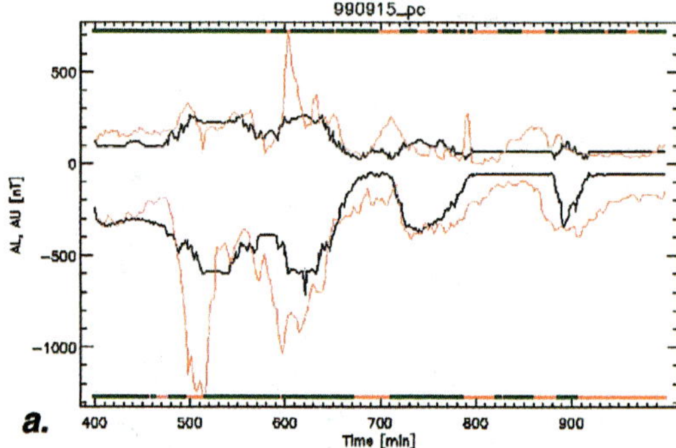

Solar wind ejecta event, April 6-7, 2000

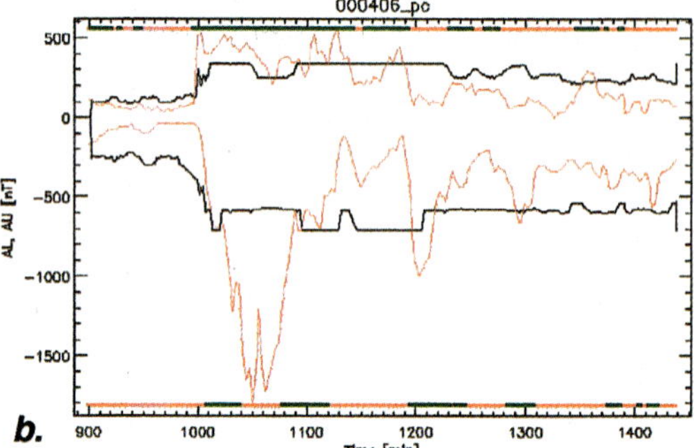

Storm, September 22-23, 1999

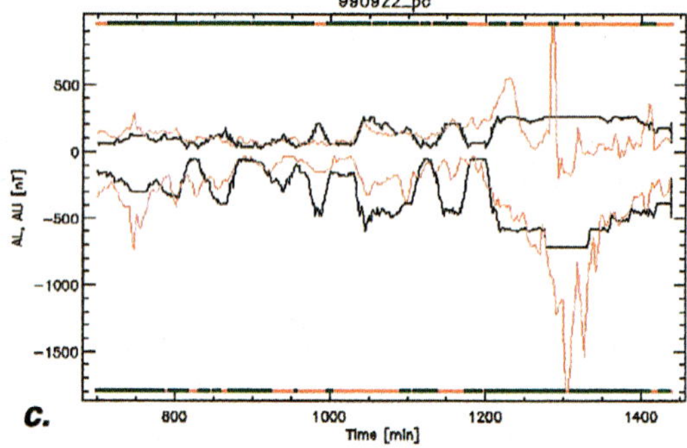

Fig. 3. Comparison of model AL/AU indices with the Quicklook indices of the Kyoto World Data Center: a. A northward IMF B_z interval leads to reverse convection in September 15, 1999. b. A storm produced by solar wind ejecta on April 6-7, 2000. c. A CME-induced storm on September 22-23, 1999. There is fair agreement between predictions and observations for the first and third interval; the second interval is predicted poorly. Note that the artificial "clipping" of high activity is due to lack of many high-activity intervals in the database at this stage.

ELECTRODYNAMIC MODELING

Recently we have extended the model into a simple high-latitude electrodynamic model. From the geomagnetic field distribution $B_x(\lambda,\theta)$ one can obtain the equivalent current function, Ψ. Then using model Hall and Pedersen conductances, σ_H and σ_P, one can solve the two-dimensional Ohm's law for the electric potential distribution $\Phi(\lambda, \theta)$ as originally shown in the KRM method [Kamide *et al.*, 1981]:

$$\sigma_H \sin\theta \frac{\partial^2\Phi}{\partial\theta^2} + \left(\frac{\partial(\sigma_H \sin\theta)}{\partial\theta} + \frac{\partial\sigma_P}{\partial\lambda}\right)\frac{\partial\Phi}{\partial\theta} + \frac{\sigma_H}{\sin\theta}\frac{\partial^2\Phi}{\partial\lambda^2} + \left(\frac{\partial\sigma_P}{\partial\theta} - \frac{\partial(\sigma_H/\sin\theta)}{\partial\lambda}\right)\frac{\partial\Phi}{\partial\lambda} = \frac{\partial(\sin\theta\frac{\partial\Psi}{\partial\theta})}{\partial\theta} - \frac{1}{\sin\theta}\frac{\partial^2\Psi}{\partial\lambda^2}$$

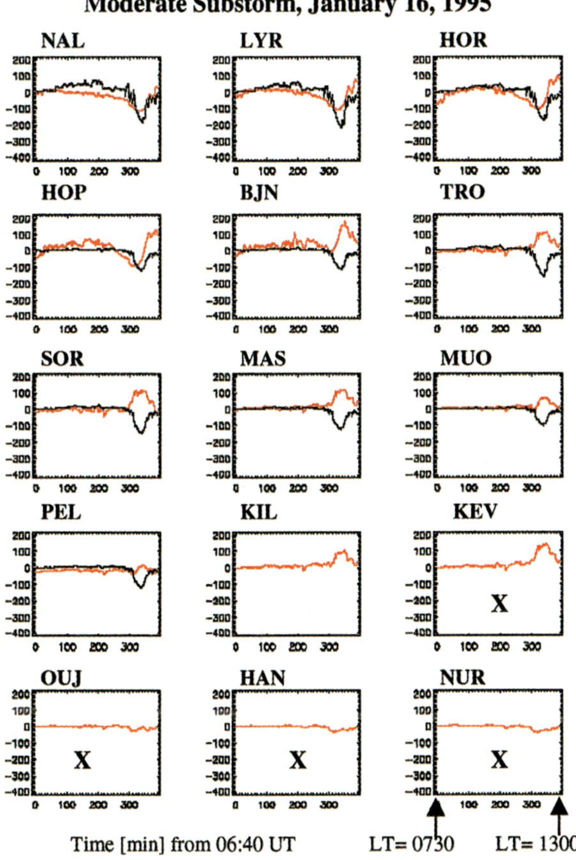

Fig. 4. Comparison of the MM210-based model field with IMAGE observations during the January 16, 1995, substorm. The array is on the dayside during the interval and the 15 magnetograms have been arranged in decreasing latitude. In the last 4 graphs, "X" indicates the locations outside the MM210 latitudinal range. There is good agreement in the poleward edge of the auroral zone and points north, but in the central auroral zone the agreement is poor: the boundary between eastward and westward electrojet is incorrectly determined (error: ~7°). In that region the sign of the modeled disturbance is opposite from observations, although the magnitude is approximately correct.

We use the Ahn *et al.* [1998] model for the Hall and Pedersen height-integrated conductances as functions of the magnitude of the horizontal magnetic field component. The time development of conductances is visually consistent with UV images of the auroral zone (e.g. from Polar) while the electric field is organized is two large-scale cells of unequal size and magnitude as shown by Weimer *et al.* [1996].

CONCLUSIONS AND OUTLOOK

As we are developing a new modeling approach for the high-latitude geomagnetic field we discover its strengths and weaknesses relative to earlier methods. The overall amplitude of the disturbance, as quantified by electrojet indices, is predicted moderately well. Also the spatial pattern agrees with substorm phenomenology. The spatial pattern of the disturbance, however, shows some significant deficiencies. These concern the placement of boundaries, such as the intra-electrojet boundary and also, although less discussed, the equatorward extent of the auroral zone. Clearly significantly more measurements need to be added to the database. We have extended the database by 100% at the time of writing and plan to continue adding more observations. In addition, we have identified 5 intervals of characteristic activity ranging from Northward B_z to a strong storm (3 of them are shown in Figure 3) to test the solar wind geoeffectiveness.

The electrodynamic part of the model is currently undergoing several changes both in the form of the equations as well as the conductance, especially to represent rapidly changing conditions such as occur in substorm expansion. Nevertheless these early results indicate that the amplitude as well as the spatial pattern of geomagnetic activity may be eventually predictable.

ACKNOWLEDGMENTS

We thank NASA/GSFC/CDAWeb for providing archive data for WIND and NOAA/SEC for providing real-time ACE data. As mentioned above the model was originally set up for real-time operation in the SCOSTEP S-RAMP campaign and ISR World Days intervals in September 1999.

REFERENCES

Ahn, B.-H., A. D. Richmond, H. W. Kroehl, B. A. Emery, O. de la Beaujardiere, and S.-I. Akasofu, An ionospheric conductance model based on ground magnetic disturbance data, *J. Geophys. Res.* 103, A7, 14,769-14,780, 1998.

Friis-Christensen, E., Y. Kamide, A. D. Richmond, and S. Matsushita, Interplanetary magnetic field control of high-latitude electric fields and currents determined from Greenland magnetometer data, *J. Geophys. Res.* 90, A2, 1325-1338, 1985.

Heppner, J.P., and N. C. Maynard, Empirical high-latitude electric field models, *J. Geophys. Res.* 92, A5, 4467-4489, 1987.

Levitin, A.E., R. G. Afonina, B.A. Belov, Y.I. Feldstein, Geomagnetic-variation and field-aligned currents at northern high-latitudes, and their relations to the solar wind parameters, *Phil. Trans. Roy. Soc. A*, 304, 1484, 253-301, 1982.

Kamide, Y., A. D. Richmond, and S. Matsushita, Estimation of ionospheric electric fields, ionospheric currents and field-aligned currents from ground magnetometer records, *J. Geophys. Res.* 86, A2, 801-813, 1981.

Kelley, M.C., *The Earth's ionosphere: plasma physics and electrodynamics*, Academic Press, San Diego, 1989.

Nishida, A., *Geomagnetic diagnosis of the magnetosphere*, Springer Verlag, New York, 1978.

Papitashvili, V.O., B. A. Belov, D. S. Faermark, Ya. I. Feldstein, S. A. Golyshev, L. I. Gromova, and A. E. Levitin, Electric potential patterns in the northern and southern polar regions parameterized by the interplanetary magnetic field, *J. Geophys. Res.* 99, A7, 13,251-13,262, 1994.

Vassiliadis, D., A. J. Klimas, D. N. Baker and D. A. Roberts, A description of solar wind-magnetosphere coupling based on nonlinear filters, *J. Geophys. Res.* 100, A3, 3495-3512, 1995.

Vassiliadis, D., V. Angelopoulos, D. N. Baker, and A. J. Klimas, The relation between northern polar cap and auroral electrojet geomagnetic indices in the wintertime, *Geophys. Res. Lett.* 23, 20, 2,781-2,784, 1996.

Weimer, D.R., A flexible, IMF-dependent model of high-latitude electric potentials having 'space weather' applications, *Geophys. Res. Lett.* 23, 18, 2549-2552, 1996.

MAGNETIC IMPULSE EVENTS AND RELATED PC 1 WAVES IN THE CUSP AND LLBL REGION OBSERVED BY A GROUND MAGNETOMETER NETWORK

H. Fukunishi[1], R. Kataoka[1] and L. J. Lanzerotti[2]

[1]*Department of Geophysics, Graduate School of Science, Tohoku University, Aramaki-aoba, Sendai 980-8578, Japan*
[2]*Bell Laboratories, Lucent Technologies, Murray Hill, NJ 07974, USA*

ABSTRACT

Using ULF wave data measured by search coil and fluxgate magnetometers installed at six Automatic Geophysical Observatories (AGOs) locating from 70° to 87° in magnetic latitude and South Pole station at 74° magnetic latitude in Antarctica, the characteristics of ULF waves associated with magnetic impulse events (MIEs) in the cusp and LLBL region have been investigated in detail. It was found that transient, narrow-band Pc 1 waves with a falling tone spectral structure often occurred related to MIEs and that one event occurred simultaneously with the hot flow anomaly (HFA) event studied by Sibeck et al. (1999). These results suggest that the rapid outward motion of the magnetopause due to HFA may induce the falling-tone Pc 1 waves. If this speculation is correct, transient Pc 1 waves related to MIEs would be used as a practical diagnostic tool for monitoring magneotspheric boundary processes.

INTRODUCTION

Impulsive, cusp latitude, magnetic ground signatures called magnetic impulse events (MIEs) have attracted a great deal of investigator's attention since mid-1980s. These impulse events have durations of ~10 min and amplitudes of a few tens to a few hundreds nT. Two candidate mechanisms of MIEs were proposed: sporadic, localized magnetic reconnection at the dayside magnetopause (e.g., Lanzerotti et al, 1986) and changes in the solar wind dynamic pressure (e.g., Sibeck, 1990; Moretto at al., 1997). In the fommer mechanism MIE have been interpreted as a possible ground signature of flux transfer events (FTEs). There is one-to-one correspondence between MIEs and traveling convection vortices (TCVs) (e.g., Friis-Christensen et al., 1988). This correspondence suggests that MIEs and TCVs are produced by antisunward-moving ionospheric Hall current loops generated by field-aligned currents. On the other hand, recent works of Sibeck at al. (1998) and Sitar et al. (1998) have suggested that a hot flow anomaly (HFA) produced by the interaction of IMF discontinuities with the bow shock is another candidate mechanism of MIEs and TCVs. Further, Pc 1 bursts in the frequency range 0.1 - 1.0 Hz were found to be closely related to MIEs (Arnoldy et al., 1996; Hansen et al, 1992; Sato et al., 1999). Andersen et al. (1996) suggested that the source region of these Pc 1 bursts is at the subsolar outer magnetosphere, earthward of the low-latitede boundary layer (LLBL), while Sato et al. (1999) suggested that MIE-related Pc 1 bursts are excited not only in the dayside outer magnetospher but also in LLBL.

In spite of a number of studies already done for MIEs and MIE-related ULF waves, there is still no consensus on their sources. To investigate the occurrence mechanisms of MIEs and MIE-related Pc 1 waves, it is necessary to observe these phenomena over the wide latitude range including the LLBL, cusp and plasma mantle. In this study we have analyzed magnetometer data obtained from the Automatic Geophysical Observatories (AGOs) covering from 70° to 86° magnetic latitude. An important finding is that transient Pc 1 waves with a falling-tone spectral structure often occurred related to MIEs. This fact suggests that these transient Pc 1 waves can be used as a practical diagnostic tool for monitoring magneotspheric boundary processes associated with MIEs.

Table 1. Geographic and geomagnetic coordinates of the AGOs and South Pole station.

Observatory	Code	Geographic Latitude	Geographic Longitude	Geomagentic Latitude	Geomagentic Longitude	UT-MLT
AGO P1	P1	83.86 S	129.61 E	80.14 S	16.75 E	3:43
AGO P2	P2	85.67 S	31.62 E	69.81 S	19.21 E	3:28
AGO P3	P3	82.50 S	30.00 E	71.78 S	40.09 E	2:10
AGO P4	P4	82.01 S	96.76 E	80.00 S	41.51 E	2:04
AGO P5	P5	77.23 S	123.51 E	86.74 S	29.44 E	2:52
AGO P6	P6	71.07 S	127.87 E	86.72 S	209.41 E	14:29
South Pole	SP	90.00 S	0.00 E	73.91 S	18.76 E	3:35

AGO MAGNETOMETER DATA

The Automatic Geophysical Observatories (AGOs) are the first large-scale network of geophysical stations in Antarctica covering from 70° to 86° magnetic latitede. Each AGO is equipped by fluxgate and search coil magnetometers, an optical all-sky imager, an imaging riometer, an ELF-VLF receiver and an HF receiver. In this study we use magnetic field data obtained at the AGO network and South Pole station. The locations of these stations are shown in Figure 1, and the geographic and geomagnetic coordinates are listed in Table 1. The AGOs and South Pole station form two meridional arrays approximately 1.6 hours apart in magnetic local time covering latitudes from equatorward of the cusp/cleft to the polar cap. The sampling frequencies of fluxgate and search coil magnetometers are 1 and 2 Hz, respectively.

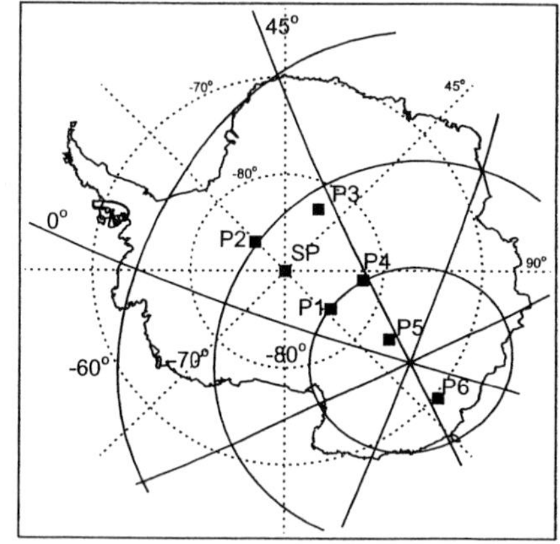

Fig. 1. Polar map showing the AGO network.

MIE-RELATED TRANSIENT PC 1 WAVES

An example of a MIE event observed at P1 and South Pole on June 24, 1997 is shown in Figure 2. The X- and Y-component magnetic field data at P1 show bipolar magnetic field perturbations with the maximum deflections of 50 and 30 nT, respectively between 1616 and 1632 UT. Much larger magnetic perpurbations are seen at South Pole with the muximum

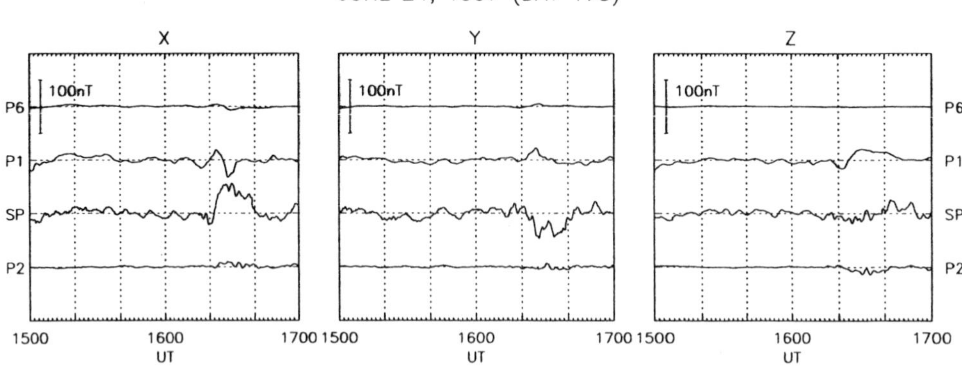

Fig. 2. Three components of fluxgate magnetometer data obtained at AGO P1, P2, P6 and South Pole on June 24, 1997. A magnetic impulse event is evident at P1 and South Pole between 1616 and 1640 UT.

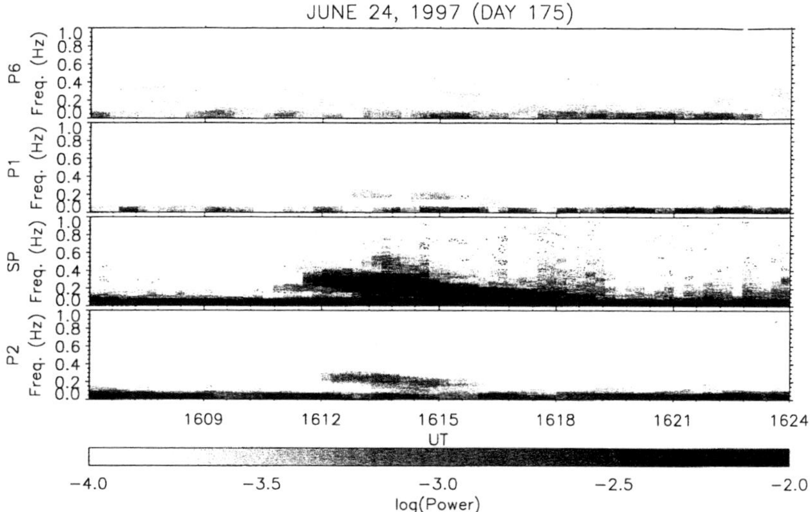

Fig. 3. Dynamic spectra of the X component obtained from the search coil magnetometer data at AGO P1, P2, P6 and South Pole in the time interval 1604 and 1624 UT on June 24, 1997.

deflections of the X and Y components of 80 and 70 nT respectively. At P2 only small-amplitude (~10 nT) Pc 5 pulsations occured during this MIE event. To inversitigate the relationship between Pc 1 activity and MIE events, dynamic spectra in the frequency range 0 - 1.0 Hz were calculated using search coil magnetometer data. Figure 3 shows the X component dynamic spectra at AGO P1, P2, P6 and South Pole in the time interval 1604 and 1624 UT on June 24, 1997. It is apparent that intense, narrow-band Pc 1 waves with a falling tone spectral structure occurred related to a MIE event at South Pole at 74° magnetic latitude (see the time interval 1611 - 1618 UT). Transient Pc 1 wave with the same falling-tone spectral structure but with much smaller amplitudes is also seen at P2 and P1 at 70° and 80° magnetic latitudes, respectively. At South Pole the frequency decreased monotonically from 0.3 Hz at 1612 UT to 0.1 Hz at 1618 UT with a rate of 0.033 Hz/min. It is important to note that this falling-tone Pc 1 event began ~5 min earlier than the MIE signature at South Pole.

Another typical example of MIE and related Pc 1 waves is shown in Figures 4 and 5. This MIE/TCV event occurred between 1130 and 1150 UT on July 24, 1996 as shown in the magnetic perturbation data in Figure 4. Sitar et al. (1998) and Sibeck et al. (1999) had already comprehensively studied this event by the analysis of

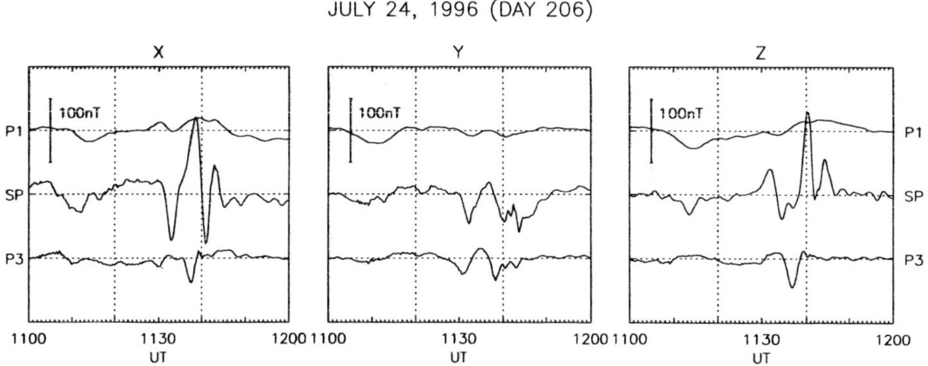

Fig. 4. Three components of fluxgate magnetometer data obtained at AGO P1, P3 and South Pole on July 24, 1996. A magnetic impulse event is evident at South Pole between 1130 and 1150 UT.

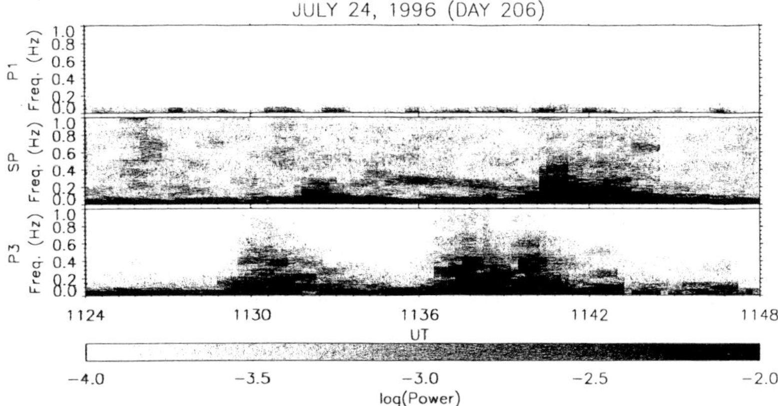

Fig. 5. Dynamic spectra of the Y component obtained from the search coil magnetometer data at AGO P1, P3 and South Pole in the time interval 1124 and 1148 UT on July 24, 1996.

a coordinated set of observations from the spacecraft (Wind, IMP-8, Interball-1, Magion-4, Geotail, Polar UVI, DMSP, GOES-8), ground incoherent scatter radar, and magnetometers in the northern hemisphere. Figure 4 shows the three components of fluxgate magnetometer data obtained at AGO P1, P3 and South Pole in the time interval 1100 - 1200 UT. A clear MIE signature is found at South Pole with the maximum deflections of 200, 70 and 170 nT in the X, Y and Z components, respectively. Figure 5 shows dynamic spectra of the Y component obtained from the search coil magnetometer data at AGO P1, P3 and South Pole in the time interval 1124 and 1148 UT. It is found again that narrow-band Pc 1 wave with a falling tone spectral structure occurred related to this MIE event (see the time interval 1134 - 1140 UT at South Pole). The frequency decreased monotonically from 0.35 Hz at 1134 UT to 0.15 Hz at 1140 UT with a rate of 0.033 Hz/min. It is also found that broad Pc 1 waves occurred at P3 in the time intervals 1129 - 1133 UT and 1136 - 1143 UT.

DISCUSSIONS

As described in the Introduction, three candidate mechanisms of MIEs/TCVs were proposed: 1) sporadic, localized magnetic reconnection at the dayside magnetopause, 2) changes in the solar wind dynamic pressure, and 3) hot flow anomalies (HFAs) produced by the interaction of IMF discontinuities with the bow shock. We examind solar wind conditions during MIE/TCV events using magnetic field and plasma data obtained from the Wind and Geotail spacecraft. In the June 24, 1997 event presented in Figures 2 and 3, the IMF Bz component was northward over 2 hours before the MIE onset and there were no rapid changes in the solar wind dymnamic pressure and the IMF orientation related to the occurrence of MIE. However, the IMF cone angle changed gradually from 50° to 85° before the MIE. On the other hand, in the July 24, 1996 event an abrupt change in the orientation of the IMF was observed at 1019 UT at Wind (Sitar et al., 1998). Corresponding to this IMF tangential discontinuity, Interball-1 located originally in the magnetosheath at 0826 LT observed evidence that the magneotpause briefly (~7 min) moves outward some 5 Re beyond its nominal position between 1132 and 1139 UT due to the formation of a HFA (Sibeck et al., 1999). As shown in Figure 5, when the falling-tone Pc 1 was observed at South Pole station between 1134 and 1140 UT, this station was located at ~0800 MLT close to the local time of Interball-1. Consequently, it is

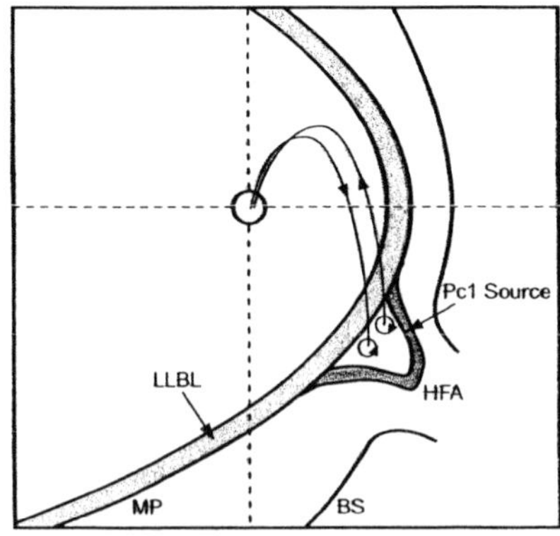

Fig. 6. Schematic diagram showing the source region of MIEs and related Pc 1 waves.

strongly suggested that Pc 1 waves with the falling-tone spectral structure are generated related to the outward motion of the magnetopause and probably in the LLBL just inside the outward inflating magnetopause bulge. Sibeck et al. (1999) showed that the magnetic field strengths fell from its nominal value to ~20 nT when the magnetopause moved ~5 Re outward over a period of ~7 min. This falling of the magnetic field strength induces a decrease in the ion cyclotoron frequency from 1.0 to 0.3 Hz. As a result, the frequency at the maximum growth rate of ion cyclotron waves excited by 1-5 keV ions in the LLBL would decrease from ~0.3 to ~0.15 Hz (cf., Cole et al., 1982; Hansen et al., 1992). The estimated falling rate is consistent with that of the observed Pc 1 waves. Based on these results, we show schematically the souce region of a MIE/TCV and MIE-retaled Pc 1 waves in Figure 6. If this proposed model is correct, we could use MIE-related Pc 1 waves as a practical diagnostic tool for monitoring magneotspheric boundary processes, particulaly for monitoring the localized deformation of the magnetopause and the LLBL due to IMF discontinuities.

CONCLUSION

It was found that associated with magnetic impulse events (MIEs), transient, narrow-band Pc 1 waves, which are characterized by a frequency range of 0.5 - 0.1 Hz, a duration of ~5 min and a bandwidth of ~0.1 Hz, occur with their maxmum amplitude usually at South Pole station at 74° magnetic latitude. It was also found that these MIE-related Pc 1 waves often have a falling tone spectral structure. It is suggested that the rapid outward motion of the magnetopause due to the formation of hot flow anomalies (HFAs) would induce these falling-tone Pc 1 waves. If this speculation is correct, transient Pc 1 waves related to MIEs would be used as a practical diagnostic tool for monitoring magneotspheric boundary processes.

ACKNOWLEGMENTS

We would like to acknowledge Prof. R. L. Arnoldy for providing search coil magnetometer data at South Pole station.

REFERENCES

Anderson, B. J., R. E. Erlandson, M. J. Engebretson, J. Alford, and R. L. Arnoldy, Source region of 0.2 to 1.0 Hz geomagnetic pulsation bursts, Geophys. Res. Lett., 23, 769-772 (1996).

Arnoldy, R. L., M. J. Engebretson, J. L. Axford, R. E. Erlandson, and B. J. Anderson, Magnetic impulse events and associated Pc 1 bursts at dayside high latitudes, J. Geophys. Res., 101, 7793-7799 (1996).

Cole, K. D., R. J. Morris, E. T. Matveeva, V. A. Troitskaya, and O. A. Pokhotelov, The relationship of the boundary layer of the magnetosphere to IPRP events, Planet. Space Sci, 30, 129-136 (1982).

Friis-Christensen, E., M. A. McHenry, C. R. Clauer, and S. Vennerstrom, Ionosphereic traveling convectiuon vortices observed near the polar cleft: A triggerd response to sudden changes in the solar wind, Geophys. Res. Lett., 15, 253-256 (1988).

Hansen, H. J., B. J. Fraser, F. W. Menk, T. -D. Hu, P. T. Newell, C. -I. Meng, and R. J. Morris, High-latitude Pc 1 bursts arising in the dayside boundary layer region, J. Geophys. Res., 97, 3993-4008 (1992).

Korotova, G. I., and D. G. Sibeck, Generation of ULF magnetic pulsations in response to sudden variations in solar wind dynamic pressure, Geophysical Monograph, 81, 265-271, 1994.

Lanzerotti, L. J., L. C. Lee, C. G. Maclennan, A. Wolf, and L. V. Medford, Possible evidence of flux transfer events in the polar isonophere, Geophys. Res. Lett., 13, 1089-1092 (1986).

Moretto, T., E. Friis-Christensen, H. Luhr, and E. Zesta, Global perspective of ionospheric traveling convection vortices: Case studies of two GEM events, J. Geophys. Res., 102, 11,597-11,610 (1997).

Sato, M., H. Fukunishi, L. J. Lanzerotti, and C. G. Maclennan, Magnetic impulse events and related Pc 1 bursts observed by the Automatic Geophysical Observatories network in Antarctica, J. Geophys. Res., 104, 19,971-19,982 (1999).

Sibeck, D. G., A model for the transient magnetospheric response to sudden solar wind dynamic pressure variations J. Geophys. Res, 95, 3755-3771 (1990).

Sibeck, D. G., N. L. Borodkova, S. J. Schwartz, C. L. Owen, R. Kessel et al., Comprehensive study of the magnetospheric response to a hot flow anomaly, J. Geophys. Res., 104, 4577-4593 (1999).

Sitar, R. J., J. B. Baker, C. R. Clauer, A. J. Ridley, J. A. Cumnock et al., Multi-instrument analysis of the ionospheric signatures of a hot flow anomaly occurring on July 24th 1996 coordinated with Polar UVI, J. Geophys. Res., 103, 23,357-23,372 (1998).

SIMULTANEOUS GROUND-BASED OBSERVATIONS OF ELECTRIC AND MAGNETIC FIELD VARIATIONS NEAR THE MAGNETIC EQUATOR FOR SPACE WEATHER STUDY

K. Yumoto[1], M. Shinohara [2], K. Nozaki [3], E. A. Orosco [4], Fr. V. Badillo [5], D. Bringas [6]
and the CPMN and WestPac Observation Groups

[1] *Dept. of Earth & Planet. Sci., Kyushu Univ., Fukuoka, 812-8581, Japan*
[2] *Solar-Terr. Environs. Lab., Nagoya Univ., Toyokawa, 442-8507, Japan*
[3] *Communication Research Laboratory, Tokyo 184-8795, Japan*
[4] *Physics Dept., University of San Carlos, Cebu City 6000, Philippines*
[5] *Manila Observatory, Manila1099, Philippines*
[6] *Muntinlupa Magnetic Observatory, National Mapping and Resource Information Authority Manila, Philippines*

ABSTRACT

In order to clarify the global nature of penetration mechanism of DP 2 electric fields from the polar to the day- and night-side equatorial ionospheres, and to understand the solar wind-Earth's magnetosphere coupling as a space weather study, we have carried out simultaneous magnetic and electric field observations using the Frequency Modulated-Continuous Wave (FM-CW) HF radar and the Circum-pan Pacific Magnetometer Network.
The DP 2 electric fields caused by the solar wind interaction with the Earth's magnetic field must be imposed on the polar ionosphere, and the ionospheric DP 2 electric fields may penetrate instantaneously from the polar ionosphere into both the day- and night-side equatorial regions.

INTRODUCTION

It is well known that daytime DP 2 magnetic variations observed on the ground are well correlated with changes of the solar wind Bz component (Nishida, 1968 a & b). Electric fields caused by the solar wind interaction with the Earth's magnetic field are believed to be imposed on the polar ionosphere and penetrate into the equatorial ionosphere (Nishida, 1968b; Kikuchi et al, 1996). DP 2 magnetic variations are clearly seen on the ground in the dayside equatorial region where the ionospheric Cowling conductivity is zonally enhanced, but are difficult to be observed in the night side because of the lower ionospheric conductivity. The magnetic dip equator is a peculiar region where the ionospheric conductivity is much larger than those at low latitudes, and therefore DP2 phenomena observed near the dayside dip equator can be a useful tool for monitoring the change of the interplanetary magnetic field (IMF) in the solar wind.
In order to clarify the global nature of penetration mechanism of DP 2 electric fields from the polar to the day- and night-side equatorial ionospheres, and to understand the solar wind-Earth's magnetosphere coupling as a space weather study, we have carried out simultaneous magnetic and electric field observations using the Frequency Modulated-Continuous Wave (FM-CW) HF radar (Nozaki et al., 1999) and the Circum-pan Pacific Magnetometer Network (CPMN; Yumoto et al., 1996; Yumoto and the CPMN Group, 2001; Tachihara et al., 1996; see the web site of http://denji102.geo.kyushu-u.ac.jp/denji/obs/obs_e.html).

OBSERVATIONS

The magnetic field data at the CPMN stations in the opposite day/night equatorial regions were obtained at Davao (7.11° N, 125.61° E, Dip= -0.6 °) in Philippines and at Gadalupe (14.0° S, 75.95° W, Dip= -2.6 °) in Peru. We measured electric field variations near the magnetic dip equator by means of a FM-CW HF radar, which was installed in University of San Carlos Talamban campus, Cebu, Philippines (10.35° N, 123.91° E, Dip= 2.6°) during the WestPac campaign of February 18-March 29, 1999. The FM-CW HF radar with ionosonde mode was operated to deduce azimuthal ionospheric electric fields of DP 2 variations near the magnetic dip equator. Radio waves are emitted every 5 min. with continuously sweeping frequency in the range of 2-19 MHz during about 3 minutes. The radar reflections of emitted radio-waves are expected to occur at ionospheric heights of 100-600 km in the dayside equatorial region. From temporal variations of the reflection heights of each 0.05 MHz bin radio-waves, we can estimate apparent upward/downward velocity (***Vd*** ; i.e. within 60 m/s, which must be ***E*** x ***B*** drift velocity in the ionosphere). Then the ionospheric electric field variation (***E***) in the east/west direction can be deduced, especially, in the ambient magnetic field (***Bo***) near the magnetic equatorial region, i.e., ***E*** = ***B*** x ***Vd***

OBSERVATIONAL RESULTS

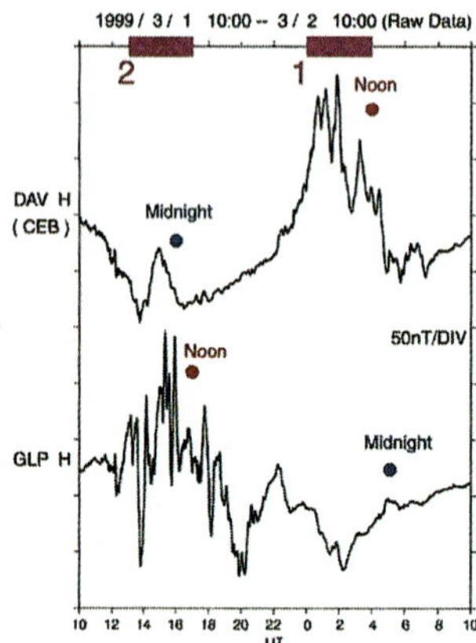

Fig. 1. Diurnal variations of the H-component magnetic fields near the magnetic dip equator stations at Davao in Philippines and at Gadalupe in Peru on March 2, 1999.

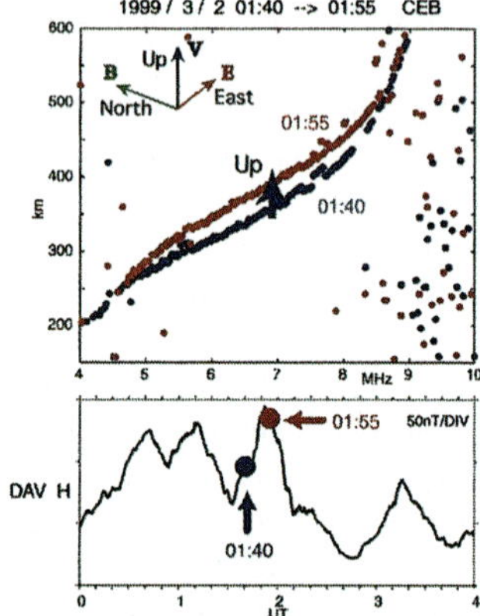

Fig. 2. (Upper) Ionograms obtained at Cebu in the noontime of 10:40-10:55 LT on March 2, 1999. (Lower) H-component DP 2 magnetic variation simultaneously observed at Davao.

Figure 1 shows diurnal variations of the H-component magnetic fields near the magnetic dip equator observed at Davao (7.11° N, 125.61° E, Dip= -0.6°) in Philippines and at Gadalupe (14.06° S, 75.95° W, Dip= -2.6 °) in Peru. We can see magnetic enhancements of Sq and DP 2 variations with several ten-minute periods during the local noontime of each stations.

The upper panel of Fig. 2 shows ionograms obtained at Cebu (10.35° N, 123.91° E, Dip= 2.6°) in the noontime interval from 01:40 to 01:55 UT (i.e., 10:40-10:55 LT) on March 2nd, 1999.

The lower panel shows the H-component DP 2 magnetic variation simultaneously observed at the Davao station at the nearly same local time. The time variations of the reflection heights at each 0.05 MHz bin radio-waves emitted every 5 min. show apparent upward velocity at each 0.05 MHz bin, which must be caused by ***E*** x ***B*** drift in the ionosphere. In the interval from 10:40 to 10:55 LT, the ionospheric electric field must be oriented in the eastward direction near the magnetic dip equator.

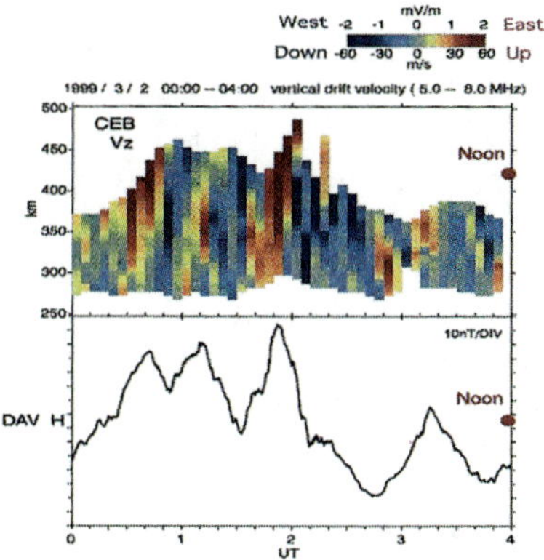

Fig. 3. (Upper) Time-height diagram of estimated azimuthal electric fields at each reflected heights at Cebu, estimated by using upward/downward velocity at each reflection heights during the interval of Fig. 2. The red and blue colors indicate east and westward electric fields, respectively. (Lower) DP 2 magnetic field simultaneously observed at Davao.

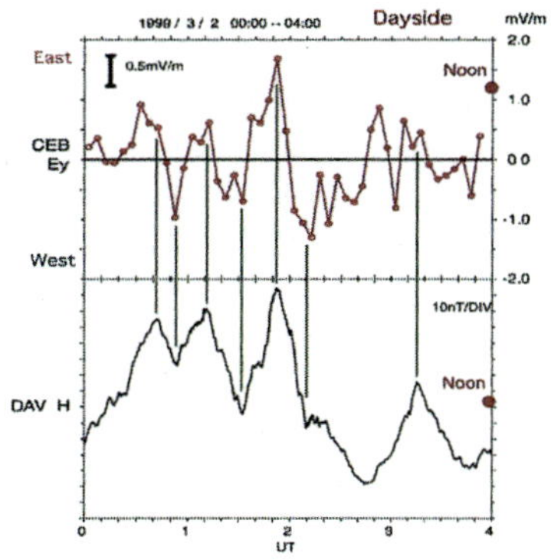

Fig. 4. The correlation chart of the averaged ionospheric electric field obtained from the upper panel of Fig. 3 and the H-component magnetic field observed simultaneously at the dayside equatorial Davao stations in the interval of 00:00 – 04:00 UT on March 2, 1999

From upward/downward velocity at each reflection heights, the azimuthal ionospheric electric fields are deduced, and plotted them as time-height diagram as shown in Fig. 3.

Upper and lower panels of the figure show the estimated azimuthal electric fields at each reflected heights at Cebu and the DP 2 magnetic field observed at Davao, respectively. The red and blue colors indicate east and westward electric fields, respectively. It is noteworthy that the estimated electric field and the observed H-component magnetic field show a good correlation to each other. We further calculated the averaged electric field in the region from lower to upper reflection heights during each 5 min. in Fig. 3, in order to see more clear relationship between the electric and magnetic fields of DP 2. Figure 4 shows the obtained correlation charts of the averaged ionospheric electric field and the H-component magnetic field observed simultaneously at the dayside equatorial stations.

The upward and downward of vertical axis indicates east and westward electric field, respectively. It is found that although there are some out-of-phase correlations in Fig. 4, the most H-component DP 2 magnetic pulses show in-phase relation to the azimuthal ionospheric electric fields in the daytime. The DP 2 magnetic variations observed on the ground are concluded to be due to the enhanced ionospheric current caused by the DP 2 electric field in the daytime equatorial region.

In order to clarify characteristics of DP 2 electric field, penetrating into the night side equatorial ionosphere, we also examined the relationship between dayside H-component DP 2 magnetic variation at Gadalupe (14.06° S, 75.95° W, Dip= -2.6 °) in Peru and nightside electric field variation at Cebu(10.35° N, 123.91° E, Dip= 2.6°) in Philippines.

Figure 5 shows correlation charts of the averaged ionospheric electric field in nighttime and the H-component magnetic field in the daytime equator. It is noteworthy that the azimuthal ionospheric electric field variations in the nighttime equatorial region show a clear out-of-phase relation to the H-component DP 2 magnetic variations in the daytime equatorial region.

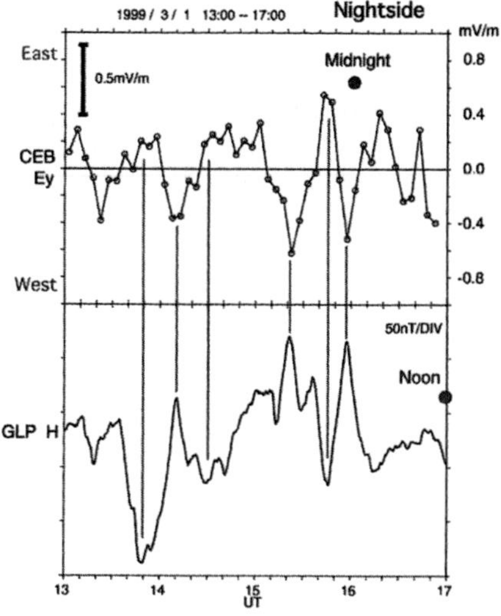

Fig. 5. The correlation chart of the averaged ionospheric electric field at Cebu, Philippines, in nighttime and the H-component magnetic field at Gadalupe, Peru, in the daytime equator in the interval of 13:00 – 17:00 UT on March 3, 1999

SUMMARY AND CONCLUSION

The peculiarities of electric and magnetic field simultaneous observations near the magnetic equator can be summarized as follows;

(1) Although there are some out-of-phase relations, the most H-component DP 2 magnetic pulses show in-phase relation to the azimuthal ionospheric electric fields observed simultaneously in the dayside equatorial region.

(2) Amplitudes of DP 2 magnetic variations observed at the nightside equatorial station are very small and/or negligible, but the azimuthal ionospheric electric field variations near the magnetic dip equator in nighttime are found to show a clear out-of-phase relation to the H-component DP 2 magnetic variations in the daytime equatorial region.

These results suggest that the ionospheric electric field variations associated with DP 2 are not in the zonal, but oriented in the dawn to dusk (or vice versa) direction in both the day- and night-side equatorial region. The DP 2 electric fields caused by the solar wind interaction with the Earth's magnetic field are imposed on the polar ionosphere, and the ionospheric DP 2 electric fields may penetrate instantaneously from the polar ionosphere into the both day- and night-side equatorial regions. This is not consistent with the magnetic zonal variation near the equator reported by Nishida (1968b), but consistent with the result of Gonzales et al. (1979). The DP 2 magnetic variations observed on the ground in the dayside equatorial region are due to the enhanced ionospheric current caused by the DP 2 electric field, while the magnetic variations on the ground in the nightside equatorial region may not be produced enough by the ionospheric electric field because of the lower ionospheric conductivity in nighttime.

It is concluded that simultaneous ground-based observations using multi-techniques are a useful tool to clarify the nature of transfer and/or penetration mechanisms of global electric fields from the polar ionosphere into the day- and night-side equatorial regions, and to understand the coupling mechanisms between the solar wind and the Earth's iono-magnetosphere for the Space Weather study.

ACKNOWLEDGMENTS

We thank all members of the CPMN and WestPac observation groups for their ceaseless supports.

REFERENCES

Gonzales, C.A., M.C. Kelley, B.G. Fejer, J.F. Vickrey, and R.F. Woodman, Equatorial electric fields during magnetically disturbed condition 2. Implications of simultaneous auroral and equatorial measurements, J. Geophys. Res., 84, 5803 (1979).

Kikuchi, T. H. Luhr, T. Kitamura, O. Saka, and K. Schlegel, penetration of the polar electric field to the equator during a DP 2 event as detected by the auroral and equatorial magnetometer and the EISCAT radar, J. Geophys. Res., 101, 17161 (1996).

Nishida, A., Geomagnetic Dp2 fluctuations and associated magnetospheric phenomena, J. Geophys. Res., 73, 1795 (1968a).

Nishida, A., Coherence of geomagnetic Dp2 fluctuations with interplanetary magnetic variations, J. Geophys. Res., 73, 5549 (1968b).

Nozaki, K., M. Shinohara, A. Saito, S. Fukao, and M. Yamamoto. Prompt report of WestPac '99 campaign observation in Cebu, Philippines, Abstracts of 1999 Japan Earth and Planetary Science Joint Meeting, held at Tokyo on June 8-11 (1999).

Tachihara, H., M. Shinohara, M. Shimoizumi, O. Saka, and T. Kitamura, Magnetometer system for studies of the equatorial electrojet and micro-pulsations in equatorial regions, J. Geomag. Geoelectr., 48, 1311 (1996).

Yumoto, K., and the 210° MM Magnetic Observation Group, The STEP 210° magnetic meridian network project, J. Geomag. Geoelectr., 48, 1297 (1996).

Yumoto, K., and the CPMN Group, Characteristics of Pi 2 magnetic pulsations observed at the CPMN stations: A review of the STEP results, 53, in press (2001).

GLOBAL POSITIONING SYSTEM STUDIES OF IONOSPHERIC IRREGULARITIES: A TECHNICAL REVIEW

T. L. Beach

Air Force Research Laboratory, Space Vehicles Directorate, 29 Randolph Road, Hanscom AFB, MA 01731, USA

ABSTRACT

Global Positioning System (GPS) techniques for studying the ionosphere have come to the fore within the past ten years or so. In that time frame: (1) the GPS constellation became fully operational, (2) GPS receivers became widely available, and (3) data from GPS receiver networks appeared on the Internet. Traditionally GPS studies of the ionosphere have focused on total electron content (TEC) measurements and the undisturbed ionosphere. More recently, researchers have begun to exploit the potential to study the disturbed ionosphere using GPS satellite signal observations. Broadly speaking, two GPS techniques are used to study disturbances: delta-TEC techniques and scintillation techniques. This article outlines both techniques and summarizes recent examples from the literature for high, low, and middle latitudes. We will discuss promising multi-instrument techniques and identify some areas where more ionospheric information can potentially be extracted from the data. Throughout the discussion we will also point out the limitations of GPS techniques for studying disturbances. Some of these limitations occur because the available GPS hardware and software were designed with other purposes in mind. In such cases we shall consider possible remedies. Other limitations are more fundamental to beacon satellite techniques. Ground-based GPS diagnostics will never match radar installations for wealth of ionospheric data, for example. What GPS techniques for monitoring disturbances offer is a relatively inexpensive means to proliferate measurements with the goal of resolving issues like spatial versus temporal variations of irregularities.

INTRODUCTION

Global Positioning System (GPS) satellite signals have been considered as tools for ionospheric research almost since the system's inception in the 1970s. Ranging errors due to ionospheric group delay are one of the largest potential sources of GPS navigation error so the designers built dual-frequency measurements into the system to monitor the group delay in real time. As a result, total electron content (TEC) measurements are readily available from dual-frequency GPS receivers along the various satellite lines of sight.

Researchers also recognized that GPS would respond to ionospheric irregularities, e.g. through the well-known phenomenon of ionospheric scintillation. Within the past ten years researchers have begun to use GPS links as tools to study ionospheric irregularities on a systematic basis. Many of those studies have focused on TEC fluctuations, since those measurements are readily available from commercial dual-frequency GPS receivers and data that are routinely collected, but specially designed GPS scintillation monitors have begun to be used more widely.

The purpose of this paper is to provide a brief review of current techniques and results in the area of ground-based monitoring of ionospheric irregularities. It will provide background on the two most common types of measurements—delta-TEC and scintillation measurements—and highlight recent examples from the literature of their use. The remainder of the paper proposes an agenda for improving the research potential of GPS techniques.

TECHNIQUES

Delta-TEC Techniques

The process of obtaining TEC measurements from GPS is well known. Usually, one derives TEC from a linear combination of phase measurements for the L1 (1.57542 GHz) and L2 (1.22760 GHz) broadcast signals (e.g., Wanninger *et al.*, 1994). Technical challenges to GPS TEC measurements are also well known: 2π ambiguities in phase, cycle slips, inter-channel biases. Even if TEC measurements along slant paths were fully accurate and absolute, the issue of relating those slant TEC values to overhead TEC values would remain.

Some of these challenges may not be so critical for studying deviations from a smooth background. A 2 TECU (TEC unit; 1 TECU = 10^{16} electrons/m^2) drop when the GPS line of sight enters an equatorial plasma depletion is still detected as a 2 TECU drop even if the offsets in phase, interchannel bias, etc. are not fully worked out. The important factor is that those offsets remain constant, or relatively so, during the period of measurement.

On the other hand, slant-to-vertical TEC conversion is more problematic when the ionosphere is disturbed. Unless the disturbance is homogeneous in character over large horizontal distances, such a conversion is likely to be nonsensical. It is even possible to imagine ionospheric disturbances and geometries where the vertical TEC does not vary but the slant TEC does.

It may be important to know the absolute TEC along the slant path in addition to the variations in some cases. For example, the quantity ΔTEC/TEC may be relevant to the geometry or physics of a particular disturbance observation. (For relating irregularities to scintillation, ΔTEC alone is probably more important.) If we need to know the absolute TEC background against which variations occur, we must return to determining the offsets in phase, etc. Still, in most cases slant-to-vertical conversion is to be recommended against for irregularity studies.

Many authors study TEC fluctuation measures related to dTEC/dt (e.g., Musman *et al.*, 1997; Aarons *et al.*, 2000). In this case, the average background TEC is eliminated and the high frequency components of ΔTEC are emphasized. Studying dTEC/dt works well for detecting irregularities in practice. Beach and Kintner (1999) have more discussion comparing dTEC/dt measures with the variance of TEC. One complicating factor of dTEC/dt is that it tends to emphasize data glitches; the data analysis routines must exclude non-physical outliers.

The major practical limitation to studying TEC variations is the lack of high sample rate data. Most surveying and solid-earth geophysics applications do not require—or desire—large quantities of high rate data from fixed receivers. For example, many of the popular International GPS Service for Geodynamics (IGS) network of receivers record 1 sample of GPS data every 30 seconds. In the mid-1990s it was also hard to find a commercial, civil dual-frequency receiver capable of recording GPS data faster than 1 sample per second. Recently, the range of available hardware capabilities has improved somewhat.

Scintillation also impacts the ability of dual-frequency receivers to make TEC observations of ionospheric irregularities. As Bhattacharyya *et al.* (2000) emphasize, "TEC fluctuations" as calculated from phase measurements are really differential phase fluctuations that respond to phase scintillation (i.e., diffraction related effects) as well as actual TEC changes. The distinction is mostly prominent at small scales; the effects of diffraction do not show up strongly in low-rate data. Secondly, amplitude scintillation can interrupt dual-frequency measurements. Codeless (i.e., civil) GPS receivers pay a penalty in link margin over code correlation, making dual-frequency measurements more susceptible to scintillation than single-frequency techniques.

The advantages of TEC measurements of irregularities are that commercially available receivers and standard file formats (i.e., RINEX) are compatible with these techniques. These fortunate circumstances are not available for scintillation measurements, as will be discussed below.

Scintillation Techniques

Scintillation measurements—measurements of the amplitude and phase fluctuations caused by signal propagation through ionospheric irregularities—are relatively straightforward to make for beacon satellites. GPS introduces complications because the broadcast signal is not a simple RF beacon but a spread-spectrum signal. Other differences from conventional scintillation measurements will be addressed later in this section.

GPS is a code-division multiple access (CDMA) spread spectrum system. Instead of pure carrier signals, each GPS satellite broadcasts signals that are modulated by pseudorandom noise codes. The codes allow for precise ranging between the GPS satellite and receiver and permit all GPS satellites to share the same spectrum allocation. (Here we mostly discuss the Coarse/Acquisition, or C/A, code on L1; the Precise, or P, code on L1 and L2 is generally unavailable to civil GPS users due to "Anti-Spoofing (AS)" policy.)

For scintillation measurements the CDMA nature of the GPS signals means an additional degree of freedom in receiving the signals and additional complexity in the receiver. Not only must the GPS receiver "tune in" the Doppler-shifted broadcast frequency, it must also correlate the incoming signal's pseudorandom noise code with a locally generated replica. The added degree of freedom is the time offset between the received code and the locally generated code.

The process of detecting and locking onto an incoming CDMA signal and tracking it, both in frequency (or phase) and in code offset, is well known (e.g., Van Dierendonck, 1996). One can also obtain statistical estimates of amplitude and phase measurements based on the correlator outputs (e.g., Van Dierendonck *et al.*, 1993; Van Dierendonck, 1996). The problem is that GPS amplitude and phase measurements are not readily accessible in standard GPS hardware and very little non-standard hardware exists to make scintillation measurements for research. Testing GPS receivers for scintillation research poses challenges as well, since simulating a GPS signal requires more complex and expensive hardware than an RF signal generator.

The limitations of available receivers especially show up in measuring amplitude scintillation. Most receivers or GPS chip sets that provide signal strength information provide scant documentation about what quantity they actually report. The standard RINEX file format provides little information about signal strength, coding it in a single decimal digit with arbitrary units (Hofmann-Wellenhof *et al.*, 1994). Experience has shown that there is no consistency in how—or even whether—signal strength recording is implemented in RINEX from manufacturer to manufacturer; one must rely on manufacturers' proprietary data file formats instead. Finally, data rates of available receivers or receiver installations are often not sufficiently high to resolve amplitude scintillation.

At least two groups have developed "GPS scintillation monitors" as a result of these concerns. Van Dierendonck *et al.* (1993) developed one based on a Novatel receiver design. This receiver—called the Ionospheric Scintillation Monitor (ISM)—usually outputs signal statistics for both amplitude and phase, although some raw data are available. Another scintillation monitor, developed at Cornell University, is based on a unique GEC Plessey (now Mitel) open-architecture receiver development kit (Beach and Kintner, 2001). The latter receiver records raw data and is primarily suitable for amplitude, not phase, measurements in its present form. (Modifications to the Cornell scintillation monitor are ongoing to provide processed output and better phase data.) When raw data are available and are synchronized to GPS timing, scintillation drift measurements are possible (Kil *et al.*, 2000).

A practical problem is the ability to sustain GPS scintillation monitor designs, since open-architecture GPS development systems are a niche market and access to proprietary systems may change through mergers and acquisition. Probably the best hope for maintaining some stability in scintillation monitor design will come from a "hybrid" approach: mating a conventional GPS receiver with separate RF/correlator sections and tracking loops. The hybrid approach would allow for simplified tracking loop design that concentrates on monitoring amplitude and phase. The navigation and housekeeping functions would solely reside in the conventional GPS receiver, which would mainly be used to aid the separate, specialized tracking loops in initial signal acquisition.

Other limitations are more fundamental to the GPS scintillation technique. The problem of satellite motion is one. Most beacon satellites for scintillation studies have been in low earth orbit (LEO) or geostationary orbit (GEO). For the LEO satellites, their radio line of sight moves very rapidly through the ionosphere so the ionospheric motion is traditionally neglected. For GEO satellites the line of sight and ionospheric puncture point are essentially fixed. GPS satellites are in 20,200-km altitude orbits, though, and the line-of-sight motion falls into an intermediate regime. For example, it is easily possible for eastward motion of the GPS ionospheric puncture point to be of the same order as the eastward drift of equatorial ionospheric irregularities (Kintner *et al.*, 2001).

Scintillation studies, especially amplitude scintillation, also provide limited ionospheric information when the ionosphere is not disturbed. For more discussion of the limitations of scintillation techniques, and of GPS techniques in general, we turn to geophysical examples.

EXAMPLES

Due to the highly condensed nature of this review, we focus on relatively recent work and comparatively few examples. To save space, citations in the following passage should carry the sense of "and references therein" when appropriate. Historically, L. Wanninger and others in the early 1990s were among the first to investigate GPS TEC fluctuations as indicative of ionospheric irregularities. Here we will start with a short tour through recent examples of GPS studies in different ionospheric regimes ranging from high latitudes to low latitudes.

At high latitudes, Aarons *et al.* (2000) compare TEC fluctuation measures, binned in latitude and universal time (UT), with similarly binned UV imager data from the Polar satellite. The results indicate good correlation of E- and F- region "microscale" structure with 100-200 km altitude particle precipitation. In this case, the authors took advantage of the fact that GPS provides TEC measurements over multiple lines of sight to extend the latitudinal coverage of a single ground station. Separately, GPS scintillation measurements at high latitudes have also been performed, although L-band scintillations are not as strong at high latitudes as in the equatorial anomalies.

Examples of mid-latitude irregularities with concurrent GPS measurements are, naturally, somewhat harder to find. Recently, Kelley *et al.* (2000) presented airglow observations of highly structured disturbances over Puerto Rico during strong geomagnetic activity. Nearby GPS TEC measurements showed significant variations correlated with the disturbances; TEC also demonstrated that the dark areas in the 630-nm airglow structure corresponded to density depletions. (Absolute TEC measurements also showed an increase as the magnetic storm developed.)

Another mid-latitude example, although of traveling ionospheric disturbances (TIDs) not irregularities, is that of Saito *et al.* (1998). Here the authors take advantage of a network of more than 900 dual-frequency receivers in Japan to demonstrate fairly high-resolution delta-TEC "imaging" of a southwesterly propagating TID. The authors have also shown good correspondence between the delta-TEC maps and 630-nm airglow images on a campaign basis (A. Saito, personal communication, 2000).

At low latitudes there are many examples to choose from, but we highlight just two. Musman *et al.* (1997) demonstrate a technique for visualizing the motion of equatorial irregularities using GPS that takes advantage of: (1) dTEC/dt measurements along several lines of sight, and (2) the field-aligned nature of equatorial irregularities. Kintner *et al.* (2001) include as part of their paper a somewhat novel technique that may hold some promise for extracting more ionospheric information out of GPS scintillation measurements using spaced receivers.

The Kintner *et al.* (2001) technique plots the 3-dB half-width of the signal strength autocorrelation function (τ_{3dB}) of one station against the offset of the peak of the two-station cross-correlation function on a scatter plot (for times when significant cross correlation exists). Specifically, the cross-correlation peak yields a net drift, v_{drift}, and the comparison converts τ_{3dB} to an effective velocity, r_{Feff}/τ_{3dB}, where r_{Feff} is an "effective Fresnel radius." (Note: $r_{Feff} = \sqrt{\lambda z}$, where $\lambda = 0.19$ m is the GPS L1 wavelength and z is the line of sight distance to 350-km altitude in the ionosphere.) Figure 1 illustrates a sample plot of this type, a composite plot for several satellites.

How well correlated the scatter plots are probably tells us something about the state of the irregularities—i.e., uncorrelated data can indicate rapidly developing irregularities. The slope of the line may also tell us about the distance to the irregularities that are causing the scintillation. We have observed that the slope values vary somewhat, though they usually fall in the neighborhood of 3-4.5 in plots formatted as shown. Rino and Owen (1980) also looked at similar plots—with similar conclusions—but the technique seems to have fallen by the wayside.

These selected examples demonstrate the most favorable aspects of using GPS for studying ionospheric irregularities: simultaneous data collection from multiple lines of sight, proliferating receivers, taking advantage of available receiver networks, and using GPS timing to synchronize data collection. The challenges of GPS studies lie in relating the limited amount of quantitative information available from GPS measurements to relevant ionospheric physics. (Note: Evans (1977) has a good review of the capabilities and limitations of beacon satellite measurements.) Generally speaking, the direct measurements that can be derived from ground-based GPS observations of ionospheric irregularities are the following: (1) azimuth and elevation of observations, (2) scintillation and ΔTEC parameters, (3) signal amplitude and phase spectra and time series, and (4) TEC values. We should bear in mind, however, that recovering all of this information might require multiple receiver types at one site. There are few, if any, receivers currently available that can perform all of these measurements.

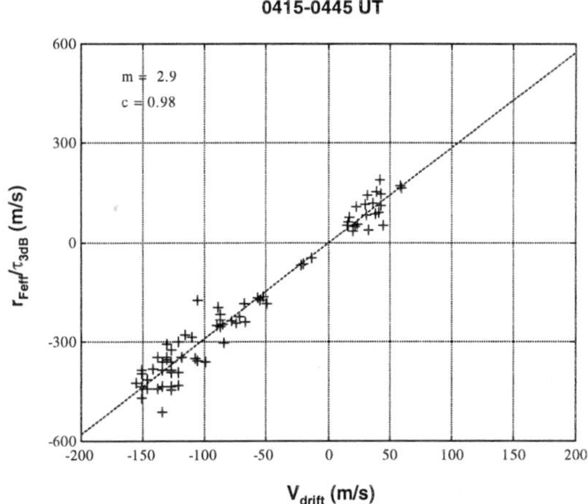

Fig. 1. Scatter plot of autocorrelation width vs. cross-correlation offset—both expressed in velocity units (see text)—for 19 Oct 1996 from Ancon, Peru. Plot also indicates slope, *m*, and correlation coefficient, *c*, of fitted (dashed) line. Note: τ_{3dB} artificially made negative when $v_{drift} < 0$ for purpose of line fitting.

The best approach to rectifying these inherent limitations is to develop relationships between GPS observable quantities and other quantities of interest based on physical constraints. These constraints can come from measurements from other sensors, including other GPS receivers, and the known physics of the problem, e.g. sufficiently mature models of the instability process. These relationships can even include interrelationships between GPS measurements, such as Beach and Kintner (1999) and others have investigated between TEC fluctuation measures and the S4 scintillation index. The Conclusions include a list of possible topics to tackle.

CONCLUSIONS

GPS measurements of ionospheric irregularities are relatively simple to perform although they may require unique receiver technology or alternative modes of operation of existing receivers. Several examples in the recent literature, including those cited in this review, have demonstrated that GPS receivers can detect ionospheric irregularities that are detected by other means. Now the community should turn to research efforts for maximizing the potential for scientific return of GPS sensors, to include some receiver technology development (e.g., the development of "hybrid" scintillation receivers as discussed under scintillation measurements above).

Much of the research proposed would be to develop and explore new techniques for GPS measurement, given all of the observable quantities available. For example, once TEC, ΔTEC, and scintillation measurements are routinely available from the same receiver, how can we combine those measurements to yield new information? We also need better understanding of correlations between time series data from multiple receivers of the same type. We need to identify which sets of different sensors best complement GPS for irregularity studies, especially other sensors that are as inexpensive and portable as GPS receivers. The state of the art in GPS-based studies is to detect and localize irregularities; we now need to begin to look deeper for identifiable characteristics within them.

Some short-term, suggested research goals follow:

- Investigate the spaced receiver technique for comparing amplitude scintillation cross-correlation offset and autocorrelation width as discussed above; this technique may provide information on disturbance altitude.

- Develop "triangulation" techniques with multiple receivers and field line constraints to localize irregularities.
- Perform more work using regular TEC measurements to identify "precursors" to irregularity development and then to monitor and characterize the subsequent irregularities using the same receivers (e.g., Mendillo *et al.*, 2000).
- Determine the feasibility of monitoring drifts of high-latitude features (e.g., patches) with GPS receivers, probably using a delta-TEC technique.

Probably the best way to develop these techniques is to demonstrate their feasibility on a campaign basis as an adjunct to multi-instrument studies. Then the techniques should be used systematically as confidence builds. Again, the best usage of ground-based GPS ionospheric sensors will probably be in long-term studies with good geographical coverage. (Space-based receiver techniques will also be needed to complement the ground-based data in the long run.) The systematic approach—developing appropriate hardware and then developing standard techniques—should be quite useful to the long-term study of space weather.

ACKNOWLEDGEMENTS

This work was performed under AFOSR Task 2310G9. Helpful conversations with J. Aarons, J. Makela, and A. Saito are appreciated.

REFERENCES

Aarons, J., B. Lin, M. Mendillo, K. Liou, M. Codrescu, Global Positioning System Phase Fluctuations and Ultraviolet Images from the Polar Satellite, *J. Geophys. Res.*, **105**, 5201 (2000).

Beach, T. L., and P. M. Kintner, Simultaneous Global Positioning System Observations of Equatorial Scintillations and Total Electron Content Variations, *J. Geophys. Res.*, **104**, 22,553 (1999).

Beach, T. L., and P. M. Kintner, Development and Use of a GPS Ionospheric Scintillation Monitor, *IEEE Trans. Geosci. Remote Sensing*, **39**, 918 (2001).

Bhattacharyya, A., T. L. Beach, S. Basu, and P. M. Kintner, Nighttime Equatorial Ionosphere: GPS Scintillations and Differential carrier Phase Fluctuations, *Radio Sci.*, **35**, 209 (2000).

Evans, J. V., Satellite Beacon Contributions to Studies of the Structure of the Ionosphere, *Rev. Geophys. Space Phys.*, **15**, 325 (1977).

Hofmann-Wellenhof, B., H. Lichtenegger, and J. Collins, *Global Positioning System: Theory and Practice (3rd ed.)*, Springer-Verlag, New York (1994).

Kelley, M. C., F. Garcia, J. Makela, T. Fan, E. Mak, C. Sia, and D. Alcocer, Highly Structured Tropical Airglow and TEC Signatures during Strong Geomagnetic Activity, *Geophys. Res. Let.*, **27**, 465 (2000).

Kil., H., P. M. Kintner, E. R. de Paula, and I. J. Kantor, Global Positioning System Measurements of the Ionospheric Zonal Apparent Velocity at Cachoeira Paulista in Brazil, *J. Geophys. Res.*, **105**, 5317 (2000).

Kintner, P. M., H. Kil, T. L. Beach, and E. R. de Paula, Fading Time Scales Associated with GPS Signals and Potential Consequences, *Radio. Sci.*, **36**, 731 (2001).

Mendillo, M., B. Lin, and J. Aarons, The Application of GPS Observations to Equatorial Aeronomy, *Radio Sci.*, **35**, 885 (2000).

Musman, S., J.-M. Jahn, J. LaBelle, and W. E. Swartz, Imaging Spread-*F* Structures using GPS Observations at Alcântara, Brazil, *Geophys. Res. Let.*, **24**, 1703 (1997).

Rino, C. L., and J. Owen, The Time Structure of Transionospheric Radio Wave Scintillation, *Radio Sci.*, **15**, 479 (1980).

Saito, A., S. Fukao, and S. Miyazaki, High Resolution Mapping of TEC Perturbations with the GSI GPS Network over Japan, *Geophys. Res. Let.*, **25**, 3079 (1998).

Van Dierendonck, A. J., J. Klobuchar, and Q. Hua, Ionospheric Scintillation Monitoring using Commercial Single Frequency C/A Code Receivers, in *Proceedings of ION GPS-93*, pp. 1333-1342, The Institute of Navigation, Arlington, VA (1993).

Van Dierendonck, A. J., GPS Receivers, in *Global Positioning System: Theory and Applications (Vol I)*, ed. B. W. Parkinson and J. J. Spilker, pp. 329-407, AIAA, Washington, DC (1996).

Wanninger, L., E. Sardón, and R. Warnant, Determination of the Total Ionospheric Electron Content with GPS—Difficulties and Their Solution, in *Proceedings of the International Beacon Satellite Symposium*, pp. 13-16, University of Wales, Aberystwth, UK (1994).

EQUATORIAL PC5 ASSOCIATED WITH MOVING CURRENT VORTICES IN THE HIGH-LATITUDE IONOSPHERE

T. Motoba[1], T. Kikuchi[2], H. Lühr[3], H. Tachihara[4], T.-I. Kitamura[5], and T. Okuzawa[1]

[1] *University of Electro-Communications, Tokyo, Japan* [2] *Communications Research Laboratory, Tokyo, Japan* *GeoForschungsZentrum Potsdam, Potsdam, Germany* [4] *National Space Development Agency of Japan, Tokyo, Japan* [5] *Kyushu University, Fukuoka, Japan*

ABSTRACT

An equatorial Pc5 (period of about 7 min) was recorded with the azimuthal phase difference at longitudinally separated equatorial stations, Mokolo (MOK; Cameroon, geographic latitude = 10.44°, geographic longitude = 13.48°) in the morning sector and Peradenia (PRD; Sri Lanka, 7.28°, 80.64°) at noon. The Pc5 amplitude was enhanced at both equatorial stations, which has been recognized as the ground magnetic signatures of equatorial electrojet currents amplified by a polar electric field propagating to the equator. The Pc5 at MOK was correlated with that at PRD with a correlation coefficient of 0.7, while it had a time lag of 80 s; i.e. the phase signal propagated from the noon to morning sector at a speed of 0.84°/s. In the IMAGE magnetometer array located at morning high latitudes, magnetic fluctuations with the same period as the equatorial Pc5 moved westward at a speed of 0.13°/s (5.6 km/s) at about 67° in magnetic latitude. The equivalent current pattern deduced from the IMAGE magnetometer data indicated small-scale ionospheric current vortices centered at approximately 70°, which could be generated by a few pairs of upward and downward field-aligned currents. At mid/low latitudes, a summer/winter asymmetry in magnetic variations was remarkable; the magnetic fluctuations in the summer hemisphere showed a more correlative waveform with the equatorial Pc5 and larger amplitude than those in the winter hemisphere. These observational features allow us to regard the magnetic fluctuations at the mid/low latitude caused by ionospheric currents depending on the ionospheric conductivity. In conclusion, the azimuthal phase difference in the equatorial Pc5 is caused by ionospheric currents extending from current vortices moving anti-sunward in the high latitude ionosphere.

INTRODUCTION

Some types of geomagnetic fluctuations at the dayside equator are amplified by the Cowling effect in the equatorial ionosphere and often appear almost simultaneously with those at auroral latitudes (e.g., Araki, 1977; Kikuchi *et al.*, 1996). These dayside equatorial properties have been thought as contributions of a polar electric field penetrating instantaneously to the equator (Kikuchi and Araki, 1979). Thus, the dayside equatorial magnetometer data are useful to identify whether a polar originating electric field expands to the global ionosphere. The dayside pulsations in the Pc5 frequency range around the dip equator are often observed and their amplitudes are also extremely enhanced in comparison with those at low latitudes (Trivedi *et al.*, 1997). For the global coherent Pc5 event observed during the storm-time of March 24, 1991, Reddy *et al.* (1994) found that both the equatorial and subauroral stations recorded similar Pc5 magnetic oscillations and that those were correlative with the electric field oscillation in the equatorial electrojet measured by the coherent backscatter radar.

These observations were also interpreted in terms of the equatorward transmission of the Pc5-related electric field originated in the polar region. On the other hand, it has been known that daytime impulsive/continuous magnetic pulsations appeared at high latitudes usually propagate in the azimuthal direction and some of the events are organized as the vortical current structures moving longitudinally in the ionosphere (McHenry *et al.*, 1990; Lühr and Blawert, 1994). These longitudinal propagations may be caused by the Kelvin-Helmholtz instability (KHI) or the passage of pressure pulses on the magnetopause. However, little is recognized about the equatorial magnetic behavior corresponding to a variety of magnetic disturbances propagating azimuthally at high latitudes.

In a case study using the high time resolution magnetometer data (3 s sampling for equatorial stations, 10 s sampling for high latitude stations), we present a dayside Pc5 with an azimuthal phase difference observed at two separated dip equatorial stations on November 8, 1993 and clarify relationships between the equatorial and high latitude Pc5.

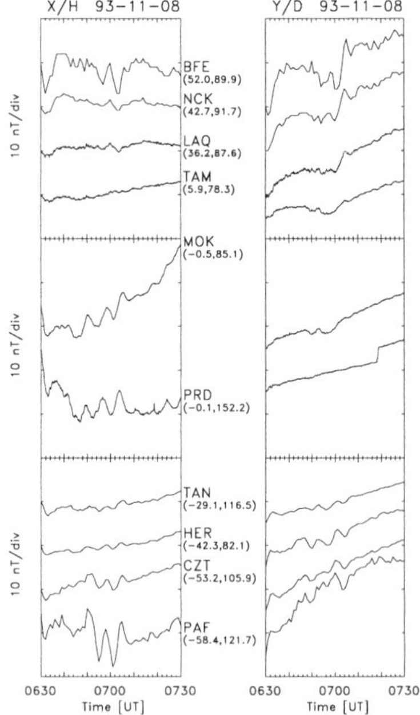

Fig1. Pc5 pulsation observed at low latitudes.

OBSERVATION

Equatorial Magnetometer Observations

Figure 1 shows the unfiltered magnetic X/H and Y/D data from the Equatorial Magnetometer Network (Tachihara *et al.*, 1996) and the INTERMAGNET during a period of 0630–0730 UT on November 8, 1993. The upper, middle and lower panels indicate northern mid/low-latitude stations, dip equatorial stations, and southern mid/low-latitude stations, respectively. In the middle panel, during 0647 to 0707 UT, Pc5 range magnetic oscillation lasting three wave cycles (period of about 7 min) in the H component was clearly seen at Mokolo (MOK: MLT = UT+1.5, dip angle $\alpha = -0.7°$) and at Peradenia (PRD: MLT = UT+5.5, $\alpha = -0.5°$) located on the longitudinally separated dip equator. The peak-to-peak amplitude at both dip equatorial stations was about 5 nT. The cross-correlation analysis (Figure 2) showed that the maximum correlation coefficient was 0.70 and the time lag between MOK and PRD was about 80 s (the phase at MOK delayed behind that at PRD by 72°). The azimuthal phase shift at the dip equator implied that the equatorial Pc5 propagated longitudinally at an apparent speed of 0.84 °/s from the noon to the morning sector (westward).

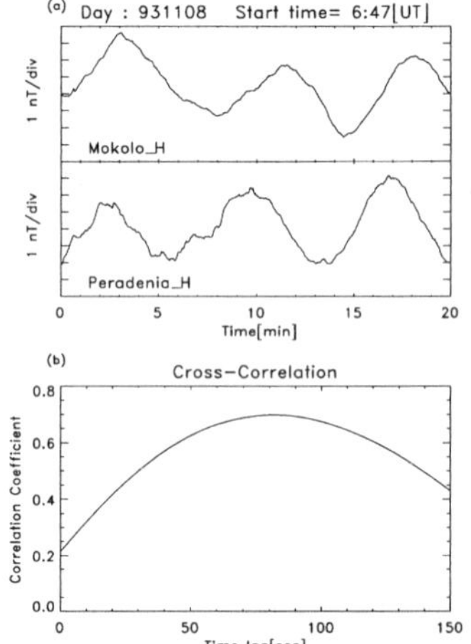

Fig 2. H component at MOK and PRD and the correlation coefficient between both stations.

At the northern (winter) mid/low-latitude stations, the H and D components at BFE, NCK, LAQ, and TAM (corrected geomagnetic latitude: $\theta = 5°$ to $52°$) did not observe magnetic variations corresponding to the equatorial Pc5. Especially, there were little magnetic fluctuations at the northern-lowest-latitude station TAM. In contrast to the northern hemisphere, the X/H component at TAN, HER, CZT, and PAF ($\theta = -29°$ to $-59°$) located in the southern (summer) hemisphere detected Pc5 magnetic fluctuations corresponding to the equatorial Pc5. Generally, the contribution of ionospheric conductivity caused by the solar radiation would be significant in summer rather than in winter. So, asymmetry of magnetic signatures between the summer and winter hemisphere suggests that Pc5 at the mid/low latitudes was excited by polar originating ionospheric currents, as pointed out for the SC/SI signatures at the mid/low latitudes (Yumoto *et al.*, 1996).

High Latitude Magnetometer Observations

Figure 3 shows the magnetic X and Y components obtained from higher-latitude stations ($\theta > 66°$) of the IMAGE magnetometer chain [Lühr *et al.*, 1998] (MLT = UT+1 to +3) for the period of 0630 to 0730 UT. The Pc5 magnetic variations with the same period as the equatorial Pc5 were observed in both X and Y components at morning high latitudes. The X component showed the largest amplitude (150–200 nT) at 73°–74° and the phase in the X component reversed between higher and lower latitude than approximately $\theta = 70°$. On the other hand, the Y component variations at all stations were almost in-phase and the maximum amplitude was about 150 nT at BJN (71°). The azimuthal phase velocity at 66° latitudes was 5.6 km/s (0.13 °/s) in the westward direction (azimuthal wavenumber: m = 6.6).

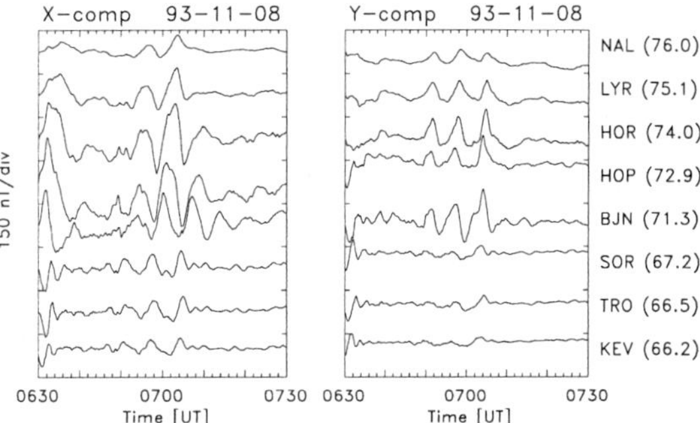

Fig.3. X and Y components recorded by the IMAGE magnetometers.

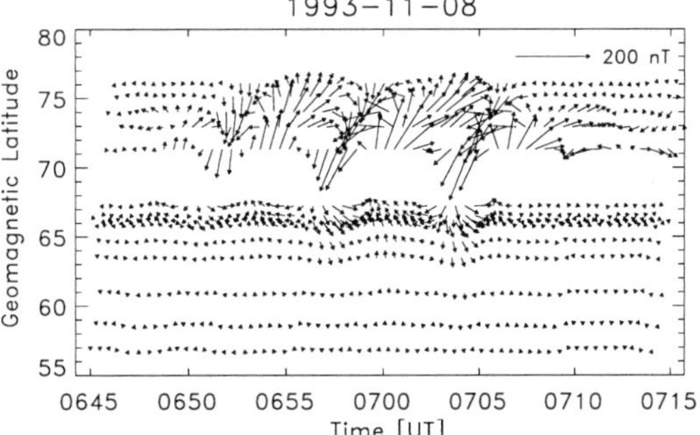

Fig.4. Equivalent current vectors deduced from the IMAGE magnetometer data.

In order to obtain the two-dimensional structure of magnetic variations, we examined the equivalent ionospheric current under an assumption that the ground magnetic variations were produced by ionospheric Hall currents. Figure 4 shows a time series of equivalent current vectors in the IMAGE chain during 0645 to 0715 UT. The six current vortices propagated over the IMAGE chain with their centers located at around 70° magnetic latitudes.

DISCUSSION AND CONCLUSION

We have presented the Pc5 event characterized by the azimuthal phase difference and enhanced amplitude at the longitudinally separated dip equatorial stations. The equatorial enhancement and the summer/winter asymmetry at mid/low latitudes in ground magnetic variations suggest that ionospheric currents play a major role during the Pc5 event. In the high latitude ionosphere at the same local time the multiple current vortices propagated in the anti-sunward direction. In conclusion, the azimuthal phase difference in the equatorial Pc5 is caused by ionospheric currents extending from current vortices moving longitudinally in the high latitude ionosphere.

Although both equatorial and auroral magnetic variations propagated in the same direction, the apparent phase velocity in the equatorial region was much faster than that in the auroral region. This may be related to a rotation of the vortical current configuration originating in the high-latitude ionosphere. For westward moving current vortices, if the line connecting the leading and trailing vortex centers rotated clockwise, the peak of the electric fields induced in the ionosphere would shift westward at the equator. This leads the apparent phase velocity at the equator to be faster than that at high latitudes.

The current structures of the event presented here are similar to those reported previously by McHenry *et al.* (1990) that are associated with the KHI at the inner edge of the low latitude boundary layer (LLBL), but the vortex centers of our event appear at lower latitude by about 5°. For the impulsive traveling convection vortices (TCV) events often associated with solar wind pressure changes, several theoretical and observational studies have suggested that TCVs are excited at the magnetopause or in the LLBL (e.g., Glassmeier, 1992, references therein). However, Yahnin and Moretto (1996) confirmed that the source region of the TCV was located deep inside the magnetosphere (for example, the central/boundary plasma sheet) using the low-altitude satellite data. Similarly, the MHD simulation by Slinker *et al.* (1999) also indicated that the double-vortex centers located at about 70° were mapped to near 7 Re in the magnetosphere. Consequently, it is suggested that the multiple current vortices centered at roughly 70° presented here are excited by the field-aligned currents generated well within the closed field-line region of the magnetosphere rather than the magnetopause or LLBL.

ACKNOWLWGMENTS

The Internal Monitor for Auroral Geomagnetic Effects (IMAGE) magnetometer network is a joint project of Finish Meteorological Institute, Sodankylä Geophysical Observatory, GeoForschungsZentrum Potsdam, Polish Academy of Sciences, University of Tromsö and Technical University of Braunschweig. The Equatorial Magnetometer Network was conducted by Kyushu University during the international STEP period. We gratefully acknowledge the INTERMAGNET and STEP projects for mid/low-latitude magnetometer data.

REFERENCES

Araki, T., Global structure of geomagnetic sudden commencements, *Planet. Space Sci.*, 25, 373, (1977).

Kikuchi, T. and T. Araki, Horizontal transmission of the polar electric field to the equator, *J. Atmos. Terr. Phys.*, 41, 927 (1979).

Kikuchi, T., H. Lühr, T. Kitamura, O. Saka, and K. Schlegel, Direct penetration of the polar electric field to the equator during a DP2 event as detected by the auroral and equatorial magnetometer chains and the EISCAT radar, *J. Geophys. Res.*, 101, 17161 (1996).

Glassmeier, K., Traveling magnetospheric convection twin-vortices: Observation and theory, *Ann. Geophys.*, 10, 547 (1992).

Lühr, H. and W. Blawert, Ground signature of traveling convection vortices, in *Solar Wind Sources of Magnetospheric Ultra-Low-Frequency Waves, Geophys. Monogr. Ser.*, vol. 81, edited by M. J. Engebretson, K, Takahashi, and M. Scholer, pp.231, AGU, Washington, D. C. (1994).

Lühr, H., *et al.*, Westward moving dynamic substorm features observed with the IMAGE magnetometer network and other ground-based instruments, *Ann. Geophys.*, 16, 425 (1998).

McHenry, M. A., C. R. Clauer, E. Friis-Christensen, P. T. Newell, and J. D. Kelly, Ground observations of magnetospheric boundary layer phenomena, *J. Geophys. Res.*, 95, 14995 (1990).

Reddy, C. A., S. Ravindran, K. S. Viswanathan, B. V. Krishna Murthy, D. R. K. Rao, and T. Araki, Observation of Pc5 micropulsation related electric field oscillations in the equatorial ionosphere, *Ann. Geophys.*, 12, 565 (1994).

Slinker, S. P., J. A. Fedder, W. J. Hughes, and J. G. Lyon, Response of the ionosphere to a density pulse in the solar wind: simulation of traveling convection vortices, *Geophys. Res. Lett.*, 26, 3549 (1999).

Tachihara, H., M. Shinohara, M. Shimoizumi, O. Saka, and T. Kitamura, Magnetometer system for studies of the equatorial electrojet and micropulsations in equatorial region, *J. Geomag. Geoelectr.*, 48, 1311 (1996).

Trivedi, N. B., B. R. Arora, A. L. Padilha, J. M. Da Costa, S. L. G. Dutra, F. H. Chamalaun, and A. Rigoti, Global Pc5 geomagnetic pulsations of March 24, 1991, as observed along the American sector, *Geophys. Res. Lett.*, 24, 1683 (1997).

Yahnin, A. G., and T. Moretto, Traveling convection vortices in the ionosphere map to the central plasma sheet, *Ann. Geophys.*, 14, 1025 (1996).

Yumoto, K., *et al.*, North/South asymmetry of sc/si magnetic variations observed along the 210° magnetic meridian, *J. Geomag. Geoelectr.*, 48, 1333 (1996).

SYSTEM PHASE BIAS ESTIMATION OF THE CHUNG-LI VHF RADAR

R. M. Kuong[1], Y. H. Chu[2], S. Y. Su[2], and C. L. Su[2]

[1] *Chung-Shan Institute of Science and Technology, Long-Tan, Taiwan, R.O.C.*
[2] *Institute of Space Science, National Central University, Chung-Li, Taiwan, R.O.C.*

ABSTRACT

A new scheme was proposed in this article to estimate the system phase bias of the interferometry array of the Chung-Li VHF radar inherently induced by phase imbalance between different modules of the array. The system phase bias of the ionospheric array of the Chung-Li VHF radar specifically designed for the purpose of observing ionospheric electron density irregularities has been successfully calibrated by using the radar returns from highly field-aligned sporadic E (Es) irregularities incorporating with IGRF95 model. These irregularities can be affected by space weather events through changes of the global electric field. By comparing the angles of arrival of the Es echoes received simultaneously by ionosphere and interferometer arrays, the system phase biases of interferometer array are determined. The results show that the phase bias is about 66° between receiving channels 1 and 2, and is about 56° between receiving channels 3 and 2. Once the interferometer array is well calibrated, the locations of meteor echoes in the scattering volume can thus be determined accurately.

INTRODUCTION

The global electric field is in a certain way controlled by high latitude electric fields which map down from the magnetosphere (Kelley, 1989). Changes in the electric field should have an influence on the generation of plasma instabilities in the E-region all over the globe (Schunk and Sojka, 1996). Knowledge of the created irregularities can be improved by the radar interferometer technique. A new calibration scheme for this technique is applied in this paper. Space weather effects are, thus, expected to be indirectly seen in the E-region irregularities when studied with the interferometer technique. Middle atmosphere and ionosphere are the important regions in the terrestrial atmosphere in the transfer energy from sun to lower part of atmosphere. The understanding of the responses of these two regions to the solar variations plays is crucial to study the space weather phenomena and solar-terrestrial relationship. The dynamic information of the middle atmosphere and ionosphere can be inferred from the drifts of refractivity irregularities observed by using the ground-based radars. The interferometer technique is used extensively to determine the positions of the discrete targets in the echoing region in terms of the phase differences between the echoes received by different pairs of the antenna modules. Practically, it is inevitable that the observed phase difference of the echoes will contain the inherent phase bias of radar system. As a result, the estimation of the angle of arrival will be incorrect if the radar system is not well calibrated and the system phase bias does not remove totally from the observed radar returns. Calibration algorithms of system phase bias have been proposed by a number of scientists. For example, a technique using an RF-sensor synchronized to the radar oscillators, which is hooked directly to each antenna module and the input signal, has been applied at the Chung-Li VHF radar for the measurements of lightning echoes (Rottger et al., 1995, 2000). The so-called self-survey method was proposed by Valentic et al. (1995) to estimate phase error for a meteor radar by receiving a given radio signal transmitted from a specific radio source with known location determined by the global positioning system (GPS). The other method, proposed by Palmer et al. (1996), is the radio star method. The radio star Cygnus is employed as a given source for the purpose of system phase calibration at MU radar. One of the limitations for the radio star method is that a high gain antenna is required so that the signal-to-noise ratio (SNR) of the received radar signal is large enough for the need of the calibration of the system phase. The radio star method is difficult to implement to the interferometer array of the Chung-Li VHF radar because of its low antenna gain and small aperture. In this article, a new method for the determination of system phase bias of a low-gain antenna array for the observation of meteor trail is proposed. The basic idea is that the electron density irregularities at the scale of 3-meter associated with ionosphere sporadic E (Es) layer can be considered as calibration sources because of their field-aligned property and fairly narrow spatial distribution in the echoing region.

CHARACTERISTICS OF THE CHUNG-LI VHF RADAR

The Chung-Li VHF radar consists of three independent and identical modules, and each of the modules contains its own transmitter, receiver, signal processor, antenna array, and other essential system units. The operational frequency of this radar is 52 MHz (corresponding to 5.77 m wavelength), and the peak transmitted power is 180 kW. The maximum duty cycle is 2% and the pulse length can be set arbitrarily from 1 μs to 999 μs. The configuration of antenna array of the Chung-Li radar is shown in Figure 1. As indicated, the whole antenna array consists of three sub-arrays with different geometric shapes and experimental purposes. The largest

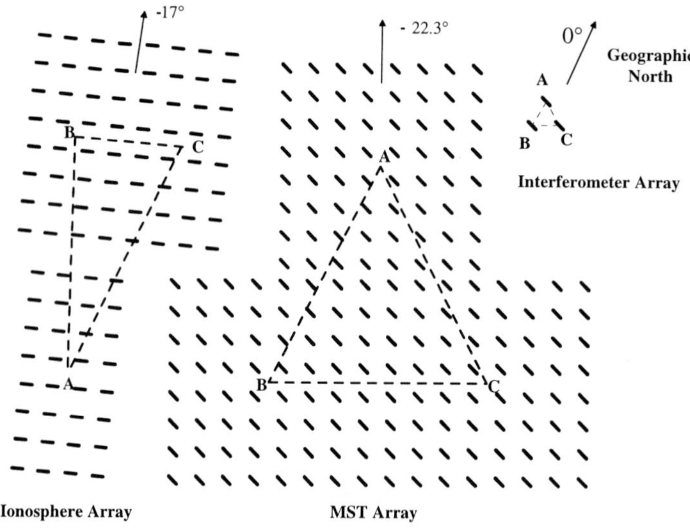

Fig. 1. The configuration of the antenna array of the Chung-Li VHF radar

antenna sub-array among them is arranged as an equal lateral triangle and employed to detect echoes from fluctuations of the atmospheric refractivity that occur in Mesosphere, Stratosphere, and Troposphere. This sub-array is called the MST array. The second sub-array located left-hand side of the MST array is called the Ionosphere array that is used to observe the plasma irregularities in the ionosphere. The smallest sub-array, which consists of three Yagi antenna elements, is employed to detect the echoes of meteor trail and is called the meteor array (or interferometer array). Note that in conducting the phase calibration experiment, the radar wave is transmitted by the Ionosphere array and received alternately by the Ionosphere array and the meteor array through a switch. The three Yagi antennas of interferometer array are arranged in a delta shape with the base line of 5 meters, slightly shorter than the radar wavelength of 5.77 meters. The phase differences of the returned signals between different pairs of the three Yagi antennas are used to determine the arrival angles of meteor in zenith and azimuth direction. In order to enhance the backscatter and also increase the range resolution of the echoes, the Barker code with 7 bits is applied to the radar pulse. Once the radar returns are received, the cross spectra of the echoes for each pair of the Yagi antennas are computed. The phase difference of the received echoes between any pair of the Yagi antennas of the interferometer array can thus be deduced by measuring the mean phase around the peak of the coherence (or amplitude) of the cross spectrum.

PHASE BIAS ESTIMATION

The phase calibration experiment was conducted during the period from April 21 to April 24, 1997. The strong Es echoes was detected on April 23 and lasted early morning of April 24. Figure 2 shows the range-time-intensity plots of the Es echoes detected by ionospheric sub-array (lower panel) and interferometer sub-array (upper panel), respectively. As shown, the echo intensity for ionosphere array is about twice stronger than that for interferometer array.

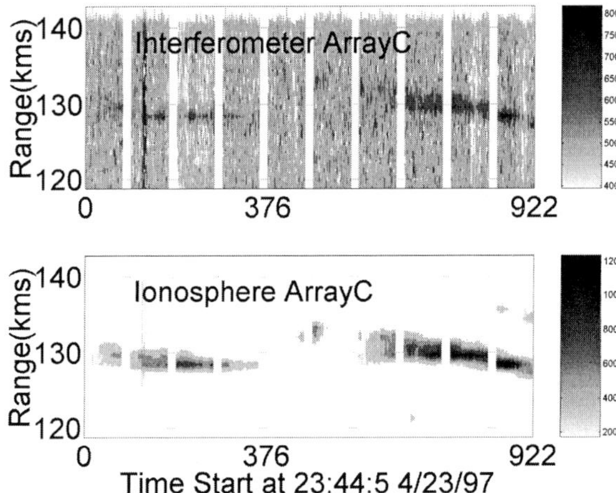

Fig. 2. Range-time-intensity contour plots of Es echoes received by interferometer and ionospheric arrays

Obviously, the two sub-arrays observed the same Es irregularities simultaneously. According to our experiences, the identifications of the desired echoes of meteor trails and Es irregularities from unwanted signals will have much higher sensitivity in frequency domain than that in the time domain. Therefore, in order to calibrate the system phase bias of interferometer sub-array, the time series of the Es echoes at every range gate were transformed into frequency spectra first, and then the complex normalized cross spectra for different pairs of antenna modules were computed. The phase difference can thus be obtained by calculating the corresponding phases of the spectral components of the cross spectra. Note that interferences from other unknown sources are always the serious problems in processing the radar returns. We find that for the present case serious interferences from other radio sources exist in the observed echoes when interferometer array is employed to receive the atmospheric echoes due to its wide beam width and low transmitted power. The other advantage of the processing radar signals in frequency domain is that the interference signal can be effectively suppressed to obtain more realistic and reasonable spectra for interferometer application. Once the phase information of the echoes are obtained, the angle of arrival of the target can be calculated By comparing the angles of arrival calculated from the Es echoes received simultaneously by ionosphere and interferometer arrays, the system phase biases of interferometer array can be determined. The observed data φ_{32}-φ_{12} and θ-ϕ of Es echoes detected by interferometer array are plotted in Figure 3, in which the data

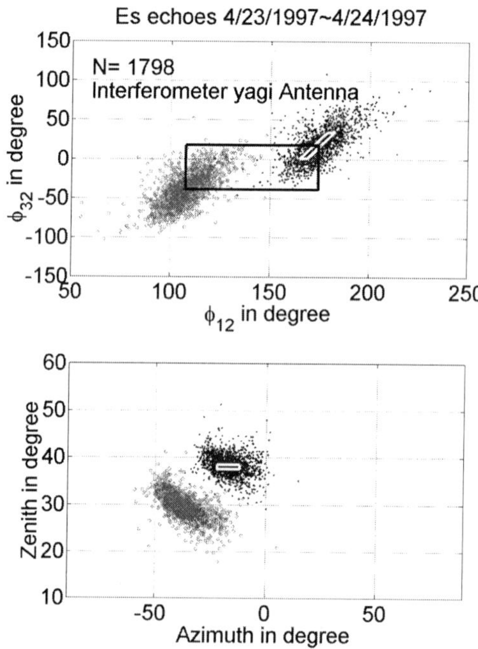

Fig. 3. Scatter diagrams of phase differences (a) between ϕ_{32} and ϕ_{12} and (b) on the plane with zenith-azimuth coordinate

points (opened circle) gathered into the left (in upper panel) and lower left (in lower panel) groups are the observed phase differences before calibration, while the data points (cross) gathered into the right (in upper panel) and upper right (in lower panel) groups are the phase differences after calibration. The small boxes surrounded by white line centered in the calibrated data group represent the theoretical phase difference calculated on the basis of the predicted Es echoing region from IGRF95 model. The result shows that the system phase bias in φ_{32} and φ_{12} are, respectively, about 66° and 56°. In addition, Figure 3 also shows that the Es echo area is locally confined around the predicted echoing region, as expected. Once the system phases are adjusted from the observed Es echoes, the true zenith and azimuth angles of the Es irregularities in the echoing region can be determined accurately. The lower panel of Figure 3 compares the data points before and after system phase adjustment, where 0° in abscissa represents the direction of antenna beam perpendicular to the magnetic field line at 105 km. The spatial distribution of Es echoes is plotted in Figure 4. It shows that the distribution of Es echoes received by interferometer array is primarily located within the main beam of ionosphere array, which was employed for the transmission of radar wave. To verify the correctness of deduced system phase bias, the spatial

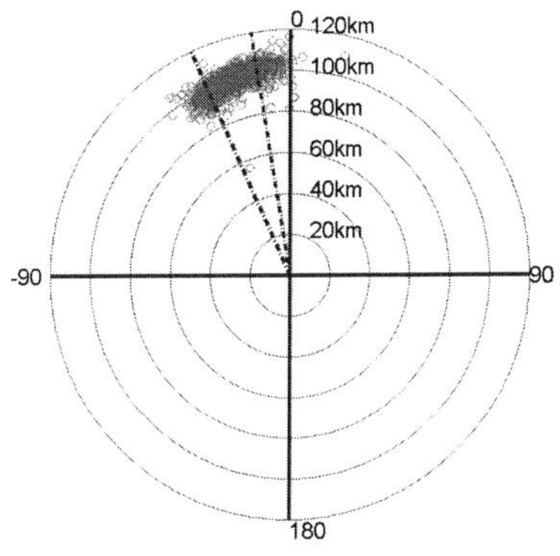

Fig. 4. Spatial distribution of Es echoes detected on April 23-24 1997

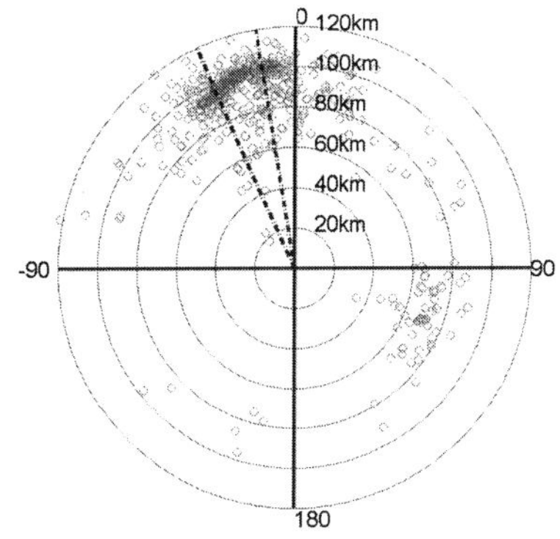

Fig. 5. Spatial distribution of meteor echoes detected on April 23-24 1997

distribution of meteor echoes obtained during the experiment is computed and collected, as shown in Figure 5. The meteor echoes locate primarily within a region that is exactly the main beam of ionosphere antenna array for the illumination of radar wave.

CONCLUSIONS

A new method employed to calibrate the system phase bias of interferometer array of the Chung-Li VHF radar is proposed in this article. Due to very low antenna gain of small interferometer array in the Chung-Li VHF radar, the frequency domain analysis is chosen to pick up the true echoes from the radar returns contaminated by the interference signal. The results show that higher sensitivity of echo detection can be obtained comparing with that of time domain analysis. The small interferometer array suffers from the serious interference problem because of the very wide antenna beam width. By using the IGRF95 model, the expected echoing region of the highly field-aligned Es irregularities is computed. The system phase bias of the interferometer sub-array of the Chung-Li VHF radar can be calibrated by comparing the observed phase differences between the Es echoes detected simultaneously by interferometer and ionospheric sub-arrays, where the latter has been well-calibrated. The results show that the system phase biases for the pair of antenna modules 2 and 3 and antenna module pair 1 and 2 are, respectively, about 66° and 56°.

ACKNOWLEDGEMENTS

This work was supported by the National Science Council of the Republic of China under the grant NSC89-2111-M-008-029-A10

REFERENCES

Kelley, M.C., The earth's ionosphere, Academic, San Diego, California (1989)

Palmer R. D., S. Vangal, M. F. Larsen, S. Fukao, T. Nakamura and M. Yamamoto, Phase calibration of VHF spatial interferometry radars using stellar sources, Radio science, 31, No.1, p.147-156 (1996)

Rottger, J., C.H.Liu, C.J.Pan, and S.Y.Su, Characteristics of lightning echoes observed with VHF ST radar, Radio Sci., 30, 1085-1097 (1995)

Rottger, J., S.Y.Su, C.J.Pan, C.H.Liu, and C.H.Wu, SDI-FDI detection of lightning echoes with the Chung-Li VHF radar: Three-dimensional interferometry, Proc. Ninth Workshop on Techn. Scient. Aspect MST Radar, SCOSTEP, 51-54, (2000)

Schunk, R.W., and J.J.Sojka, Ionosphere-thermosphere space weather issues, J. Atmos. Terr. Phys., 58, 1527-1574 (1996)

Valentic, T. A., S. K. Avery, and J. P. Avery, Self-survey calibration of meteor radar antenna arrays, IEEE Geosci. Remote Sens., in press (1995)

ROCSAT Program Session

EFFECTS OF LIGHTNING ON THE MIDDLE AND UPPER ATMOSPHERE: SOME NEW RESULTS

D.D. Sentman[1], H.C. Stenbaek-Nielsen[1], E.M. Wescott[1], M.J. Heavner[2], D.R. Moudry[1] and F. São Sabbas[1]

1. *Geophysical Institute, University of Alaska, Fairbanks, AK 99775, USA*
2. *Los Alamos National Laboratory, Los Alamos, NM 87845, USA*

ABSTRACT

Recent observations of the effects of lightning on the middle and upper atmosphere have revealed several new dynamical effects of sprites and yielded initial estimates of the total internal energy, including energy residing in nonradiative vibrational states. High speed images (1000 fps) of sprite development and decay show evidence of modification of the local atmosphere by sprites, which may in turn affect the form of subsequent sprites occurring in the same region. Small, relatively long lived (~100s ms) stationary "balls" are often observed within the decaying regions of some sprite tendrils. Calculations of diffusion time scales for the balls are roughly consistent with thermal diffusion, but the relatively long lifetime of the balls relative to nearby tendrils suggests that other processes, perhaps chemical in nature, are operative to extend the lifetime of the balls. Other contrasting features within sprites include an invariably spatially diffuse top and highly structured tendril bottoms. Following the suggestion of Williams [2000], we test as a possibly useful plasma parameter the number of electrons per electron-neutral collision volume for being potentially able to account for the clear separation between the highly structured tendril lower region and the diffuse upper regions of sprites. Models of spectrographic measurements of sprites obtained in 1998 suggest that substantial energy resides in nonradiating vibrational states of both molecular nitrogen and oxygen, in addition to the energy associated with the N2(1PG) optical emissions. Estimates of the internal energy of sprites are ~1 GJ for a large event.

INTRODUCTION

The transient optical excitation of the mesosphere by lightning (sprites) came as an unexpected surprise to atmospheric electricians when first reported by Franz *et al.* [1990]. Subsequent investigations from the Space Shuttle [Boeck *et al.*, 1998], from aircraft [Sentman *et al.*, 1993; Sentman *et al.*, 1995], and from the ground [Lyons *et al.*, 1994, 1996] using a variety of instruments have shown that sprites occur globally and are associated with mesoscale convective complexes (MCCs), intense large scale thunderstorm systems Observations of sprites and correlated electromagnetic measurements [Bocippio *et al.*, 1995], have revealed a causative relationship to positive cloud-to-ground lightning. Various theories have been advanced to account for this connection, including a quasi-electrostatic mechanism [Pasko *et al.*, 1996,

1997], an electromagnetic mechanism [Cho and Rycroft, 1998], and runaway electrons [Roussel-Dupre *et al.*, 1998]. Spectroscopic observations [*Hampton et al.,* 1996; *Mende et al.*, 1995; *Heavner*, 2000] show that most of the optical emissions in sprites are from molecular nitrogen, $N_2(1P)$ leaving the nitrogen in the N_2 exited A state. Recent reviews are given by Rowland [1998] and Rodger [1999].

In this paper we present several new observations of sprites and results obtained by the University of Alaska as part of a NASA-sponsored EXL98 aircraft campaign, and the 1999 Sprites99 balloon campaign, balloon campaign, and follow-up analyses.

OBSERVATIONS

High Speed Images

Most of the spatial phenomenology of sprites has been determined using intensified television technology, with interlaced frame rates of 30/sec, or deinterlaced field rates of 60/sec (e.g., Franz *et al.* [1990], Sentman *et al.* [1993, 1995], Lyons *et al.* [1994]). These video systems are limited in their temporal resolution to 16.7 ms, but high speed photometer measurements (e.g., Armstrong *et al.*, 1998) reveal that the major features of sprites develop over time scales much shorter than this, typically a few ms. Imaging measurements have only recently achieved the required speeds to resolve sprite development at this time scale. The first high speed image sequences of sprites were obtained by Stanley et al. [1999], and showed that they start as small breakdown regions at an altitude of ~75 km and subsequently developed simultaneously upward and downward from this "ignition" point.

High speed images of sprites were subsequently obtained by the University of Alaska [Stenbaek-Nielsen *et al.*, 2000] using a digital, low-light-level, 1000 frame per second intensified CCD imager developed by the Geophysical Institute and operated at the University of Wyoming Infrared Observatory (WIRO) at Jelm Mountain south of Laramie, Wyoming, during August 1999. The observatory is at an altitude of 9650 feet and has unobstructed views ranging from the Canadian border area to the north, the Mississippi river to the East, and into Texas and New Mexico to the South. Over a three week period in August 1999 several hundred sprites were observed from this location and at a second location at Bear Mt., SD. On the night of August 18, 1999 a very active thunderstorm over Nebraska-Kansas (Figure 1) provided an opportunity to capture numerous sprites with excellent spatial resolution.

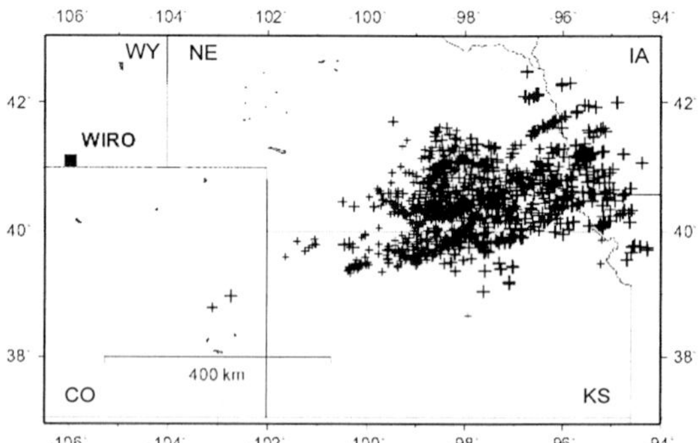

Figure 1. Geographic distribution of lightning strikes, 4-7 UT, August 18, 1999, as recorded by the National Lightning Detection Network (NLDN). Observations presented here were made at WIRO.

Sample image sequences from the high speed camera at WIRO are presented in Figures 2 and 3. The digital images are 256x256 pixels at 8 bit (256 gray levels) and the field of view is 6.4 x 6.4 deg. The instrument wavelength pass band is 500-900 nm with maximum sensitivity at 700 nm. All observations were made unfiltered and at 1000 frames per second. At this data rate the imager saturates at 3.0 MR (at 700 nm). Since most, if not all, sprite events substantially saturated the imager at onset, the maximum brightness is considerably above 3 MR.

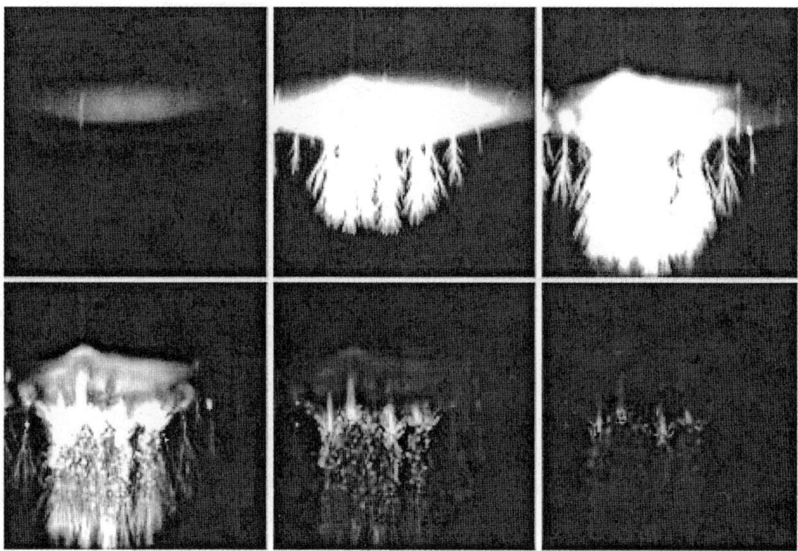

Figure 2. Six consecutive high-speed images, each 1 ms apart, of a large sprite 06:15:07.67 UT on 18 August 1999. The example illustrates some of the typical sprite features: The large horizontal, fairly featureless structure prominent in the two first images is the sprite halo. It often precedes the sprite event. The sprite then develops from an altitude near that of the halo with tendrils going down and branches going up. In this example most of the activity is in the tendrils.

Figure 2 shows a typical large sprite obtained on the night of 18 Aug 1999 using the high-speed imager. The duration of the event shown in Figure 2 is only 6 ms, compared to more typical durations of 10-30 ms. Sprites are often immediately preceded by the appearance of a diffuse "sprite halo" at an altitude of 70-80 km [*Barrington-Leigh et al.*, 2000; *Wescott et al.*, 2000], and this behavior is evident in Figure 2. The sprite develops tendrils branching both upwards and downwards from the onset region near the altitude of the sprite halos. A columniform sprite (C-sprite) [*Wescott et al.*, 1998] is evident in the first frame. In the event shown in Figure 2, as well as in other events, most of the spatial structure is in the tendrils on the bottom side of the sprite. The sprite is fully developed within 2 ms of onset and then begins to fade. In our high-speed imager recordings the tendrils typically reach more than 90% of their full downward extent in less than 1 ms, i.e., essentially the full sprite appears within one video frame, demonstrating that the characteristic time scale for development is less than 1 ms.

Lightning information from the National Lightning Detection Network (NLDN) provided the position of the associated cloud to ground strike at 40.422N, -98.817E, which is 620 km from WIRO. The direction to this point is in the center of the observed sprite. Using the 620 km distance we derived the altitude of the center of the sprite to lie at 84 km, and its full horizontal width is about 75 km. The top of the optical emissions is at 94 km, and the bottom at 38 km.

A second example, observed at 05:24:22 UT and shown in Figure 3, exhibits a more complex development and decay sequence. Here we show two sequences of 1 ms images, separated by 98 ms. Initially, a large sprite occurs in the middle of the imager field of view (top row). The main sprite fades rapidly, but very faint activity continues in the region of the sprite itself. After almost 100 ms low-level activity picks up in the area to the left of the sprite with some filamentary structure visible in the images. Eventually, additional sprite structures appear near the left edge of the high-speed imager field of view (start of the

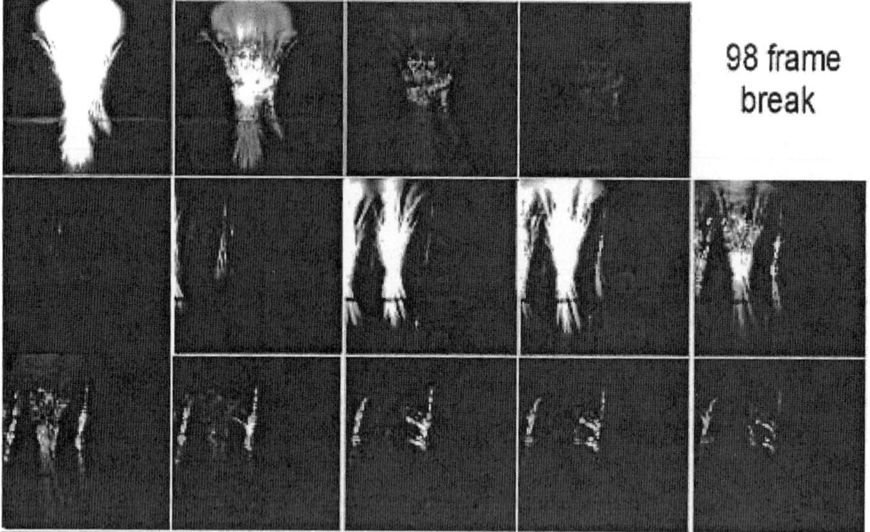

Figure 4. Two sequences from a sprite event at 05:24:22.829 UT, 18 August 1999. The images in each sequence are 1 ms apart, and the sequences are separated by 99 ms. Note the branches going up, in contrast to the tendrils going down in Figure 2, but directed towards the sprite just left of center.

second sequence of images). A filamentary structure in the center of the field of view gradually brightens and after 5 ms (first image bottom row), develops branches towards the sprite to the left, which at this time is rapidly fading. Again, the branches appear to start from small, localized ball-like features at or near the surface of the volume occupied by the sprite. The branches also appear to terminate at the surface of the other, older sprite.

The sprite presented in Figure 3 is an example of an event in which a decayed sprite appears to be "reactivated" by subsequent nearby sprites. This suggests that sprites can modify the local electrical properties of the atmosphere to lower the threshold for subsequent sprite development, and that the changes last substantially longer than the optical sprite events themselves. We also frequently observed new, off-vertical or even horizontal tendrils, such as are evident in the third row of Figure 3. These oblique structures seem to originate from the surface of the volume occupied by the first sprite. These off-vertical tendrils presumably follow the local electric field, and could be evidence of either a local modification by the previous sprite of the electrical conductivity that distorts an external field, or the presence of residual electrical charge generating an additional local field.

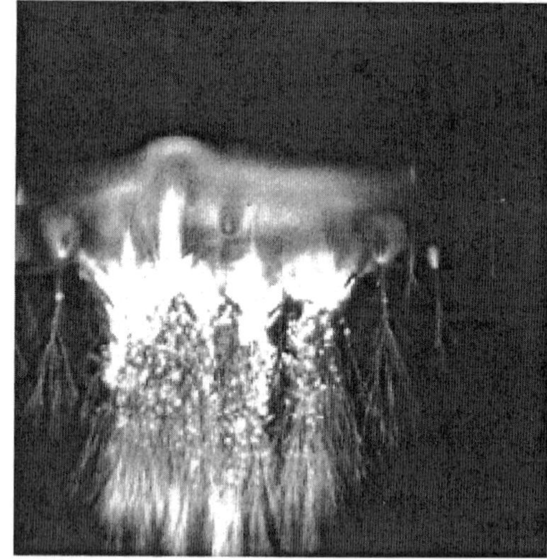

Figure 3. A single frame from the sequence of images in Figure 2, showing the presence of long-lived balls at the juncture of tendril forks.

Long Lived Stationary "Balls" Within Sprite Tendrils

The downward propagating tendrils often lead to the formation of secondary small ball-like structures [*Sentman et al.*, 1996], as noted in Figure 2. One of the frames showing this clearly is shown in Figure 4, where the small isolated balls are visible at the left and right, and numerous others are distributed throughout the volume occupied by the tendrils. The filamentary tendril structures, once developed, seem to remain spatially stationary. The balls do not develop during the tendril development phase, but appear several ms after the tendril onset. They decay in brightness much slower than the main sprite structures, often remaining visible for more than 100 ms. This is longer than the duration of both the causative low-altitude lightning stroke and the mesospheric electrical relaxation time, as well as nearby or adjacent tendril structures. This in turn suggests that the main sprite event initiates secondary localized processes within the balls that are different from processes within the tendrils, and perhaps chemical in nature [Stenbaek-Nielsen *et al.*, 2000]. The presence of such optically emitting balls at the juncture of branch structures of sprite tendrils implies the existence of a local energy enhancement of some kind at these locations. In the absence of an ongoing electrical process to replenish their energy source, these features might otherwise be expected to die with the parent sprite.

We estimate the expected characteristic transport times for thermal diffusion and electron cooling by elastic collisions with neutrals. The electron thermal diffusion coefficient is given by $D_{en} = \lambda_{en}^2 / \tau_{en}$, where λ_{en} is the electron-neutral mean free path and τ_{en} is the electron-neutral collision time. The diffusion coefficient permits estimating the escape time τ_{esc} from a region of characteristic size L as $D_{en} = L^2 / \tau_{esc}$. The electron-neutral mean free path is given by $\lambda_{en} = 1/N_n\sigma_{en}$, where N_n is the neutral number density and σ_{en} is the electron-neutral collision cross section. The electron-neutral collision time is given by $\tau_{en} = \lambda_{en}/a_{th}$, where a_{th} is the electron thermal speed. Combining these, we obtain the approximate electron diffusion escape time from a region of size as $\tau_{esc} = \tau_{en}L^2/\lambda_{en}^2 = L^2/\lambda_{en}a_{th}$. An upper limit for the electron cooling time by interaction with neutrals is estimated to be $\tau_{cool} \sim \tau_{en}m_n/m_e$, where m_n and m_e are the mean neutral mass and electron mass, respectively, where we consider only elastic collisions. Including effects of inelastic collisions would result in a shorter cooling time.

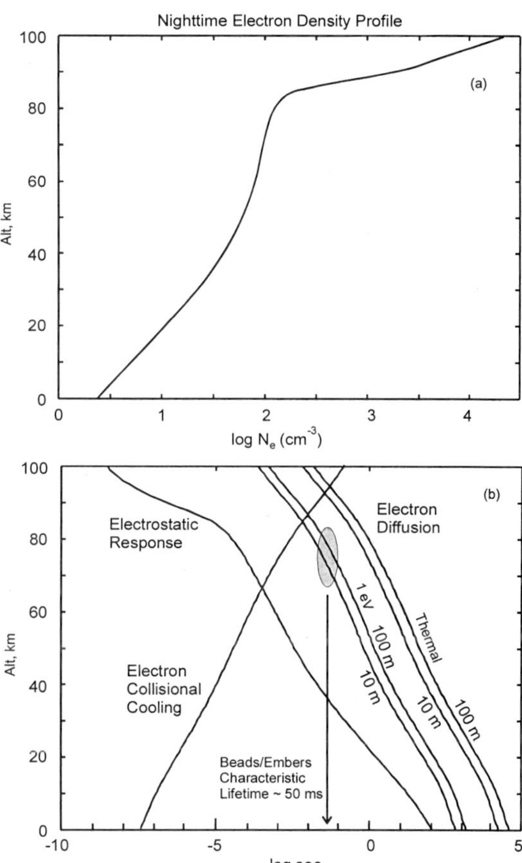

Figure 5. Ambient nighttime electron density profile used in calculation models. (b) Electrostatic relaxation (response) time, electron collisional cooling time, and electron diffusion times. The diffusion times are computed assuming two temperatures and two diffusion distances.

Figure 5 summarizes several characteristic time scales associated with electrical and transport processes within the atmosphere as functions of altitude. We have used a standard model for the neutral atmosphere number densities and temperatures and the nighttime electron density profile shown in Figure 5(a), and assumed a constant electron-neutral collision cross section of 10^{-15} cm^2. [We note in passing that with this model the fractional electron density is only 10^{-12}-10^{-14}, so the ambient nighttime medium is extremely weakly ionized. However, with roughly 10^4-10^5 electrons in a Debye sphere within the ambient mesosphere region of sprite occurrence, a micro-physical plasma description of collective behavior using the Boltzmann equation (not provided here) is in principle feasible.] In Figure 5(b) are shown cooling and diffusion time scales derived for this model. From the basic atmospheric and electron density parameters the ambient electrical conductivity and the associated electrostatic relaxation (response) time are computed. Four electron thermal diffusion time profiles are computed, assuming ambient thermal electron temperature and 1 eV electron energy, and diffusion scale distances of 10 m and 100 m, respectively. Also shown is the electron elastic collisional cooling time. The altitude-duration regime (70-80 km, 10-100 ms) occupied by the balls is indicated by the oval in Figure 5b. The profile most nearly matching the observed altitude-duration regime of beads is electron-neutral collisional diffusion. From these results we conclude that the lifetime of the balls is broadly consistent with electron thermal diffusion. Their long lifetime relative to nearby tendrils may be a reflection of differences in electron energy in the respective structures, i.e., the balls have longer lifetime because the characteristic electron energy of the balls is lower than that of tendrils.

A Plasma Parameter Delineating Regions of Discrete Tendrils and Diffuse Hair

A persistent feature in virtually all well defined sprites, such as in Figure 4, is the diffuse "hair" in the topmost 5-10 km of the structures between 85-95 km, often separated from the highly structured tendril regions at lower altitudes by a dark band of depressed optical intensity [Sentman *et al.*, 1997]. Until recently, no theories have been advanced to account for the underlying physical mechanism behind the difference between the diffuse top region of a sprite and the structured tendril region.

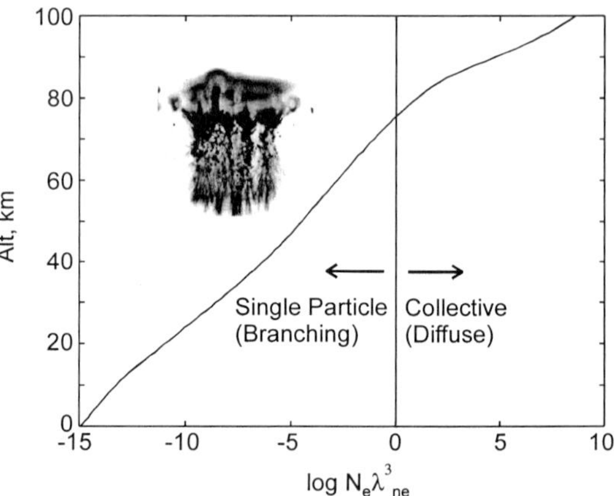

Figure 6. Number of electrons in a collision sphere vs. altitude.

Williams [2000] has proposed that the plasma parameter consisting of the number of electrons per thermal electron-neutral collision volume may be an important determinant separating the diffuse and structured regions. The underlying concept is that at low altitudes electron avalanches originating from single thermal seed electrons are widely separated, by many mean free paths, from other avalanches that may similarly develop, so the avalanches will tend to develop as discrete structures, such as observed in the tendrils. By way of contrast, at high altitudes where there are many electrons within a collision volume avalanches initiated by any single electron would immediately be joined within a single collision time by electrons within the collision volume, thus producing a homogenized glow, such as is observed in the diffuse "hair" of the sprite, instead of a discrete structures.

In Figure 6 is plotted the number of electrons in a collision sphere vs. altitude, with the sprite of Figure 4 included at scale. Although the heuristic concept remains to be developed quantitatively, it is intriguing to note that the Williams [2000] parameter behaves as suggested., steeply increasing from sub-unity values at altitudes below 75-80 km, below which sprites are highly structured, to large values above this height, where the emissions are diffuse.

Nonradiative Energy in Sprites

Based on spectral measurements and optical intensities obtained from aircraft during the EXL98 sprite campaign, Heavner *et al.* [2000] concluded that the total energy stored in N_2 electronically excited states in a typical sprite is ~26 MJ. This figure is obtained from the total intensity of a sprite as observed in white light, combined with spectral observations that show that most of the optical emissions arise from the N_2(1P) group of transitions from the electronically excited B state to the lower energy A state. These transitions occur primarily at red and near infrared wavelengths (hence the name "red sprite"). There is only a very weak ionized component, although the earliest stages of sprite development undoubtedly involve ionization of neutrals. The total internal energy of the excited neutrals includes in addition to the electronic excitation energies a much larger amount of energy stored in nonradiative vibrational states within the ground electronic states of both N_2 and O_2. Using an assumed value of 0.3 eV for the vibrational energy of the ground electronic state associated with sprite emissions, the ratio of total internal energy (vibrational + electronic excitation) to the energy emitted optically is ~20. For a sprite with an optical emission energy of 50 kJ [Sentman *et al.*, 1995], this corresponds to approximately 1 GJ [Heavner *et al.*, 2000]. Uncertainties in the various parameters entering into this calculation are estimated to produce a corresponding multiplicative uncertainty in the energy on the order of 2-5. Hence, the total internal energy of a "typical" sprite is 200 MJ – 5 GJ.

SUMMARY

We have presented several examples of recent observations of sprites imaged a 1 ms time resolution and discrete "balls" of persistent (~100 ms) of luminous emissions that often occur in sprite tendril structures. The images show several new features, including evidence for local modification of the mesosphere by sprites. The relatively long lifetime of the luminous balls is broadly consistent with thermal diffusion lifetimes. However, the fact that the lifetimes are longer than adjacent tendrils suggests that either the internal energy density of the two structures is different, or perhaps balls possess an internal energy source for optical emissions not present in, or distinct from, those in the tendrils, such as from chemical processes. The potential for chemical processes within sprites has been recognized from almost the earliest reports [Sentman *et al.*, 1993, 1995]. Based on spectral and optical measurements, estimates of the total internal energy of a "typical" sprite yield values of ~1 GJ, with generous error estimates ranging between factors of 2-5 above and below this.

ACKNOWLEDGMENTS

Research at the University of Alaska was performed under National Aeronautics and Space Administration grant NAS5-5125.

REFERENCES

Armstrong, R. A., J. A. Shorter, M. J. Taylor, D. M. Suszcynsky, W. A. Lyons, and L. S. Jeong, Photometric measurements in the SPRITES '95 & '96 campaigns of nitrogen second positive (399.8 nm) and first negative (427.8 nm) emissions, *J. Atmos. Terr. Phys.*, **60**, 787, 1998.

Barrington-Leigh, C., U. S. Inan and M. Stanley, Elves: Identification of sprites and elves with intensified video and broadband array photometry, *J. Geophys. Res.*, submitted 2000.

Boccippio, D.J., E.R. Williams, S.J. Hackman, W.A. Lyons, I.T. Baker, and R. Boldi, Sprites, Extremely-Low-Frequency transients, and positive ground strokes, *Science*, **269**, 1088-1091, 1995.

Boeck, W.L., O.H. Vaughan, Jr., R.J. Blakeslee, B. Vonnegut and M. Brook, The role of the space shuttle videotapes in the discovery of sprites, jets and elves, *J. Atmos. Terr. Phys.*, **60**, 669-678, 1998.

Cho, M., and M.J. Rycroft, Computer simulation of the electric field structure and optical emission from cloud-top to the ionosphere, *J. Atmos. Solar-Terr. Phys.*, **60**, 871-888, 1998.

Franz, R.D., R.J. Nemzek and J.R. Winckler, Television images of a large upward electrical discharge above a thunderstorm system, *Science*, **249**, 48-51, 1990.

Hampton, D. L., M. J. Heavner, E. M. Wescott, and D. D. Sentman, Optical spectral characteristics of sprites, *Geophys. Res. Lett.*, **23**, 89, 1996.

Heavner, M.J., D D Sentman, D R Moudry, E M Wescott, C L Siefring, J S Morrill, and E J Bucsela, “Sprites, Blue Jets, and Elves: Optical evidence of energy transport across the stratopause,” in *Atmospheric Science Across the Stratopause* (D. Siskind, Editor), AGU Monograph Series, American Geophysical Union, Washington, D.C., 2000.

Heavner, M. J., *Optical spectroscopic observations of sprites, blue jets, and elves: Inferred microphysical processes and their macrophysical implications*, PhD thesis, University of Alaska Fairbanks, Alaska, 2000.

Lyons, W.A., Sprite observations above the U.S. High Plains in relation to their parent thunderstorm systems, J. Geophys. Res., 101, 29641-29652, 1996.

Mende, S. B., R. L. Rairden, G. R. Swenson, and W. A. Lyons, Sprite spectra: N_2 1 PG band identification, *Geophys. Res. Lett.*, **22**, 2633, 1995.

Morrill, J.S., E.J. Bucsela, V.P. Pasko, S.L. Berg, M.J. Heavner, D.R. Moudry, W.M. Benesch, E.M. Wescott and D.D. Sentman, Time resolve N2 triplet state vibrational populations and emissions associated with red sprites, *J. Atmos. Solar-Terr. Phys.*, **60**, 811-830, 1998.

Pasko, V. P., U. S. Inan, and T. F. Bell, Sprites as luminous columns of ionization produces by quasi-electrostatic thunderstorm fields, *Geophys. Res. Lett.*, **23**, 649, 1996.

Pasko, V. P., U. S. Inan, T. F. Bell, and Y. N. Taranenko, Sprites produced by quasi-electrostatic heating and ionization in the lower ionosphere, *J. Geophys. Res.*, **102**, 4529, 1997.

Rodger, C. J., Red sprites, upward lightning and VLF perturbations, *Rev. Geophys.*, **37**, 317, 1999.

Roussel-Dupre, E.,E. Symbalisty, Y. Taranenko and V. Yukhimuk, Simulations of high altitude discharges initiated by runaway electrons, *J. of Atmos. Solar-Terr. Phys.*, **60**, 917-940, 1998.

Rowland, H.L., Theories and simulations of elves, sprites and blue jets, *J. Atmos. Solar-Terr. Phys.*, **60**, 831-844, 1998.

Sentman, D. D. and E. M. Wescott, Upper atmospheric optical flashes observed from an aircraft, *Geophys. Res. Lett., 20*, 2857, 1993.

Sentman, D. D. and E. M. Wescott, Red sprites and blue jets: thunderstorm excited optical emissions in the stratosphere, mesosphere, and ionosphere, *Phys. Plasmas, 2*, 2415, 1995.

Sentman, D.D., E.M. Wescott, M.J. Heavner, and D.R. Moudry, Observations of sprite beads and balls, *EOS Trans. Am. Geophys. Union*, **77**(46), F61, 1996

Sentman, D.D., E.M. Wescott, M.J. Heavner, and D.R. Moudry, “Horizontal Banded Structure in Sprites,” *EOS Trans. Am Geophys. Union*, **78**(46), F71, 1997.

Stanley, M., P. Krehbiel, M. Brook, C. Moore, W. Rison, and B. Abrahams, High Speed Video of initial sprite development, *Geophys. Res. Lett.*, **26**, 3201, 1999.

Stenbaek-Nielsen, H.C., D.R. Moudry, E.M. Wescott, D.D. Sentman, and F. São Sabbas, Sprites and possible mesospheric effects, *Geophys. Res. Lett.* (in press), 2000.

Wescott, E. M., D. D. Sentman, M. J. Heavner, D. L. Hampton, W. A. Lyons, and T. Nelson, Observations of ‘Columniform’ sprites, *J. Atmos. Solar Physics*, **60**, 733, 1998.

Wescott, E. M., H. C. Stenbaek-Nielsen, D. D. Sentman, D. R. Moudry, and F. S. Sabbas, Triangulation of sprites, associated halos and their relation to causative lightning and micro-meteors, *Geophys. Res. Lett.*,. (in press) 2000

Williams, E.R., Transition from a lightning-like constricted discharge to a uniform glow: Dependence on air density and altitude,” *EOS Trans. A. Geophys. Union* (to appear), Fall Meeting American Geophysical Meeting, Paper A12B-10, San Francisco, 2000.

FINE STRUCTURE OF SPRITES AND PROPOSED GLOBAL OBSERVATIIONS

S. B. Mende[1], H. U. Frey[1], R. L. Rairden[2], Han-Tzong Su[3], R-R Hsu[3], T. H. Allin[4], T. Neubert[4], E. A. Gerken[5], and U. S. Inan[5]

[1]University of California, Berkeley, CA 94720, USA
[2]Lockheed-Martin Research Laboratories, Palo Alto, CA 94304, USA
[3]Physics Department, National Cheng Kung University, Tainan, Taiwan 70101
[4]Danish Meteorological Institute, 2100 Copenhagen, Denmark
[5]STAR Laboratory, Stanford University, Stanford, CA, USA

ABSTRACT

In order to understand sprite processes we have to explain the phenomena from spatial scales of a few meters to the scale of thunderstorm cells. The intricate small-scale vertical structuring of sprites or the so called beads are particularly difficult to understand. From a two-station triangulation featuring observations from Kitt Peak, Arizona and Socorro, New Mexico it was possible to make high resolution observations of the sprite structure when the sprite events occurred within the field of view of the narrow field imager. In several cases the lower altitude luminous filamentary structures of columniform sprites (C sprites) consisted of slant directed, nearly vertically aligned columns of intense pinpoint like beads. The distance of the sprites from the observer was measured and the altitude and vertical spacing of the beads were estimated. The distribution of beads showed that the most frequently observed bead spacing is between 0.6 and 1 km. The vertical and horizontal size of the bright luminous beads was about 80 m or less. The bead spacing showed a trend to increase with altitude and the e folding distance or altitude "scale-height" of bead spacing was found to be 20 and in another case 25 km. In order to make systematic observations of the large-scale sprite morphology a satellite based instrument the Imager for Sprites and Upper Atmospheric Lightning (ISUAL) instrument is planned to fly on the Taiwanese satellite, ROCSAT 2. The instrument will consist of an imager and two bore-sighted photometers. The imager will locate the sprites near the earth limb and make global synoptic measurements while the photometers will measure the spectral and temporal properties of sprites and other upper atmospheric luminous phenomena in a number of different wavelength regions uninhibited by atmospheric absorption.

INTRODUCTION

Optical emissions constitute observational evidence of strong electrical coupling between the troposphere (altitude range of 5-15 km) and the mesosphere/lower ionosphere (altitude range 60 to 100 km). Sprites are the brightest and most frequently observed luminous coupling phenomena between troposphere and mesosphere/lower ionosphere.

Sprites were categorized according to their morphological appearance. The most commonly reported sprites are the so called "carrot sprites" which have a strong luminous center regions, tapering towards lower altitudes while showing a diffuse "hair" region above the body. Near the bottom, the carrots are often accompanied by thin hairline discharges or tendrils. Wescott et al. [1998] observed sprites consisting of thin relatively uniform vertical columns so called columniform or "C" sprites. From limited analysis they concluded that C-sprites follow positive flashes with currents ranging from about 23-100 kA. They measured spectra and they suggest that there are spectral differences between C sprites and other type of sprites. We do not know yet whether the morphological distinctions in the horizontal/vertical structuring have significant relevance to qualitative differences in the underlying electro-dynamic processes. For example it is possible that the tendency to develop full carrot like features with hair and tendrils are the result of strong electric fields and accompanied by substantial discharge while less intense discharges exhibit the features of C sprites. Most sprites are largely elongated in the vertical direction presumably because discharges line up with the predominantly vertical electric field.

Substantial structuring in the horizontal direction was also observed by Sentman et al., [1996] who report occurrences of small, often isolated, patches of bright luminosity that are often distributed throughout the volume or within the immediate vicinity of the sprite at altitudes of 60-80 km. They also reported on observing isolated patches, with scale sizes not exceeding few km, resembling plasma beads and balls and may occur wherever the local breakdown criteria is met within the complex three dimensional structure of nodes and anti-nodes of the radiation field created the underlying lightning stroke. Winckler [1998] also reported the observation of beads and in their magnified images for example one can discern beads in their Figure 6b. Wescott et al. [1998] also showed example of a few beads in their Figure 5. Very high resolution spatial studies of sprites have been carried out during the Sprite 1998 campaign [Gerken et al., 2000] revealing extensive intricate structural complexity of sprites.

During the 1999 summer campaign simultaneous observations were made with high resolution cameras from Kitt Peak, Arizona and Socorro, New Mexico. The two station observations permitted to obtain triangulated quantitative measurements, which allowed us to examine the properties of the beads in more detail. We were able to measure the sizes and altitude separation of beads occurring underneath "C" sprites.

The detailed morphology of the discharges, their intrinsic structural richness, the conditions which leads to high probability of occurrence and the tendency for sprites to recur in the same region has not been explained. Recent high time resolution investigations have uncovered some of the temporal properties of Sprites and that some ionization does occur in the initial period of sprite formation which subsequently gives way to pure excitation [Armstrong et al., 2000].

Most prior studies used ground or aircraft based observations. The observations were inhibited by the dense lower atmosphere, which produces substantial wavelength selective absorption. For a significant statistical study of the global distribution of the frequency of occurrence brightness time-profile and wavelength content a spacecraft based observing program is needed. The Imaging of Sprites and Upper Atmospheric Lightning (ISUAL) satellite program is designed to fulfill this need.

DESCRIPTION OF THE OBSERVATIONS

High resolution data were taken at Kitt Peak, with special emphasis to search for luminous clustering or bead production phenomenon. From simultaneous data taken at Socorro, New Mexico, we were able to triangulate sprites and obtain accurate range and altitude information. The instrument used at Kitt Peak was a pair of intensified cameras. One of the cameras, a 300 mm focal length lens intensified CCD camera, had a focal plane consisting of a 25 mm diameter S-20 photo-cathode intensifier tube. This was fiber-optically coupled to an interline transfer CCD. The instrument included a built in video processor, which added annotation on each video frame, encoding time, gain setting and frame integration duration. For sprite viewing the camera was operated at standard video rates (30 frames per second). Another bore sighted camera, had a 50 mm lens and also used a 25 mm S-20 intensifier tube. In this camera a Charge Injection Device (CID) was optically coupled to the intensifier. Data from a time code generator were superimposed on both video signals, which were separately recorded on Hi-8 videocassettes. A team of researchers representing the Lockheed Martin Research Laboratories, the National Chen Kung University of Taiwan, and the Danish Meteorological Institute of Denmark operated the instruments. The University of California, Berkeley provided the most field instruments and coordinated the observations.

The Socorro instruments, a long focal length telescope [Gerken et al., 2000] and the bore sighted wide field (50 mm focal length camera provided by the Lockheed Martin Palo Alto Research Laboratories) were operated by Stanford University. Several spectacular sprite events were recorded on Aug 8, 1999. Some of these events were analyzed in some detail. Several stars were identified in the wide FOV camera and in the Socorro wide field frame to provide a basis for triangulating the sprite at 220:04:22.3. Another sprite was observed in the Kitt Peak wide field cameras at 220:04:22.4 only 100 msec later which was not inside the telephoto field of view. From the data sets concerning several stars it was possible to construct a generalized expression which associated a value of azimuth and elevation for any pixel coordinate x and y. The tips of three sprite clusters were identified from the similarities of the images from both stations and azimuth and elevation of the tips of the sprites were obtained. These data combined with the station coordinates were used in a triangulation program previously developed for barium ion cloud position determination. The triangulated range (distance) from the Kitt Peak observers were found to be about 680+-10 km for the triangulated tips of the sprites for the sprite occurring an 8/8/1999 day 220 04:22:09 UT.

Table 1. The Results of the Triangulation for Sprite Observed at 220-04-22-09

	Lat (km)	Lon (km)	Range(km)	Alt (km)
Kitt Peak	26.38	251.097	680	75.97
Socorro	26.11	251.17	895	76.02

The National Lightning Detection Network (NLDN) data was examined for the 04:09:22 second periods and we found the events which had positive lightning discharges and they were found closely consistent with the triangulated results.

Stars were also identified in the images taken by the 300 mm telephoto and a linearized relationship between azimuth, elevation and pixel x, y space was found in the telephoto view. Greater relative accuracy was obtained by the use of the telephoto lens. Using the range distance obtained from the wide field triangulation (See table 1) the azimuth and elevation of each pixel was determined as single points in space and the true altitude was estimated above the oblate spheroid earth surface. Thus the altitude of each Sprite feature including the altitude of the beads was determined.

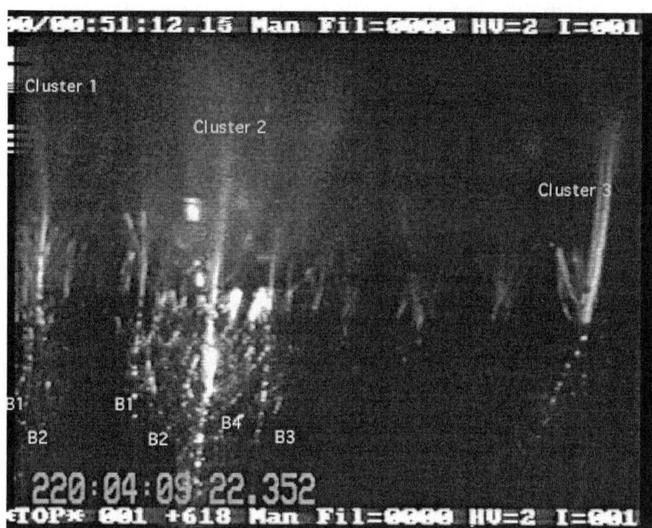

Fig. 1. Telephoto Sprite image. The various features were labeled as clusters and branches.

Figure 1 shows the example of a sprite. The upper portion of the so called C sprites are bright and continuous and they abruptly end at the same lower altitude (74.5 km +- 1.5 km) with the notable exception of the brightest branch of Cluster 2 which appears to come lower. Below that altitude one can see distinct "beads". For the purposes of the analysis presented here we arbitrarily identified 3 clusters and branches within the clusters.

The diameter of the beads is close to instrument resolution, although some of the stars visible on this image (e.g. just above the left branch of Cluster 3) are somewhat sharper than the beads associated with cluster 3. Each CCD pixel in the 300 mm telephoto system was estimated to be 83 meters in diameter at 680 km distance.

The bead spacing distances were sorted into 0.2km bins from 0 to 2.4 km. The distribution is double peaked at spacings of 0.6 and 1 km with 12 events in each spacing range. The apparent minimum at 0.8 km may not be significant. Figure 2 shows the statistical distribution of the altitude differences between consecutive beads from all branches of all sprites in figure.

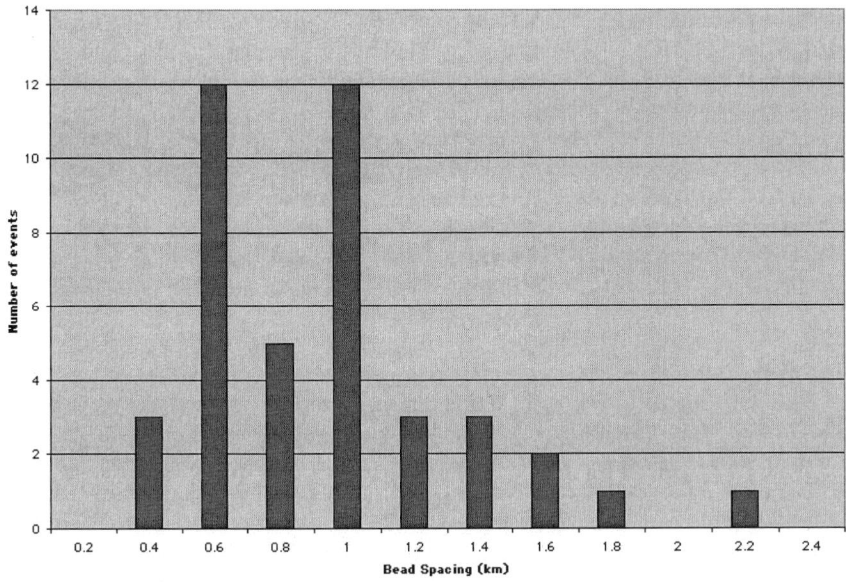

Fig. 2. Distribution of the altitude spacing between beads.

It was interesting to investigate whether the bead spacing varies systematically with atmospheric pressure. Figure 3 is a scatter plot of the bead altitude vs. bead spacing for all the beads measured on Figure 2. The symbols represent the branches of the bead in Figure 1. Although the scatter is quite large there is a trend that the bead spacing is increasing with altitudes. This trend is more or less present in each branch and can be seen as an overall trend.

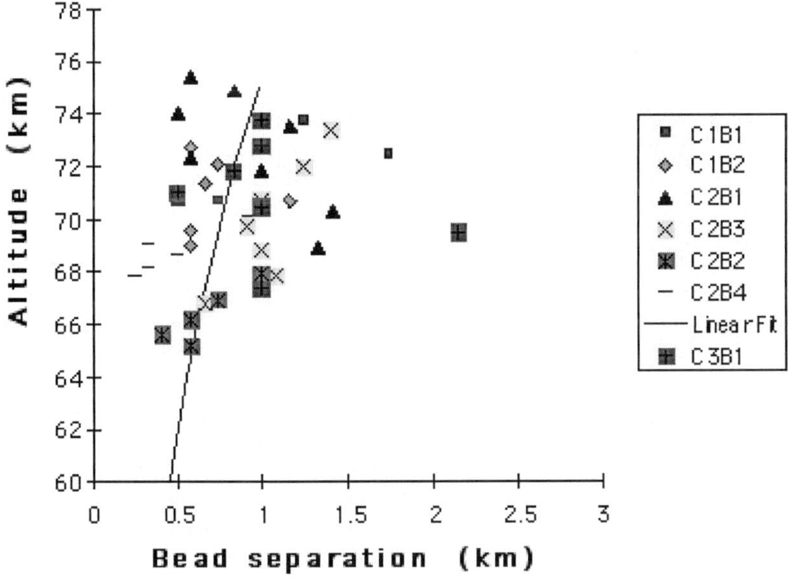

Fig. 3. Bead Separation as a function of altitude.

One would expect spacing to increase with pressure/altitude as the mean free path increases. If the bead spacing were proportional to the mean free path then the plot of log of the spacing against altitude should have a slope of 1/H where H is the scale height. The curve representing the regression analysis of the natural logarithm of the bead spacing as a function of altitude is labeled as "Linear fit" on Figure 3. Linear regression analysis showed a value of 20 km for the "bead scale height" for the event at 04:09:22.

A similar sprite exhibiting extensive bead structure was observed at 04:34:27 on the same night. This sprite was also observed both at Kitt Peak and at Socorro. The bead separation "scale height" in this case was found to be 25 km which considering the scatter of the data can be regarded as relatively good agreement with the 20 km obtained from the 04:09:22 event.

DISCUSSION

High spatial resolution triangulated observation of sprites revealed interesting properties of the beads under C sprites. These beads are spatially well defined and appear to form near vertical strings at the bottom of c sprites below 74 km. The vertical spacing of the beads on average is less than one km. We examined the dependence of the bead spacing with altitude and found that there is a slight variation with altitude and the bead separation tends to increase slowly with altitude. The "scale height" or the logarithmic variation of bead spacing with altitudes, was about 20 km in one case and 25 km in another. The scale height of the atmosphere at sprite altitude is about 6 km. This shows that the bead formation phenomena are not solely dependent on the atmospheric mean free path. The externally impressed electric field electric field decreases with altitude probably as an inverse square or inverse cube law. One would therefore expect a longer "bead scale height" than the atmospheric scale height.

As the precursor lightning discharge occurs in the thundercloud at a large distance below, in the mesosphere the externally superimposed quasi- electrostatic field should not have a fine structure. The fine scale structure exhibited by beads show strong light emission modulations indicating localized fine structure. One approach to explain sprite structure has been associated with the larger scale variations with gravity wave induced changes of the neutrals [Pasko et al., 1997]. However the wavelength of directly observed gravity wave induced variations are multi kilometer in scale. Another possible structural variable is the pre-existent ambient electron density that could exhibit fine structuring because cosmic ray produced contribution to the electron density. Such electron source is indeed a requirement for generating runaway electrons which are needed in the current explanation of seeding thunderstorm associated gamma ray events [Russell-Dupre and Gurevich, 1996; Russell-Dupre et al., 1998]. Structuring of sprites have also been explained as caused by the interference of electric fields produced by the electromagnetic pulse (EMP) of an intricate current system within the cloud during the lightning event [Valdivia et al., 1997; Valdivia et al., 1998]. However the observed time delay between the precursor lightning, the elv which is caused by EMP and the sprite is inconsistent with such an "instantaneous" production mechanism of sprites.

If we assume that sprites are in a relatively uniform exponential atmosphere and that they are produced by the superposition of a quasi-static electric field induced by the cloud to ground discharge then the quasi-static electric field (downward directed, negative) is smooth and diminishes with altitude according to some model. The critical field, which is required for either excitation or ionization, falls off exponentially with altitude following the atmospheric density. The quasi-static electric field can reach the magnitude of the critical field since the critical field falls off faster than the electric field. At the lowest altitude where the electric field exceeds the critical excitation field the ambient electrons will excite the air molecules and faint tendril glows could be expected. At the point where the electric field is above the critical value for ionization ions and electrons will be produced. We associate beads and lower boundary of the bright region in the C sprites where the excitation and a bright intensity enhancement will be seen which are associated with the body of the C sprite.

Although it has not been specifically measured it is safe to assume that the bulk of light emission in the beads is the N_2 first positive red emission [Mende et al, 1995; Hampton et al., 1996]. The excitation potential of this emission is about 8 – 10 eV. The most remarkable property of the beads is their relatively short altitude extent and their relatively constant spacing along the vertical direction. The sprite intensity was measured in the wide field of view image by comparison of the sprites to the luminosity of observed stars in the same field of view. It was found that each digitized video unit is equivalent to 2 kR of red light. Some of the Sprite features were found to be over the maximum of 256 units. Thus a representative sprite intensity was >500 kR. The beads are considerably brighter than this (in the 1.5 – 2 MR range) and therefore require electron multiplication hence the association of beads with the regions where ionization takes place. Substantial electron multiplication (~40) is required to produce the required intensities. At 80 km the needed electron multiplication is 400 increasing rapidly with decreasing atmospheric density. In summary we expect that substantial electron multiplication and subsequent charge generation and possible space charge separation would take place to provide the observed large intensity variation characteristic of the bead appearance. Regions of positive space charge would push the local excitation/ionization potential above the critical value producing bright beads while regions of negative space charge produce dark regions.

The situation is analogous to a situation where a rigid insulator surrounded by vacuum is bombarded by electrons of energy higher than the secondary emission critical potential. The ionization i.e. secondary electron production produces a positively charged region with the electrons carrying negative charge to the adjacent regions. This action would result in small regions of positive charge with intense local excitation with adjacent regions having retarding electric field, which slows the electrons below the critical potential. Thus the size of the bright regions would be defined by the mobility of the ions, (which would be zero in the example of a rigid insulator) and the size of the dark regions would be related to the mobility of the electrons. During the sprite's ionizing phase (few msec) the electrons with a kinetic energy of 10 eV travel a distance which is of the order of about 1 km which is consistent with the bead separation distance.

Understanding of the physics of sprite production would be greatly improved if we knew what atmospheric conditions favor occurrences of sprites and other luminous phenomena. A systematic program to observe the global distribution of sprites would be of great value. To make systematic and global observations of sprites and related phenomena a satellite based instrument the Imager for Sprites and Upper Atmospheric Lightning (ISUAL) instrument is planned to fly on the ROCSAT 2 Taiwanese satellite. A satellite-based instrument will provide global coverage and will allow observations unhindered by atmospheric extinction or cloud problems. The instrument will consist of an imager and two bore-sighted photometers. The imager will image sprites (and other luminous phenomena) near the earth limb spatially separating them from tropospheric lightning. The photometers will measure the spectral and temporal properties of each sprite in a number of different wavelength regions. By correlating the measured global sprite distributions with other atmospheric data we hope to gain insight into the physics of sprite production.

ACKNOWLEDGMENTS

The authors wish to acknowledge encouragement and helpful discussions with Jyh-Long Chern and Lou Lee of the National Chen Kung University of Taiwan. The observations were performed at the Kitt Peak National Observatory and the authors are indebted to the Observatory and its staff for their assistance.

REFERENCES

Armstrong, R. A., D. M. Suszcynsky, W. A. Lyons, T. E. Nelson, Multi-color photometric measurements of ionization and energies in sprites, *Geophysical Research Letters*, **27**, 653-6 (2000).

Gerken, E. A., U. S. Inan, and C. P. Barrington-Leigh, Telescopic Imaging of Sprites, Stanford University pre-print (2000).

Hampton, D. L., M. J. Heavner, E. M. Wescott, D. D. Sentman, Optical spectral characteristics of sprites, *Geophysical Research Letters*, **23**, 89-92 (1996).

Mende, S. B., R. L. Rairden, G. R. Swenson, W. A. Lyons, Sprite spectra, N/sub 2/ 1 PG band identification, *Geophysical Research Letters*, **22**, 2633-6 (1995).

Pasko, V. P., U. S. Inan, Y. N. Taranenko, T. F. Bell, Heating, ionization and upward discharges in the mesosphere due to intense quasi-electrostatic thundercloud fields, *Geophysical Research Letters*, **22**, 365-8 (1995).

Pasko, V. P., U. S. Inan, T. F. Bell, Sprites as evidence of vertical gravity wave structures above mesoscale thunderstorms, *Geophysical Research Letters*, **24**, 1735-8 (1997).

Pasko, V. P., U. S. Inan, T. F. Bell, Y. N. Taranenko, Sprites produced by quasi-electrostatic heating and ionization in the lower ionosphere, *Journal of Geophysical Research*, **102**, 4529-61 (1997).

Roussell-Dupre, R. A., A. V. Gurevich, On runaway breakdown and upward propagating discharges, *Journal of Geophysical Research*, **101**, 2297 (1996).

Roussel-Dupre, R., E. Symbalisty, Y. Taranenko, V. Yukhimuk, Simulations of high-altitude discharges initiated by runaway breakdown, *Journal of Atmospheric and Solar-Terrestrial Physics*, **60**, 917-40 (1998).

Sentman D. D., E. M. Wescott, M. J. Heavner, D. R. Moudry, Observations of sprite beads and balls, Abstract A71B-7, Suppl. To EOS, 77, F61 (1996).

Valdivia, J. A., G. Milikh, K. Papadopoulos, Red sprites: lightning as a fractal antenna, *Geophysical Research Letters*, **24**, 3169-72 (1997).

Valdivia, J. A., G. M. Milikh, Papadopoulos, K. Model of red sprites due to intracloud fractal lightning discharges, *Radio Science*, **33**, 1655-68 (1998).

Wescott, E. M., D. D. Sentman, M. J. Heavner, D. L. Hampton, W. A. Lyons, T. Nelson, Observations of 'columniform' sprites, *Journal of Atmospheric and Solar-Terrestrial Physics*, **60**, 733-40, 1998.

Winckler J. R., Optical and VLF radio observations of sprites over a frontal storm viewed from O'Brien observatory of the university of Minnesota., J. A. T. P., 60, 679-688 (1998).

SPATIAL AND TEMPORAL STRUCTURES OF SPRITES AND ELVES OBSERVED BY ARRAY PHOTOMETERS

H. Fukunishi, Y. Watanabe, A. Uchida, and Y. Takahashi

Department of Geophysics, Graduate School of Science, Tohoku University, Aramaki-aoba, Sendai 980-8578, Japan

ABSTRACT

To observe rapid spatial and temporal structures of sprites and elves, we developed high-speed array photometers using multianode photomultipliers and operated these photometers and image intensified CCD cameras at Yucca Ridge Field Station (40.7°N, 104.9°W), Colorado during the SPRITES campaigns in the summer seasons from 1996 to 1999. We also operated these instruments at Dodaira Observatory (36.0°N, 139.2°E) and Sendai (38.3°N, 140.9°E) in Japan in December 1998 and January 1999, and succeeded for the first time in observing sprites and elves above the coast of the Sea of Japan near the Hokuriku region in the winter season. The obtained space-time structures of sprites and elves are consistent with those predicted from the quasi-electrostatic (QE) model and the electromagnetic pulse (EMP) model, respectively.

NTRODUCTION

Sprites are transient luminous events which occur at altitudes typically from 50 to 90 km and usually in association with intense positive cloud-to-ground lightning (Sentman et al., 1995). The brightest part of a sprite called "head" consists of a cluster of luminous columns and beneath the head faint "tendrils" extend downward to altitudes of ~50 km, changing color from red at the head to blue at their lowest extremity. Recently Barrington-Leigh et al. (2001) have presented clear evidence on the existence of a diffuse "halo" capping a cluster of columns. The quasi-electrostatic (QE) model and the theoretical analysis of electrical breakdown properties at different altitudes expect these structures (e.g., Pasko et al., 1997, 1998). On the other hand, elves are ring-shaped diffuse glows expanding at altitudes of ~90 - 100 km (Fukunishi et al., 1996; Inan et al., 1996, 1997). These features are explained by the electromagnetic (EMP) model in which elves are induced by electromagnetic pulses launched by impulsive return stroke currents (Inan et al., 1996; Veronis et al., 1999; Uchida, 1999).

Sprites and elves are transient glows with durations of a few ms to tens of ms and durations of less than 1 ms, respectively. Therefore, standard-speed low-light video cameras with a time resolution of 16.7 ms can not detect their transient spatial and temporal structures. While photometers have much higher time resolution, standard photometers could not measure spatial structures of optical emissions. To observe both rapid spatial and temporal structures, we have developed a new photometer called a multi-anode array photometer (MAP).

INSTRUMENTATION

The characteristics of the developed photometer are summarized in Table 1. This photometer consists of a camera lens, a linear multi-anode photomultiplier tube with 16 photocathodes and 16 amplifiers. Thus, the MAP has 16 fields-of-view arrayed vertically as shown in Figure 1. We used two array photometers with red and blue filters, respectively to measure the N_2 first positive and second positive emissions separately to estimate the energy of electrons exciting optical emissions from the ratio of $N_2 1P$ to $N_2 2P$. Analogue output signals from two

Table 1. Characteristics of developed multi-anode array photometers.

Photomultiplier tube	Hamamatsu R5900U-01-L16
Number of fields-of-view	16 (vertical)
Individual field-of-view	0.67° (vertical) x 10.6° (horizontal)
Total field-of-view	10.7° (vertical) x 10.6° (horizontal)
Red filter	560 - 800 nm ($N_2$1P measurement)
Blue filter	380 - 500 nm ($N_2$2P measurement)
A/D converter	12 bits, 32 ch.
Sampling rate	20 kHz (50 μs)
Data length of each event	800 ms (400 ms before triggering)

array photometers are converted into digital signals by a 32-channel A/D converter and are recorded on a hard disk. Each transient luminous event is selected by a trigger pulse from a trigger photometer with a wider field of view (~8°x14°). The trigger level is changeable by a command from a PC. In order to obtain photometer data over a 400-ms interval before the trigger onset, we use a pre-trigger mode of the A/D converter with a FIFO memory.

OBSERVATIONS

We operated array photometers and image intensified CCD cameras at Yucca Ridge Field Station (YRFS, 40.7°N, 104.9°W), Colorado during the SPRITES campaigns which were carried out every summer from 1996 to 1999. All instruments were installed on a single mount to point to the same direction as shown in Figure 2. Using real-time information on cloud-to-ground lightning discharges provided by the National Lightning Detection Network (NLDN), we pointed all instruments to the locations of intense positive CGs. The geometry for observing lightning-induced luminous events above the U.S. High Plains is shown in Figure 3. We also operated these instruments at Dodaira (36.0°N, 139.2°E) and Sendai (38.3°N, 140.9°E) in Japan in December 1998 and January 1999, and succeeded for the first time in observing sprites and elves above the coast of the Sea of Japan near the Hokuriku region in the winter season.

SPATIAL AND TEMPORAL STRUCTURES OF SPRITES

Optical observations with array photometers and CCD cameras demonstrated that sprites can be classified into two categories, columniform sprites and carrot sprites, which exhibit different spatial and temporal structures. The array photometer data of a typical columniform sprite observed at YRFS, Colorado at 06:02:24 UT on July 24, 1996 are shown in Figure 4. In this figure the wave form of the causative sferics is shown in the bottom panel, while signals from the channels 6 to 14 of the array photometer are shown in the upper panel as a function of time in milliseconds from the onset of the causative sferics. The altitudes of the individual channels

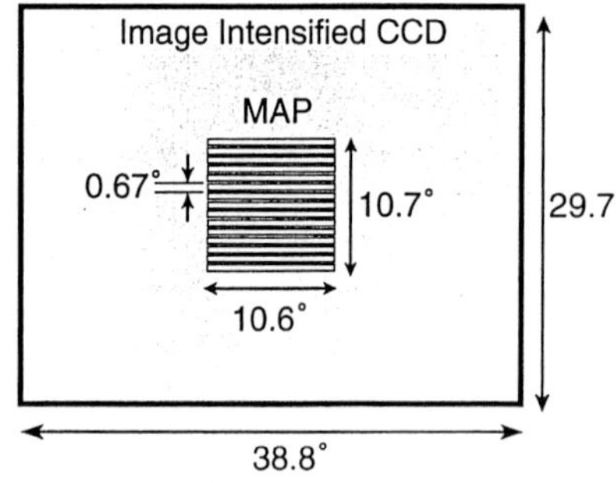

Fig. 1. Fields-of-view of the array photometer and the image intensified CCD camera.

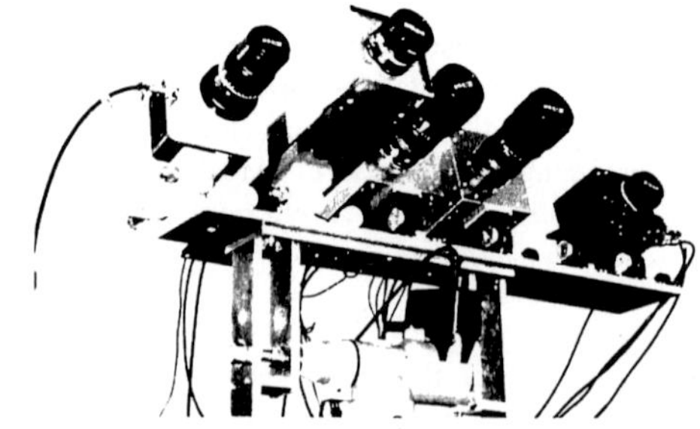

Fig. 2. View of the optical system used for observing sprites and elves.

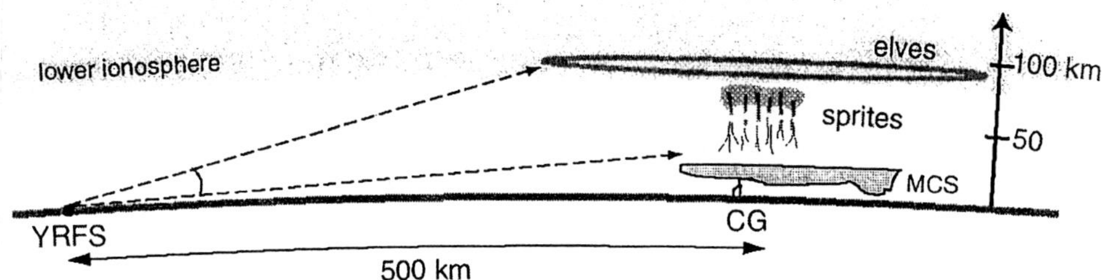

Fig. 3. Geometry for observing lightning-induced luminous events above the U.S. High Plains.

given on the right side of Figure 4 are estimated using the distance from observation site to the location of the causative sferics and the elevation angle of each channel field-of-view. The initial intensity enhancement starts at the top altitude around 90 km just after the onset of sferics (see channels 11 - 14). This intensity enhancement is due to the onset of an elve with an apparent downward motion as indicated by arrow (1). Following the onset of elve, a large luminosity enhancement characterized by downward development is seen in channels 8 - 10 as indicated by arrow (2). This enhancement is due to sprite emission from the head part of sprite. Then emission from the tendril is identified in channels 6 and 7 as indicated by arrow (3). The luminosity of the head part is found to be enhanced again ~1 ms later as seen in channels 8 and 9.

The array photometer data of a typical carrot sprite observed at YRFS at 04:22:20 UT on July 24, 1996 is shown in Figure 5 in the same format as Figure 4. Note that VLF data plot given in the bottom panel indicates no evidence on occurrence of sferics within the displayed time interval. The causative sferics occurred 87 ms before the onset of this sprite event (see the time delay from the onset of the causative sferics displayed at the bottom). This event started with a downward development in the tendril portion as shown by arrow (1) in channels 6 and 7. Then a large luminosity enhancement occurred in the head portion with an upward development as shown by arrow (2). Further, the second upward enhancement occurred with a time delay of ~0.6 ms in the head portion as shown by arrow (3).

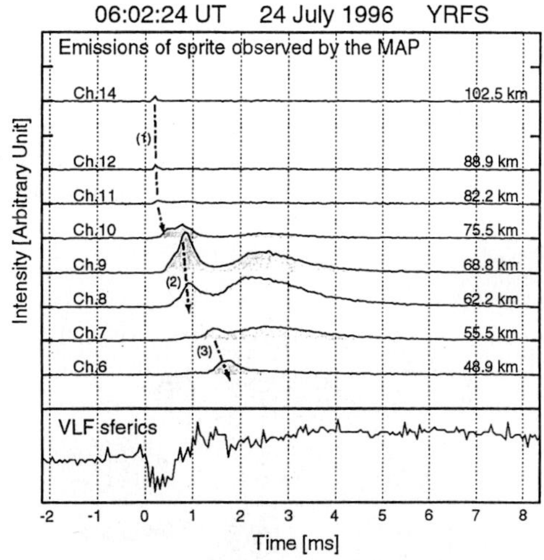

Fig. 4. Array photometer data plot for a typical columniform sprite observed at YRFS, Colorado.

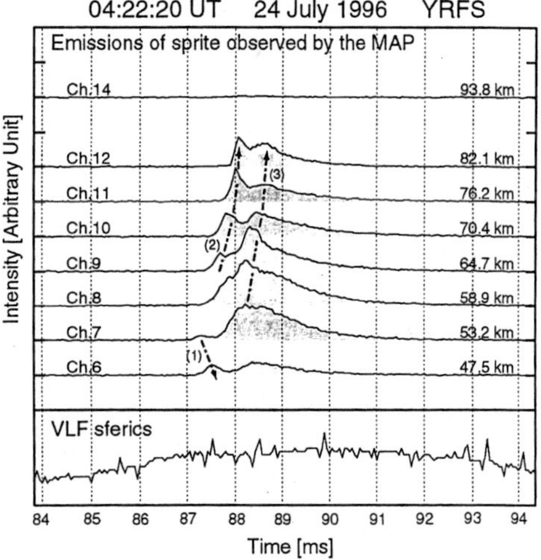

Fig. 5. Array photometer data plot for a typical carrot sprite observed at YRFS, Colorado.

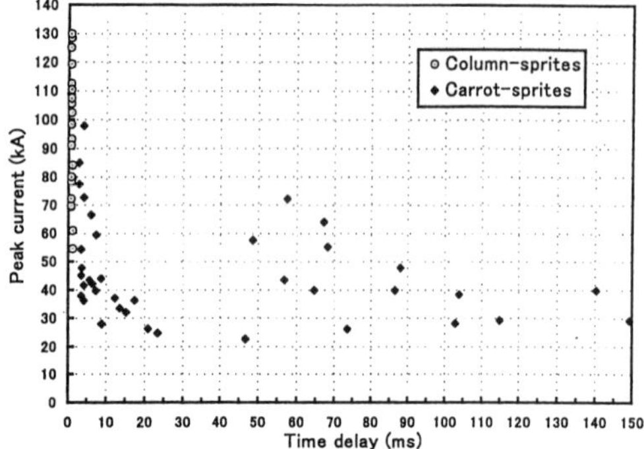

Fig. 6a. Relationship between the peak current of causative lightning discharge and the time delay from the onset of sferics to the onset of sprites.

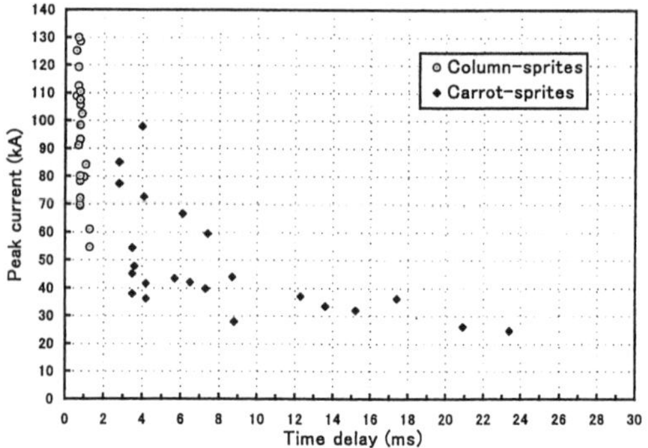

Fig. 6b. Extended display of Fig. 6a for the time delay from 0 to 30 ms.

As demonstrated by two examples of typical columniform and carrot sprites in Figures 4 and 5, respectively, the time delay from the onset of causative sferics to the onset of sprites depends on the type of sprites. The time delay of the columniform sprites is very short (~1 ms), while that of the carrot sprites is long with large scattering from 2 to 150 ms. Another important difference between the columniform sprites and the carrot sprites is found in the peak current of causative cloud-to-ground lightning discharges (CGs). The peak current of CGs for the columniform sprites is larger than that for the carrot sprites. These characteristics are clearly demonstrated in Figures 6a and 6b. Note that columniform sprites occur within ~1 ms after the onset of sferics. On the other hand, carrot sprites are found to be classified into two groups, i.e., sprites with time delays of ~0 - 30 ms and ~40 - 150 ms, respectively. Further, there is a tendency that the peak current becomes smaller as the time delay becomes longer for the sprites with time delays less than 30 ms.

SPATIAL AND TEMPORAL STRUCTURES OF ELVES

Similar to the typical sprite event shown in Figure 4, columniform sprites always accompany preceding elves. However, elves often occur without following sprites. An example of an intense elve event observed at YRFS, Colorado at 05:49:46 UT on August 19, 1999 is shown in Figure 7. Here the field-of-view of the array photometer is superimposed on the image of a ring-shaped elve. The distance from YRFS to the location of the causative CG lightning is 517 km. This event was observed by two array photometers with their sensitivity to red (560 - 880 nm) and blue (380 - 500 nm) emissions, respectively. The obtained array photometer data of this event are displayed in Figure 8, in which the intensity of red emissions and the ratio of blue to red emissions are represented by solid lines and dotted lines, respectively. It is apparent that the blue/red ratio is very high at the initial phase of the elve. An apparent downward motion seen in the top to bottom channels is due to a lateral expansion of the elve emission ring. From the blue/red ratio, we can estimate the energy of electrons inducing elve optical emissions. Here we assumed that electrons are monoenergetic, and that red emissions are due to N_2 first positive band and N_2^+ Meinel band, while blue emissions are due to N_2 second positive band and N_2^+ first negative band. The result is shown in Figure 9. Estimated peak energy of electrons is ~20 - 40 eV with its maximum at a distance ~100 km apart from the location of the causative CG discharge.

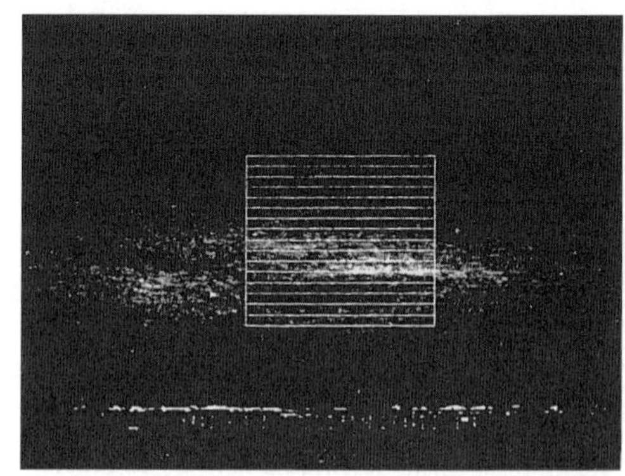

Fig. 7. Example of an elve observed at YRFS at 05:59:46 UT on August 19, 1999.

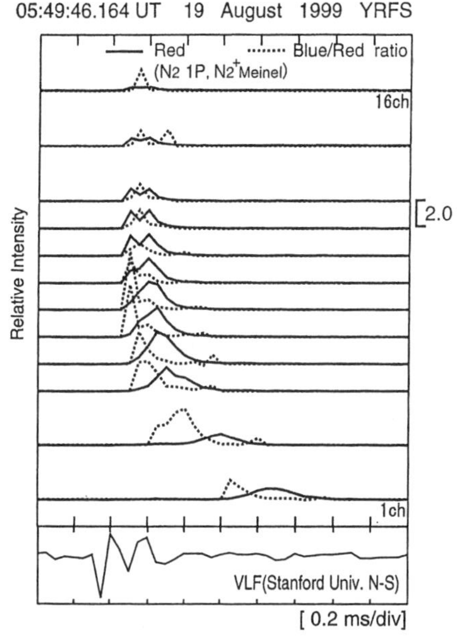

Fig. 8. Array photometer data plot for a typical elve shown in Fig. 7.

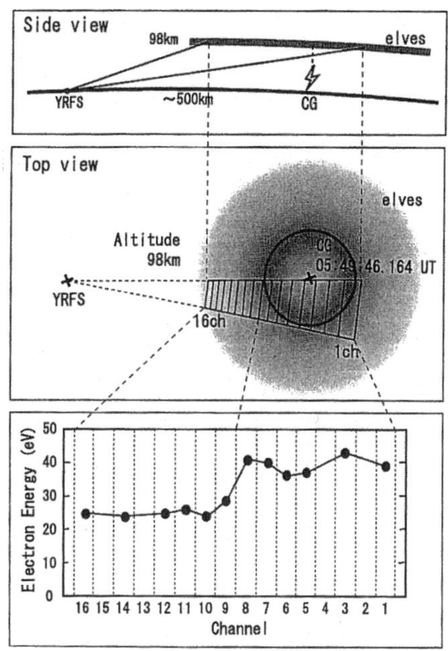

Fig. 9. Field-of-view of the array photometer in the vertical plane (top) and the horizontal plane (middle) and the estimated electron energy (bottom).

SPRITES AND ELVES OBSERVED IN JAPAN

Sprites have been observed during the summer months in North America, Africa (Boeck et. al. 1995), South and Central America (Heavner et al., 1995), and Australia (Dowden et al., 1997). Most of these sprites are induced by positive cloud-to-ground lightning discharges (CGs). From an observational fact presented by Suzuki (1992) that the occurrence probability of positive CGs is very high (~30 %) during the winter months in the Hokuriku region, Japan, we have planned to confirm whether sprites and elves occur associated with these positive CGs or not. We operated array photometers and image intensified CCD cameras at Dodaira Observatory (36.0°N, 139.2°E) and Tohoku University, Sendai (38.3°N, 140.9°E) in Japan in December 1998 and January 1999 and at Maebashi (36.4°N, 139.1°E) in January 2000. The locations of these observation sites are shown in Figure 10. We have succeeded in observing many sprites and elves during these campaigns. Sprites and elves occurred associated with the passage of a cold front aligned on the coast of the Sea of Japan as shown in Figure 10. A CCD camera image of a typical sprite event observed on January 27, 1999 is shown in Figure 11.

Fig. 10. Locations of observation sites superposed on a cloud map on January 27, 1999.

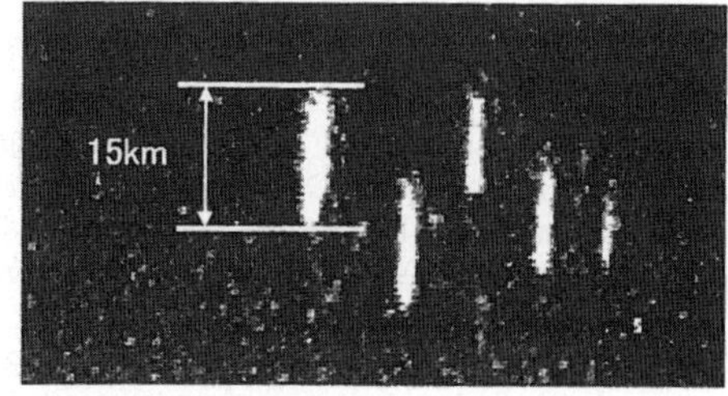

Fig. 11. Example of wintry sprites observed above Hokuriku region at 19:01 UT on January 27, 1999.

DISCUSSION AND SUMMARY

Array photometer observations demonstrated that sprites could be classified into two categories, columniform sprites and carrot sprites, which exhibit different space-time structures. Columniform sprites are always accompanied by preceding elves and are characterized by an initial downward development starting around 80-90 km altitude. On the other hand, carrot sprites are characterized by an initial upward development starting around 50 km altitude and rare occurrences of preceding elves. It was also found that the delay time from the onset of the causative sferics to the emission peak of sprites is very short (~1 ms) for columniform sprites, while it ranges from 2 to 150 ms for carrot sprites. These differences can be explained by a difference in the time constant of charge removal by cloud-to-ground discharges in the quasi-electrostatic (QE) model, as suggested by Pasko et al. (1996) and as discussed further by Watanabe (1998).

The image intensified CCD camera observed the ring-shaped images of elves, while the array photometer observed the time delays from high elevation channels toward low elevation channels. These characteristics are consistent with a lateral expansion of the emission ring produced by a lightning-induced electromagnetic pulse (EMP), as pointed out by Veronis et al. (1999). Two color observations using the two sets of the array photometers demonstrated that electrons producing elves have the peak energies of ~20 - 40 eV in the initial phase of elves and at a distance ~100 km apart from the center of the emission ring. These results suggest that the most effective heating is induced by electric field produced during the risetime of EMP.

ACKNOWLEGMENTS

We would like to acknowledge Drs. W. A. Lyons and T. E. Nelson and Prof. U. S. Inan for their support in the SPRITES campaign at YRFS, Colorado.

REFERENCES

Barrington-Leigh, C. P., U. S. Inan, and M. Stanley, Identification of sprites and elves with intensified video and broadband array photometry, J. Geophys. Res., 106, 1741-1750 (2001).

Boeck, W. L., O. H. Vaughan Jr., R. Blakeslee, B. Vonnegut, M. Brook, and J. M. McKune, Observations of lightning in the stratosphere, J. Geophys. Res, 100, 1465-1475 (1995).

Dowden, R., S. Hardman, J. Brundell, J. Bahr, Z. Kawasaki, and K. Nomura, Red sprites observed in Australia, IEEE Antennas Propag. Mag., 39, 106 (1997).

Fukunishi, H. Takahashi, K. Kubota, K. Sakanoi, U. S. Inan, and W. A. Lyons, Elves: lightning-induced transient luminous events in the lower ionosphere, Geophys. Res. Lett., 23, 2157-2160 (1996).

Heavner, M. J., D. L. Hampton, D. D. Sentman, and E. M. Wescott, Sprites over Central and South America, Eos Trans. AGU, 76(46), Fall. Meet. Suppl., F115 (1995).

Inan, U. S., W. A. Sampson, and Y. N. Taranenko, Space-time structure of optical flashes and ionization changes produced by lightning EMP, Geophys. Res. Lett., 23, 133-136 (1996).

Inan, U. S., C. P. Barrington-Leigh, S. Hansen, V. S. Glukhov, T. F. Bell, and R. Rairden, Rapid lateral expansion of optical luminosity in lightning-induced ionospheric flashes referred to as 'elves', Geophys. Res. Lett., 24,583-24,586 (1997).

Pasko, V. P., U. S. Inan, and T. F. Bell, Sprites as luminous columns of ionization produced by quasi-electrostatic thundercloud fields, Geophys. Res. Lett., 23, 649-652 (1996).

Pasko, V. P., U. S. Inan, T. F. Bell, and Y. N. Taranenko, Sprites produced by quasi-electrostatic heating and ionization in the lower ionosphere, J. Geophys. Res., 102, 4529-4561 (1997).

Pasko, V. P., U. S. Inan, and T. F. Bell, Spatial structure of sprites, Geophys. Res. Lett., 25, 2123-2126 (1998).

Sentman, D. D., E. M. Wescott, D. L. Osborne, D. L. Hampton, M. J. Heavner, Preliminary results from the Sprites94 aircraft campaign: Red sprites, Geophys. Res. Lett., 22, 1205-1208 (1995).

Suzuki, T., Long term observation of winter lightning on Japan Sea Coast, Res. Lett. Atmos. Electr. 12. 53-56 (1992).

Uchida, A., A study on space-time structures of elves and emission processes, Master's thesis in Japanese (1999).

Veronis, G., V. P. Pasko, and U. S. Inan, Characteristics of mesospheric optical emissions produced by lightning discharges, J. Geophys. Res, 104, 12,645-12,656 (1999).

Watanabe,Y., A study on space-time structures of sprites based on photometric observation, Master's thesis (1998).

OBSERVATION OF ANGEL SPRITES

H. T. Su[1], R. R. Hsu[1], Alfred B. Chen[1], S. F. Chen[1], S. B. Mende[2], R. L. Rairden[3], T. H. Allin[4], and T. Neubert[4]

[1]Physics Department, National Cheng Kung University, Tainan 70148, Taiwan
[2]Space Science Laboratory, University of California, Berkeley, CA 94720, USA
[3]Lockheed-Martin Research Laboratories, Palo Alto, CA 94304, USA
[4]Danish Meteorological Institute, 2100 Copenhagen, Denmark

ABSTRACT

Angel sprites are column-like sprites, which are capped with diffuse hair regions and trailed by tendrils. This paper presents some interesting properties of the angel sprites observed in the SPRITES'99 campaign from Kitt Peak, Arizona. From the observed images, it is noted that there are faint columniform glows preceding the angel sprites, but not for other types of sprites. For the angle sprites, the top of the leading glows and the top of the tendrils roughly match in altitude, but the leading glows extend to lower altitude below the tendrils. At its brightest stage, the sprite streamers are topped with diffused hair regions and trailed by branches of luminous beads and smaller, fainter and slant side branches. These luminous beads and the side branches are the main components of the tendrils. Some sprites occurred at the same location separated in time by hundreds of milliseconds to a few minutes. In these recurrent sprite events, the earlier sprites seem to have some influence on the dynamical behavior of the later sprites.

INTRODUCTION

Sprites are often described as an electric discharge or breakdown, which occurs at altitudes of 30-90 km above large thunderstorm systems. The first image of sprites was accidentally recorded using a low light level camera by Franz *et al* (1990) on the night of 22 September 1989. Since then, the sprites research groups around the globe have staged many ground and aircraft sprite campaigns and many important properties of sprites have been deduced (Boeck *et al.*, 1992; Sentman *et al.*, 1995; Mende *et al.*, 1995; Lyons *et al.*, 1996; Winckler *et al.*, 1996; Stanley *et al.*, 1999; Cummer and Stanley, 1999).

So far, a widely accepted theoretical model for producing sprites is the quasi-electrostatic field model proposed by Pasko *et al.* [1995, 1997]. According to this model, a large positive cloud-to-ground lightning discharge (+CG) induces a large electric field between the cloud top and the ionosphere. Sprites are produced by air breakdown generated by this field. The electric field accelerates electrons, which cause inelastic collisions with air molecules and initiate optical emissions when the electrons have acquired sufficient energy to excite molecules. Although most the sprites are associated with +CG, there are strong evidence that sprites can be induced by –CG (Barrington-Leigh and Inan, 1999).

Sprites have been categorized according to their morphological features. The most commonly reported sprites are the "carrot sprites" which have a intense, luminous center body, tapering towards lower altitudes while showing a diffuse "hair" region above the body. Near the bottom, the carrots are often accompanied by thin-hairline discharges or tendrils. Wescott *et al.* (1998) observed sprites consisting of thin relatively uniform vertical columns called the columniform or "C" sprites. After analyzing their data, they concluded that c-sprites follow positive flashes with currents ranging from about 23 to 100 kA. They also noted that there are spectral differences between c-sprites and other type of sprites. However, it is still not clear whether the morphological distinctions in the horizontal/vertical structures have any significant relevance to qualitative differences in the underlying electrodynamical processes. Most sprites are elongated in the vertical direction presumably because their discharges tend to occur along the predominantly vertical electric field.

As the spatial resolution of the sprite images becomes higher, finer structures of spites have been widely reported. Sentman *et al.* (1996) reported sprites, which have isolated patches, with scale sizes not exceeding a few km, resembling plasma beads and balls. These sprites occur wherever the local breakdown criterion is met for the complex three dimensional structure of nodes and anti-nodes of the radiation field created by the underlying lightning stroke. Winckler (1998) also reported the presence of beads in their sprite images. Wescott *et al.* (1998) showed the occurrence of beads in their observed sprite images. Very high resolution imaging of sprites have been carried out during the Sprites'98 campaign (Gerken *et al.*, 2000). The obtained images revealed the intricate complexity in the spatial structures of sprites.

GROUND OBSERVATION AND RESULTS

The data reported in this article were taken from Kitt Peak, Arizona in the summer of 1999. The instruments used on Kitt Peak were an intensified CCD and an intensified CID (Charge Injection Device) cameras. Both cameras have standard adapters to be equipped with 50 mm or 300 mm telephoto lens and are bore-sighted. When equipped with 50 mm lens the field-of-view is 24 degrees by 18 degrees, while with telephoto lens, the FOV is 4 degrees by 3 degrees. The CCD camera was originally developed for the Tether Optical Phenomena (TOP) experiment and was operated on two Shuttle missions STS-46 and STS-75 in 1992 and 1996, respectively. In this camera, a 25-mm diameter S-20 photo-cathode intensifier tube set on the focal plane of this camera was fiber-optically coupled to an interline transfer CCD. The instrument included a built-in video processor, which added annotation on each video frame, encoding time, gain setting and frame integration duration. For sprite observation both cameras were operated at standard video rates (30 frames per second). Data from a timecode generator were superimposed on both video signals which were separately recorded on Hi-8 video cassettes (Fig. 1).

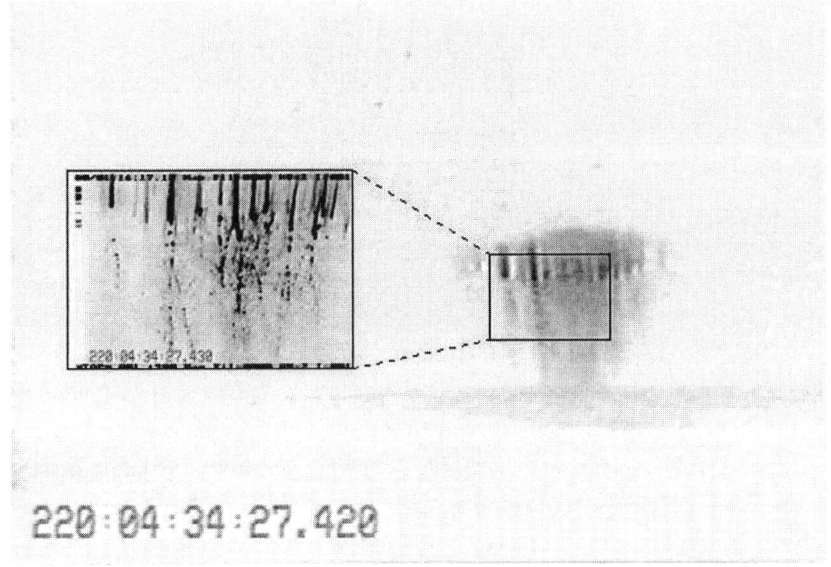

Fig. 1 Wide-angle view (right) and telephoto view (left) of a sprite event recorded at 04:34:27UT on 8 August 1999.

In the SPRITES'99 campaign, 5 angel sprites were recorded and Fig. 2 is a collage of 2 image frames for the event shown in Fig. 1. For the angel sprites we observed, the main streamers are all topped with diffused hair regions and trailed by branches of luminous beads and smaller, fainter and slant side branches. The time separations for the two image frames shown in Fig. 2 is 33 ms. In the upper image, the faint leading glows clearly can be discerned and their locations seem to be correlated closely with the brighter streamers seen in the lower image. To confirm whether these leading glows are real signals or instrumental artifacts, IDL routines were used to plot the histogram for each image. After filtered out the signals, which are weaker than the mean noise intensity plus three standard deviations, the leading glows still stand out. Furthermore, the leading glows are not all aligned vertically. Thus they cannot be the bleedings of the imaging system caused by the CCD saturating bright sprites.

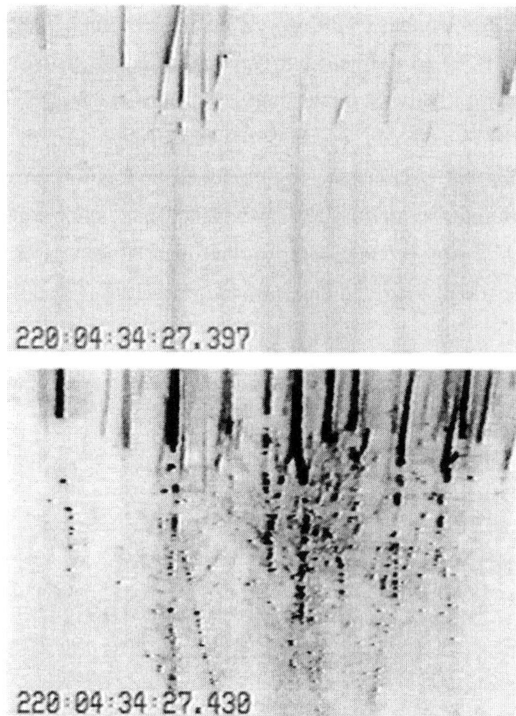

Fig. 2. A collage of two image frames for the event shown in Fig. 1. This sprite is classified as the angel sprite since the column-like streamers are topped with diffused hair region, and trailed by branches of luminous beads and smaller, fainter and slant side branches. The time separation between these two frames is 33ms. In the upper image, a few faint leading glows can be discerned and they seem to be correlated with the brighter streamers seen in the lower image.

Also in our observed sprite events, some groups of sprites re-emerged later at the same position within hundreds of milliseconds to a few minutes. For the sprites occur in the same area, the earlier sprites seem to have some influence on the dynamical behaviors of the later sprites (see Fig. 3).

DISCUSSION

From the SPRITES'99 data, all the angel sprites have leading columniform glows but not for c-sprites and carrot sprites. According to the NLDN data, the event shown in Fig. 1 has a peak current of 270 kA, which is much higher than the case of c-sprites reported by Wescott *et al* (1998). Also for another angel sprite event, which occurred at 09:22UT on 8 August1999, the NLDN peak current is even higher than 330 kA. Thus it is found that the lightning discharges leading to the angel sprites are more energetic than those of the c-sprites, which suggests that the leading glows might be a signaturs of the precursor discharges.

Also for the angel sprites, the leading glows always appear bellow the following main streamers and the top of the glows nearly corresponds to the lower ends of the streamers. Furthermore, the top of the leading glows corresponds to the top of the tendrils, but the glows extend below the tendrils.

At their brightest stage, there are diffuse hairs above the main streamers and branches of luminous beads and smaller, fainter and slant side branches below them. From the telephoto images, it can be concluded that the beads and the side branches are the components of the tendrils. We also noted that the distance between the bottom of the streamers and the branching points of beads is nearly half of the length of the main streamers. The reason for the angel sprites to show this kind of scaling behavior is still unknown.

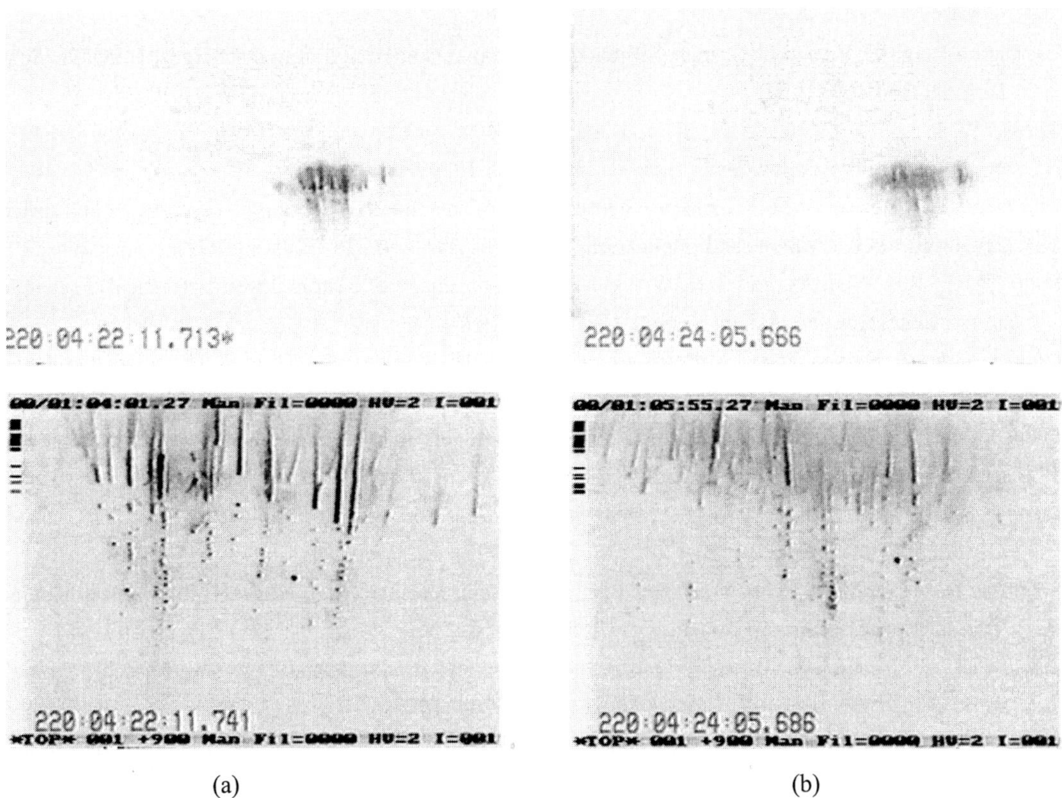

Fig. 3 Example of two angel sprite events with a time separation of 2 minutes. The second sprite appeared to be more diffused. Again, the top panels are wide-angle views and the bottom panels are telephoto views.

For the successive sprites, which occurred at the same location, the forms of the later sprites tended to be less well defined and the amount of halo between the main streamers became more prominent. It is likely that the electrical properties of the atmosphere in the vicinity of the sprites are altered by the earlier sprites, and they persist up to a few minutes. Thus for the subsequent sprites, their energy might be spread over a larger volume and produces greater amount of diffusive halo as seen in Fig. 3(b).

ACKNOWLEDGEMENT

We thank the logistic supports provided by the Kitt Peak National Observatory. H.T Su and R.R. Hsu would like to acknowledge the financial support of the NSPO, Taiwan under contract number NSC88-NSPO(B)-ISUAL-FA07-01.

REFERENCES

Barrington-Leigh C. P., and U. S. Inan, Sprites triggered by negative lightning discharges, *Geophys. Res. Lett.*, 26, 3605-08 (1999).

Boeck, W. L., O. H. Vaughan Jr., R. J. Blakeslee, B. Vonnegut, and M. Brook, Lightning induced brightening in the airglow layer, *Geophys. Res. Lett.*, 19, 99-102 (1992).

Cummer, S. A., and M. Stanley, Submillisecond resolution lightning currents and sprite development:observations and implications, *Geophys. Res. Lett.*, 26, 3205-8 (1999).

Franz, R. C., R. J. Nemzek, and J. R. Winckler, Television image of a large upward electrical discharge above a thunderstorm system, *Science* 249, 48-50 (1990).

Gerken, E. A., U. S. Inan, and C. P. Barrington-Leigh, Telescopic imaging of sprites, *Geophys. Res. Lett.*, 27, 2637–2640 (2000).

Lyons, W. A. Sprite observations above the U.S. High Plains in relation to their parent thunderstorm systems, *J. Geophys. Res.*, 101, 29641-52 (1996).

Mende, S. B., R. L. Rairden, G. R. Swenson, and W. A. Lyons, Sprite Spectra; N2 1 PG band identification, *Geophys. Res. Lett.*, 2 2633–2636 (1995).

Sentman, D. D., E. M. Wescott, D. L. Osborne, D. L. Hampton, and M. J. Heavner, Preliminary results from the Sprites94 aircraft campaign: 1. Red sprites, *Geophy., Res., Lett.*, 22, 1205-1208 (1995).

Stanley, M., P. Krehbiel, M. Brook, C. Moore, W. Rison, and B. Abrahams, High speed video of initial sprite development, *Geophys. Res. Lett.*, 26, 3201-3204 (1999).

Pasko, V. P., U. S. Inan, Y. N. Taranenko, and T. F. Bell, Heating, ionization and upward discharges in the mesosphere due to intense quasi-electrostatic thundercloud fields, *Geophys. Res. Lett.*, 22, .365-8 (1995).

Pasko, V. P., U. S. Inan, T. F. Bell, and Y. N. Taranenko, Sprites produced by quasi-electrostatic heating and ionization in the lower ionosphere, *J. Geophys. Res.*, 102, 4529-61 (1997).

Wescott, E. M., D. D. Sentman, M. J. Heavner, D.L. Hampton, W.A. Lyons, and T. Nelson, Observations of Columniform sprites, *J. Atmos. Terr. Phys.*, 60, 737-740 (1998).

Winckler, J. R., W. A. Lyons, T. E. Nelson, and R. J. Nemzek, New high-resolution studies of sprites, *J. Geophys. Res.*, 101, 6997-7004 (1996).

The Atmospheric Correction Algorithm of ROCSAT-1/OCI Data

Shih-Jen Huang[1,3],Gin-Rong Liu[1], Tang-Huang Lin[2], and Tsung-Hua Kuo[1,2]

[1] *Institute of Space Science, National Central University, Chung Li, Taiwan 320, ROC*
[2] *Center for Space and Remote Sensing Research, National Central University, Chung Li, Taiwan 320, ROC*
[3] *Department of oceanography, National Taiwan Ocean University, Keelung, Taiwan 202, ROC*

ABSTRACT

In this study, the Normalized Difference Vegetation Index (NDVI) and the band ratio of the total radiance at channels 670nm and 865nm were used to determine the sea surface albedo. The air mass character parameter and aerosol optical depth were then assessed by a simulated process. The pixel-by-pixel aerosol scattering radiance and water-leaving radiance are the main goals to retrieve in this study. As the Ocean Color Imager (OCI) is similar to the Sea-viewing Wide Field-of-view Sensor (SeaWiFS), a set of images were acquired both by SeaWiFS and OCI at the same temporal and spatial parameters. Their respective models –SeaWiFS Data Analysis System (SeaDAS) and Ocean Color Imager TRANsmittance / radiance computation code (OCITRAN) were employed to retrieve the water-leaving radiance so we could compare and evaluate the accuracy of OCITRAN. The results showed a high correlation (R>0.76) between the two models, proving that the OCITRAN algorithm established by this study is adaptable.

INTRODUCTION

The observation of the ocean color can be used to assess the global chlorophyll concentration and distribution, which is one the most important factors in global changes. It is well known that the chlorophyll inside phytoplankton plays a key role in converting carbon dioxide to oxygen via photosynthesis processes. Both gases contribute greatly to the greenhouse effect, energy balance and ozone concentration maintenance. Therefore, efforts have been made in the past decades to convert the satellite observed water-leaving radiance to chlorophyll concentration and distribution assessments. Thus, new ocean color sensors and instruments have been manufactured in recent years.

The OCI mounted on the ROCSAT-1 satellite was launched in January 1999, and has six channels (443nm, 490nm, 512nm, 555nm, 670nm and 865nm), which like SeaWiFS onboard the SeaStar satellite operates in a similar electromagnetic spectrum. The ROCSAT-1 is a low-earth orbit (LEO) science satellite situated at an altitude of 600 km with an inclination of 35 degrees, and has an orbital period of 97 minutes. The OCI is a push-broom scanner owning a 700-kilometer field-of-view with 896 pixels in each line. The 64 pixels at the center has a spatial resolution of 400 meters, while the remaining pixels is at 800 meters. The primary mission of the OCI is to provide observational data in the analysis of the marine environment. Since the OCI's channels are located at the visible and near IR regions, its data are easily influenced by the atmosphere. To ensure the best accuracy in OCI image interpretations, a method in retrieving the water-leaving radiance from the OCI's total radiance is required.

Table 1. The parameters of OCI and SeaWiFS images

	OCI	SeaWiFS
Observation time	Jun. 8 1999 0909z	Jun. 8 1999 0910z
Coverage Area	Mozambique Strait	Southeastern Africa
Inclination	35°	98.25°
Altitude(km)	600	705
Period(min)	96.6	98.9
Spectral bands(nm)	 B1 443-453 B2 480-500 B3 500-520 B4 545-565 B5 660-680 B6 845-885	B1 402-422 B2 433-453 B3 480-500 B4 500-520 B5 545-565 B6 660-680 B7 745-785 B8 845-885
Nadir resolution(m^2)	800x800	1130x1130
Swath width(km)	702	2801
Total pixels per scan line	896	1285
Observation method	Scanner	Push broom
Tilt	No	-20° ~ 20° step 2°
Launch date	Jan. 1999	Aug. 1997

METHODOLOGY

In a clear sky, the total radiance observed by the ocean color sensors comes from aerosol and molecular scattering in the atmosphere, the solar radiance reflected by the sea surface, and the constituents (e.g. chlorophyll, plankton) of the sea. Because the OCI operates within the visible and near-infrared (NIR) channels, the atmospheric effect is ubiquitous, and tends to degenerate the information obtained from the ocean. According to previous study experiments analyzing the Coastal Zone Color Scanner (CZCS) and SeaWiFS data, the atmospheric effect is significant enough in making accurate interpretations difficult (Gordon and Wang, 1994).

Due to the similarity between OCI and SeaWiFS, (Note: the two instruments operate within different channels) we attempt to do devise a new method by utilizing some of the concepts used by SeaDAS, which is the standard operational package for the atmospheric correction of SeaWiFS. However, we have remodeled the concepts by adding many new ideas to construct a new atmospheric correction model (OCITRAN) for the OCI data. The OCI does not provide the 412nm and 765nm channels, as does SeaWiFS (see also Table 1) (Li et al., 1999). SeaDAS uses the 765nm and 865nm channels as an important reference to assess the atmospheric effect, hence its algorithm can not be directly employed to process the OCI data. Therefore, an atmospheric correction procedure was designed to process the OCI data in this study. First, the albedo at 865nm and the band ratio of 865nm and 555nm were used to mask the land and cloud-contaminated pixels, in order to pick out the clear sky pixels over the ocean (Saunders and Kriebel, 1988). In addition, a sun glint examination procedure (Cox and Munk, 1954a,b) was used, since the OCI does not provide the sun glint avoidance function. In this study, the pixels, which have probability values larger than 1.5, were excluded (McClain and Yeh, 1994).

After filtering out the sun glint contaminated pixels and retaining only the clear sky pixels, the Rayleigh radiance was calculated by the multi-scattering algorithm in a pixel-by-pixel basis. To calculate the aerosol scattering radiance, the NDVI and band ratio algorithms were both employed in this study to estimate the sea surface albedo at 670nm and 865nm in advance (Figure 1). First, a Lowtran-simulated database of the satellite-observed radiance for different sea surface albedo at these two bands were generated (Kneizys et al., 1988). With this information in hand, the relation between the albedo and

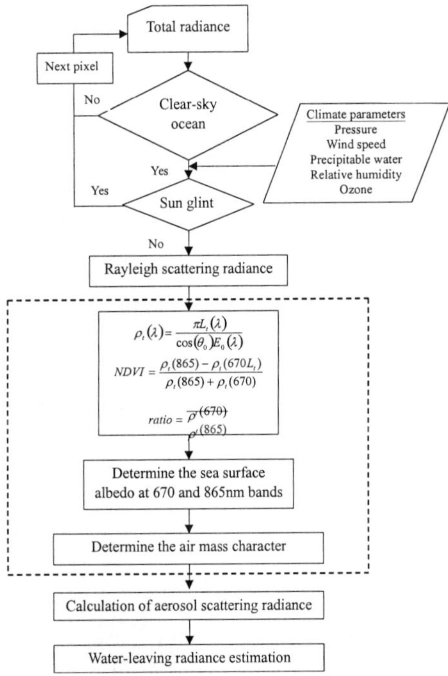

Fig. 1. The flowchart of the atmospheric correction algorithm for the OCI data. The steps in the dashed rectangle is the algorithm to determine the air mass character.

radiance of the two bands can be established by using either the NDVI or band ratio values. Multiple pairs (each pair consists of one assessed by the NDVI method, and the other by the band ratio method) of the albedo existed under different atmospheric and sea surface conditions, and the pair which had the closest albedo values were chosen and averaged to determine its mean. Afterwards, the air mass character, which is used for representing the marine air quality (Gathman, 1983), and the aerosol optical depth, could also be determined. The aerosol scattering radiance can be further calculated with the Angstron relationship. Once the Rayleigh and aerosol scattering radiance are obtained, the water-leaving radiance is finally retrieved by our OCITRAN procedure.

DATA

One set of SeaWiFS and OCI data covering the Mozambique Strait area were employed to evaluate OCITRAN's accuracy. Both data were observed on June 8 1999 at 0910z and 0909z, respectively (see also Table 1). The tested area was within 35°E ~ 42°E and 14°S ~ 24°S. The investigation was to compare the retrieved water-leaving radiance with the two images, which were processed by the SeaDAS and OCITRAN models respectively. The processed results are discussed in the following section. Because the two images were acquired at almost the same time, the retrieved radiance can provide objective evidence to evaluate OCITRAN's practicability.

DISCUSSION

The comparison between the two retrievals shows generally a higher water-leaving radiance appearing around the seashore zone. The ocean area southeast of Mozambique had a lower value than the adjacent

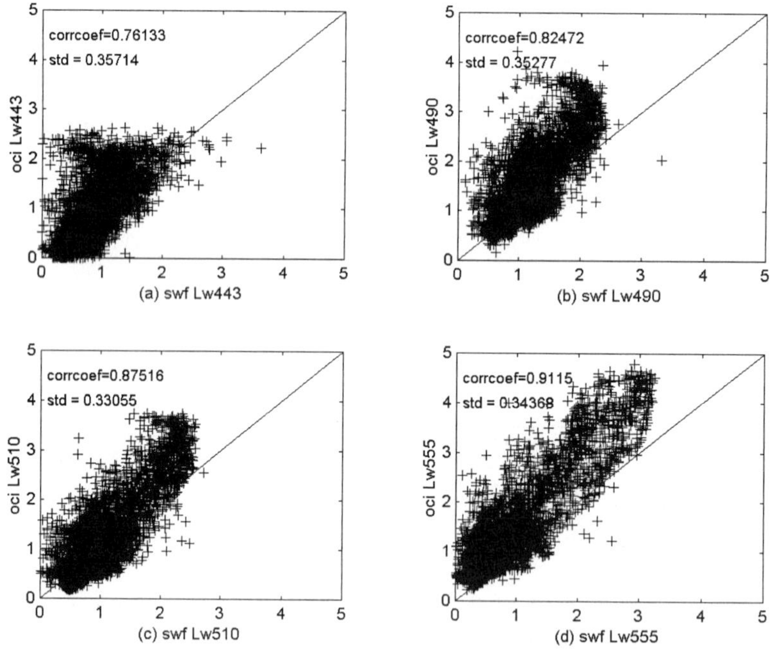

Fig. 2. Comparison of water-leaving radiance derived from the OCITRAN(OCI) and SeaDAS(SeaWiFS) models. Units are in $mw/cm^2/\mu m/sr$.

ocean areas, which was witnessed throughout all the bands, especially in the shorter wavelength bands. The analysis also showed that the area contained a higher chlorophyll concentration.

In Figure 2 a comparison of the retrieved water-leaving radiance of SeaDAS (SeaWiFS) and OCITRAN (OCI) is shown. The retrieved values vary from 0.01~3.62 *mw/cm²/μm/sr* at 443nm, and the correlation coefficient and standard derivation are 0.76 and 0.36, respectively. At 490nm, the values are 0. 11~3. 72 *mw/cm²/μm/sr*, and the correlation and standard derivation are 0. 82 and 0. 35, respectively. At 510nm and 555nm, the values are 0.02 – 3.73 and 0.02 – 4.78 *mw/cm²/μm/sr*, the correlation at 0.88 and 0.91, and the standard deviation at 0.33 and 0.34. Generally, the OCI-retrieved radiance is slightly higher than the SeaWiFS-derived radiance. Previous studies showed that this bias disappeared when the same SeaWiFS images were processed by the SeaDAS and OCITRAN models (Liu et al., 1999; Liu and Huang, 2000). This inconsistency requires further investigations in the future. Despite the downside, the rest of the results revealed a very good consistency between the SeaDAS(SeaWiFS) and OCITRAN(OCI) retrievals.

CONCLUSION REMARKS

Although the OCI does not provide the 765nm channel, which is very important in the SeaDAS procedure, the atmospheric correction can still perform well when the NDVI and band ratio algorithm is adopted in this study. The water-leaving radiance retrieved from OCITRAN(OCI) agrees quite well with the SeaDAS(SeaWiFS) retrieval where correlation coefficient was at least 0.76. Notwithstanding, comparison with field measurement data is expected in further studies in improving the OCITRAN's accuracy. Meanwhile, a bio-optic procedure in delineating a more accurate chlorophyll concentration pattern is encouraged to extend the OCI data applications.

ACKNOWLEDGMENTS

The OCI data used in this study were provided by the ROCSAT-1 Science Data Distribution Center of Ocean Color Imager, supported by the National Space Program Office (NSPO) of the National Science Council (NSC), and operated by the National Taiwan Ocean University. The research is supported by the NSPO under the NSC of the ROC under grant NSC89-NSPO-A-OCI-019-01-02. In addition, we would like to extend our special thanks to Mr. Chih-kang (Charlie) Liang for his careful proofreading in improving the English writing.

REFERENCES

Cox, C. and W. Munk, Measurement of the Roughness of the Sea Surface from Photographs of the Sun's Glitter, *J. of Opt. Soc. Am.*, 44, 838-850 (1954a).

Cox, C. and W. Munk, Statistics of the Sea Surface Derived from Sun Glitter, *J. of Mar. Res.*, 13,198-277 (1954b).

Gathman, S.G., Optical Properties of the Marine Aerosol as Predicted by the Navy Model, *Opt. Eng.*, 22, 57-62 (1983).

Gordon H. R. and M. Wang, Retrieval of Water-leaving Radiance and Aerosol Optical Thickness over the Oceans with SeaWiFS: a Preliminary Algorithm, *Appl. Opt.*, 33, 3, 443-452 (1994).

Kneizys, F. X., E.P. Shettle, L.W. Abreu, G.P. Anderson, J.H. Chetwynd, W.O. Gallery, J.E.A. Selby, and S.A. Clough, User guide to LOWTRAN 7, *AFGL-TR 880177 Environ. Res. Papers*, 1010 (1988).

Li, H.W., C.R. Ho, N.J. Kuo, C.T. Chen, T.S. Fang, W.P. Tsai, S.M. Lin, W.C. Su, L.H. Leu, W.K. Hu, and

H.A. Chen, The Scientific Project for ROCSAT-1/OCI, *TAO, Supplementary issue,* 85-98 (1999).

Liu, G.R., S.J. Huang, and T.H. Kuo, The Atmospheric Correction of ROCSAT-1 OCI Imagery—part II: OCITRAN-2", *TAO*, 10, 4, 886-872 (1999).

Liu, G.R. and S.J. Huang, The Atmospheric Correction of ROCSAT-1 OCI Imagery using near infrared radiance, *Proceedings-Ocean Conference on Weather Analysis and Forecasting 2000, Taipei, 10-12 July 2000,* 121-124 (2000).

McClain, C. R. and E. N. Yeh, Sun glint flag sensitivity study, *NASA Tech. Memo. 104566(13),* S.B. Hooker and E.R. Firestone Eds., NASA Goddard Space Flight Center, Greenbelt, Maryland, 51pp (1994).

Saunders, R.W. and K.T. Kriebel, An improved method for detecting clear sky and radiances from AVHRR data, *Int. J. Remote Sensing,* 9, 1, 123-15 (1988).

FIRST MEASUREMENT OF SCINTILLATION AND ATTENUATION OF 19.5 GHZ BEACON SIGNAL FOR EXPERIMENTAL COMMUNICATION PAYLOAD OF ROCSAT-1

Yen-Hsyang Chu[1], Shun-Peng Shih[1] and Ging-Shing Liu[2]

[1] *Institute of Space Science, National Central University, Chung-Li, Taiwan, R.O.C.*
[2] *National Space Program Office, Hsin-Chu, Taiwan, R.O.C.*

ABSTRACT

The ROCSAT-1 has been launched successfully on January 27, 1999. With employing transportable and fixed ground terminals to receive 19.5 GHz beacon signal of the experimental communication payload (ECP), the Ka band propagations experiment were carried out to study the atmospheric effects, including rain particles, clouds, water vapor content, melting hydrometeors, atmospheric refractivity fluctuations, and so on, on the satellite signal at Ka band over Taiwan area. Preliminary measurement shows that even during precipitation-free environment the variation range of 19.5 GHz beacon signal amplitude can be as large as 18 dB. Data analysis indicates that strong scintillations with peak-to-peak amplitude fluctuations greater than 5 dB are often observed under the cloudy condition. In addition, the frequency of amplitude fluctuation is higher at the low elevation angle than that at large elevation angle. A comparison between amplitude variation of beacon signal and sky noise temperature at 19.5 GHz measured by a ground-based radiometer is also made in this paper.

INTRODUCTION

It is generally recognized that the electromagnetic waves at the Ka band are susceptible of the weather-related precipitation, impairing the quality of the Earth-satellite communication. Except for the rain attenuation through the absorption and scattering effects, the absorption by oxygen molecules and water vapor, scattering and diffraction by the atmospheric refractivity irregularities, and abnormal refracting by the stratification of atmospheric structures play the important roles in the degradation of the earth-satellite communication link. In order to design an optimal earth-satellite communication link over Taiwan area, the environmental effects influencing the quality of satellite communication mentioned above should be investigated thoroughly. The first satellite of Republic of China, ROCSAT-1, was successfully launched on January 27, 1999. It is a low-earth orbit (LEO) experimental satellite, orbiting the Earth at an altitude of 600 km with an inclination of 35° and an orbital period of 97 minutes. The frequency of occurrence that the beam downlink telemetry signals of collected data to local receiving stations is about 6 times per day. One of three scientific payloads mounted on ROCSAT-1 is Experimental Communication Payload (ECP) designed for the communication and propagation experiments at Ka band. The frequency of the downlink beacon signal of ECP is at Ka band, more specifically, at 19.5 GHz. The major objectives of the Ka band propagation experiment of ROCSAT-1 are (1) to study the characteristics of rain attenuation at the frequency of Ka band, (2) to understand the behavior of scintillation caused by atmospheric refractivity fluctuations and rain over Taiwan area, (3) to carry out the site diversity experiment to measure the diversity gain for the compensation of rain attenuation effect, (4) to conduct the multi-path and low-elevation propagation using the beacon signal of the ROCSAT-1 satellite

With employing transportable and fixed ground terminals to receive the beacon signal of ECP, combined with a number of ground-based instruments, such as optical raingauge, radiometer, VHF radar, weather station, etc., the atmospheric effects on LEO satellite signal at Ka band are investigated, including effects of rain attenuation, atmospheric absorption, scintillation, depolarization, and so on. For more information on the details of ECP and the characteristics of the corresponding ground-based instruments, see Liu et al. (1999) and Shih and Chu (1999). In this paper, the preliminary results of the measurement of ECP beacon signal are presented, including scintillation and atmospheric attenuation of the beacon signal. A comparison between predicted and measured ranges of amplitude variation of the beacon signal is also made. It shows that measured ranges are generally greater than predicted ones by about 5 dB, indicating that atmospheric effect on the amplitude fluctuation of the signal at Ka band for a LEO satellite should be taken into consideration.

INSTRUMENTATION FOR Ka BAND PROPAGATION EXPERIMENTS

In order to conduct Ka band propagation experiment, a number of ground-based instruments are implemented. Three high resolution optical raingauges were implemented, respectively, at Chung-Li, Hsin-Chu, and Tai-nan for routine measurements of rainfall rates since 1997. Portable stand-alone radiometer at 19.5 GHz and automatic weather station were employed to measure background noise temperature and background weather condition at the experimental site. A disdrometer was set up in 1999 on the campus of National Central University at Chung-Li to measure the raindrop distribution. A VHF radar located on the campus of National Central University is used to measure related atmospheric parameters, including the profile of three-dimensional wind field, vertical distribution of precipitation aloft, and the rain height. The key ground-based instruments for Ka band propagation experiments are transportable and fixed ground terminal for receiving simultaneously beacon signal of ECP at 19.5 GHz. With employing these two terminals, a number of important experiments can be carried out, such as, rain attenuation experiment, scintillation experiment, site diversity experiment, depolarization experiment, and low-elevation angle experiment.

PRELIMINARY RESULTS OF MEASUREMENTS

As mentioned above, the main goal of Ka band propagation experiment is to estimate the atmospheric effects on LEO satellite signal at Ka band from the measurements of amplitude variations of beacon signal. Because ROCSAT-1 orbits the Earth at the altitude of about 630 km at the speed of around 7.2 km/s, the received amplitude variations of beacon signal will be influenced by the characteristics of transmission antenna on board, including antenna gain pattern, axial ratio, pointing direction of apex, and so on. Fig.1 shows the antenna gain pattern of transmission antenna of beacon signal of ECP. As indicated, the antenna pattern is very corrugated and rough. In order to remove the effect of fluctuations of transmission antenna pattern, a predicted amplitude variation of beacon signal due to satellite movement should be obtained before true atmospheric effects on the satellite signals are abstracted from the measured ones. With the

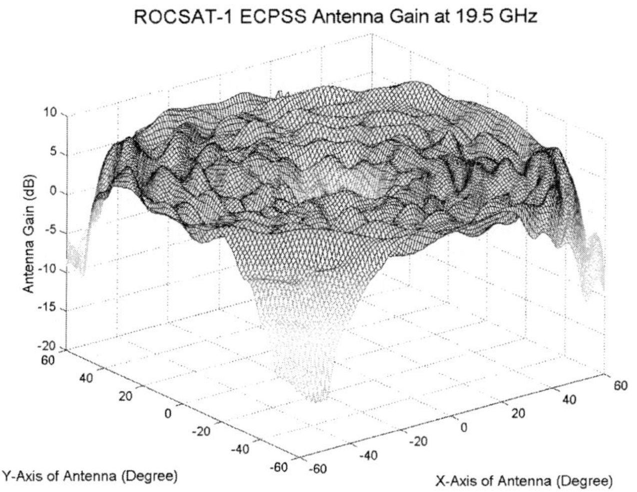

Figure 1 Transmission antenna pattern of beacon signal of ECP of ROCSAT-1.

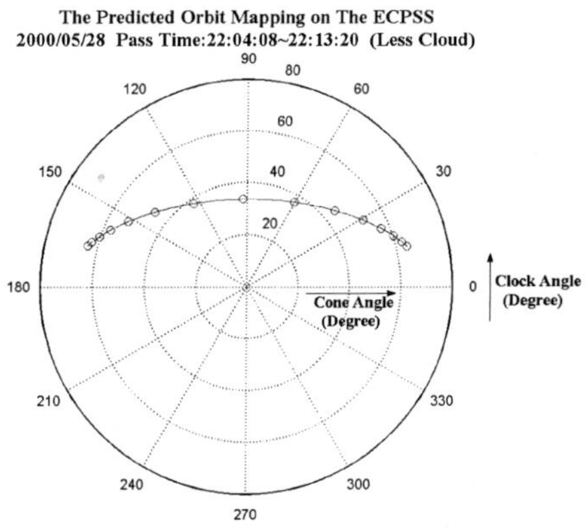

Figure 2 Projection of track of ground terminal on transmission antenna plane

help of orbital information of the satellite, the clock and cone angles of the satellite can be calculated, from which the projection of antenna apex of ground terminal on the transmission antenna plan on board can be calculated accordingly. The predicted power of beacon signal can thus be obtained if the system gain of ground terminal is well calibrated and the losses of free space and gaseous absorption are considered. Figure 2 shows an example of the trajectory of antenna apex of ground terminal projected on the plan of transmission antenna on board. After considering the free space loss, the atmospheric gaseous absorption, and system gain of ground terminal, the predicted amplitude variations of beacon signal can be obtained. The upper panels of Figs 3 and 4 present two cases of measured power variations (solid curves) of left-hand polarized beacon signal made by Transportable Ground Terminal (TGT) at Tai-nan, in which predicted beacon power variations with (upper dashed curves) and without (lower dashed curves) atmospheric loss are also shown. The atmospheric loss is converted from the brightness temperature observed by a stand-alone radiometer at 19.5 GHz that is operated simultaneously during the period of receiving beacon signal. The characteristics of the radiometer, see Shih and Chu (1999). The middle panels of Figs.3 and 4 display the zenith variation of apex of TGT antenna and the bottom panels indicate the status of contact between satellite and TGT. It is clear that basically the measurements are in good agreement with the predictions, implying that the characteristics of transmission antenna on board are not impaired during the launch of satellite. In addition, substantial variations of measured beacon power are seen in Figure 3 and 4, suggesting significant scintillations exist in the signals. Inspecting the beacon power variations shown in Figs.3 and 4 indicates that the peak-to-peak amplitude of the scintillation can be

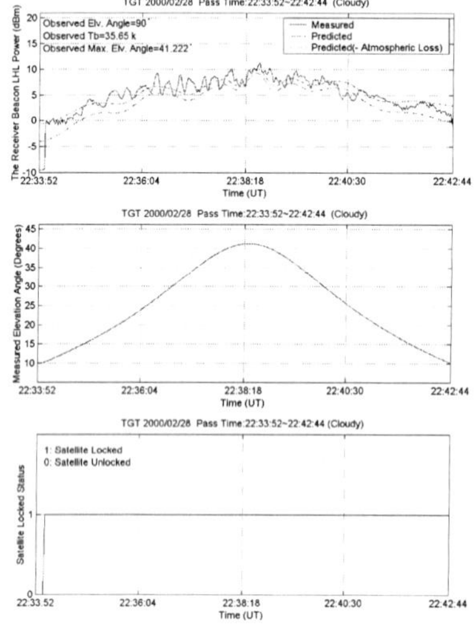

Figure 3 Example of measured power variations of beacon signal of ECP.

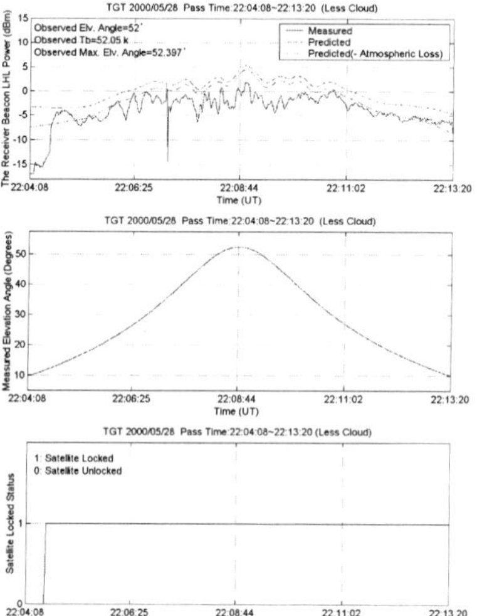

Figure 4 Same as Figure 3, but for another case.

as large as more than 6 dB. The frequency of the scintillation is generally between 0.1 and 1 Hz. Moreover, it appears that the oscillation frequency of the scintillation is elevation angle dependent. Namely, it is higher at large elevation angle and lower at small elevation angle, opposite to the results from the

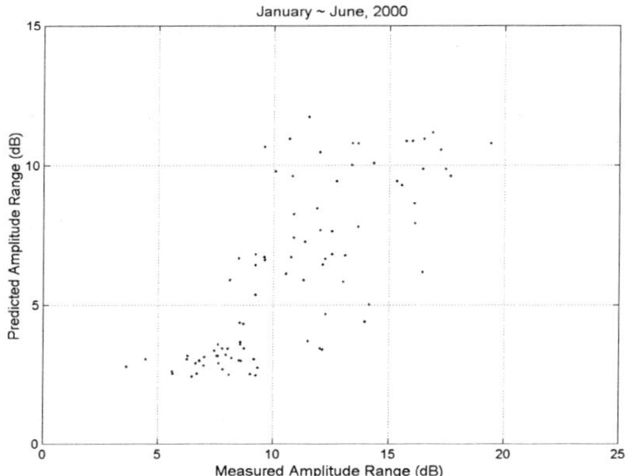

Figure 5. Scatter diagram of predicted versus measured amplitude levels of beacon signal

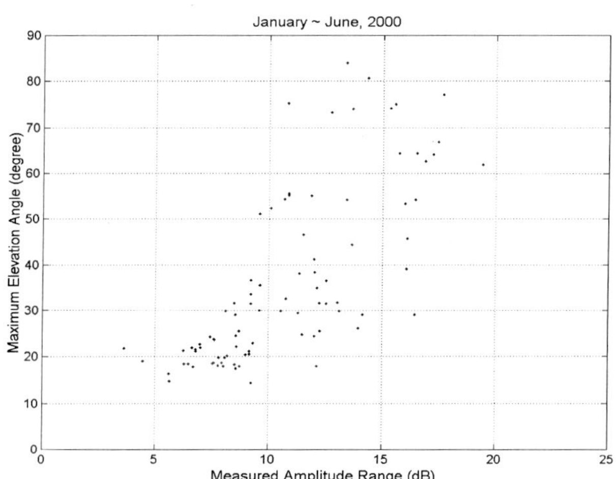

Figure 6 scatter diagram of measured amplitude range versus zenith angle

measurements of geostationary satellite. More data are required to ascertain this feature. Figure 5 show the comparison between predicted and measured amplitude levels of the beacon signal. Statistically, measured levels are greater than predicted ones by about 5 dB in average, showing that other propagation effects, including depolarization and cloud absorption, may play significant roles for the change in amplitude of beacon signal. Figure 6 is the scatter diagram of measured amplitude range versus maximum zenith angle of TGT antenna apex. It is clear that measured range of amplitude variation of beacon signal is proportional to the maximum zenith angle. This feature implies that the path length of satellite signal propagation will be a factor influencing the fading range of the satellite signal.

CONCLUSION

The preliminary results of Ka band propagation experiment conducted by 19.5 GHz beacon signal transmitted by ROCSAT-1 and transportable ground terminal (TGT) are presented in this report. It shows that basically measured beacon signal powers are in good agreement with the predicted ones, suggesting that the power pattern of on board beacon transmitting antenna is not impaired during the launch procedure of ROCSAT-1. Observations show that pronounced scintillations exist in the time series of beacon power variation. The peak-to-peak variation of the scintillation can be as large as more than 6 dB, and the oscillation frequency of the scintillation is generally between 0.1 and 1 Hz. In addition, the range of the 19.5 GHz beacon power variation is a function of elevation angle of ROCSAT-1 seen from the ground. The higher the elevation angle is, the larger the range will be. More measurements are required to explore the propagation effects due to atmosphere on Ka band LEO satellite signal.

ACKNOWLEDGEMENTS

This work was supported by National Space Program Office, Republic of China, under grants NSC88-NSPO-A-ECP-08-001 and NSC89-NSPO-A-ECP-008-01.

REFERENCES

Liu, G.S., S.J.Yu, J.K.Ho, J.Lin, and P.Chen, Experimental communication payload project of the ROCSAT-1 satellite, *TAO*, Supplementary Issue, 127-144 (1999)

Shih, S.P., and Y.H.Chu, Ka band propagation experiments of experimental communication payload (ECP) on ROCSAT-1 – Preliminary results, *TAO*, Supplementary Issue, 145-164 (1999)

A NEW METHOD OF RETRIEVING WATER VAPOR CONTENT USING GROUND-BASED RADIOMETER WITH SINGLE BAND

Shun-Peng Shih and Yen-Hsyang Chu

Institute of Space Science, National Central University, Chung-Li, Taiwan, R.O.C.

ABSTRACT

Using a single-frequency ground-based microwave radiometer with steerable horn antenna at frequency 19.5 GHz, we measure sky radiometric temperature over Taiwan area during the period from May 1997 to September 1999. A method of retrieving integrated precipitable water vapor content (IPWV) from the observed brightness temperature made by a single-frequency radiometer at 19.5 GHz is developed and examined. Comparing the retrieved and measured IPWVs shows a good agreement between them, validating the method proposed in this article.

INTRODUCTION

The microwave radiometer has been proved to be an effective instrument in the remote sensing of atmosphere. The applications of ground-based microwave radiometer to the measurements of meteorological parameters have been widely made in the community of atmospheric remote sensing for years (e.g., Janssen, 1993, and references therein). It is possible to establish a useful relation between the magnitude of the radiation intensity from a specific terrestrial or atmospheric object and corresponding parameter of interest. Once the relation is established, the desired parameter can thus be obtained from the measurements of microwave radiometer. Since the pioneered work of Westwater (1965), atmospheric temperature profiles were retrieved successfully by many scientific workers from the radiometry measurements of gaseous emission in the use of various inversion techniques (Snider, 1972; Westwater et al., 1975; Decker et al., 1978; Askne and Westwater, 1982). The precipitable water vapor and cloud liquid can also be retrieved from the observation of microwave radiometer operated at selected frequencies (Guiraud et al., 1979; Snider et al., 1980). The retrieved accuracy of precipitable water vapor made by microwave radiometer is believed to be the same as or even better than those estimated by radiosonde (Guiraud et al., 1979). In addition to the atmospheric parameters, atmospheric attenuation of radiowave due to absorptions of atmospheric gases, i.e., water vapor and oxygen molecule, can also be estimated from the observed brightness temperature made with a microwave radiometer. The atmospheric attenuation is an essential parameter in designing the earth-satellite communication link (Ippolito, 1986). Westwater et al. (1990) measured the atmospheric attenuations from the conversion of observed brightness temperatures at 20.6, 31.65, and 90 GHz and found that measurement and theory generally agree to within 30%.

A portable microwave radiometer with steerable horn antenna is employed for the measurement of brightness temperature in this study. This ground-based radiometer is a single linearly polarized, superheterodyne, and Dicke-switched instrument. The central frequency of the radiometer is 19.5 GHz with IF bandwidth of 500 MHz. The half-power beam width of the horn antenna is approximately 6 degrees and the sensitivity is 0.5 K for an integration time of 1 sec. The linear polarized radiation, either horizontal or vertical, is capable of being received through orienting the feed horn vertically or horizontally. The horizontal or vertical voltage signal from the radiometer is fed into the low pass filter/amplifier which determines the integration time of the radiometer. The low pass filter used for the measurements has a various cutoff frequency for the integration time between 4 and 1024 ms. The horizontal or vertical signal from the low pass filter is sent to a printer/recorder for viewing the data in real time and to a data collection board connected to the bus of the host personal computer that controls the data acquisition. The data collection board digitizes the radiometer output with 12 bits precision. The positions of the antenna in azimuth and elevation are recorded at each time when the radiometer signals are sampled.

In reality, the noise temperature measured by a radiometer contains not only true sky noise temperature, but also the system temperatures. Therefore, proper calibration of the radiometer is required to derive constants for use in a radiometer equation to estimate the absolute sky noise temperature. The 19.5 GHz radiometer employed in this experiment has a Dicke reference load and an avalanche noise diode that serves as an internal calibration source. In addition to the internal calibration, the radiometer is also calibrated using external loads before the deployment of field experiment. The hot and clod calibrations are, respectively, performed in this experiment by using 0°C (273 K) ice water and -196°C (77 K) liquid nitrogen. These two points permit an absolute calibration of the radiometer.

RETRIEVAL OF INTEGRATED PRECIPITABLE WATER VAPOR

According to the radiative transfer theory, the brightness temperature measured by a ground-based microwave radiometer is the integration of the radiation emitted by the absorptive gases in the antenna beam. Once the brightness temperatures are measured, the corresponding total atmospheric attenuation due to gaseous absorption can be converted directly.

It is well known that oxygen molecule and water vapor both give rise to the radiowave absorption, computation shows that in Taiwan area the contribution of water vapor to the absorption is much more significant than that of oxygen molecule by a factor of about 12, where the meteorological data employed for the calculation are taken from Pan-Chiao rawinsonde station over the period from 1st August 1994 to 30 November 1999. This result implies that oxygen absorption can be reasonably neglected in retrieving water vapor content from radiometer-observed brightness temperature.

Ideally, the radiometer-measured attenuation should be identical to the calculated one if the expression of specific attenuation coefficient is accurate enough and the total amount of water vapor content responsible for the measured and calculated attenuations is the same. Therefore, with the help of a relevant approximation equation connecting the gaseous absorption and water vapor concentration, it is possible to estimate the integrated precipitable water vapor content (IPWV) from the brightness temperature measured by a single-frequency ground-based radiometer. An approximation equation as a complicate function of temperature, pressure and water vapor concentration has been recommended by ITU to calculate the gaseous attenuation. However, the ITU-recommended equation is still so complicate that the use of it to retrieve water vapor content from the measurement of single-frequency radiometer is difficult and impractical. By appropriately adjusting the constants in the ITU-recommended approximation equation (ITU Recommendation, 1997), a very simple approximation expression at the frequency of 19.5 GHz is proposed as follows

$$\alpha_w = 0.00036 + 2.1432608\text{x}10^{-3}\rho_w \quad \text{(dB/km)} \tag{1}$$

where ρ_w is the water vapor concentration in unit of g/m^3. An examination of the accuracy of the approximation equation as shown in (1) is made and the results are presented in Fig.1, where 69 months radiosonde data from January 1, 1994, to September 30, 1999 are employed for the examination. As indicated, the attenuations A_a calculated from the approximation equation agree very well with those estimated from ITU-recommended equation with RMS of less than 0.0061. This good

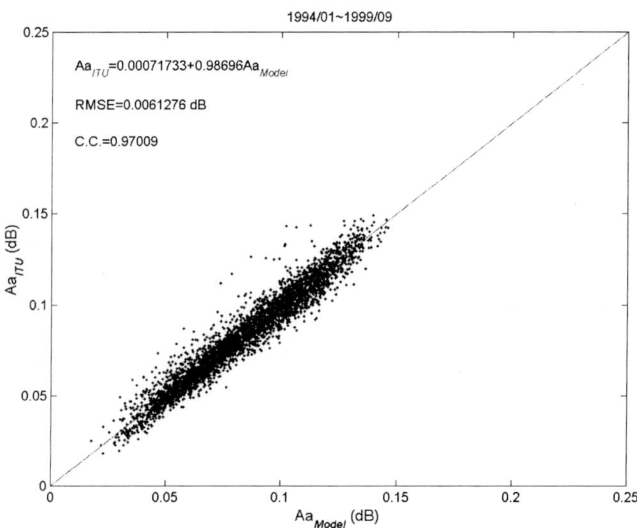

Fig. 1 Comparison between gaseous attenuations calculated by approximation equation shown in Eq.4 and ITU-recommended equation.

agreement validates the usefulness of Eq.1. Note that A_a is defined as

$$A_a = \int_0^{r_0} \alpha_w(z)dz \quad \text{(dB)} \tag{2}$$

where r_0 is the maximum height that the water vapor extends up to and can be detected by the radiosonde. Assume that the concentration of water vapor decreases with height exponentially in the form of

$$\rho_w = \rho_{ws} \exp(-\frac{z}{H_w}) \quad (g/m^3) \tag{3}$$

where ρ_{ws} and H_w are, respectively, surface concentration (g/m^3) and scale height (km) of water vapor concentration. Note that the integrated precipitable water vapor content (IPWV) h_v can be estimated in accordance with following equation

$$h_v = \frac{M_v}{10\rho_L} = \frac{\int_0^{r_0} \rho_w dz}{10\rho_L} \quad (cm) \tag{4}$$

where ρ_L is the density of liquid water (equal to approximately 1 g/cm^3), ρ_w is water vapor concentration (in unit of g/m^3), r_o (in unit km) is the highest altitude that radiosonde can reach, and M_v (g km/m^3) is the total mass of atmospheric water vapor contained in a vertical column of unit cross section. It is obvious that in practice the measured profile of ρ_w may not necessarily decrease with height exponentially. However, data analysis indicates that h_v estimated from real profile of ρ_w recorded by radiosonde is very close to that calculated by using exponential model, as shown in Fig.2. Therefore, it is justified to estimate h_v in terms of exponential decrease model.

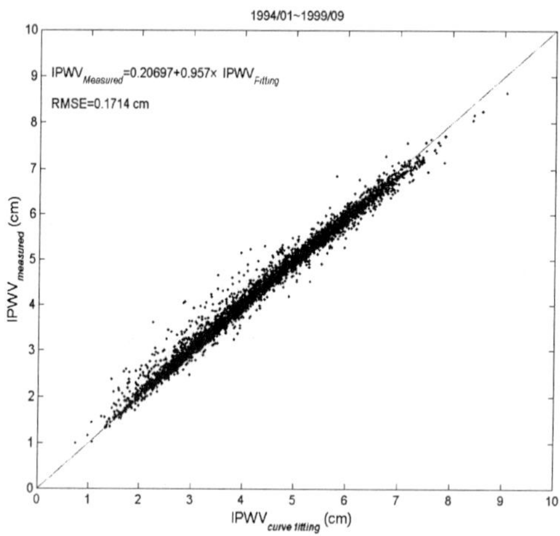

Fig. 2. Comparison between integrated prcipitable water vapor IPWV estimated from measured profile of water vapor density and calculated by best fitting exponential model to the measured profile

Substituting Eq.3 into Eqs.1 and 2, we have

$$A_a = Br_0 + C\rho_{ws} H_w (1 - e^{-r_0/H_w}) \quad (dB) \tag{5}$$

where B=0.00036 dB km^{-1} and C=2.1432608x10^{-3} dB·m^3·g^{-1} km^{-1}. From Eqs.4 and 5, the IPWV h_v can thus be derived as follow

$$h_v = \frac{A_a - Br_o}{10C\rho_L(1 - e^{-r_0/H_w})} \text{ (cm)} \tag{6}$$

Note that the mean values of r_o and H_w over Taiwan area are, respectively, 10 and 2.2 km, producing that Br_o=0.0036 dB and exp(-ro/Hw)~0.0106. In view of the fact that A_a is in general greater than Br_o by 1-2 orders of magnitude and 1-exp(-ro/Hw)~0.99, Br_o and exp(-ro/Hw) can be neglected reasonably in the use of Eq.6 to calculate IPWV from observed zenith opacity τ_o by letting τ_o be equal to A_a as defined in Eq.2. Fig.3 presents the comparison of IPWVs' estimated from radiosonde data and

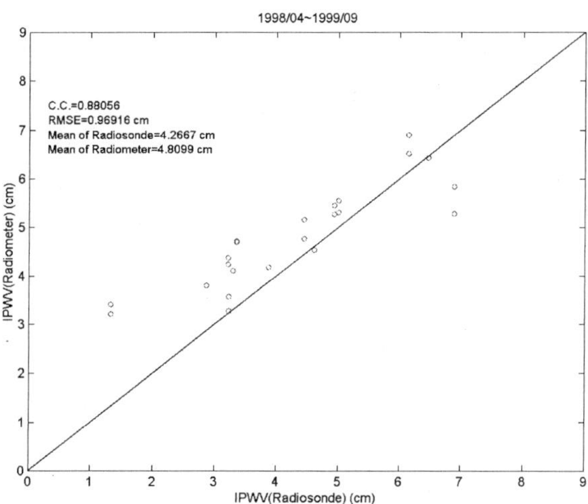

Fig. 3. Comparison of IPWVs estimated from radiosonde data and retrieved from zenith opacity measured by a 19.5 GHz radiometer.

retrieved from zenith gaseous attenuation observed by a 19.5 GHz radiometer, where the data employed for the comparison are collected in the clear-air condition to avoid the effects of both cloud and precipitation on observed brightness temperature. As shown, the retrieved IPWVs are in good agreement with the measured ones with RMS error of about 0.97 cm. Note that Pan-Chiao rawinsonde station locates about 25 km northeast from the Chung-Li. It is reasonable to speculate that water vapor contents over Pan-Chiao and Chung-Li will be different due to spatial inhomogeneity. We believe that this is the major cause for a few points with large scatter from the line with slope of 1. Therefore, the method of retrieving IPWV from the measurement of single-frequency radiometer proposed in this section is applicable.

CONCLUSION

With the help of ITU-recommended equation, a method of retrieving integrated precipitable water vapor content using zenith opacity measured by a single-frequency radiometer at 19.5 GHz is developed in this article. A comparison of retrieved IPWV and observed IPWV by radiosonde shows good agreement between them, validating the proposed method.

ACKNOWLEDGEMENT

This work was supported by the National Space Program Office of the Republic of China (R.O.C.) on Taiwan under grants NSC89-NSPO-A-ECP-008-01 and NSC89-2111-M-008-029-A10.

REFERENCES

Askne, J., and E.R.Westwater, A review of ground-based remote sensing of temperature and moisture by passive microwave radiometers, IEEE Trans. Geosci. Remote Sens., GE-24, 340-352 (1982)

Decker, M.T., E.R.Westwater, and F.O.Guiraud, Experimental evaluation of ground-based microwave radiometric sensing of atmospheric temperature and water vapor profiles, J. Appl. Meteorol., 17, 1788-1795 (1978)

Guiraud, F.O., J.Howard, and D.C.Hogg, A dual-channel microwave radiometer for measurement of precipitable water vapor and liquid, IEEE Tran. Geosci. Electron., GE-17, 129-136 (1979)

Ippolito Jr., L.J., Radiowave propagation in satellite communication, Van Nostrand Reinhold Company Inc., New York, pp.241 (1986)

ITU Recommendation ITU-R P.676-3 (1997)

Janssen, M., Atmospheric remote sensing by microwave radiometry, John Wiley & Sons, Inc., New York, pp.572 (1993)

Snider, J.B., Ground-based sensing of temperature profiles from angular and multi-spectral microwave emission measurements, *J. Appl. Meteorol.*, 11, 958-967 (1972)

Snider, J.B., F.O.Guiraud, and D.C.Hogg, Comparison of cloud liquid content measured by two independent ground-based systems, *J. Appl. Meteorol.*, 19, 577-579 (1980)

Van Vleck, J.H., and V.F.Weisskopf, On the shape of collision broadened lines, *Rev. Mod. Phys.*, 17, 227-236 (1945)

Westwater, E.R., Ground-based passive probing using the microwave spectrum of oxygen, *Radio Sci.*, 69D, 1201-1211 (1965)

Westwater, E.R., J.B.Snider, and A.V.Carlson, Experimental determination of temperature profiles by ground-based microwave radiometry, J. Appl. Meterolo., 14, 524-539 (1975)

Westwater, E.R., J.B.Snider, and M.J.Falls, Ground-based radiometric observations of atmospheric emission and attenuation at 20.6, 31.65, and 90.0 GHz: A comparison of measurements and theory, *IEEE Trans. Antennas Propagat.*, 38, 1569-1580 (1990)

COSMIC Research Program Session

MODELING, TRACKING AND INVERTING THE TROPOSPHERIC RADIO OCCULTATION SIGNALS

S. V. Sokolovskiy

University Corporation for Atmospheric Research, 3300 Mitchell Lane, Boulder, CO, 80301, USA

ABSTRACT

Radio occultation signals propagating through the moist lower troposphere are subjected to strong fluctuation due to the multipath. This causes problems for tracking such signals by means of routine phase-locked loop technique, as well as for their inversions into refractivity by means of routine calculation of the bending angles from Doppler. In this paper we discuss: 1) modeling of realistic radio occultation signals with the use of high resolution tropical radiosondes; 2) open-loop tracking technique capable of capturing the complicated structure of such signals; 3) radio holographic inversion technique capable of solving for the multipath propagation.

INTRODUCTION

The Global Positioning System Meteorology (GPS/MET) experiment demonstrated high potential of radio occultation (RO) technique in the Earth's atmosphere (Kursinski et al., 1997, Rocken et al., 1997). But, at the same time, this experiment revealed some problems of RO technique in the lower troposphere. The complicated and very sharp structures in refractivity induced mainly by humidity result in multipath propagation and strong fluctuation in phase and amplitude of RO signals. In the GPS/MET experiment RO signals were often not tracked for a long enough time (due to the loss of lock by receiver) to retrieve the refractivity profiles down to the Earth's surface. The tracked RO signals were often significantly corrupted at the end and that resulted in large errors in the retrieved refractivity. Statistically those errors include negative bias which is larger in tropics than at high latitudes. Understanding the source of those errors is difficult with the use of observational data only, because true refractivity in the atmosphere is never known accurately enough. For understanding the source of those errors it was necessary to model realistic lower tropospheric RO signals. Then those signals were used for choosing and testing the tracking technique capable of their processing without a corruption. Finally, the signals were used for testing two radio holographic inversion techniques capable of solving for the multipath propagation. The main results are discussed in the following sections.

MODELING OF RADIO OCCULTATION SIGNALS

For modeling of RO signals we use high resolution tropical radiosondes. An example of the vertical refractivity profile $N(z)$ calculated from the tropical radiosonde data is shown in Figure 1 by solid line (dashed line shows the exponential profile $N_0(z)$ which was used for comparison and for validation purposes). The structure of the profile $N(z)$ is rather complicated due to humidity. It includes super-refraction ($dN/dz \lesssim -160$ N-units/km) and sub-refraction ($dN/dz > 0$) gradients. As seen from the panel insight Figure 1, the irregularities with the vertical scales of about 100 m and larger are well resolved by the radiosonde. In our modeling of RO signals we assume the 3D refractivity field as to be horizontally homogeneous, i.e., spherically symmetric. This assumption is not fully realistic, especially for the irregularities with small vertical scales. However, we use this assumption due to the following reasons. 1) There is a certain evidence of anisotropy of the irregularities in refractivity in the troposphere. 2) Given vertical refractivity profile the assumption of horizontal homogeneity results in the largest fluctuation in phase and amplitude of RO signal due to the multipath. The "worst case" signals are useful for testing tracking technique in order to be confident in its stable operation under real conditions. 3) Inversions of RO signals normally use the assumption of the spherical symmetry in refractivity. Thus the use of this assumption in the forward modeling of RO signals allows to separately study the inversion errors resulting from the multipath propagation, by not mixing them with the errors resulting from the horizontal inhomogeneity in refractivity. More realistic simulations with 3D (2D) refractivity models have to be done in the future.

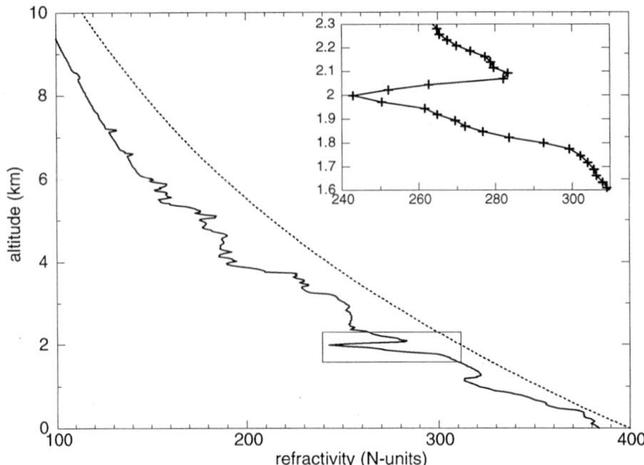

Fig.1. Solid line: refractivity profile $N(z)$ calculated from the high resolution tropical radiosonde (Island Majuro, $7.1N$, $171.4E$, 12UTC, October 1, 1995). Dashed line: model refractivity profile $N_0(z) = 400\exp(-z/8\text{km})$.

Calculation of RO signals, induced by refractivity profiles as shown in Figure 1 and observed in low Earth orbit (LEO), i.e., at a distance of several thousand kilometers from ray tangent point, by means of ray tracing would require precise calculation of large amount of multiple rays arriving at receiver and their summation with the individual phases and amplitudes. The Fresnel zones for some of those rays may overlap. Thus the ray tracing is not applicable. To calculate RO signals we replace the continuous refractivity in the atmosphere by a large number of phase screens normal to the direction of initial wave propagation. Then we solve Helmholtz equation in vacuum between those screens by using boundary conditions on the screens. This is done by Fourier expansion of the complex amplitude of electromagnetic field on output of a screen, association of each harmonic with the plane wave in space, and summation of those plane waves on input to the next screen (successively for all screens). This technique is similar to the "split-step" method widely used for modeling wave propagation problems in random media (Martin, 1993). In our simulations we use 2000 screens spaced by 1 km and we apply 1 m vertical discretization on each screen (substantiation of these parameters may be found in (Sokolovskiy, 2000a)). We assume transmitter infinitely far away (incident plane wave) and receiver moving with velocity of 3.2 km/sec along the straight-line trajectory normal to the direction of initial wave propagation at a distance of 3000 km from the Earth's limb (this approximately corresponds to observations from LEO of about 700 km altitude when transmitter is in LEO plane). We use L1 GPS frequency of 1.57542 GHz.

Figure 2 shows amplitude of RO signals, calculated for $N_0(z)$ and $N(z)$, as function of altitude on the observation trajectory (in our simulations this altitude coincides with the altitude of tangent point of the straight line transmitter-receiver). Negative altitudes correspond to observations below the straight line tangent to the Earth's limb where RO signal is observed due to diffraction of radio waves in the Earth's atmosphere (in case of smooth refractivity profile it may be interpreted as bending of radio wave trajectories). As seen, diffraction of radio waves by small-scale irregularities in refractivity results in: 1) strong fluctuation in amplitude of RO signal; 2) propagation of RO signal down to substantially lower observation altitudes than in case of smooth refractivity profile. Thus tracking of RO signals in the moist troposphere is important down to low enough observation altitudes (because neglecting the information contained in RO signal at those altitudes may result in inversion errors).

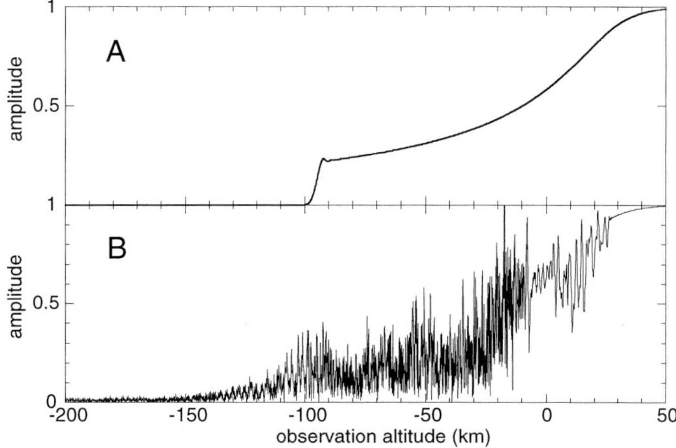

Fig.2. Amplitude of RO signals simulated with the use of $N_0(z)$ (A), and $N(z)$ (B).

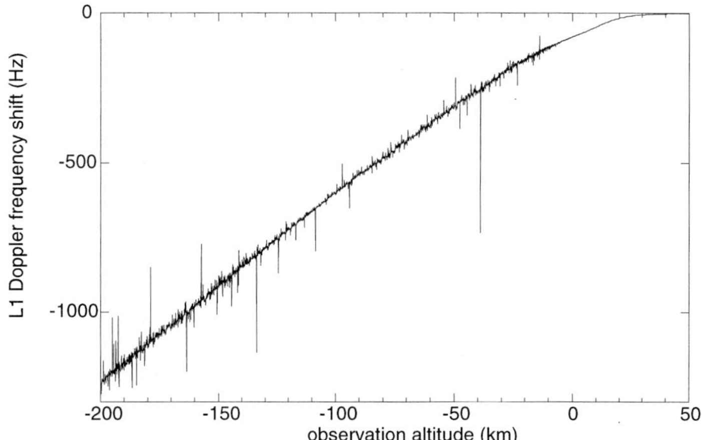

Fig.3. Doppler frequency shift of RO signal simulated with the use of $N(z)$.

Figure 3 shows Doppler frequency shift of RO signal calculated for $N(z)$ (Doppler calculated for $N_0(z)$ is so smooth as amplitude (A) in Figure 2 and it is not presented). As seen, after the signal descends into the moist troposphere below ~7 km (which corresponds to ~-10 km of the observation altitude), the structure of Doppler becomes very complicated with a lot of fluctuation. As seen from Figures 1-3, severe multipath propagation and diffraction do not allow for explicit mapping of the irregularities in refractivity into the irregularities in amplitude and phase (such mapping is rather clear in the case of single path propagation). Under the conditions of multipath propagation the whole RO signal may be treated as a radio hologram.

Figure 4 shows two examples of the spectra of complex RO signal simulated with the use of $N(z)$. The spectra shown in Figure 4 were calculated within the time window of 1.28 sec (which corresponds to 4.096 km window along the observation trajectory) above (A) and below (B) -10 km observation altitude (in each case the signal was downconverted to eliminate mean Doppler). As seen, the spectral bandwidth of RO signal substantially increases up to ~50 Hz after the signal descends into the moist troposphere. The structure of RO signal spectrum in the moist troposphere (B) is very complicated with a lot of local spectral maxima corresponding to multiple tones (rays) arriving at receiver.

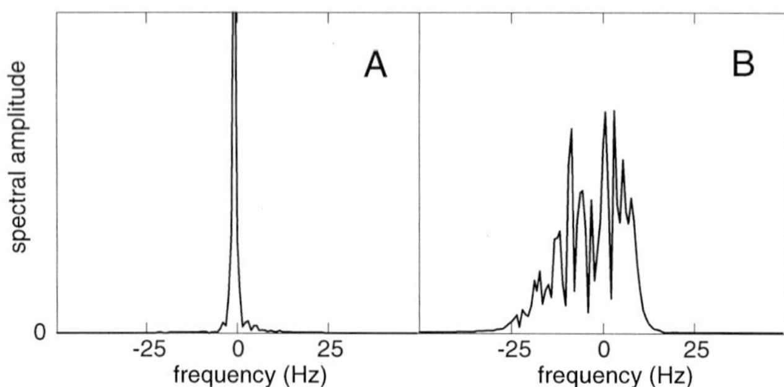

Fig.4. Spectra (absolute value of the complex spectral amplitude) of RO signal simulated with the use of $N(z)$.

Random phase acceleration of RO signal in the moist troposphere is very large, of about ~1 kHz/sec, which is substantially larger than ~ 300 Hz/sec (which corresponds to the phase path acceleration ~6 g for L1 frequency) admitted for stable operation of a generic GPS receiver equipped with phase-locked loop (PLL) (Kaplan, 1996). Thus tracking of RO signals with the use of PLL is questionable in the moist troposphere.

OPEN-LOOP TRACKING OF RADIO OCCULTATION SIGNALS

It is important to choose a proper tracking technique for RO signals in the moist troposphere in order to avoid loss or corruption of the signals due to their complicated structure. PLL is an optimal tracking technique under the assumption of a quasi-monochromatic signal. However, (see previous section) the moist tropospheric RO signal may contain a lot of multiple tones which must be properly resolved for radio holographic inversions of those signals (see next section). The importance of tracking technique also follows from the GPS/MET experiment. During the observational "prime time" 2 (June-July 1995) the PLL tracking technique (which allows different tuning options) had been modified (Kuo et al., 2000). As a result, the tropical occultations were tracked down to substantially lower altitudes and the negative bias after inversions was about twice smaller than for other observational "prime times".

A tracking technique which is not limited by any assumptions about the signal structure (which had been previously applied for planetary occultations) is open-loop (OL) tracking or, in other words, raw sampling of the complex signal and the following post-processing of the sampled In-phase (I) and Quadrature (Q) components. Direct sampling of raw RO signal would result in low signal-to-noise ratio (SNR) due to aliasing of noise into the sampling frequency band. To minimize the noise aliasing the signal must be downconverted to as low mean frequency as possible and then low-pass filtered prior to sampling. Figure 5 shows a layout of RO signal in the frequency domain. On top of carrier frequency, f_c, there is a Doppler shift, $f_{vac} \lesssim 40$ kHz, related to the movement of transmitter and receiver in vacuum, which can be predicted very accurately based on the predicted orbits of GPS and LEO. Next, there is a Doppler shift, $f_{atm} \lesssim 1$ kHz, related to both movement of the satellites and the refraction of radio waves in the atmosphere. Finally, the spectrum of RO signal has spread Δf_{tro} ~50 Hz due to the diffraction by small-scale laminated refractivity irregularities in the troposphere. The main question is: how well the f_{atm} can be predicted prior to an occultation. To answer this question f_{atm} was calculated for given satellite orbits (GPS and LEO of 700 km altitude), for a large number of refractivity profiles represented by NCEP T62 numerical weather prediction (NWP) model (Sela, 1980) for all latitudes and different weather conditions. It appeared that the spread in f_{atm} is rather small, so that the frequency of RO signal can be predicted (modeled) prior to an occultation with accuracy $\lesssim$15 Hz with the use of the refractivity climatology which depends on latitude, and with accuracy $\lesssim$20 Hz with the use of the all-latitude refractivity climatology. An additional frequency mismodeling of the order of several Hz may be introduced by transmitter and receiver clock errors (however, GPS clock dither is cancelled now). Thus, after the downconversion with the use of the frequency model f_{mod} based on the predicted orbits and refractivity climatology the RO signal occupies a double-sided frequency band $\lesssim$100 Hz. Thus it can be low-pass filtered with Δf =100 Hz bandwidth and sampled at 100 Hz rate to preserve its information content. In fact the sampling rate may be even lower (but not lower than the spread of the spectrum, 50 Hz), then it would require an additional downconversion in the post-processing to eliminate aliasing in the spectrum (Sokolovskiy, 2000b).

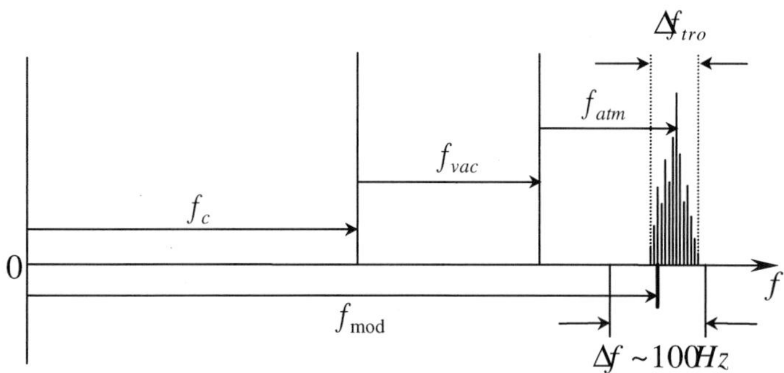

Fig.5. Lay-out of RO signal in frequency domain.

The small weather-related spread of mean Doppler frequency shift of a RO signal, which allows for its prediction (modeling) with accuracy of about 15-20 Hz and thus for the efficient low-pass filtering, has a simple explanation. This spread is related to the spread in arrival angles of multiple tones received in LEO and thus it substantially depends on the distance from the receiver to limb. The larger is the distance the smaller is the angular size of the atmosphere as viewed from the receiver, and thus the smaller is the spread in arrival angles (for given atmospheric properties). For the distance of 3000 km this spread is about 10 times smaller than the spread in bending angle of ray trajectories given altitude of their tangent point. While the spread in frequency of a RO signal decreases with the increase of the distance to limb, the duration of the signal, on opposite, increases.

In the post-processing the downconverted, low-pass filtered and sampled complex RO signal (I and Q components) must be upconverted with the use of the same Doppler model, $f_{\rm mod}$, which was used for in-real-time downconversion, corrected for the solution of transmitter and receiver clocks. After the upconversion the complex signal introduces the complex electromagnetic field along the receiver trajectory (LEO) and it can be used directly for radio holographic inversions (see next section). When the continuous (accumulated) phase is necessary it can be reconstructed by resolving cycle ambiguities in raw phase $\phi = \arctan(Q/I)$. This includes re-sampling the complex signal at higher rate (this is possible according to sampling theorem (Proakis and Manolakis, 1992) when the sampling rate is not smaller than the spectral bandwidth of the downconverted signal), successive comparison of ϕ_i for the adjacent samples, i and $i+1$, and adding 0 or $\pm 2\pi$ to minimize $|\phi_{i+1} - \phi_i|$. The Doppler frequency shift can be calculated by differentiation of the accumulated phase.

RADIO HOLOGRAPHIC INVERSIONS OF RADIO OCCULTATION SIGNALS

A generic inversion technique for RO signals includes calculation of the vertical refractivity profile $N(r)$ from ray bending angle α as function of its impact parameter a under the assumption of the local spherical symmetry in refractivity (Abel inversion) (Eshleman, 1973). However, the calculation of $\alpha(a)$ from raw observation data (phase and amplitude as functions of the positions and velocities of the satellites) can be done by different methods. In case of the single ray propagation, which mostly occurs above the moist troposphere, $\alpha(a)$ can be calculated directly from the Doppler frequency shift of RO signal (i.e., basically, from the slope of phase front), without the use of amplitude (Vorob'ev and Krasil'nikova, 1994). In the presence of multipath propagation this technique results in errors which may be large in the moist tropical troposphere (typically $\alpha(a)$, formally calculated from Doppler, is ambiguous in the presence of multipath). At present two radio holographic techniques solving for the multipath propagation, which had been previously applied for planetary atmospheres, are being tested for the Earth's troposphere.

The first technique uses back propagation (BP) of the complex electromagnetic field (i.e., solution of Helmholtz equation in vacuum for a given boundary condition) from the receiver trajectory closer to the atmosphere (Marouf, et al., 1986, Karayel and Hinson, 1997, Gorbunov and Gurvich, 1998, Gorbunov et al., 2000). The back propagation, which is basically straight line continuation of rays, preserves their impact parameters. Thus, although the back propagated electromagnetic field does not coincide with the true field, it preserves the function $\alpha(a)$. This allows for the calculation of the function $\alpha(a)$ on some auxiliary trajectory where the multipath is not expected (or, at least, is expected to be substantially reduced). The back propagation method not only allows to solve for the multipath, but it allows the sub-Fresnel vertical resolution which may be theoretically as good as the wavelength. The main problem of BP method is that there is no an apriori criterion to find the position of the auxiliary trajectory which is located in one ray region. Under certain conditions (super-refraction) it may be not possible at all (Gorbunov et al., 2000).

Another technique uses spectral analysis of the received complex signal within some small aperture sliding along the observation trajectory (Lindal et al., 1987, Hocke et al., 1999, Gorbunov et al., 2000) (we call this technique as sliding spectral (SS)). Local spectral maxima in each aperture are associated with plane waves (rays) arriving at that aperture. The corresponding spatial frequency defines the arrival angle, and, with account for the position of the center of the aperture, it also defines the impact parameter of the ray. Thus, the SS technique allows to independently calculate bending angles and impact parameters for multiple rays arriving at receiver and to reconstruct $\alpha(a)$. The main problem of this technique (this problem allows different approaches) is detecting the local maxima in the spectrum of RO signal which may be rather complicated (e.g., as shown in Figure 4(B)).

We suggest an approach for SS technique which does not use any detection of the local spectral maxima at all. Instead the whole spectral content of RO signal in each aperture is taken into account. Each spectral component (harmonic) is associated with ray. The spectral power is assigned as an additional parameter to that ray. When the aperture is sliding along the observation trajectory all rays are stored and then sorted according to the increase in their impact parameter. Then $\alpha(a)$ is calculated by sliding averaging of the accumulated array of rays where each ray is weighted proportional to its power. This approach by the definition provides an unambiguous function $\alpha(a)$, however, at the expense of the vertical resolution. But, it appears, that in the presence of the severe multipath propagation induced by multiple super-refraction layers this approach provides more feasible results than BP.

Figure 6 shows $\alpha(a)$ calculated by BP method (solid line) for two positions of the auxiliary trajectory. For comparison dashed line shows the "geometric optical" $\alpha(a)$, i.e., obtained from $N(z)$ by means of ray tracing (super-refraction layers formally result in singularities of $\alpha(a)$). It appears that for the RO signal simulated with the use of the refractivity profile such as $N(z)$ in Figure 1 under the assumption of the spherical symmetry it is not possible to find the position of the auxiliary trajectory which provides feasible unambiguous $\alpha(a)$.

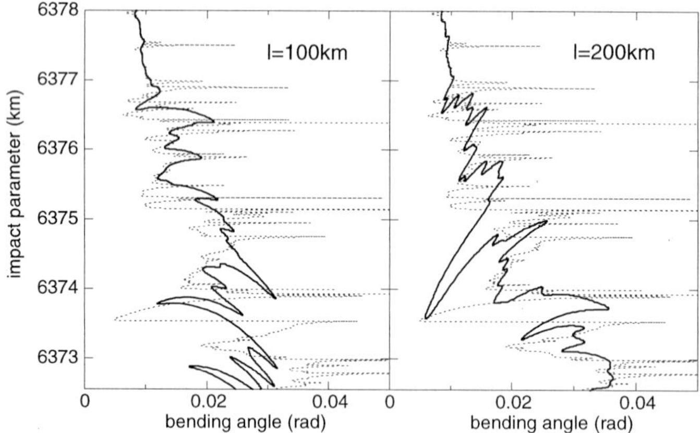

Fig.6. Solid line: bending angle calculated by BP method (l is the distance from the auxiliary trajectory to limb). Dashed line: "geometric optical" bending angle.

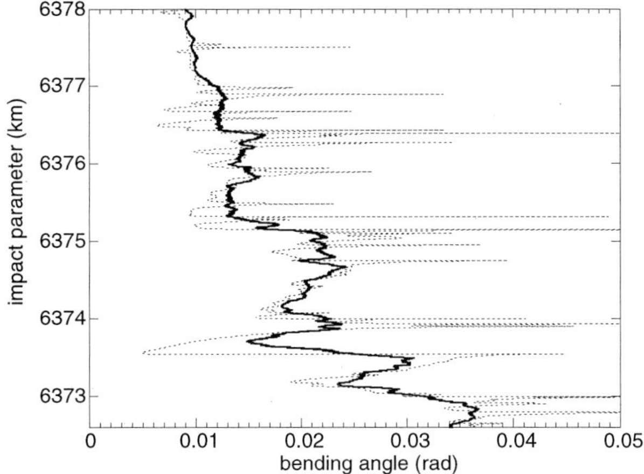

Fig.7. Solid line: bending angle calculated by SS method. Dashed line: "geometric optical" bending angle.

Figure 7 shows $\alpha(a)$ calculated by SS method which takes into account the whole spectral content of RO signal as discussed above (solid line). Comparison to "geometric optical" $\alpha(a)$ (dashed line) indicates that the discussed option of SS method provides feasible reconstruction of $\alpha(a)$ with the vertical resolution of about 0.5 km.

Figure 8 shows the refractivity profile retrieved by Abel inversion with the use of $\alpha(a)$ calculated by SS method (solid line) compared to true $N(z)$ (dashed line). Even though sharp structures in $N(z)$ are not resolved, the retrieval errors are much smaller than in the case when $\alpha(a)$ is calculated directly from Doppler (in that case either very strong smoothing of Doppler or some ad hoc technique to correct ambiguities of $\alpha(a)$ are necessary).

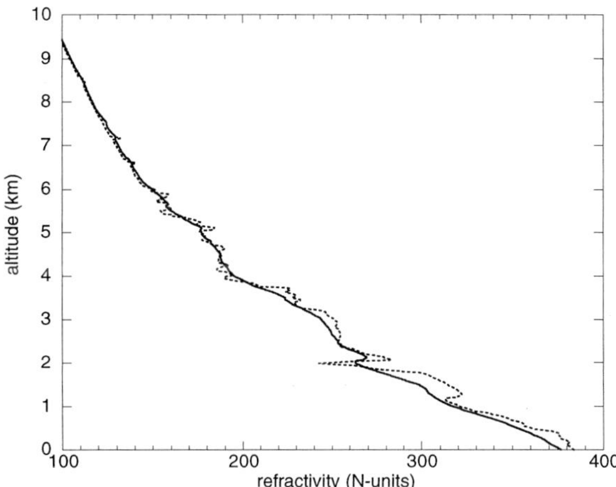

Fig.8. Solid line: refractivity retrieved by Abel inversion from the bending angle calculated by SS method. Dashed line: true refractivity.

As it can be noticed in Figure 8, the inversion errors include some negative bias below 2-4 km. This bias is beleived to be caused by the super-refraction layers in the profile $N(z)$. Negative bias of about the same magnitude is obtained if to apply Abel inversion to "geometric optical" $\alpha(a)$. It must be noted that in case of the super-refraction (resulting in that the rays with certain altitude of tangent point are trapped, i.e., they do not start or end outside the atmosphere) Abel inversion formally does not provide exact solution of the inverse refraction problem.

For validation of the effect of super-refraction on the inversion bias we performed the following simulations. We modeled RO signals by replacing the spherically symmetric atmosphere with a single phase screen (this approximation strictly is not applicable in the troposphere, but we use it as a model only). In case of the refractivity profile such as shown in Figure 1 the single screen approximation results in so severe multipath propagation as in the case of multiple phase screens, and the structure of RO signal is so complicated as in Figures 2,3. But, in case of the single phase screen there is no such a phenomenon as super-refraction (i.e., trapping). For the single phase screen BP method provides exact solution of the inverse diffraction problem, i.e., reconstruction of the phase on the screen, which is trivial. SS method resulted in random inversion errors in the reconstructed phase on the screen of about the same fractional magnitude as the errors in reconstructed refractivity after multiple phase screen propagation. But, in case of the single phase screen there was no evidence of the inversion bias. Also, we modeled RO signals with the use of multiple phase screens for random isotropic irregularities in refractivity whose magnitude was large enough to provide strong fluctuation in amplitude and Doppler of RO signal. In that case the inversion with the use of SS method resulted in random retrieval errors without evidence of bias. Thus the super-refraction introduces a problem for RO technique in the lower troposphere. One possible way to overcome this problem would be direct inversion of the complex RO signal into the refractivity without the intermediate calculation of $\alpha(a)$. However, such inversion is not known at present.

The bias in Figure 8 is clear overestimation of the RO inversion bias due to the assumption of the spherical symmetry in refractivity applied in the forward problem. For more realistic evaluation of the inversion errors in the moist troposphere with the existing RO inversion techniques it is necessary to perform simulations with more realistic 3D(2D) refractivity models.

As it was already noticed, diffraction by laminated irregularities in refractivity results in the propagation of radio waves down to substantially lower observation altitudes than in case of smooth refractivity profile. It appears that the information contained in RO signal at those altitudes is important for radio holographic inversions. The $\alpha(a)$ and $N(z)$ in Figures 7,8 were reconstructed by SS method with the use of the simulated RO signal down to -160 km observation altitude (using the signal down to lower altitudes did not noticeably change the results). However, if to use the RO signal down to -80 km observation altitude only (which is about the mean altitude where receiver stopped tracking due to loss of lock during "prime time" 2 of the GPS/MET experiment) then the negative bias is substantially larger than in Figure 8. This indicates that tracking of RO signals in the moist troposphere is important down to low enough observation altitudes to minimize the inversion errors.

CONCLUSIONS

Simulations of realistic RO signals with the use of high resolution tropical radiosondes allowed to estimate the structure of those signals. The amplitude and phase are subjected to very strong fluctuation due to diffraction of radio waves by the small-scale laminated irregularities in refractivity induced mainly by water vapor in the troposphere. For observations in LEO the spectral bandwidth of RO signals may be as large as $\sim$50 Hz and their random phase acceleration may be much larger than $\sim$300 Hz/sec (i.e., $\sim$6 g, which is admitted for stable operation of a generic GPS receiver with PLL). The RO signals propagate down to substantially lower observation altitudes than in case of smooth refractivity profiles. Information content of RO signals at those altitudes is important for radio holographic inversions. Tracking RO signals down to low enough observation altitudes, -150 -(-200) km, must be done in OL mode that does not use a feedback (the feedback in PLL often results in tracking failure under the conditions of large random phase acceleration and low SNR). Also, OL tracking is the only possible technique for rising occultations. Prediction of the mean frequency of RO signal with the use of refractivity climatology for OL tracking may be done with accuracy of about $\pm$15-20 Hz. This allows for efficient noise filtering prior to sampling of RO signal. Under the conditions of severe multipath propagation induced by multiple super-refraction layers in the moist troposphere BP inversion technique encounters problem due to impossibility to find position of the auxiliary trajectory in one ray region and to reconstruct unambiguous bending angle as function of impact parameter. Under such conditions the SS technique, which takes into account the whole spectral content of RO signal, provides more feasible results and allows for the reconstruction of refractivity with vertical resolution of about 0.5 km. Super-refraction layers introduce negative bias in inversions. However, this bias is much smaller than the bias observed in the GPS/MET experiment which is believed to be caused mainly by the corruption of RO signals after PLL tracking (the magnitude of the bias was substantially different when different options of the PLL tracking firmware were applied). For more realistic evaluation of the errors of RO technique in the lower troposphere simulations with more realistic 3D(2D) refractivity models must be performed.

ACKNOWLEDGMENTS

This work was performed as part of development of the Constellation Observing System for Meteorology Ionosphere and Climate (COSMIC) Data Analysis and Archive Center (CDAAC) at University Corporation for Atmospheric Research (UCAR), funded by National Science Foundation (NSF) under the Cooperative Agreement No. ATM-9732665.

REFERENCES

Eshleman, V.R., The Radio Occultation Method for the Study of Planetary Atmospheres, *Planet. Space Sci.,* 21, 1521 (1973).

Gorbunov, M.E., and A.S. Gurvich, Microlab-1 Experiment: Multipath Effects in the Lower Troposphere, *J. Geophys. Res.*, 103(D12), 13819 (1998).

Gorbunov, M.E., A.S. Gurvich, and L. Kornblueh, Comparative analysis of radioholographic methods of processing of radio occultation data, *Radio Sci.*, 35(4), 1025 (2000).

Hocke, K., A.G. Pavelyev, O.I. Yakovlev, L. Barthes, and N. Jakowski, Radio occultation data analysis by the radio holographic method, *J. Atmos. and Solar-Terr. Phys.,* 61, 1169 (1999).

Kaplan, E.D., *Understanding GPS, Principles and Applications*, Artech House, Inc., 554 pp. (1996).

Karayel, E.T, and D.P. Hinson, Sub-Fresnel-Scale Vertical Resolution in Atmospheric Profiles from Radio Occultation, *Radio Science*, 32(2), 411 (1997).

Kuo, Y.-H., C.Rocken, S.Sokolovskiy, E.R.Kursinski, D.Chu, and L.Lee, Constellation Observing System for Meteorology, Ionosphere, and Climate - COSMIC: an Overview, *4th Symp. on Integrated Observing Systems,* 9-14 Jan., 2000, Long Beach, CA, pp. 86-92 (2000).

Kursinski, E.R., G.A. Hajj, J.T. Schofield, R.P. Linfield, K.R. Hardy. Observing Earth's Atmosphere with Radio Occultation Measurements using the Global Positioning System. *J. Geophys. Res.,* 102(D19), 23429 (1997).

Lindal, G.F., J.R. Lyons, D.N. Sweetnam, V.R. Eshleman, D.P. Hinson, G.L. Tyler, The atmosphere of Uranus: Results of radio occultation measurements with Voyager 2, *Journal of Geophysical Research,* 92, 14987 (1987).

Marouf, E.A., G.L. Tyler, and P.A. Rosen, Profiling Saturn's rings by radio occultation, *Icarus,* 68, 120 (1986).

Martin, J.M., Simulation of Wave Propagation in Random Media: Theory and Applications, in *Wave Propagation in Random Media (Scintillations)*, edited by Tatarskii V.I., A. Ishimaru, and V.U. Zavorotny, pp. 463-486, SPIE Press, Bellingham WA, USA, and the Institute of Physics Publishing, Bristol, UK (1993).

Proakis, J.G., and D.G. Manolakis, *Digital Signal Processing: Principles, Algorithms and Applications*, Macmillan Publishing Company, N.Y., 969 pp. (1992).

Rocken, C., R. Anthes, M. Exner, D. Hunt, S. Sokolovskiy, R. Ware, M. Gorbunov, W. Schreiner, D. Feng, B. Herman, Y.-H. Kuo, and X. Zou, Analysis and Validation of GPS/MET Data in the Lower Troposphere, *J. Geophys. Res.*, 102(D25), 29849 (1997).

Sela, J.G., Spectral Modeling at the National Meteorological Center, *Monthly Weather Review*, 108(9), 1279 (1980).

Sokolovskiy, S.V., Modeling and Inverting Radio Occultation Signals in the Moist Troposphere, *Radio Sci.*, 36(3), 441 (2001).

Sokolovskiy, S.V., Tracking of Tropospheric Radio Occultation Signals from Low Earth Orbit, *Radio Sci.*, 36(3), 483 (2001).

Vorob'ev, V.V., and T.G. Krasil'nikova, Estimation of the accuracy of the atmospheric refractive index recovery from Doppler shift measurements at frequencies used in the NAVSTAR system, *Izv. Russ. Acad. Sci., Atmos. Oceanic Phys.,* Engl. Transl., 29, 602 (1994).

ACTIVE LIMB SOUNDING OF ATMOSPHERIC REFRACTIVITY AND DRY TEMPERATURE PROFILES BY GPS/MET OCCULTATION

Yuei-An Liou and Cheng-Yung Huang

Center for Space and Remote Sensing Research, and Institutes of Space Sciences and Hydrologic Sciences National Central University, Chungli 32054, Taiwan. Email: yueian@csrsr.ncu.edu.tw

ABSTRACT

In this paper, we present a newly developed algorithm for inferring atmospheric profiles of refractivity and dry temperature profiles from GPS occultation data. Our approach differs from the previous algorithms presented in the literature in that the solutions are solved through a 3-dimensional vector analysis rather than a 2-dimensional approach. The retrieved atmospheric profiles are compared with those predicted by ECMWF (European Centre for Medium-Range Weather Forecasts) and NCEP (National Centers for Environmental Prediction). The comparison shows that our algorithm infers reasonably well with the profiles from altitudes of ~5 to 40 km. The profiles deviate from the expected one because of ignorance of relatively high humidity near the Earth's surface. The task of reducing the deviation could be achieved through an iterating approach of solution finding between the retrieval algorithm and atmospheric models and shall be further explored in future.

INTRODUCTION

The radio occultation technique has been used in planetary science for decades to obtain reliable and accurate temperature profiles of the other planets' atmospheres (Phinney et al. 1968). It counts on the fact that radio waves are bent and delayed due to the gradient of atmospheric refractivity along-ray-path. With the advent of Global Positioning System (GPS), it becomes possible to retrieve the refractivity and temperature profiles of the atmosphere from the occultation data (Ware et al. 1996), which are also named as GPS/MET occultation data. However, the problem becomes more complicated in the lower troposphere due to the presence of water vapor, which together with temperature determine the refractivity of the atmosphere. It becomes an underdetermined issue if one intends to simultaneously recover the temperature and water vapor profiles from a single occultation. One must simplify the problem by a certain way such as assuming the amount of water vapor is negligible. Such an assumption is acceptable at higher altitudes, but fails near the Earth's surface. As an initial effort, we develop a computationally efficient algorithm for retrieving atmospheric refractivity and temperature profiles from the GPS/MET occultation data by ignoring the water vapor.

MEASUREMENT OF DOPPLER-SHIFTED FREQUENCY FROM GPS/MET OCCULTATION

Basic observables of the GPS satellite from which the bending angle is derived are the phase delays. In GPS/MET occultation, the phases are measured and recorded as functions of time. Knowing the position of the transmitter (GPS satellite) and the receiver (ground- or LEO-based GPS receivers) and their clocks, the delay due to the intervening media can be isolated. Both the GPS frequencies are used to calibrate the ionospheric effects so that the extra neutral atmospheric delay is isolated. The extra neutral atmospheric delay is used to derive the extra Doppler shift introduced by the medium. Assuming a spherically symmetric atmosphere in the surroundings of occultation, the extra Doppler shift, Δf in a three dimensional coordinate system can be written as (Feng and Herman 1999)

$$\frac{c}{f}\Delta f + (\vec{v}_g \bullet \hat{e} - \vec{v}_l \bullet \hat{e}) = (\vec{v}_g \bullet \hat{e}_g - \vec{v}_l \bullet \hat{e}_l) \qquad (1)$$

where f is the operating frequency, c is the velocity of light, $\hat{e}_g$ and $\hat{e}_l$ are unit vector of signal transmitted by GPS and that of signal received by LEO's respectively and $\hat{e}$ is the unit vector of length from GPS to LEO, $\vec{v}_g$ and $\vec{v}_l$ are the 3-dimensional velocity vector of GPS satellite and LEO satellite, respectively. Any vector can be expressed as

$$\vec{u}_l = u_l^r \hat{r} + u_l^\theta \hat{\theta} + u_l^\varphi \hat{\varphi} \qquad (2)$$

where u_l^r, u_l^θ, and u_l^φ represent the three components in spherical coordinates. Then the vectors of interest $\hat{e}_g$ and $\hat{e}_l$ are solved through vector analysis without any mapping transform performed so that this solution finding method is named as "3-dimensional (3D) vector analysis." The 3D is stressed because equation (1) must be solved through 3D to 2D mapping as seen in the literature (Kursinski et al. 1997, 2000; Feng and Herman 1999).

In our approach we consider the two impact distances as two vectors $\vec{a}_g$ and $\vec{a}_l$. According to this new consideration the vector $\vec{a}_g$ is coplanar with $\vec{r}_g$ and $\vec{r}_l$ and perpendicular with $(\vec{r}_g - \vec{a}_g)$. The two vectors $\vec{r}_g$ and $\vec{r}_l$ represent the distance vectors from GPS to Earth center, and from Earth center to LEO, respectively. Then the following two relations for $\vec{a}_g$ must hold (Figure 1)

$$(\vec{r}_g - \vec{a}_g) \bullet \vec{a}_g = 0 \qquad (3\text{-}1)$$

$$(\vec{r}_l - \vec{a}_l) \bullet \vec{a}_l = 0 \qquad (3\text{-}2)$$

Given equations (2) and (3), the two impact distance vectors can be described as functions of $\vec{r}_g$ and $\vec{r}_l$. Then, together with the assumption $|\vec{a}_g| = |\vec{a}_l|$, equation (1) can be expressed as a one-parameter function of impact distance $|\vec{a}_g|$. The evolution for the solution from one iteration to the next is accomplished by Taylor expansion. The partial derivatives are performed with respect to the impact parameters. The major advantage of using this 3-dimensional vector analysis is in that the 3D to 2D mapping process can be easily avoided.

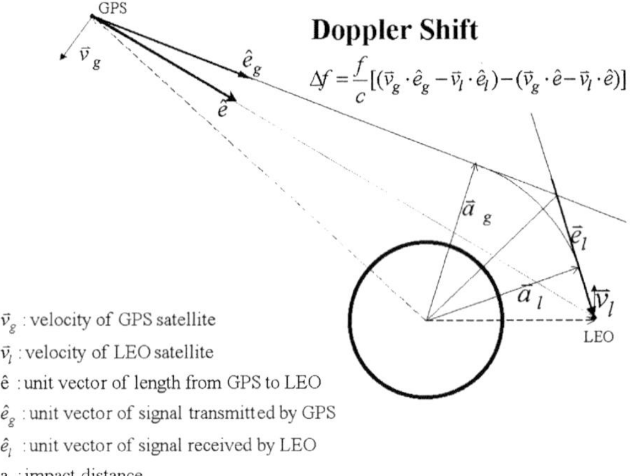

Figure 1. Geometry of GPS/MET occultation. It shows the relation between vectors $\vec{a}_g$, $\vec{a}_l$, $\vec{e}_g$ and $\vec{e}_l$.

RETRIEVED REFRACTIVITY AND DRY TEMPERATURE PROFILES

Figure 2 shows the refractivity profiles from our retrieval algorithm (3D), the UCAR scheme (Rocken *et al.* 1997), and predictions from NCEP and ECMWF. The corresponding temperature profiles are shown in Figure 3. A comparison of refractivity profiles shows a remarkable agreement whereas a comparison of dry temperature profiles shows some deviation below ~5 km and above 40 km. The deviation of the dry temperature profile near the Earth's surface is because of ignorance of relatively high humidity near the Earth's surface. On the other hand above 40 km the deviation is probably due to the different treatment of the upper boundary condition, data runaways and noise. Several temperature profiles with wavelike structures at tropospheric and stratospheric heights are shown while the periodic structures at upper stratospheric heights could be caused by residual errors of the ionospheric correction method. The periodic temperature fluctuations at heights below 40 km are most likely caused by atmospheric waves like gravity waves and equatorial waves. In the lower part of the troposphere, weak signal-to-noise ratio and multipath undermines the applicability of the Abel equation, causing larger errors in refractivity and dry temperature retrievals (Feng and Herman 1999; Kursinski et al. 2000). The multipath problem primarily associated with water vapor limits the applications of the GPS/MET sounding in the lower troposphere. Further studies are still desired to effectively improve the GPS/MET retrievals in the lower troposphere.

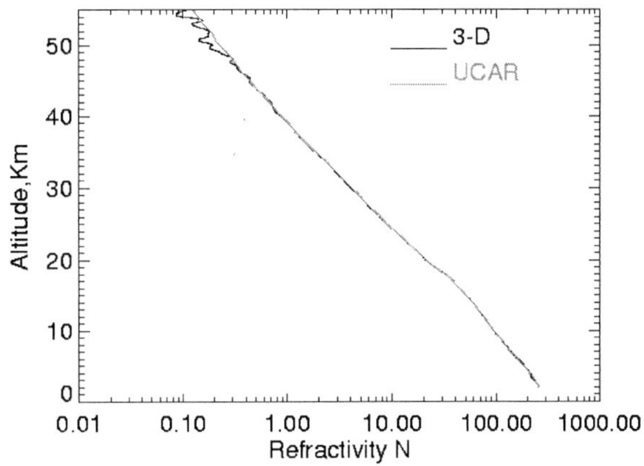

Figure 2. Refractivity profiles derived from the 3-D scheme (solid line) and UCAR's method (dashed line). Occultation was taken at 4°N latitude and 80°W longitude at 6:00 p.m. on June 19, 1995. (Occ_ID: occ_0494_02.04_95.170).

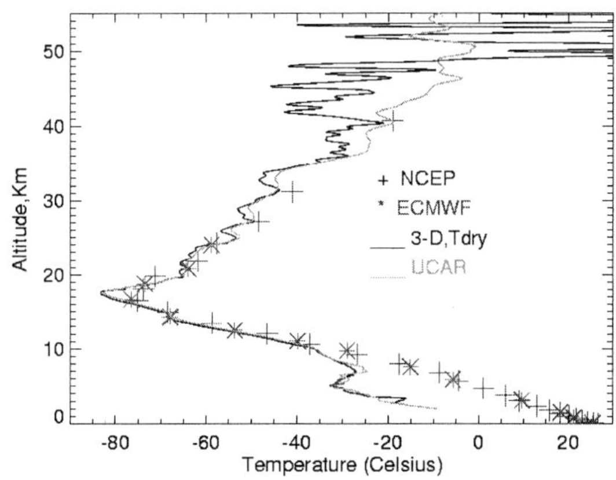

Figure 3. Dry temperature profiles derived from the 3-D scheme (solid line) and UCAR's method (dashed line). Temperature profiles predicted by NCEP (+), and ECMWF (*) at the same location and time as the occultation taken.

CONCLUSIONS

Results from the 3D vector analysis method agree with those derived from the UCAR scheme, and predicted by the ECMWF and NCEP in the altitude range from 5 km to 40 km. The 3D vector analysis method is computationally efficient and could be more appropriate for operational purposes than the 2D retrieval algorithms proposed in the literature. A further refinement of the data analysis and development of ionospheric correction methods of higher order may improve the quality of dry temperature profile at higher altitude. The task of reducing the deviation from the reference could also be achieved through an iterating approach of solution finding between the retrieval algorithm and atmospheric models and will be further explored in future.

ACKNOWLEDGEMENTS

We are grateful to National Science Council of Taiwan and Office of Naval Research of USA for their financial support under grants NSC 89 2111-M-008-025-AP3 and N00014-00-1-0528, respectively.

REFERENCES

Feng, D. D. and M. Herman, Remote Sensing the Earth's Atmosphere Using the Global positioning System (GPS) — The GPS/MET Data Analysis, *J. Atmos. Oceanic Technol.,* 16, 989 (1999).

Kursinski, E. R., G.A. Hajj, K.R. Hardy, J.T. Schofield, and R. Linfield, Observing Earth's atmosphere with radio occultation measurements, *J. Geophys. Res.*, 102, 23429 (1997).

Kursinski, E. R., G. A. Hajj, S. S. Leroy, and B. Herman, The GPS radio occultation technique, *Terr. Atmos. Oceanic Sci.*, 11, 53 (2000).

Phinney, R. A., and D. L. Anderson, On the radio occultation method for studying planetary atmospheres, *J.Geophys. Res.,* 73, 1819 (1968).

Rocken, C., R. Anthes, M. Exner, D. Hunt, S. Sokolovskiy, R. Ware, M. Gorbunov, W. Schreiner, D. Feng, B. Herman, Y.-H. Kuo, and X. Zou, Analysis and validation of GPS/MET data in the neutral atmosphere. *J. Geophys. Res.*, 102, 29849 (1997).

Ware, R., M. Exner, D. Feng, M. Gorbunov, K. Hardy, B. Herman, H. K. Kuo, T. Meehan, W. Melbourne, C. Rocken, W. Schreiner, S. Sokolovskiy, F. Solheim, X. Zou, R. Anthes, S. Businger, and K. Trenberth, GPS sounding the atmosphere from low earth orbit: Preliminary results, *Bull. Am. Meteor. Soc.*, 77, 19 (1996).

A STUDY ON THE COSMIC ELECTRON DENSITY PROFILE

H. F. Tsai[1], D. D. Feng[2], J. Y. Liu[1], C. H. Liu[1], and W. H. Tsai[1]

[1]*Institute of Space Science, National Central University, Jungli City, Taoyuan 320-01, Taiwan*
[2]*Institute of Atmospheric Physics, University of Arizona, Tucson, AZ 85721, USA*

ABSTRACT

This paper shows a simulation process of retrieving ionospheric electron density from the Global Positioning System/Meteorology (GPS/MET) to examine the accuracy of the retrieval procedure for the COSMIC project[3] during solar minimum and maximum periods. Results show that the error of the maximum electron density (NmF_2) is under 25%, and the error is greater in the solar maximum than in the solar minimum period. The mean error of the height of NmF_2 (i.e. hmF_2) is less than about 13 km and seems to have no correlation with solar activity.

INTRODUCTION

To understand ioniospheric construction on the issue of space weather, many scientists have studied the radio-occultation retrieval technique for the Global Positioning System/Meteorology (GPS/MET) experiment to obtain vertical profiles of ionospheric parameters during solar minimum (*Rius et al.*, 1997; *Hajj and Romans*, 1998; *Schreiner et al.*, 1999). For the COSMIC project, the ionospheric parameters and their accuracy should be computed and examined not only during solar minimum but also solar maximum. In this study, we retrieve ionospheric electron density profiles from precise ephemeris in GPS/MET Level 2 data and excess phase measurements obtained by ROSAP[4], which simulates the ray path and excess phase between GPS and LEO (Low Earth Orbit) satellites according to the ephemeris and the NeUoG global ionospheric model[5]. Finally, we analyze the error percentage in electron density to the NeUoG model during 1995 and 2000 (solar minimum to solar maximum), which allows us to examine accuracy of the retrival procedure and the relationship between the error and solar activity.

DATA RETRIEVAL

The GPS/MET Level 2 data contain excess phase measurements and precise ephemeris (coordinates and velocities) of GPS and LEO satellites. To analyze the computation error of the procedure, ROSAP is adopted to simulate the excess phase measurements in length Φ to replace the original excess phase measurements. The extra Doppler frequency can be written by

[3] A Taiwan-US joint project for Constellation Observing System for Meteorology, Ionosphere, and Climate

[4] A ray tracing code of Radio Occultation Simulator for Atmospheric Profiling.

[5] Electron density (Ne) from the University of Graz global model, which is included in ROSAP.

$$f_d \equiv -\frac{f}{c}\frac{d\Phi}{dt} = -\frac{f}{c}v_d$$

where f is the GPS carrier frequency, c is the light speed, and $v_d \equiv d\Phi / dt$. Assuming that ionosphere is local spherical symmetric, the extra Doppler frequency can be written as (*Hajj and Romans*, 1998)

$$f_d = \frac{f}{c}\left[\mathbf{v}_t \cdot \mathbf{k}_t - \mathbf{v}_r \cdot \mathbf{k}_r - (\mathbf{v}_t - \mathbf{v}_r)\cdot \mathbf{k}\right], \quad (1)$$

where $\mathbf{k}_t$ and $\mathbf{k}_r$ are the unit vectors of the asymptotes of the ray path at GPS and LEO; $\mathbf{k}$ is the unit vector from GPS to LEO; $\mathbf{v}_t$ and $\mathbf{v}_r$ are the components of GPS velocity $\mathbf{v}_G$ and LEO velocity $\mathbf{v}_L$ on the occultation plane (Figure 1). The relationship between these velocities are

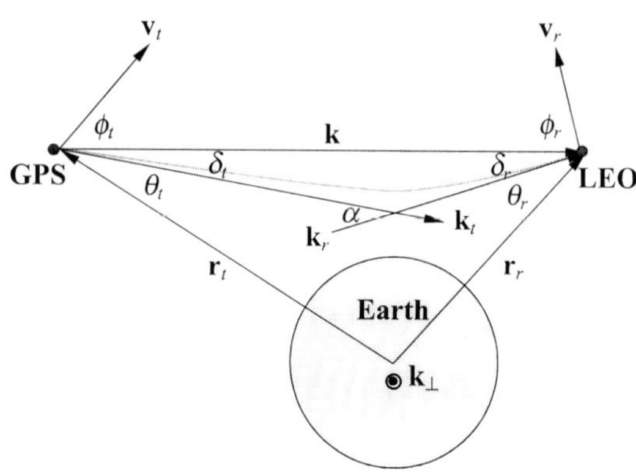

Fig. 1. The geometry of the GPS/MET ray trace (gray curve) and its asymptotes ($\mathbf{k}_t$ and $\mathbf{k}_r$) at GPS and LEO on the occultation plane.

$$\mathbf{v}_t = \mathbf{v}_G - \mathbf{k}_\perp(\mathbf{v}_G \cdot \mathbf{k}_\perp)$$
$$\mathbf{v}_r = \mathbf{v}_L - \mathbf{k}_\perp(\mathbf{v}_L \cdot \mathbf{k}_\perp)$$

where $\mathbf{k}_\perp$ is the unit normal vector to the plane. According to Bouguer's rule, the impact distance can be represented as

$$a = n(\mathbf{r}_t)|\mathbf{r}_t \times \mathbf{k}_t| = n(\mathbf{r}_r)|\mathbf{r}_r \times \mathbf{k}_r| \quad (2)$$

where n is ionospheric phase refractive index; $\mathbf{r}_t$ and $\mathbf{r}_r$ are the orbital radii of GPS and LEO. *Hajj and Romans* (1998) showed that for the GPS/MET experiment, if $n(\mathbf{r}_t) = n(\mathbf{r}_r) = 1$, the derived electron density could be overestimated by no more than 0.5% of the true density.

Note that ϕ_1, ϕ_2, δ_1, δ_2, θ_1, and θ_2 are the acute angles between $\mathbf{k}$ with $\mathbf{v}_t$, $\mathbf{v}_r$, $\mathbf{k}_t$, $\mathbf{k}_r$, $\mathbf{r}_t$, and $\mathbf{r}_r$ (Figure 1). Therefore, when δ_1 and δ_2 approach to zero, the bending angle ($\alpha = \delta_t + \delta_r$) and the impact distance a have analytic solutions as

$$\alpha = -\frac{v_d(r_t \cos\theta_t + r_r \cos\theta_r) + (\pm v_t \sin\phi_t \mp v_r \sin\phi_r)(r_t \sin\theta_t - r_r \sin\theta_r)}{v_t r_r \sin\phi_t \cos\theta_r + v_r r_t \sin\phi_r \cos\theta_t} \quad (3)$$

$$a = r_r \sin\left(\theta_r + \frac{\pm v_t \sin\phi_t (r_t \sin\theta_t - r_r \sin\theta_r) + v_d r_t \cos\theta_t}{v_t r_r \sin\phi_t \cos\theta_r + v_r r_t \sin\phi_r \cos\theta_t}\right). \quad (4)$$

The sign (±) associated with v_t (or v_r) is determined by the direction of $\mathbf{v}_t$ (or $\mathbf{v}_r$). Upper sign is chosen when v_t (or v_r) is upward (see Figure 1), and vice versa. Based on the Abel inversion integral formula, the phase refractive index is (*cf. Phinney and Anderson*, 1968; *Fjeldbo et al.*, 1971; *Tricomi*, 1985; *Hajj and Romans*, 1998)

$$n(a) = \exp\left(-\frac{1}{\pi}\int_a^\infty \frac{\alpha(x)}{\sqrt{x^2-a^2}}dx\right)$$
$$= \exp\left(-\frac{1}{\pi}\int_a^{a_0} \frac{\alpha(x)}{\sqrt{x^2-a^2}}dx\right)\exp\left(-\frac{1}{\pi}\int_{a_0}^\infty \frac{\alpha(x)}{\sqrt{x^2-a^2}}dx\right) \tag{5}$$

where a_0 is the upper boundary of the impact distance a and therefore the first exponential expression can be solved. After comparing to ionospheric Chapman function, we adopted exponential extrapolation to find the unknown α above a_0 in the second exponential expression in form of (*Davies*, 1990)

$$Ne(y) = Ne(a_0)\exp\left(\frac{a_0 - y}{H}\right), \tag{6}$$

where Ne and H are the electron density and the scale height, respectively. According to Appleton-Hartree formula (*Davies*, 1990)

$$n(y) = 1 - \frac{40.3Ne(y)}{f^2}, \tag{7}$$

and Abel integral formula (*Fjeldbo et al.*, 1971)

$$\alpha(x) = 2x\int_x^\infty \frac{dn/dy}{n(y)\sqrt{y^2-x^2}}dy, \tag{8}$$

the refractive index n in Eq. 5 can be obtained. Finally, the ionospheric electron density profile can be solved by Appleton-Hartree inversion formula

$$Ne(a) = \frac{(1-n(a))f^2}{40.3}. \tag{9}$$

PROFILING RESULTS

Figure 2 illustrates five cases of the bending angle profiles and electron density profiles during 1995 and 2000. The solar radio flux, F10.7, during the solar minimum (1995-1997) and the solar maximum (2000) are also displayed. The solid/dashed curves indicate the experimental value from Eq. 9 and the reference value directly from the NeUoG model, respectively. The data sampling rates are 0.1 Hz (Figures 2a-2c) and 1 Hz (Figures 2d-2e), respectively.

The bending angle lying between ±0.01° suggests that the assumption of δ_1 and δ_2 near zero is reasonable. The maximum electron density at F_2 region, NmF_2, in Figure 2a is less than 10^{11} el m^{-3}, and the electron density is overestimated. For Figure 2b, the electron density is overestimated under the height of NmF_2, hmF_2, and underestimated around and above hmF_2. The curves in Figures 2d-2e are much smoother than those in Figures 2a-2c, which suggest higher data sampling rate. In solar maximum (F10.7=160), the bending angle is up to 0.01°, which is 10 times the value of Figure 2a, and NmF_2 is up to 8×10^{11} el m^{-3} (Figure 2e).

Table 1 summarizes the differences of experimental and reference values in NmF_2 and hmF_2 (ΔNmF_2 and

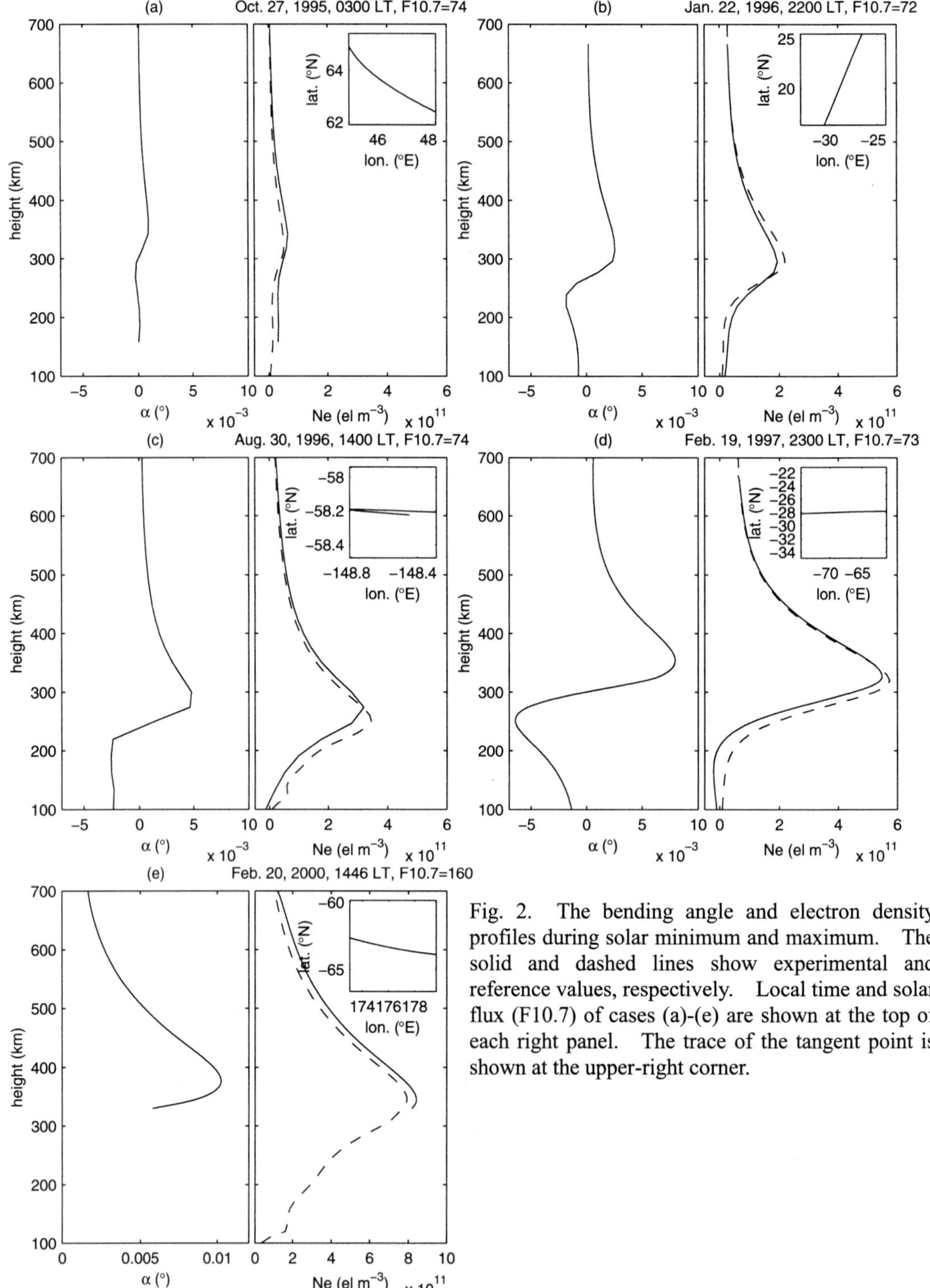

Fig. 2. The bending angle and electron density profiles during solar minimum and maximum. The solid and dashed lines show experimental and reference values, respectively. Local time and solar flux (F10.7) of cases (a)-(e) are shown at the top of each right panel. The trace of the tangent point is shown at the upper-right corner.

ΔhmF_2) and the associated error percentage. It is found that in the solar minimum, ΔNmF_2 ranges between 1.3×10^{11} el m^{-3} and 2.6×10^{11} el m^{-3}, while in the solar maximum, ΔNmF_2 is as large as 4.9×10^{11} el m^{-3}. Since NmF_2 is much higher in solar maximum, its error percentage 6% is relatively smaller. Moreover, ΔhmF_2 lies between 4 km and 24 km and its mean error is less than about 13 km. Thus, the relationship between ΔhmF_2 variation and solar activity seems to be not clear.

Table 1. The difference of NmF_2, its error percentage, and the difference of hmF_2 between experimental and reference values during solar minimum and maximum. Cases 1-5 correspond to Figures 2a-2e.

	Case	ΔNmF_2 (el m^{-3})	Error (%)	ΔhmF_2 (km)
solar min.	1	1.3×10^{10}	25	22.7
	2	2.5×10^{10}	11	4.0
	3	2.5×10^{10}	7	24.0
	4	2.6×10^{10}	5	6.5
max	5	4.9×10^{10}	6	5.8
	Mean	2.8×10^{10}	-	12.6

DISCUSSION AND CONCLUSIONS

The simulation of ionospheric electron density profile retrieved from GPS/MET ionospheric data by using Abel integral technique and exponential extrapolation demonstrates that the error percentage of NmF_2 is greater in solar maximum than in solar minimum and ΔhmF_2 seems to have no clear relationship with the solar activity.

The errors may result from both of simulated excess phase measurements, which may include errors from ROSAP and the NeUoG model, and the retrieval procedure, which caused by assumption and computation. Since the reference and experimental values are derived from the NeUoG model directly and indirectly, the errors from the NeUoG model should be excluded. Therefore, after neglecting the errors from ROSAP, the errors from the retrieval procedure must be smaller than the results. Meanwhile, the ionosphere is not in spherical symmetric distribution (*cf. Davies*, 1990), and the trace of tangent points usually crosses many longitudes and latitudes (see Figure 2), so the local spherical symmetric assumption may produce major errors.

ACKNOWLEDGMENTS

We thank Dr. Stig Syndergaard for providing ROSAP program and thank Dr. William Schreiner in UCAR for providing the GPS/MET Level 2 ionospheric data. This work was partially supported by National Science Council in Taiwan Grants NSC 89-2119-M-008-005.

REFERENCES

Davies, K., *Ionospheric Radio*, Peter Peregrinus Ltd., London, UK (1990).

Fjeldbo, G. F., A. J. Kliore, and V. R. Eshleman, The neutral atmosphere of Venus as studied with the Mariner V radio occultation experiments, *Astront. J.*, 76, 123 (1971).

Hajj, G. A., and L. Romans, Ionospheric election density profiles obtained with the Global Positioning System: Results from the GPS/MET experiment, *Radio Sci.*, 33, 175 (1998).

Phinney, R. A., and D. L. Anderson, On the radio occultation method for studying planetary atmospheres, *J. Geophys. Res.*, 73, 1819 (1968).

Rius, A., G. Ruffini, and L. Cucurull, Improving the vertical resolution of ionospheric tomography with GPS occultations, *Geophys. Res. Lett.*, 24, 2291 (1997).

Schreiner, W. S., S. V. Sokolovskiy, C. Rocken, and D. C. Hunt, Analysis and validation of GPS/MET radio occultation data in the ionosphere, *Radio Sci.*, 34, 949 (1999).

Tricomi, F. G., *Integral Equations*, Dover, Mineola, NY (1985).

SPACE GEODESY AND CLIMATE CHANGE STUDIES USING COSMIC MISSION

C. K. Shum [1] , Christopher Cox [2] , and Benjamin F. Chao [3]

[1] *Laboratory for Space Geodesy and Remote Sensing Research, Department of Civil and Environmental Engineering and Geodetic Science, The Ohio State University, 470 Hitchcock Hall, 2070 Neil Ave. Columbus, Ohio 43210-1275 U.S.A. Email: ckshum@osu.edu*

[2] *Raytheon Technical Services Corporation, 7701 Greenbelt Rd, Suite 400 Greenbelt, Maryland 20770, U.S.A.*

[3] *Space Geodesy Branch, NASA Goddard Space Flight Center Greenbelt, Maryland 20771, USA*

ABSTRACT

This study provides a preliminary assessment of the potential contribution of COSMIC to space geodesy and climate change studies. COSMIC is a 6-satellite constellation and a joint Taiwan-US (NSPO/UCAR) mission to be launched in 2005 with a 5-year mission life time, for research in climate change, meteorology, space weather, and space geodesy. This study complements earlier analyses conducted by Chao et al. [2000a, 2000b] on COSMIC's contribution to gravity field improvement and geodetic applications. It is anticipated that COSMIC will improve the current knowledge of surface pressure fields with an accuracy of 1 hPa and temporal resolution of several hours over a spatial resolution of a few hundred km, especially over Southern Ocean and the Antarctica continent pole-ward up to 72^0S latitude (COSMIC inclination coverage). It is not clear that COSMIC would significantly contribute to secular gravity observations. However, COSMIC is anticipated to improve estimates of the seasonal and annual time-variable gravity observations with finer temporal and spatial resolutions due primarily to its 6-satellite constellation and its orbital inclination at 72^0. The use of COSMIC data together with GRACE and GOCE data will allow improved separations of atmospheric effect from the observed time-variable gravity signals of oceanic, hydrological and glaciology origin, and will improve the accuracy of long-wavelength time-variable gravity field observations.

INTRODUCTION

Accurate knowledge of the Earth's mean and time-variable gravity field is encompassed in a number of U.S. National Research Council, NASA, and other studies [e.g., NASA Coolfont Report, 1991; NRC Report, 1997]. Momentum and mass transports among the atmosphere, oceans, solid Earth, hydrosphere, and cryosphere produce measurable changes in the Earth's gravity field, its rotation, oceanic and atmospheric circulations and sea level. These phenomena are considered to be possible consequences of global climate changes, and as indicators of natural hazard occurrences. Some of these phenomena, such as Earth rotation and temporal changes in zonal harmonics of the Earth's gravity field [e.g., Cheng et al., 1997], have been accurately measured using satellite geodetic techniques. Conventional ground-based tracking data (e.g., SLR, DORIS, GPS), spaceborne GPS, satellite altimetry, and surface and airborne gravity data have been used to determine Earth's gravity field. By the middle of the next decade, the approved gravity mapping missions, GFZ CHAMP (launched in 1999) [Reigber et al., 1996], NASA/GFZ GRACE (2001 launch) [Tapley et al., 1996], and ESA's GOCE [Rummel et al., 1999] missions are expected to provide more than three-fold improvement over the current knowledge of the Earth's gravity field model, e.g., EGM96 [Lemoine et al., 1998].

Taiwan's National Space Program Office (NSPO), the University Corporation for Atmospheric Research's (UCAR), and a number of other organizations and Universities are jointly undertaking a satellite mission, the Constellation Observing System for Meteorology, Ionosphere and Climate (COSMIC), to be launched in 2005 [Kuo et al., 1999]. The current plan for COSMIC is to launch a constellation of 6 mini-satellites using the

Orbital Science launcher in the year 2005. These mini-satellites are anticipated to have circular orbits with altitudes of about 800 km and inclination of 72^0, and the satellites are to be placed in orbital planes with the ascending nodes separated by about 30^0. The primary scientific objectives for COSMIC currently include atmospheric limb-sounding using high-low GPS radar occultation to extract troposphere and ionosphere delays [e.g., Kursinski et al., 2000] for research and operational opportunities in climate change studies, meteorology, space weather and space geodesy [Kuo et al., 1999]. The contributions of COSMIC to space geodesy include the improvement in the Earth's mean and temporal gravity field solutions in the long wavelength component, orbit determination procedure towards accurate near-real time (3 hours) computation and prediction of the orbits of the COSMIC satellites with an accuracy of 0.1 mm/sec in the absolute velocity component for computation of atmospheric refraction and GPS-ray bending angle measurements from GPS limb-sounding, and global surface pressure field for improved separation of atmospheric mass signals from signals of other geophysical origins (hydrology, glaciology, ocean), e.g., from GRACE measurements. These studies are reported by Chao et al. [2000a, 2000b], Pavlis [1998], Shum and Jekeli [1998], and Hwang [2000]. Most of the studies assumed an earlier COSMIC mission configuration of 8-satellite constellation and the possibility of a period of tandem-flying satellites in pairs at lower altitudes (500 km) during the first phase of the mission.

This paper discusses a complementary study of potential contribution of COSMIC to focus on climate sensitive time-variable gravity signal mapping at seasonal and secular time scales, and speculated on the potential improvement of surface pressure field measurements. This study is also intended to further refine the space geodetic science objectives of the COSMIC mission.

COSMIC MISSION

The COSMIC constellation of satellites is based on an improved version of the highly successful MicroLab-1 GPS-Met atmospheric occultation satellite mission launched in April, 1995 [Exner et al., 1996]. The current satellites have a mass of around 40 kg (includes 7 kg fuel). The mission has a nominal mission life-time of five years. The cross-sectional area of each satellite along the flight path is around 25.4 cm by 81.3 cm. The two sun-tracking solar panels are fixed with respect to the satellite motion and are of length 96.5 cm. The spacecraft attitude is controlled by using 3-torque-rods and gravity gradient stabilized and nadir pointing. The knowledge of the center of mass of the spacecraft is around 1 cm. Each satellite carries a geodetic quality, 24-channel, 6 antenna dual-frequency JPL Blackjack type GPS receiver. The current mission plan will launch the 6 mini-satellites using the Orbital Science launcher, to altitudes of 800 km with each satellite separate by approximately 30^0 in their nodal planes during the second year of the mission. The inclinations of the satellites are about 72^0.

COSMIC CONTRIBUTIONS IN TIME-VARIABLE GRAVITY FIELD MAPPING

Satellite measurements can detect the climate-sensitive signals manifesting in the form of time-variable gravity field, which is the focus of a wide area of interdisciplinary research studies [Chao et al., 2000c]. Cheng et al. [1997] and Cox et al. [2000] reported studies of using satellite laser ranging (SLR) to measure the composite signals of time-variable gravity field at the seasonal and interannual scales and over long wavelengths ((>5000 km). The GRACE gravity mapping mission is anticipated to measure time-variable gravity signals with monthly time scales and a spatial scale longer than 350 km. In this study, we have assessed the contributions of the COSMIC 6-satellite constellation at inclinations of 72^0 to the improved mapping and separation of temporal gravity field signals at the seasonal and interannual scales.

Though very accurate, GRACE measurements will have the limitations of monthly or longer temporal sampling, and hence potentially significant signal aliasing problems, which are challenging to the processing and interpretation of the data. A 6-satellite constellation from COSMIC at a higher altitude than GRACE (800 km vs. 500 km), even without the benefit of 3-axis accelerometers to correct for non-gravitational forces [e.g., Sehnal et al., 1997], could provide improved temporal resolution at the long wavelength component of the time-variable gravity field.

Fig. 1 shows an example of time-variable gravity field solutions using multiple satellite tracking data for the solution of J_2, J_3, and J_4, representing the longest wavelength components of the signal from 1986-1999 [Cox et al., 2000]. The satellite data include SLR to Lageos 1/2, Starlette, BE-C, Etalon 1/2, Ajisai, Stella, TOPEX/POSEIDON (T/P), and Westpac. Fig. 2 shows estimates of solution uncertainties for the annual time-variable gravity using all the satellites, without Stella (98.6^0 inclination, 800 km altitude satellite), and without

T/P (1354 km altitude, and 66^0 inclination). It shows that T/P (with satellite-laser-ranging and global DORIS radiometric tracking) contributes significantly to the annual gravity solution, while Stella, although a polar satellite ($\pm 81.4^0$ latitude coverage as compared to T/P's $\pm 66^0$ latitude coverage) contributes much less. This is attributable to the fact that DORIS tracking is essentially global for T/P, while the SLR tracking is quite spare for Stella.

Fig. 3 shows a 30-day solutions for C4,4 (a sectoral geopotential harmonic coefficient) over one year in 1994 with all satellites, and without contribution from T/P. Also plotted (thick continuous curve) is the anticipated atmosphere+ocean+hydrology from the NCEP (atmospheric, courtesy of A. Au) and POCM (oceanic, courtesy of T. Johnson, Johnson et al., 2001) general ocean circulation model outputs, and the associated error bars for both solutions (all satellites and without T/P). Fig. 3 shows that the solution with T/P is more accurate compared to the solution without contribution from T/P. Fig. 4 further shows that the uncertainty of the C4,4 solutions with T/P is almost one order of magnitude better than the solutions without T/P.

The significant contribution from T/P to temporal gravity solutions is primarily due to (1) the dense tracking from DORIS, and (2) T/P's distinct inclination in the solution. Compared to T/P, COSMIC will have more global coverage in tracking (inclination of COSMIC is 72^0 and with high-low GPS tracking) and more temporal resolutions (6-satellite constellation) and is therefore expected to contribute significantly in the long wavelength time-variable gravity field recovery.

IMPROVEMENT OF GLOBAL PRESSURE FIELDS

Fig. 5 shows the surface pressure differences between operational NCEP and ECMWF atmospheric general circulation model (GCM) outputs with 6-hour intervals, or 4-times daily averaged data. The surface pressure error is largest over Antarctica and over the Southern Ocean (>6 Pha). This comparison provides an uncertainty estimate of surface pressure prediction using these data products. The accuracy for more frequent sampling (<6 hours) would be much worse. Accurate (<1 Pha) pressure fields are required for separation of atmospheric mass variations from oceanic, glaciological and hydrological signals, for proper use of measurements such as GRACE for time-variable gravity field mapping.

COSMIC's contribution is from the adequately temporally sampled occultation profiles globally in the form of atmospheric refraction and GPS-ray bending angles. One method is to assimilate bending angles and refractions into atmospheric GCMs (e.g., Kuo et al. [1999]) to provide improved prediction of surface pressure fields. Alternatively, another possible method is to empirically determine surface pressure over Antarctica using COSMIC retrieved water vapor profiles over the Southern Ocean near Antarctica and assuming that there is essentially no water vapor over the continent [Bromwich et al., 2001]. Although it is not clear how close to surface the pressure profile could be determined, it is speculated that COSMIC will contribute to global pressure field observations using either or both of the above scenarios.

SUMMARY

This study complements earlier analyses conducted by Chao et al. [2000a, 2000b] on COSMIC's contribution to gravity field improvement and geodetic applications. It is unclear that COSMIC would improve secular gravity field solutions. However, COSMIC is anticipated to improve estimates of the seasonal and annual temporal gravity observations with improved resolution due primarily to its 6-satellite constellation and its distinct orbital inclination (72^0). It is speculated that COSMIC will improve the current knowledge of surface pressure fields with accuracy of 1 hPa and temporal resolution of several hours over a spatial resolution of a few hundred km, over Southern Ocean and much of Antarctica. The use of COSMIC data together with GRACE and GOCE data will allow improved separation of atmospheric effects from the observed time-variable gravity signals of oceanic, hydrological and glaciology origin, and will improve accuracy of long-wavelength time-variable gravity field observations. This paper also intends to further refine the space geodetic science objectives of COSMIC mission.

ACKNOWLEDGEMENT

We appreciate the travel grant provided by NSPO to attend the Workshop. This work benefited from earlier analysis and discussions with Erricos Pavlis and Steve Klosko, to whom we gratefully acknowledged.

REFERENCES

Bromwich, D., C. Shum, and B. Kuo, Combining GPS occultation data and EOS sensor to improve Antarctica meteorology, EOS/IDS Meeting, Miami, FL., 2001.

Chao, B., C. Hwang, C. Liu, E. Pavlis, C. Shum, C. Tseng, and M. Yang, white paper on Space Geodesy with COSMIC, May 22, 2000.

Chao, B., E. Pavlis, C. Huang, C. Liu, C. Shum, C. Tseng and M. Yang, COSMIC: Improving Earth's Gravity Field Model and Other Geodetic Applications, Terrestrial Atmospheric Oceanic Sciences (TAO), 11, 365-378, 2000a.

Chao, B. F., C. W. Huang, C. C. Liu, E. C. Pavlis, C. K. Shum, C. L. Tseng, and M. Yang, Space Geodesy with COSMIC, White Paper, Submitted to National Space Program Office, May 2000b.

Chao, B. F., V. Dehant, R. S. Gross, R. D. Ray, D. A. Salstein, M. M. Watkins, and C. R. Wilson, Space Geodesy Monitors Mass Transports in Global Geophysical Fluids, in press, *EOS, Trans. Amer. Geophys. Union,* 2000c.

Cheng, M., C. Shum, and B. Tapley, Determination of long-term changes in the Earth's gravity field from satellite laser ranging observations, Jl. Geophys. Res., 102(B10), 22377-22390, October 1997.

Cox, C., B. Chao, and S. Klosko, Temporal gravity field solutions using geodetic satellites, EGS Symposium, Nice, France, 2000.

Exner, M. et al., GPS-Met, Spring AGU Meeting, Baltimore, MD, USA, May 1996.

Hwang, C., Gravity Recovery Using COSMIC GPS Data: Application of Orbital Perturbation Theory, Report of the Dept. of Civil and Environmental Engineering and Geodetic Science, The Ohio State University, Columbus, 2000.

Johnson, T. J., C. R. Wilson, and B. F. Chao, Non-tidal oceanic contributions to gravitational field changes: Predictions of the Parallel Ocean Climate Model, *J. Geophys. Res., 106*, 11,315-11,334, 2001.

Kuo, Y., B. Chao, and L. Lee, A Constellation of Microsatellites Promises to Help in a Range of Geoscience Research, EOS Trans. Amer. Geophys. Union, 80, 467-471, 1999.

Kursinski, E. R., G. Hajj, S. S. Leroy, and B. Herman, The GPS radio occultation technique, *Terr. Atmos. and Ocean. Sc., 11*, 53-114, 2000.

Lemoine, F. and 14 others, The development of the NASA GSFC and the National Imagery and Mapping Agency (NIMA) geopotential model EGM96, NASA/TP-1998-206861, NASA Goddard Space Flight Center, July 1998.

National Research Council, *Satellite Gravity and the Geosphere: Contributions to the study of the solid Earth and its fluid envelope*, J. O. Dickey (Ed.), Washington, D.C., 1997

Pavlis, E., Gravitational modeling contributions of an international multidisciplinary space mission: COSMIC, *EOS, Trans. Amer. Geophys. Union, 79*, S57, 1998.

Reigber, C., Z. King, R. Konig, and P. Schwintzer, CHAMP, a minisatellite mission for geopotential and atmospheric research, Spring AGU Meeting, Baltimore, MD, USA, May 1996.

Rummel, R. et al., Gravity Field and Steady-State Ocean Circulation Mission, Earth Explorer Mission Selection Workshop Report, SP-1233(1), European Space Agency, Granada, Spain, October 12-15, 1999.

Sehnal, L., L. Pospisilova, R. Peresty, and A. Kohlhase, Accelerometric measurements of satellite dynamics, Proc. 12th International Symposium on Space Flight Dynamics, ESOC, Darmstadt, Germany, 2-6 June 1997.

Shum, C., and C. Jekeli, Contribution of COSMIC Mission to Geodesy and Geodynamics, *Proc. COSMIC Science Workshop*, Taiwan, February, 1998.

Tapley, B., C. Reigber, and W. Melbourne, Gravity recovery and climate experiment (GRACE) mission, Spring AGU Meeting, Baltimore, MD, USA, May 1996.

Zhao, C.Y., The sensitivity of atmospheric temperature retrieved from radio occultation technique to orbit errors of GPS and LEO satellites, GeoForschungsZentrum Potsdam, Scientific Technical Report STR98/23, 1998.

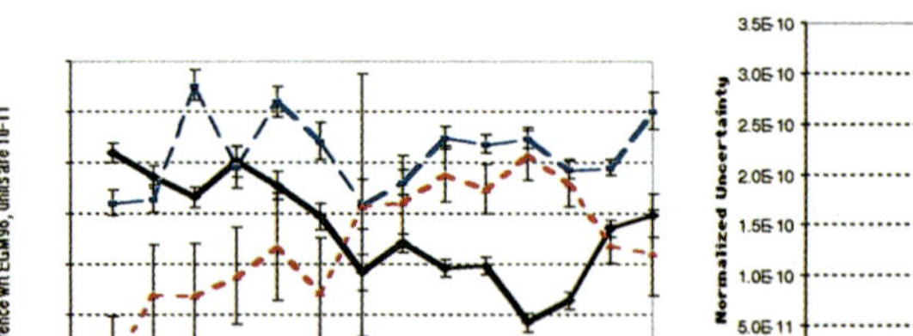

Figure 1. Annual and Seasonal Zonal Solutions From multiple satellites [Cox et al., 2000].

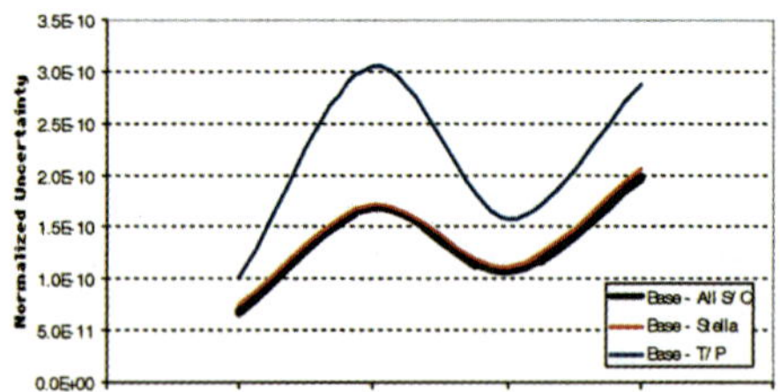

Figure 2. Solution uncertainties using all satellites, and without T/P or Stella.

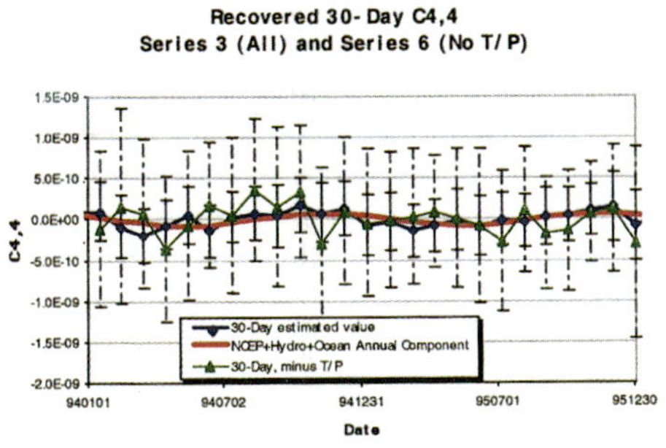

Recovered 30-Day C4,4
Series 3 (All) and Series 6 (No T/P)

Fig. 3. C4,4 solutions using all satellites and without T/P, compared with NCEP model output

Recovered 30-Day C4,4 Uncertainty
Series 3 (All) and Series 6 (No T/P)

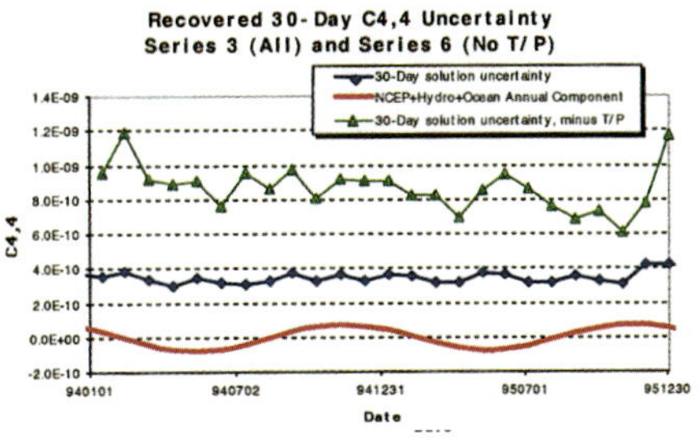

Fig. 4. C4,4 solution uncertainty using all satellites and without T/P, compared with NCEP model output

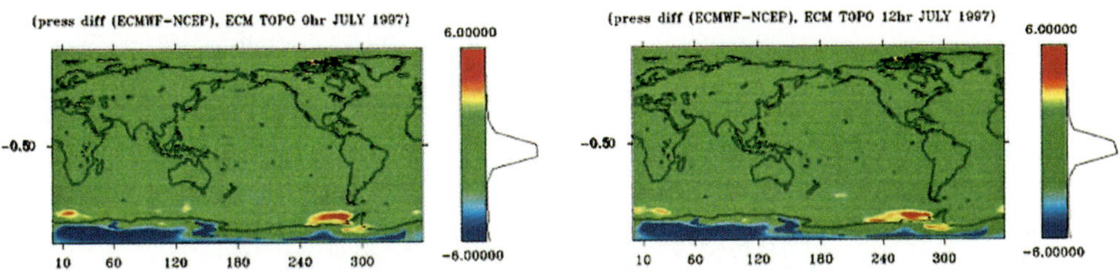

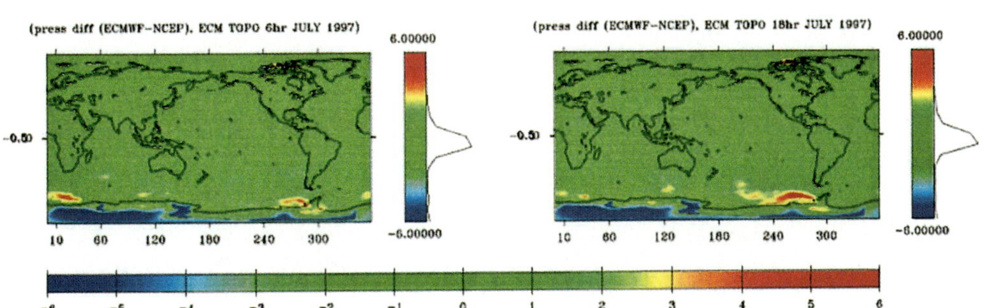

Fig. 5. Surface pressure differences between NCEP and ECMWF models for the four times daily measurements. Pressure errors are largest over Antarctica and the Southern Ocean (> 6 hPa).

GLOBAL IONOSPHERE DYNAMICS INFERRED FROM TOPSIDE SOUNDING

S. A. Pulinets, and I. A. Safronova

Institute of Terrestrial Magnetism, Ionosphere and Radiowave Propagation, Russian Academy of Sciences, (IZMIRAN), Troitsk, Moscow Region, 142190, RUSSIA

ABSTRACT

Topside sounding from onboard the satellites is probably one of most powerful remote sensing techniques for ionospheric research. Two Russian (Soviet Union) satellites Intercosmos-19 and Cosmos 1809 gave opportunity to create the richest database on global distribution of the near-earth plasma. Some of major results from both missions are reviewed in the paper. We can mention among them the global longitudinal effect explanation and main ionospheric trough model creation including the trough dynamics during magnetic storms. The global reaction of the ionosphere on magnetospheric disturbances was studied also using as global mapping of the critical frequency, so the changes of the ionosphere vertical structure inferred from the topside profiles. The fine wavy structure of the equatorial anomaly was revealed from the topside sounding data as well.

TECHNICAL PARAMETERS

One can find the detailed information about the Russian topside satellites and topside sounders installed onboard in Pulinets (1989). Here we will only mention the main information necessary for the following discussion. The topside sounder IS-338 was installed onboard satellites Intercosmos-19 and Cosmos 1809. It was the first digital topside sounder, working on 338 fixed frequencies from 0.3 up to 15.95 MHz with the frequency step 25 kHz within the range 0.3 – 1.5 MHz and 50 kHz for the rest of the frequency band of the sounder. Intercosmos-19 satellite was launched on 27 of March 1979 at the elliptical orbit of 470 km – 980 km with inclination 74°, and was active until the middle of 1982. The Cosmos 1809 satellite was launched on 28 of December 1986 and topside sounder on it was active until the middle of 1988. The orbit of the satellite was quasi-circular (960-980 km) with inclination of 82.5°.

GLOBAL LONGITUDINAL EFFECT AND MAIN TROUGH

One of the main features of the global structure of the ionosphere revealed by Intercosmos-19 was the strong and steady longitudinal effect that was observed at all latitudes (Karpachev, 1995, 1996). It was shown that the longitudinal effect at middle latitudes has a steady character in the F2-peak electron density NmF2 and electron temperature Te, and a less stable one in the F2-peak height hmF2. The new version of the IRI model (IRI-95) reproduces the longitudinal variations in N_mF2 very well but is much worse in h_mF2 and is completely inadequate

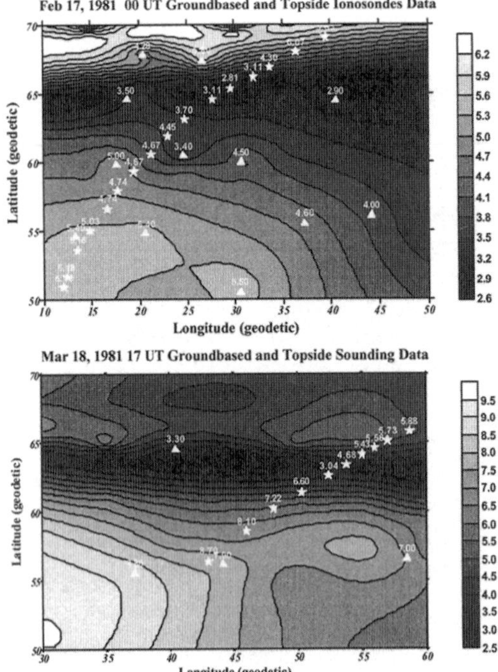

Fig. 1 Examples of model distribution of the critical frequencies for COST-251 ionospheric model with main trough included, in comparison with experimental data of topside sounding (stars) and ground-based sounding (triangles) for quiet (top) and disturbed (bottom) geomagnetic conditions)

in reproducing the observed longitudinal variations in T_e. The calculations of the vertical drift from h_mF2 values, by means of the model developed in IZMIRAN (Karpachev and Gasilov, 1998), show that the longitudinal variations in N_mF2 and h_mF2 are caused, mainly, by the neutral wind effect and, partially, by the longitudinal variations of thermospheric parameters with prevailing contribution from the neutral temperature. The HWM-93 neutral wind model inadequately reproduces the vertical drift variations for the above-mentioned conditions. The neutral wind components were determined recently from longitudinal variations of h_mF2 scaled from topside ionograms (Karpachev and Gasilov, 2000).

As a result of many years of Intercosmos-19 satellite data analysis, a model of the main ionospheric trough (MIT) was developed based on data from the Intercosmos-19 and Cosmos-900 satellites (over 1500 orbits) (Karpachev *et al.*, 1996). It is valid for nighttime (18 to 06 MLT), winter and equinox and Kp-indices from 0 to 8. It covers the topside ionosphere up to 1000 km and describes the trough minimum position depending on longitude, altitude, magnetic local time MLT and Kp-index. The model is presented in analytical form as well as a nomogram versus MLT and Kp and a nighttime segment of a circle (in polar coordinates) with the radius as a function of the Kp-

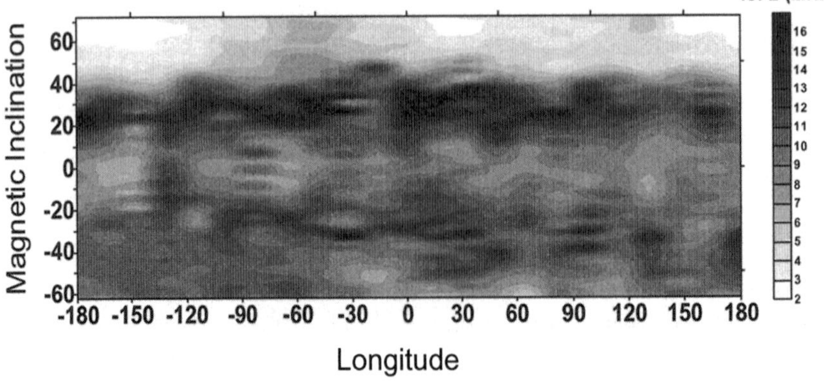

Fig. 2 Example of global distribution of the peak density in the vicinity of equatorial anomaly for summer night conditions by Intercosmos-19 topside sounding data. High solar activity

index. The effective Kp-index taken for the preceding 4.4 hours is used in the model. The Karpachev's MIT model was used to insert the trough into existing models of the ionosphere. Figure 1 demonstrates the comparison of the model distribution of critical frequencies with the experimental values from Intercosmos-19 satellite sounding as well as with the groundbased vertical soundings. Two distributions are shown for the COST-251 model (Mikhailov *et al.*, 1998) with MIT introduced for European region.

In addition to the main ionospheric trough (MIT), the ring ionospheric trough (RIT) was identified and studied (Karpachev, 2000) which, contrary to the MIT, correlates poorly with the Kp-index. It is observed only during the storm recovery phase at invariant latitudes near 55° and is connected to a residual ring magnetospheric current and to steady red auroral arcs (SAR-arcs).

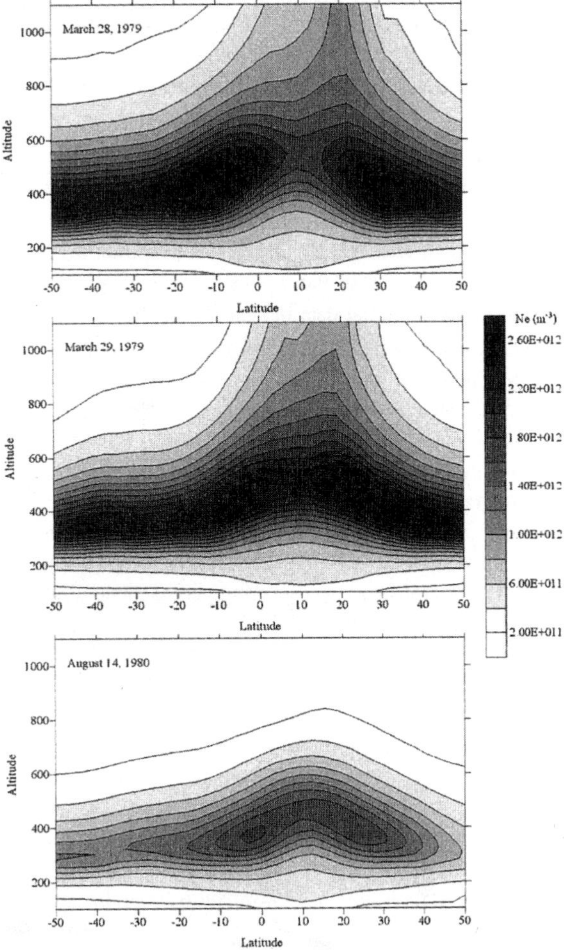

Fig. 3 Quasi-longitudinal cross sections of the ionosphere for geomagnetic disturbed (top and middle panels) and quiet conditions (bottom panel).

EQUATORIAL ANOMALY

In the vicinity of the equatorial anomaly, the longitudinal effect has a more complex character. In the majority of cases (more than 80 %) the longitudinal variations have a regular wavy character with a period of about 75° in longitude.

The average amplitude of this feature is 2-3 MHz on the dip equator and 3-4 MHz on the crests of the Appleton anomaly (Deminova, 1995a). The longitudinal variations of foF2 on the crests of the anomaly are usually in opposite phase relative to those along the dip equator. On the average, the extrema of these wave structures settle down on the same longitudes. The changes of the average longitudinal position of these extrema during nighttime (21-05 LT) and from season to season was not prominent. Along the dip equator, the minima of foF2 were observed, on average, to be at longitudes close to 45°, 130°, 200° and 330°. The minimum close to 330° is the deepest and is especially prominent during the June solar-solstice. At the transition from the crests of the anomaly to the middle latitudes, such wavy structure gradually disappears and at invariant latitudes above 30° it practically does not exist. The critical frequency scaled from topside ionograms versus magnetic inclination global distribution in the vicinity of equatorial anomaly for summer night conditions is presented in Figure 2.

DISTURBED CONDITIONS

Topside sounding from a satellite reveals how the ionosphere reacts to geomagnetic storms on a global scale. A schematic pattern of a global ionospheric response to a magnetic disturbance was constructed using the strong storm of 3 - 4 April 1979 as an example for further study (Karpachev *et al.*, 1995). In general, the MIT position's response to a southward turning of B_Z corresponds well to the equatorial edge dynamics of auroral diffuse precipitation. Under isolated B_Z southward turnings, the height of the equatorial night-time F-layer decreases by 50 - 100 km, and when B_Z turns northward, the low latitude F-layer rises. The latter effect is most pronounced at 03 LT (Deminova, 1995b). Such vertical movements of ionospheric plasma occur under the action of an additional electrical field of a magnetospheric origin which, in turn, is caused by the turning of B_Z. About one hour after this turning, it can penetrate from the magnetosphere into the plasmasphere. Such strong changes of the height of the F2-layer at B_Z turns could only be detected from topside-sounder satellite data. Ground-based sounding data indicated much smaller change in the height of the F2 layer at B_Z turns due to the strong recombination in the bottom part of the F2 layer. Large-scale internal gravity waves arriving at equatorial latitudes, from the auroral oval, cause an intensification of the equatorial anomaly in both the daytime and nighttime.

Not only the peak parameters, but the whole vertical profile dynamics during geomagnetic disturbances could be studied from topside sounding data as well. The sequence of topside profiles along the satellite orbit could be presented in the form of quasitomographic reconstruction of the vertical structure of the ionosphere in combination with the bottom side vertical profile models (Nava *et al.* 2000). One can see in Figure 3 how the equatorial ionosphere expands during geomagnetic disturbances (top and middle panels) in comparison with the quiet conditions (bottom panel).

CONCLUSION

The aim of the paper was to show using few examples how powerful still is the lost and forgotten topside sounding technique. No one of the modern fashion methods such as GPS TEC, tomography or radio occultation can provide such information on global scale. That's why the special attempts were undertaken to save the topside database and provide it for common use at IZMIRAN under NASA grant NRA 98-OSS-03(5.2) sponsorship (http://antares.izmiran.rssi.ru/projects/IK19). Due to new processing software and algorithms these data are revised now and new information about the ionosphere dynamics is still expected.

REFERENCES

Deminova G. F. Wave-like structure of the longitudinal variations of the nighttime equatorial anomaly, *Geomagnetism and Aeronomy*, 35, No. 4, 169 (1995 a)

Deminova G. F. Modifications in the nighttime low-latitude ionosphere after southward turning of the IMF, *J. Atmos. Terr. Phys.*, 57, 1459 (1995 b)

http://antares.izmiran.rssi.ru/projects/IK19

Karpachev A. T., Distribution of electron concentration in the high latitude topside ionosphere of a Southern hemisphere for nighttime summer conditions, *Geomagnetism and Aeronomy*, 35, No. 6, 82 (1995)

Karpachev A. T., Distribution of electron concentration near F2 layer maximum in Northern hemisphere for nighttime summer conditions, *Geomagnetism and Aeronomy*, 36, No. 3, 86 (1996)

Karpachev A. T., G. F. Deminova, S. A. Pulinets, Ionospheric changes in response to IMF variations, *Journal of Atmospheric and Terrestrial Physics*, 57, 1415 (1995)

Karpachev A. T., M. G. Deminov, V. V. Afonin, Model of the mid-latitude ionospheric trough on the base of Cosmos-900 and Intercosmos-19 satellites data, *Adv. Space Res.* 18, No. 6, 221 (1996)

Karpachev A. T., N. A. Gasilov, Vertical drift variations with longitude in the midlatitude nighttime summer ionosphere calculated from H_mF2, *Geomagnetism and Aeronomy*, 38, No. 5, 38 (1998)

Karpachev A. T., The characteristics of the ring ionospheric trough, *Geomagnetism and Aeronomy*, 40, No. 6, 1 (2000)

Karpachev A. T., N. A. Gasilov, Extraction of horizontal and meridional components of the neutral wind from longitudinal variations of hmF2, *Geomagnetism and Aeronomy*, 40, No. 4. 79 (2000)

Mikhailov A.V, S.A.Pulinets, V.V.Mikhailov, A.T.Karpachev and V.N.Shubin, An attempt to include midlatitude trough to the foF2 instantaneous mapping model MQMF2-IM, in *Proceedings of the 2-nd COST 251 Workshop "Algorithms and Models for COST 251 Final Product"*, edited by A. Vernon, Turkey, pp.100-103 (1998)

Nava B., Radicella S.M., Pulinets S., Depuev V. Modelling Bottom and Topside Electron Density and TEC with Profile Data from Topside Ionograms, *Adv. Space Res.* 27, No.1, 31 (2001)

Pulinets, S.A., Prospects of topside sounding, in *WITS Handbook, 2*, edited by C. H. Liu, pp. 99-127, SCOSTEP publications, Urbana, Illinois (1989)

Summary Session

DOES SPACE WEATHER REALLY MATTER ON THE GROUND?

Ingrid Sandahl

Swedish Institute of Space Physics, P.O. Box 812, SE-981 28 Kiruna, Sweden

ABSTRACT

During recent years there has been an increasing number of observations showing that various space factors have an influence on the biogeosphere. Several effects have been reported, including potential fluctuations in pipelines and power lines, correlations with weather phenomena, correlation between the solar cycle and mortality, and between magnetic disturbances and heart problems. For several of these effects the mechanisms behind the correlations are poorly understood. Measurements of different input parameters, such as the properties of the solar wind and cosmic ray intensity, are needed to study these mechanisms. It is less obvious when it is useful to have a space weather prediction service. In some cases it is better to rely on space weather proof design of technical systems. The paper will discuss requirements on space weather measurements for monitoring and prediction for different applications, and when predictions may be worth the effort.

INTRODUCTION

Man has been living at least reasonably happily for tens of thousands of years without even knowing that there is such a thing as space weather. It has of course been suspected; astrology was once considered an important field, but there has for a long time been agreement among scientists that astrology is superstition. Now, however, scientists are beginning to tell politicians and the public about all the dangers coming from out there and are arguing that resources should be spent on studying how to predict these dangers. If future generations are not going to give us the same verdict as we are now giving the astrologers, we should have really good reasons for claiming our case.

The issue here is not whether solar terrestrial physics and space weather effects should be studied – of course they should. It is rather a question about the responsibility that scientists have to give the public balanced information about the dangers of space weather and realistic expectations regarding what can be achieved with predictions.

In this paper we will first make some general observations about predictions. We will then examine some space weather effects on the ground and in the atmosphere and discuss how useful predictions may be in these cases. I have chosen some examples of space weather effects that may be less well known in the space science community.

GENERAL CONSIDERATIONS ABOUT SPACE WEATHER PREDICTIONS

There are three different approaches regarding how to handle space weather:

1. You do not take any preventive measures, just accept occasional space weather induced failures.
2. You understand space weather and design space weatherproof systems.
3. You invest in a good prediction service and take preventive action upon warning.

In order to make a wise choice between these alternatives there are a number of questions that need to be examined:

1. What is the accuracy of the prediction? The accuracy can be illustrated with the diagram in Figure 1. The diagram gives likelihoods for the four possible cases in a prediction situation. In an ideal prediction system the likelihoods are 1 in the YesYes and NoNo squares and 0 in the Unpleasant surprise and False alarm squares, but no ideal prediction systems exist and all squares will contain numbers deviating from 1 or 0.

2. How does the cost of a loss compare to the cost of making the prediction? This is a classical optimization problem. It is obvious that the costs for really elaborate predictions systems will be very great. Such systems will require monitor systems on the ground and in space with around the clock operation and data transmission, real time analysis and alarm distribution systems, development and maintenance. It will take many saved satellites and transformers to balance those costs.

3. Is it possible to prevent damage? A warning is only useful if one can take protective action.

4. Will the warning come early enough? This question is related to the previous two. Different types of preventive actions will require different amounts of time. A fast prediction system will normally be more expensive than a slow one.

One important parameter in the optimization problem is how often expensive space weather effects take place. NOAA has introduced three space weather scales (eg.Poppe, 2000). For geomagnetic storms the top value, G5, corresponds to Kp=9 and this is reached on average four

		Predicted event	
		Yes	No
Real event	No	1	0 Unpleasant surprise
	Yes	0 False alarm	1

Fig. 1. Likelihoods for different cases in an ideal prediction.

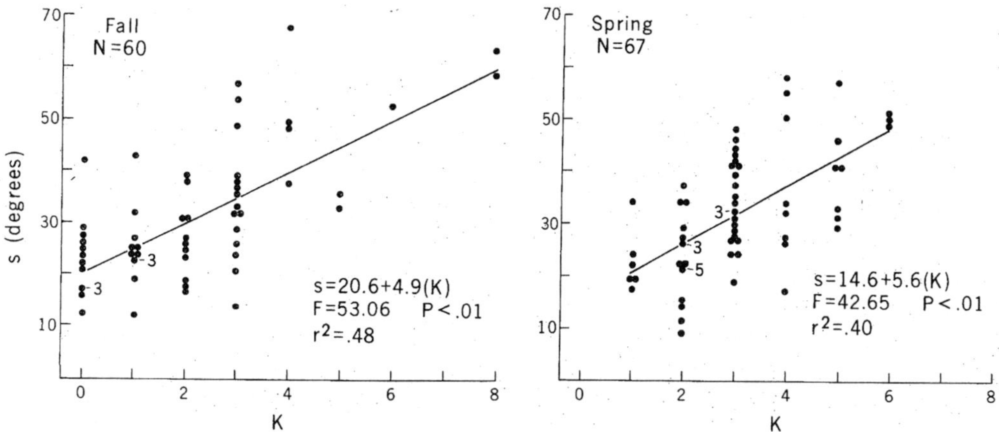

Fig. 2. Deviation in flight direction of migrating birds as a function of geomagnetic disturbance. From Moore (1977).

times per solar cycle. Severe solar radiation storms, S5, with ≥ 10 MeV proton flux greater than $10^5\ s^{-1}sr^{-1}cm^{-2}$ and severe radio blackouts, R5, both occur less than once per cycle.

EXAMPLES OF SPACE WEATHER EFFECTS

Animals

Several animal species use the geomagnetic field for orientation and navigation (Åkesson, 1990 and references therein). Some examples are magnetotactic bacteria, honey bees, sharks, birds, rodents, and whales. There are a number of different receptor mechanisms, for example involving molecules in the retina of the eye or magnetite. It has been found experimentally that the compass of European Robins and Homing Pigeons is sensitive to the inclination of the geomagnetic field. Several passerines, such as Starlings, have an inborn compass direction and birds of many species migrate alone, even in their first fall. Figure 2 shows the deviation of the flight direction of nocturnally migrating passerines as a function of geomagnetic disturbance given by the K index (Moore, 1977). Even if the spread is large there is a clear correlation.

Weather and Climate

There are now many observations of correlations between solar activity and meteorological conditions, both on time scales of days and on time scales of years and longer. An example is the correlation between the northern hemisphere land temperature and solar cycle duration shown in Figure 3 (Friis-Christensen and Lassen, 1991). The solar cycle duration is related to the solar activity; the shorter the cycle the higher the activity. The mechanisms behind these correlations are not yet understood but a promising candidate

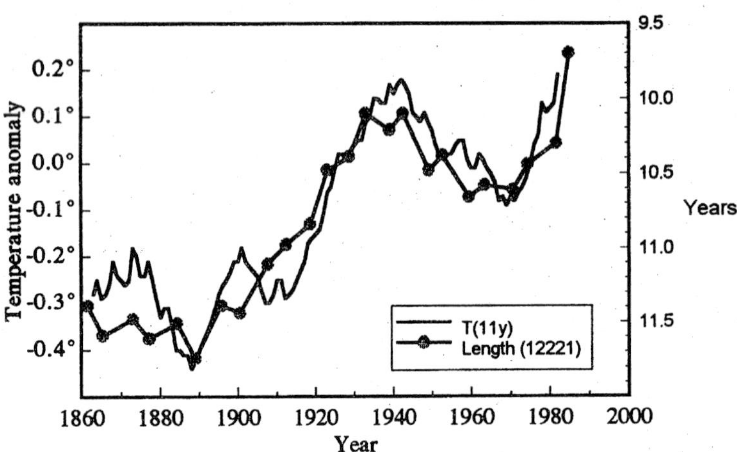

Fig. 3. Northern hemisphere land temperature as a function of solar cycle length, which is used as a proxy of solar activity level. From Friis-Christensen and Lassen (1991).

is the cosmic ray influence on cloud formation. In a recent study including also the period up until the present the temperature was found to be higher than predicted by the correlation in Figure 3 and this is thought to be due to the greenhouse effect which is working in parallel with the solar activity effect. (Thejll and Lassen, 2000)

Health

In the field of human health a number of intriguing correlations with solar activity have been found. Figure 4 is a comparison of the solar cycles to the longevity of U.S. representatives, U.K. House of Commons members and Cambridge alumni, in total a little over 20,000 people, as a function of year of birth. Only people who have died from natural causes have been included. There is a variation in the longevity with an amplitude of more than one year and a period very similar to the solar cycle period. A 20 year shift between the two curves has been taken to indicate a mechanism related to the egg cell formation of the mother.

There are also results showing effects correlated with short term disturbances. Figure 5 shows some results from superposed epoch studies using a data base of ambulance calls in Moscow 1979-81. An increased mortality from myocardial infarction was found about 1-2 days after significant southward turnings of the IMF and an even clearer increase after Forbush decreases (Villoresi at al, 1994, Halberg et al. 1999). In another study a person was equipped with the same type of heart activity monitoring equipment as is used by Russian cosmonauts. Figure 6 shows an anticorrelation between heartrate and the magnetic Z component (Chernouss et al., 2001).

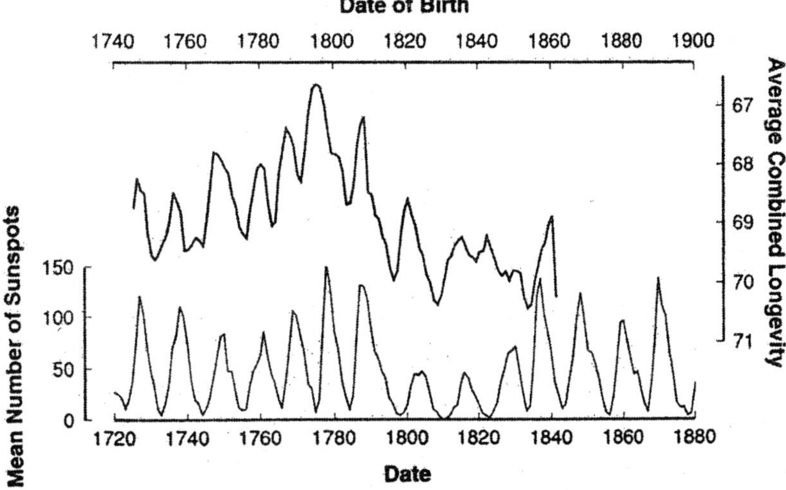

Fig. 4. Combined birth cohort longevity for U.S. representatives (5300 cases), UK House of Commons members (2336 cases), and University of Cambridge alumni (12900 cases) compared to sunspot activity. From Juckett and Rosenberg (1993).

A frequently used measure of heart stress is the heart rate variability, HRV. In a healthy heart the heart beat rate varies from one beat to the next. When the organism is under stress one response is that the heart beat rate becomes much more constant. There are examples of decreased heart beat variability during geomagnetic disturbances both on the ground (Chernouss et al., 2001) and in space (Baevsky et al., 1997).

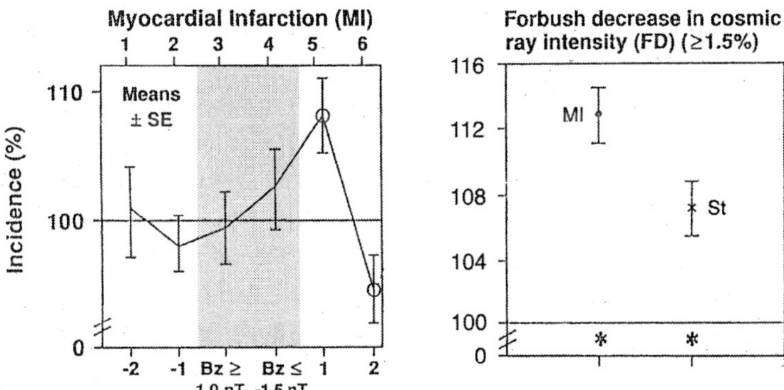

Fig. 5. To the left: Normalized mortality from myocardial infarctions in Moscow from two days before until two after a southward turning of the interplanetary magnetic field. To the right: Normalized mortality in Moscow from myocardial infarction (MI) and stroke (St) following Forbush decreases. From Villoresi et al. (1994).

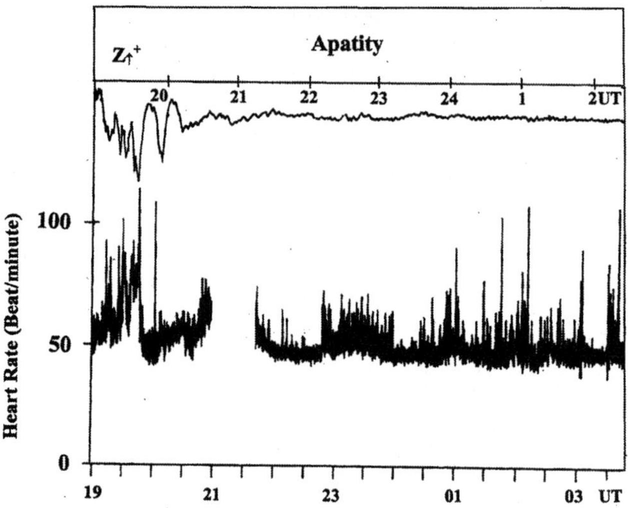

Fig. 6. Heart rate compared to the Z component of the local magnetic field. Negative bays correspond to heart rate increases. From Chernouss et al. (2001).

Radiation in Aircraft

At the cruising altitude of several airplanes the shielding by the atmosphere against cosmic radiation is significantly smaller than at ground level. During times of increased cosmic ray fluxes, especially during solar proton events, high radiation levels may be reached, affecting both humans and electronics. The radiation levels increase quickly with altitude and latitude (Dyer et al., 1990, Johansson and Dyreklev, 1998). Levels as high as 30 mSv/h have been measured, leading to a dose of more than 100 mSv during one flight. Several papers on radiation risks in aircraft are found in Kelly et al. (1999).

In order to protect flight crews and passengers against dangerous radiation there are internationallyagreed rules which flight operators, that is airlines, have to follow. In Europe these regulations are issued by the Joint Aviation Authorities, JAA. They are at present being revised and amended. The regulations dictate that a pilot has to initiate a descent as soon as certain radiation limit values are exceeded. It is now proposed that the working schedules have to be arranged so that the radiation exposure of crew members is kept below 6 mSv per year. For pregnant crew members the schedules have to be such that the dose does not exceed 1 mSv for the remainder of the pregnancy, once the operator has been notified. There are also rules regarding monitoring of radiation levels. On-board monitoring equipment is required for aircraft flying above 15,000 m. For flights between 8,000 and 15,000 meters it is proposed to introduce individual radiation dose estimates based on computer programs and internationallyagreed information on radiation levels. In comments to the proposed revision both the Swedish Radiation Protection Institute and the Swedish Airline Pilot Association state that it is necessary to have on board monitoring equipment in order to be sure to detect fast radiation level increases.

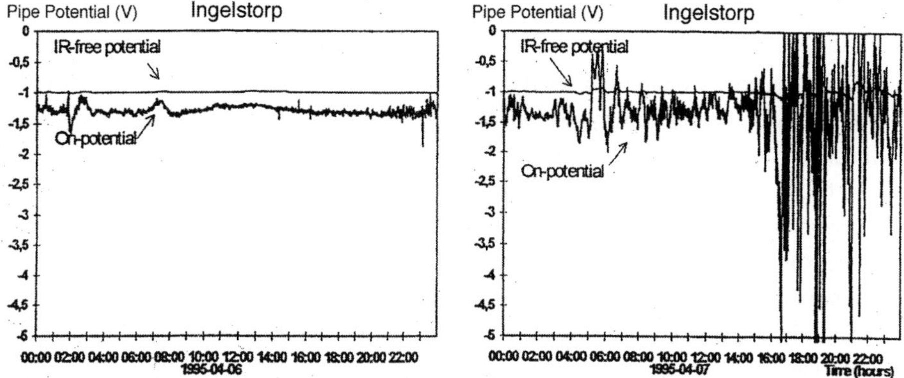

Fig. 7. Potential along a pipeline on a quiet day (1995-04-06, Ap=6) and a disturbed day (1995-04-07, Ap=100). The pipeline runs in an east-west direction. From Harde and Johansson (1996).

Geomagnetically Induced Currents, GIC

During geomagneticallydisturbed times currents are induced in long conductors on the ground, such as power lines and oil and gas pipelines. Figure 7 shows the potential along a gas pipeline in southern Sweden measured during a calm and a disturbed day (Harde and Johansson, 1996). The increased potential may lead to increased corrosion. In electric power lines the result may be saturation and destruction of transformers.

Auroral Tourism

Recently an auroral prediction service was offered to inhabitants and tourists in Northern Sweden. The predictions are distributed as SMS messenges to the customer's mobile phone a few times per day (www.forskningsturism.nu). They are based on solar and solar wind observations and are produced by the Lund division of the Swedish Institute of Space Physics. There is a growing interest in Auroral Tourism and with this a market for predictions.

Insurance

Since space weather may cause many different types of damage, the insurance industry is beginning to take an interest in the subject. Recently a booklet about Space weather was produced by the Swiss insurance company SwissRe, (Jansen et al., 2000). In the chapter "What does space weather have to do with insurances?" we find the following:

- "It is the insurance industry's responsibility to provide information and raise awareness."
- "It is the responsibility of the insured to implement risk-mitigating measures."
- "For both these reasons, early warning systems....will become more important in the future."

USEFULNESS OF PREDICTIONS

We will now look at the above examples to see if predictions would be useful. In the case of wild animals this is clearly not the case. There is no way to tell the birds that a magnetic storm is coming. In the case of weather and climate it is of great interest to understand the mechanisms, but for this realtime monitoring is not needed. Perhaps, in a distant future, space weather input will be able to increase the accuracy of meteorological predictions, but this seems to be very far away. Regarding health it is difficult to see what kind of protective measures would be in any reasonable proportion to the actual risk levels and practical difficulties of a warning system. Possibly the information could be of some use to intensive care units. Radiation in airplanes is the first of our examples where predictions can be of real value. If there is a solar proton event, routes at lower altitude and lower latitude may be used. More important here, however, is that there is direct monitoring of the radiation level on the aircraft. In the case of geomagneticallyinduced currents predictions are already being used by power companies. For example, certain service jobs are delayed when geomagnetic storms are expected and measures are taken to prevent a very high load on the power lines. In auroral tourism predictions are not only useful, they are essential. Predictions make it possible to plan the program and make an auroral holiday much more rewarding. Many persons are willing to pay much to see a good aurora. When it comes to insurance, at least SwissReclaim that predictions will become important. But at the same time this puts a high pressure on the quality of the predictions. One may run into unpleasant questions of responsibility if the predictions are not correct.

CONCLUSIONS

It is not obvious that an extensive prediction service will be economically justified. In order to find that out careful estimates are needed. Predictions to answer different special needs are, however, useful, and are already being made today. As scientists we have a responsibility to give the public balanced information about the risks of space weather and the usefulness of space weather predictions.

ACKNOWLEDGEMENTS

Several persons have generously contributed to the information in this paper. In particular I would like to thank G. Cornelissen, J.-E. Kyllönen, and K. Wigler for providing information on health effects and radiation monitoring in aircraft.

REFERENCES

Åkesson, S, Animal Orientation in Relation to the Geomagnetic Field, *Introductory paper no 59,* Department of Ecology, Lund University, Sweden, Lund (1990).

Baevsky, R. M., V. M. Petrov, G. Cornélissen, F. Halberg, K. Orth-Gomér et a., Meta-analyzed Heart Rate Variability, Exposure to Geomagnetic Storms, and the Risk of Ischemic Heart Disease, *Scripta medica*, 199 (1997).

Chernouss, S., A. Vinogradov, and E. Vlassova, Geophysical Hazard for Human Health in the Circumpolar Auroral Belt: Evidence of a Relationship between Heart Rate Variation and Electromagnetic Disturbances, *Natural Hazards*, 23, 121 (2001).

Dyer, C. S., A. J. Sims, J. Farren, and J. Stephen, Measurements of Solar Flare Enhancements to the Single Event Upset Environment in the Upper Atmosphere, *IEEE Trans. Nucl. Sci.*, NS-37, 1929 (Dec. 1990)

Friis-Cristensen, E. and K. Lassen, Length of the Solar Cycle: An Indicator of Solar Activity Closely Associated with Climate, *Science,* 254, 698 (1991).

Halberg, F., G. Cornélissen, G. Katinas, D. Bruhn, R. B. Sothern et al., From Human to Microbial Gauges of the Cosmos, Poster, 1999 Fall AGU meeting.

Harde, C. and C. Johansson, Space Weather Effects on Natural Gas Pipelines, *in AI Applications in Solar-Terrestrial Physics*, eds. I. Sandahl and E. Jonsson, ESA WPP-148, 43, (Apr. 1998).

Jansen, F., R. Pirjola, and R. Favre, *Space Weather Hazard to the Earth?*, SwissRe, Zürich, 2000. Available at www.swissre.com/e/publications/publications/society1/spaceweather.html.

Johansson, K. and P. Dyreklev, Space Weather Effects on Aircraft Electronics, *in AI Applications in Solar-Terrestrial Physics*, eds. I. Sandahl and E. Jonsson, ESA WPP-148, 29, (Apr. 1998).

Juckett, A. and B. Rosenberg, Correlation of Human Longevity Oscillations with Sunspot Cycles, *Radiation Research,*133, 312 (1993).

Kelly, M, H.-G. Menzel, P. Ryan, and K. Schuuer, (Eds.), Cosmic Radiation and Air Crew Exposure, *Radiation Protection Dosimetry*, 86(4) (1999).

Moore, F. R., Geomagnetic Disturbance and the Orientation of Nocturnally Migrating Birds, *Science,* 196, 682 (1977).

Poppe, B. B., New Scales Help Public, Techicians Understand Space Weather, *EOS Transactions,* 81, 322, 2000. See also www.sec.noaa.gov.

Thejll, P. A. and K. Lassen, Solar forcing of the Northern Hemisphere land air temperature: New data 2000, *Journal of Solar-Terrestrial Physics* 13, 1207 (2000).

Villoresi, G., Y. A. Kopytenko, N. G. Ptitsyna, M. I. Tyasto, E. A. Kopytenko et al., The influence of geophysical and social effects on the incidences of clinically important pathologies (Moscow 1979-1981), *Physica Medica,* 10, 79 (1994).

AUTHOR INDEX